Frontiers in Sedimentary Geology

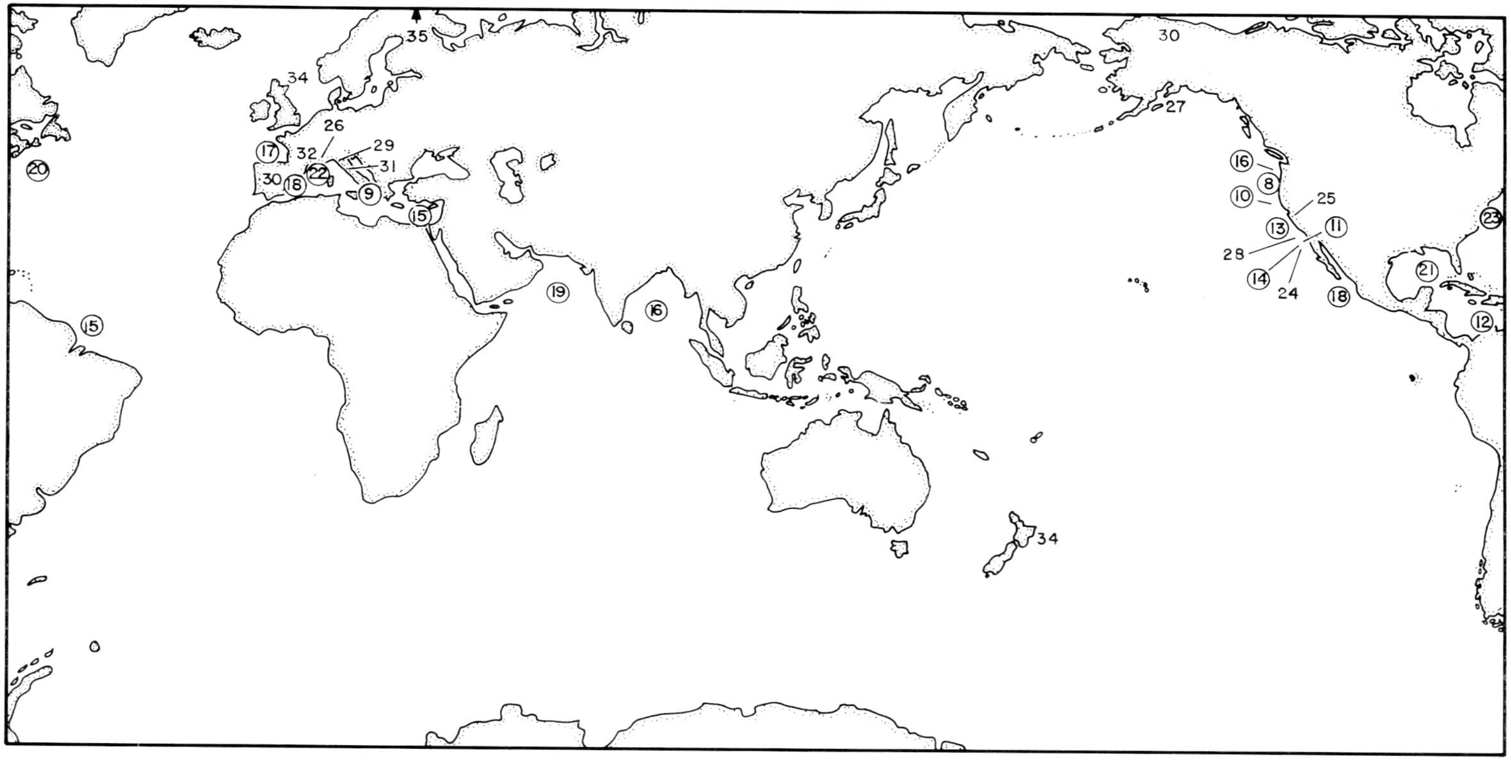

Index map to fan locations keyed to chapter numbers:
Amazon (15), Astoria (8), Bengal (16), Blanca 24, Brae 34, Butano 25, Cap-Ferret (17), Cengio 26, Chugach 27, Crati (9), Delgada (10), Ebro (18), Ferrelo 28, Gottero 29, Hecho 30, Indus (19), Kongsfjord 35, La Jolla (11), Laurentian (20), Magdalena (12), Marnoso-Arenacea 31, Mississippi (21), Monterey (13), Navy (14), Peira-Cava 32, Rhone (22), Torlesse 33, Wilmington (23)

Circled numbers are for chapters on modern fans.

Submarine Fans and Related Turbidite Systems

Edited by
Arnold H. Bouma, William R. Normark
and Neal E. Barnes

With 251 Figures

Springer-Verlag New York Berlin Heidelberg Tokyo

Arnold H. Bouma*
Gulf Research & Development Company
11111 South Wilcrest Drive
Houston, Texas 77236
U.S.A.

William R. Normark
U.S. Geological Survey
Mail Stop 999
Menlo Park, California 94025
U.S.A.

Neal E. Barnes*
Gulf Research & Development Company
11111 South Wilcrest Drive
Houston, Texas 77236
U.S.A.

*currently with Chevron Oil Field Research Company

Library of Congress Cataloging in Publication Data
Main entry under title:
Submarine fans and related turbidite systems.
(Frontiers in sedimentary geology)
Includes index.
1. Submarine fans. 2. Turbidites. I. Bouma, Arnold H. II. Normark, William R. III. Barnes, Neal E. IV. Series.
GC87.6.S92S82 1985 552′.5 85-7994

Earlier versions of 27 chapters in this book were included in the journal, Geo-Marine Letters, Volume 3, No. 2-4, published in 1984 by Springer-Verlag, New York. © 1984/1985 by Springer-Verlag, New York, Inc.

Typeset by David E. Seham Associates, Inc., Metuchen, New Jersey.
Printed and bound by Halliday Lithograph, West Hanover, Massachusetts.
Printed in the United States of America.

9 8 7 6 5 4 3 2 1

ISBN 0-387-96142-9 Springer-Verlag New York Berlin Heidelberg Tokyo
ISBN 3-540-96142-9 Springer-Verlag Berlin Heidelberg New York Tokyo

Preface

Exchange of information in the field of earth sciences is increasingly needed to stay informed about advances. However, the continuous increase in the number of journal articles and books is very noticeable, while the available time to keep up is decreasing. Such a large flow of information commonly necessitates professionals to search selectively for material and special publications in one's sub-discipline that have more specific coverage.

In addition to surveying research needs, earth scientists working in a pure or applied research environment collect and produce information that often is of interest to the much larger group of industry-employed geologists and geophysicists, to professionals employed by agencies, and to students.

To accommodate this exchange of needed information, Springer-Verlag is launching a monograph series entitled "Frontiers in Sedimentary Geology." This series will cover a number of subjects related to sediments and sedimentary rocks in a manner that both the researcher and the industrially oriented earth scientist can use constructively. Publications in this monograph series may fit one or more of the following main categories:

Topical

A topical subject will cover either the different aspects of a selected environment of deposition, or present a world tour of a particular depositional environment to demonstrate its variability and its commonalities. The author(s) or editor(s) accepts the responsibility to guide the reader as to the state of knowledge, rather than providing a set of independent chapters.

Regional

This category may deal with large areas in both a descriptive and interpretive manner, with the results being applicable worldwide. The topic may emphasize the influence of sea level variations on depositional environments, the mechanics and results of oceanographic forces on sediments such as waves and currents, or the effect of large scale and small scale tectonics on depositional basins. Each subject should stand by itself concerning the region under discussion, while the author(s) or editor(s) bears the responsibility to demonstrate its applicability to other areas.

Interdisciplinary

Unconsolidated sediments and sedimentary rocks are seldom studied using strictly sedimentological approaches. Examples of potential subjects include principles of seismic stratigraphy and acoustical facies, geotechnical approaches and the new field of geotechnical stratigraphy, diagenetic studies and changes in porosity and permeability, the principles of wire-line logs and their relationships to petrophysical and lithological characteristics, and the multi-disciplinary approach to the study of one type of sediment. Such topics should be treated very carefully to avoid loss of direction.

Any combinations of these three broad categories are likely to occur in a publication in this monograph series. However, it is our intent to publish only subjects of current interest, and to insure that sufficient new information is supplied to arouse the reader's interest.

The first volume in the monograph series "Frontiers in Sedimentary Geology" deals with submarine fans and related turbidite systems. This topic has received much interest in the past by the scientific community and more recently has found a growing audience in the industry. This edited volume is intended to update the reader on the status of current knowledge by presenting many examples of modern fans and ancient turbidites in a comparative manner. It is clear that we still treat the subject as two different sedimentary bodies with significant differences in study approaches, with the result that they cannot be compared yet as one sedimentary family. It also demonstrates the difficulty with our present models and the confusion in terminology.

This publication certainly does not close the subject, nor does it provide all the definitive answers. On the contrary, its aim is to provide documentation and to bring the topic into focus by comparing several examples and indicating present weaknesses. Its strength comes forward by viewing modern and ancient systems, and by presenting detailed studies based on recent drilling by the Deep Sea Drilling Project on the Mississippi Fan; the only submarine fan drilled so far in a systematic manner.

Arnold H. Bouma

Contents

Contributors

ERNESTO ABBATE, Istituto di Geologia, Universita de Firenza, Florence, Italy

STEVEN B. BACHMAN, Crouch, Bachman & Associates, San Diego, and Scripps Institution of Oceanography, La Jolla, California, U.S.A.

°NEAL E. BARNES, Gulf Research & Development Company, Houston, Texas, U.S.A.

*°ARNOLD H. BOUMA, Gulf Research & Development Company, Houston, Texas, U.S.A.

*JAMES M. BROOKS, Department of Oceanography, Texas A&M University, Texas, U.S.A.

*WILLIAM R. BRYANT, Department of Oceanography, Texas A&M University, Texas, U.S.A.

RICHARD T. BUFFLER, Institute for Geophysics, University of Texas at Austin, Austin, Texas, U.S.A.

CARLO CAZZOLA, Istituto di Geologia, Universita de Parma, Parma, Italy

THOMAS E. CHASE, U.S. Geological Survey, Menlo Park, California, U.S.A.

WILLIAM J. CLEARY, Department of Earth Sciences, University of North Carolina, Wilmington, North Carolina, U.S.A.

ALBINA COLELLA, Dipartimento di Scienze della Terra, Universita della Calabria, Cosenza, Italy

*°JAMES M. COLEMAN, Coastal Studies Institute, Louisiana State University, Baton Rouge, Louisiana, U.S.A.

*RICHARD E. CONSTANS, Chevron U.S.A. Inc., New Orleans, Louisiana, U.S.A.

°FRANCIS COUMES, Societé National Elf Aquitaine, Pau, France

*MICHEL CREMER, Laboratoire de Geologie et Oceanographie, Université de Bordeaux, Talence, France

°JOSEPH R. CURRAY, Scripps Institution of Oceanography, Geological Research Division, La Jolla, California, U.S.A.

°JOHN E. DAMUTH, Mobil Research & Development Corporation, Dallas, Texas, U.S.A.

D. DEFREITAS, Department of Oceanography, Texas A&M University, Texas, U.S.A.

*LAURENCE I. DROZ, Laboratoire de Geodynamique Sous-Marine, Villefranche-sûr-Mer, France

FRANS J. EMMEL, Scripps Institution of Oceanography, Geological Research Division, La Jolla, California, U.S.A.

MARY H. FEELEY, Exxon Production Research Company, Houston, Texas, U.S.A.

ROGER D. FLOOD, Lamont-Doherty Geological Observatory of Columbia University, Palisades, New York, U.S.A.

GIANNI GABBIANELLI, Istituto di Geologia, Universita di Bologna, Bologna, Italy

HENRY GOT, Centre de Recherches de Sédimentologie Marine, Université de Perpignan, Perpignan, France

STEPHAN A. GRAHAM, Departments of Applied Earth Sciences and Geology, Stanford University, Stanford, California, U.S.A.

CHRISTINA E. GUTMACHER, U.S. Geological Survey, Menlo Park, California, U.S.A.

°DAVID G. HOWELL, U.S. Geological Survey, Menlo Park, California, U.S.A.

*TOSHIO ISHIZUKA, Ocean Research Institute, University of Tokyo, Tokyo, Japan

*MAHLON C. KENNICUTT, II, Department of Oceanography, Texas A&M University, Texas, U.S.A.

*BARRY KOHL, Chevron U.S.A. Inc., New Orleans, Louisiana, U.S.A.

°V. KOLLA, Elf-Aquitaine Petroleum, Houston, Texas, U.S.A.

°FRANCO RICCI LUCCHI, Istituto di Geologia, Universita di Bologna, Bologna, Italy

°TIMOTHY R. MCHARGUE, Chevron Overseas Petroleum, Inc., San Francisco, California, U.S.A.

T. C. MACKINNON, Chevron Oil Field Research Company, La Habra, California, U.S.A.

HUGH MCLEAN, U.S. Geological Survey, Menlo Park, California, U.S.A.

ANDRES MALDONADO, Instituto de Geologia, Jaime Almera, C.S.I.C., Barcelona, Spain

*AUDREY A. MEYER-WRIGHT, Ocean Drilling Project, Texas A&M University, Texas, U.S.A.

RICHARD J. MOIOLA, Mobil Research & Development Corporation, Dallas, Texas, U.S.A.

ANDRE MONACO, Centre de Recherches de Sédimentologie Marine, Université de Perpignan, Perpignan, France

°EMILIANO MUTTI, Istituto di Geologia, Universita de Parma, Parma, Italy

°C. Hans Nelson, U.S. Geological Survey, Menlo Park, California, U.S.A.

J. C. Nelson, Department of Geology, Duke University, Durham, North Carolina, U.S.A.

°Tor H. Nilsen, U.S. Geological Survey, Menlo Park, California, U.S.A.

*°William R. Normark, U.S. Geological Survey, Menlo Park, California, U.S.A.

*Suzanne O'Connell, Lamont-Doherty Geological Observatory of Columbia University, Palisades, New York, U.S.A.

Patrick Orsolini, Elf-Aquitaine Petroleum, Houston, Texas, U.S.A.

*Mary E. Parker, Department of Geology, Florida State University, Tallahassee, Florida, U.S.A.

R. C. Pflaum, Department of Oceanography, Texas A&M University, Texas, U.S.A.

*Kevin T. Pickering, Department of Geology, Goldsmith's College, University of London, London, Great Britain

Orrin H. Pilkey, Jr., Department of Geology, Duke University, Durham, North Carolina, U.S.A.

°David J.W. Piper, Atlantic Geoscience Centre, Geological Survey of Canada, Bedford Institute of Oceanography, Dartmouth, Nova Scotia, Canada

Christian Ravenne, Institut Francais du Petrole, Rueil-Malmaison, France

Harry H. Roberts, Coastal Studies Institute, Louisiana State University, Baton Rouge, Louisiana, U.S.A.

Sergio Rossi, Istituto di Geologia Marina, C.N.R., Bologna, Italy

*Claudia Schroeder, Department of Geology, Dalhousie University, Halifax, Nova Scotia, Canada

G. Shanmugam, Mobil Research & Development Corporation, Dallas, Texas, U.S.A.

*°Charles E. Stelting, Gulf Research & Development Company, Houston, Texas, U.S.A.

*°Dorrik A.V. Stow, The University of Nottingham, Department of Geology, Nottingham, Great Britain

*William E. Sweet, Mineral Management Service, Metairie, Louisiana, U.S.A.

Paul A. Thayer, Amoco Production Company, New Orleans, Louisiana, U.S.A.

John G. Vedder, U.S. Geological Survey, Menlo Park, California, U.S.A.

Bartolomeo Vigna, Istituto di Geologia, Universita de Parma, Parma, Italy

°Roger G. Walker, McMaster University, Department of Geology, Hamilton, Ontario, Canada

*Andreas Wetzel, Geologisches Institut, Tübingen, Federal Republic of Germany

*Jean K. Whelan, Department of Chemistry, Woods Hole Oceanographic Institute, Woods Hole, Massachusetts, U.S.A.

Pat Wilde, Marine Sciences Group, University of California, Berkeley, California, U.S.A.

---

*DSDP Leg 96 Shipboard Scientist

°Participant in the COMFAN meeting

I

Submarine Fans and Related Turbidite Sequences

General Topics

CHAPTER 1

Introduction to Submarine Fans and Related Turbidite Systems

Arnold H. Bouma

Abstract

Increased research and economic interest in submarine fans and ancient turbidite sequences provided the impetus to convene a meeting of specialists to discuss the status of existing knowledge and future directions of study. This small international group, known as COMFAN (COMmittee on FANs), provided updated reviews of 23 modern and ancient turbidite systems in a special issue of *Geo-Marine Letters*, which was used as the core for this publication. The COMFAN review is augmented by a select number of general contributions (without attempting to provide a complete survey of submarine fans and related turbidite systems) and the main results of the drilling on the Mississippi Fan.

Introduction

In response to increased research and economic interest in modern submarine fans and ancient turbidite sequences, a meeting of an international group of specialists was called to share "state-of-the-art" knowledge and discuss new directions of study. The COMmittee on FANs (COMFAN) met in September 1982 at the research facilities of Gulf Research & Development Company in Harmarville near Pittsburgh, Pennsylvania, and was hosted by Gulf Oil Corporation.

During the COMFAN meeting, special attention was paid to: 1) recent or unpublished observations, 2) differences in types of studies (sedimentologic, geophysical, borehole data, outcrop mapping, etc.) and consequent interpretational differences between modern deep-sea fans and ancient turbidite sequences, 3) the value of present sedimentation models, and 4) the types of future studies needed, such as Deep Sea Drilling Project (DSDP) drilling. Although numerous individual papers and a few books [1–4] have been published concerning many of these deposits, the COMFAN members felt that succinct reviews presented in a standardized format to aid comparison of individual fan systems could form a useful contribution to turbidite research. This publication took the form of a special issue of *Geo-Marine Letters* (Volume 3, 1983–84) under the general title "Submarine Clastic Systems: Deep Sea Fans and Related Turbidite Facies" [5].

The broad interest in the results of DSDP Leg 96 drilling on the Mississippi Fan and in the *Geo-Marine Letters* special issue, which is not generally available as a separate volume, encouraged us to combine the two in this book. In addition, we have included additional chapters on various other aspects of deep-sea fan research and the petroleum potential of these deposits.

COMFAN Meeting

Participants in the COMFAN meeting were principally from the academic community, although some representatives from industry were present (see list of contributors) who emphasized other interests and approaches. By stressing recent and/or unpublished studies and focusing on the problems or discrepancies encountered, the discussions were very lively and constructive. During the first day-and-a-half, participants informally presented their recent work. Much of the data presented were so new that impromptu poster sessions created by pinning maps and data tables to the conference room walls commonly disrupted the lectures. Slightly more than half-a-day was directed to discussing the objectives of drilling on modern submarine fans, specifically the Mississippi Fan sites slated for drilling during Leg 96 of the Deep Sea Drilling Project. While considering the wide range of many problems that might be investigated by drilling, these discussions focused on setting priorities between sites and on what one might expect to find at the proposed drill sites.

The last official day of the meeting was mainly devoted to general discussions on "state-of-the-art" knowledge of the meeting, the value of presently published models, and what directions of study should be emphasized.

The informal presentations and discussions were recorded by a certified court stenographer and on video and cassette tapes. From this, a set of COMFAN notes were produced to give the participants a permanent record of the meeting.

The participants agreed that the main issues discussed during the COMFAN meeting were too timely for the geological community to be kept to the "membership" only. The transcribed meeting notes were edited and each participant was consulted to avoid mistakes, especially in geological terminology, that can creep into a courtroom style of transcript. These notes were then used as the basis for the manuscripts to be published in the special issue of *Geo-Marine Letters*.

After the meeting, the editors and several interested COMFAN participants produced a more detailed outline for the special issue and set up guidelines concerning the contents of each contribution to ensure that the papers would be compatible. In addition, it was decided to construct a wall chart to demonstrate more effectively the large variety in shape, size, and morphologic features of modern deep-sea fans and a comparative listing of a number of parameters, such as dimensions, setting, type of sediments, etc. Chapter 3 presents a discussion of the construction of this chart.

Mississippi Fan Drilling Results

Normally, a brief description of the results from any leg of the Deep Sea Drilling Program is prepared on-board ship and published in both *Geotimes* [6] and *Nature* [7]. The next step is to rewrite and edit the shipboard descriptions of the site chapters for the *Initial Reports of the Deep Sea Drilling Project*, which also include manuscripts based on the results from shore-based laboratory analyses and studies. Preparation and publication of this document by the U.S. Government Printing Office under the guidelines of the National Science Foundation normally takes years (2 to 5) before it is available to the geologic community.

The results of DSDP Leg 96 on the Mississippi Fan have received considerable attention, and oral presentations already have been given in many countries by individual shipboard participants. There have also been numerous requests, primarily from industry, to publish the major results in a timely manner. DSDP encourages such publication provided that all shipboard scientists are included in the list of contributors (in front of book). The chapters (36 to 47) dealing with the results of drilling on the Mississippi Fan will show the leading authors by name and other contributors as "Leg 96 Scientific Party" with reference to the mentioned table. The chapters will deal only with the results through November 1984. At that time, most of the laboratory analyses had been completed; interested readers should contact individual researchers for more details.

Submarine Fans and Related Turbidite Sequences

This publication uses the special issue "Submarine Clastic Systems: Deep Sea Fans and Related Turbidite Facies" [5] of *Geo-Marine Letters* as its core. A number of aesthetic modifications and a few small additions to the *Geo-Marine Letters* special issue papers were made to produce a more updated and edited book appearance. The contributions are arranged in sections grouped by tectonic setting; within sections the papers appear alphabetical.

The order of chapter sections is as follows (see Table of Contents for details):

Section I (Chapters 1–7) deals with introductory issues, external influences that affect submarine-fan development, and hydrocarbon potential, and provides an introduction to the fan comparison chart.

Section II (Chapters 8–14) contains contributions on modern fans in active margin settings.

Section III (Chapters 15–23) deals with papers on modern fans in passive margin settings.

Section IV (Chapters 24–33) presents examples of ancient turbidite sequences in active margin settings.

Section V (Chapters 34 and 35) has contributions on ancient turbidite sequences deposited in passive margin settings.

Section VI (Chapters 36–47) presents contributions on the results of the DSDP Leg 96 drilling on the Mississippi Fan and a general review of fan structure based on seismic stratigraphy.

Section VII (Chapter 48) contains a general summary.

Chapters 2 and 48 are, to a certain degree, companion papers discussing, respectively, the principle 1982 COMFAN meeting results and where we stand one year after drilling on the Mississippi Fan.

Each chapter includes its own list of references. This results in some duplication, but avoids publication delay and lightens the task of the editors in restructuring the COMFAN contributions. The affiliations of each contributor are presented in the front matter, pages xi–xiii.

Acknowledgments

The COMFAN meeting participants express their gratitude to the management of Gulf Research & Development Company for providing foresight, facilities, and funds to conduct the successful meeting in Harmarville, and to produce the COMFAN notes. Our special thanks go to the President, Dr. R. E. Wainerdi. Personnel from different drafting sections of Gulf Oil Corporation prepared many of the map illustrations to ensure reasonable compatibility. All contributors to *Geo-Marine Letters* are thankful for that enormous help. All present authors are indebted to Springer-Verlag New York, Inc. for their interest and cooperation in publishing this book in a timely fashion. Many persons besides the authors reviewed manuscripts for technical content and we are indebted to them.

References

[1] Bouma, A. H., and Brouwer, A. (eds), 1964. Turbidites. Developments in Sedimentology 3, Elsevier, Amsterdam, 264 pp.

[2] Dzulynski, S., and Walton, E. K., 1965. Sedimentary Features of Flysch and Greywackes. Developments in Sedimentology 7, Elsevier, Amsterdam, 274 pp.

[3] Whitacker, J. H. McD. (ed), 1976. Submarine Canyons and Deep Sea Fans: Modern and Ancient. Benchmark Papers in Geology 24, Dowden, Hutchinson & Ross, Stroudsburg, PA, 461 pp.

[4] Stanley, D. J., and Kelling, G. (eds), 1978. Sedimentation in Submarine Canyons, Fans, and Trenches. Dowden, Hutchinson & Ross, Stroudsburg, PA, 395 pp.

[5] Normark, W. R., Mutti, E., and Bouma, A. H. (eds), 1983/1984. Submarine Clastic Systems: Deep Sea Fans and Related Turbidite Facies. Geo-Marine Letters, v. 3, no. 2–4, pp. 53–224.

[6] Leg 96 Scientific Party, 1984. Challenger drills Mississippi Fan. Geotimes, v. 29/7, pp. 15–18.

[7] Staff Leg 96, Deep Sea Drilling Project, 1983. On the Mississippi Fan. Nature, v. 306, no. 5945, pp. 736–737.

CHAPTER 2

COMFAN: Needs and Initial Results

Arnold H. Bouma, William R. Normark, and Neal E. Barnes

Abstract

COMFAN (COMmittee on FANs) met in 1982 to discuss problems in turbidite research. Problems encountered in research on modern submarine fans include the wide variety in types of observations that make comparisons between fans difficult or suspect, as well as a lack of sedimentologic and stratigraphic information below the upper tens of meters. Ancient turbidite sequences seldom provide basinwide information for time-equivalent intervals. Comparison of modern submarine fans and ancient turbidite sequences is still tentative, and in many cases impossible, because of the incompatibility of study approaches and lack of a common scale of observations.

Introduction

COMFAN (COMmittee on FANs), formed in September 1982, was conceived to: 1) review the current status of research on modern and ancient turbidites in fan and nonfan settings, 2) identify the main reasons why comparison between modern fans and ancient turbidite sequences is still tentative, 3) identify the primary areas of confusion (i.e., nonuniform uses of terminology, process concepts, and data types), 4) present the most recent, generally unpublished, results of turbidite research by active specialists in the field, 5) attempt to recognize and define specific criteria such as facies types, morphologic features, morphometric zonations, sediment distribution, and depositional processes that can be applied to both fan and nonfan turbidite environments [1], and 6) critically review and, if necessary, modify the proposed Deep Sea Drilling Project program on the Mississippi Fan. The meeting was not designed to resolve differences between published depositional models or to propose new ones. The number of participants was kept small and the number and length of presentations on modern and ancient turbidite systems were restricted to permit active discussion during the three-day meeting.

Exploration for and production of hydrocarbons are gradually moving to deeper water environments in which turbidites form the dominant sedimentary facies. These types of deposits have been explored on land for several decades and have been regarded as difficult targets because of the problems encountered in correlating wells over short distances. In addition, little is understood about the influence on the distribution and characteristics of deep-water sand by such factors as type of continental margin, size and shape of the sedimentary basin, sand/clay ratio of the sediment source, single versus multiple point (line) source, eustatic sea level changes, and seafloor gradients. All of these factors may strongly affect the transport processes and, thus, the sorting of the sands, which in turn influences porosity and permeability.

Published models are generally based on detailed studies of one or, at most, a few modern submarine fans or ancient turbidite systems. As a result, the observer has to be extremely cautious in applying one of those models to other areas. In addition, models derived from ancient turbidite systems tend to show more differences than commonalities with modern deposits, and the application of models can be more confusing or damaging than helpful ([1]; see Chapter 6).

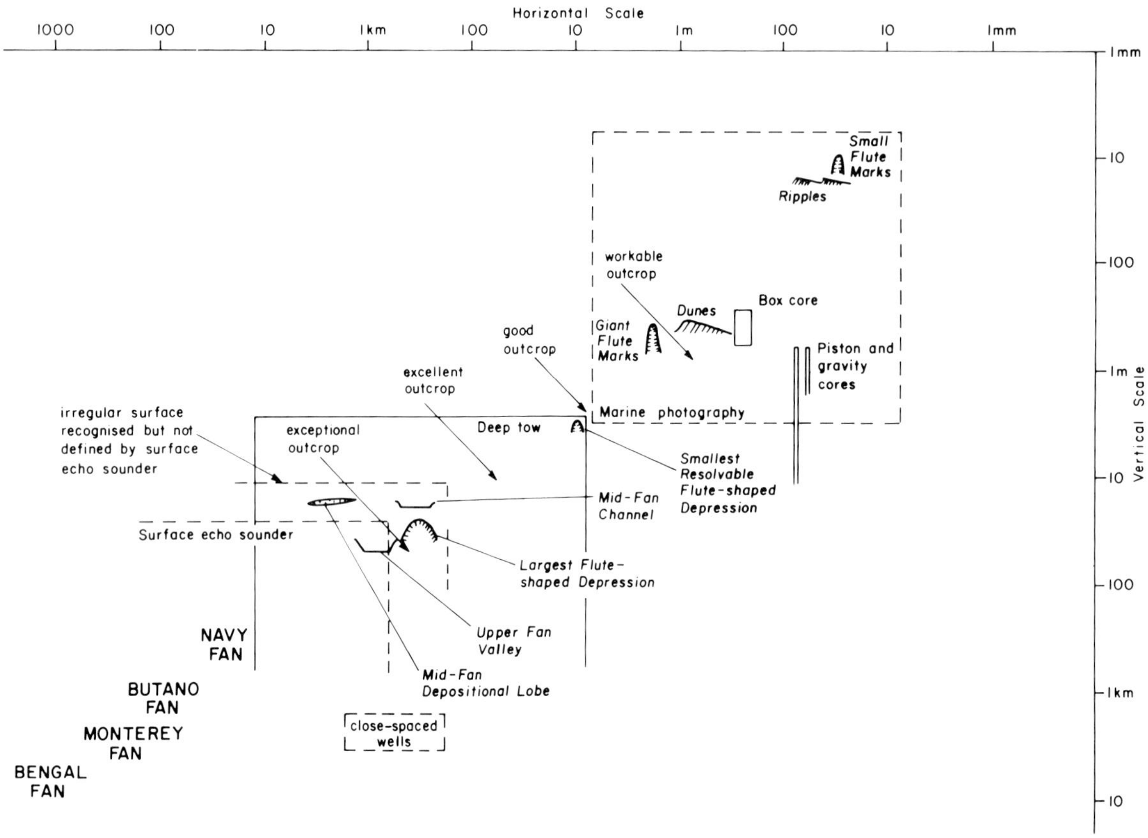

Figure 1. Horizontal and vertical scale comparison for modern submarine fan data sets with sedimentologic observations based on outcrop studies of ancient turbidites; modified from Normark and others [1]. Non-deep tow systems are capable of resolving features in areas noted by dashed lines in the block on the left.

COMFAN Issues of Discussion

Scale Relationships and Data Types

The lack of a common data set to compare modern submarine fans and ancient turbidite sequences is the result of differences in both the scale of observation and data acquisition methods [1,2]. It may be possible, if not probable, that no common data set for comparison exists. Data collection from ancient turbidite units provides details primarily on layer thickness, distribution in lateral and vertical senses over short intervals, composition, grain size, color, sedimentary structures on bedding planes, trace fossils, and paleocurrent directions. Detailed stratigraphic sections, ranging in thickness from a few meters to hundreds of meters, can be measured, and individual layers might be followed across the width of the outcrop. Biostratigraphic control does not have the same resolution as the depositional events that are represented by single layers or packets of layers. Marker beds that provide basinwide time correlations generally are rare. Erosion, masking of layers by debris or vegetation, and structural complications can prevent or make it difficult to carry out detailed correlation attempts and, thus, any reconstruction through time of the original basin configuration. Paleocurrent measurements seldom define the source area adequately.

Studies of modern submarine fans can provide a good insight into basin size, shape, gradients, surficial facies, and source area(s). The acoustic stratigraphy or acoustic facies, or both, can be identified for some fans, allowing correlation of units across the basin. The total thickness of the fan systems also can be established with acoustic techniques. The surface morphology can be defined to varying degrees depending on the survey system(s) used [1]; in general, the most detailed information from the water-sediment interface is obtained by side-scan sonar, multibeam echo-sounding, and high resolution seismic reflection systems, especially those towed near the seafloor. Detail of internal structure decreases with depth of acoustic penetration on seismic reflection records and, basically, these high-resolution techniques cannot provide the details comparable with those observed in outcrops (Fig. 1). Although sediment cores can provide details similar to those from outcrop studies, the short cores only sample the uppermost sediment. Deep drilling with continuous coring, such as the Deep Sea Drilling Project and the new Ocean Drilling Project, might eventually narrow this gap.

Figure 1 illustrates the main difficulty we face when attempting to compare modern submarine fan systems with ancient turbidite systems. Acoustical equipment that scans the seafloor from the water surface cannot adequately resolve a vertical relief that is less than 10 meters, or horizontal dimensions smaller than several hundred meters when the water depth exceeds a few thousand meters. Deep-towed side-scan sonar with 3.5 or 4.5 kHz profiler systems provide considerably more detail, but because of the slow towing speeds and the small area of observation, their use has been limited and sporadic, and information has been collected only from a few fans.

Observations on the surface and near-surface area represent the upper skin of a modern fan and, therefore, may not be representative of the principal depositional processes. Channels and their depositional fills, for example, can be many kilometers wide and tens to hundreds of meters thick.

Figure 1 also demonstrates that the small size of many phenomena described from ancient turbidite systems, as well as the thickness of well-published vertical sequences, are beyond the resolution of most of the instruments available to the marine geologist. Typical channels observed in large outcrops may be seen on side-scan sonar, but they are too small to be resolved on most subbottom reflection records.

Models and inferred transport/depositional processes for modern submarine fans are typically based on surface morphology, planimetric shape, and surficial sediments. Differential compaction of fine-grained and coarse-grained sediments, together with erosion, partial masking, and tectonic influences, prevent the reconstruction of paleosurfaces of ancient systems over large areas. Models constructed from ancient turbidite facies concentrate on stratigraphic variations and vertical sediment sequences. Lack of a common comparative factor and differences in approach to studies of modern and ancient systems, therefore, have resulted in different terminologies. The most simple and least confusing examples are the upper-middle-lower divisions for modern submarine fans and inner-middle-outer divisions for ancient turbidite systems.

Morphometric Emphasis Versus Sedimentary Facies

Submarine fans and related turbidite systems that are basically still exposed at the seafloor in present ocean basins occur as distinct morphologic features. In many cases, these features can be further divided into a number of subenvironments that are primarily defined by their morphologic characteristics.

Ancient turbidite sequences are characterized by their sedimentary rocks and vertical sequences. Consequently, the comparison between ancient and modern systems is limited by several factors [1]:

1) There is no unanimity in the application of the term "fan." This word normally suggests a fan-shaped deposit that results from long-term active turbidite-type deposition, related to a point source, and taking place in a large unconfined basin of low relief and with gentle gradients. However, most examples of modern fans commonly involve the infilling of narrow irregular depressions that prevent the development of a typical fan shape. When viewing the enclosed wall map (see Chapter 3), one observes that the shapes are strongly influenced by basin configuration and that an unrestricted basin is more the exception than the normal occurrence. These basin-shape influences must have been equally common in earlier stratigraphic times. Most ancient turbidite systems that are now exposed in outcrops were deposited in active margin settings, and the chance that the depositional basins were rather narrow and oblong is high. Those fans that developed in passive margin settings on oceanic crust had a better chance to build a more fan-shaped deposit.

2) Channels and lobes can commonly be defined on modern fans. For ancient turbidite systems, we can only describe the sedimentary facies and infer lobe morphology; therefore it is advised to describe ancient deposits without specific morphologic connotations.

3) Entire turbidite sandstone bodies can seldom be seen in outcrop. As a result, the original morphology and three-dimensional shape of an ancient turbidite environment can only be inferred.

4) Sequence analysis of vertical facies has been widely used to define the physiographic subdivision of an ancient turbidite system. Internal changes in bed thickness, sandstone/shale ratio, texture, sedimentary structures, and other characteristics define directional relationships and serve to reconstruct the shape of the ancient fan. Although researchers demonstrated the workability of the concepts from areas where detailed studies led to the development of a depositional model [3], many followers misuse the model by applying it to systems exposed in only a few outcrops.

5) Thinning- and/or fining-upward sequences and thickening- and/or coarsening-upward sequences have been widely and often successfully used to recognize channel fill and lobe deposits, respectively. A comparison of modern distributary channels and channel mouth bars in fluvially-dominated deltas was used as the basis for the sequence concept. The actual validity of such trends as environmental indicators has not been tested yet by suf-

ficiently deep cores from modern fans. The meaning of accretion and progradation processes is, therefore, completely interpretive. Throughout the literature, much confusion exists about the criteria and even the terminology that should be applied to establish these sequences. Unfortunately, this has resulted in an inconsistent usage of these concepts.

Other Important COMFAN Issues

1) It was generally agreed upon that very few investigators discuss the tectonic setting of an ancient or modern fan location and, consequently, the potential tectonic influence on source material and basin shape or other controlling factors (e.g., size, gradients) is seldom mentioned.

2) Investigators of ancient turbidite sequences may mention point or line sources for the sediments without defining the meaning of such a source designation. In particular, the term "line source" can result in confusion if it is not defined; even large deltaic sources often have a canyon (point source) equivalent for each growth stage of the fan.

3) Little study has been attempted on the mechanisms that set the material in a shallow-water source area into motion. Terms like "oversteepening," "heavy storm surge and wave action," "earthquakes," and "pore water pressure" are too freely cited without critical discussion, thus, side-stepping the real causes.

4) The COMFAN participants agreed that a "submarine fan" is in reality a channel-(levee)-overbank system in which the overbank areas either consist primarily of muds or thin-bedded turbidites [4].

5) It became apparent that small- and large-sized sediment failures in both source and fan areas are more common than heretofore realized. The causes and mechanisms have not yet been sufficiently studied.

6) Few investigators have a proper feeling for a channel and its dimensions or fill. The dimensions of channel depth or its fill range from a few tens to hundreds of meters. In an aggradational channel, the extant relief might be only tens of meters, where the fill can be hundreds of meters thick. Channel widths on modern fans are commonly measured in kilometers rather than hundreds of meters, as is the case for most ancient systems.

7) We know very little about the fills of modern fan channels. Is it always a fining-upward sequence with possibly a coarse-grained lag or bottom deposit overlain by sediment representing passive infilling? Or is there also an active depositional fill? To us, the first case refers to a conduit that is cut off from its supply of coarse material and becomes filled with fine-grained material later on, while the second case refers to an aggradational channel.

8) Channel terminations and related sediments became a heated point of discussion during the COMFAN meeting. There was a great amount of discussion on subjects ranging from descriptive and often poorly-defined terms (i.e., sheet sands, depositional lobes, and channel mouth bars) to interpretive terms (compensation cycles [5] and flow expansion), but no answers or consensus resulted.

9) The specialists working on modern fans specifically agreed that eustatic sea-level variations have a major influence on forming submarine fans, with the major activity taking place during low sea-level stands.

10) The term "lobe" now causes confusion because of non-standard usage. In ancient sequences, a lobe is basically recognized by the lack of basal channeling within the sequence. A very broad channel-fill sequence can, therefore, be mistaken for a lobe. "Lobe" also applies to specific modern features, such as suprafan, as well as to nonchannelized ancient sandstone bodies that can be the analogs of modern lobes.

Because many researchers like to use the word "lobe" rather than the complete term "depositional lobe," serious objection was raised to the term "fanlobe" for a complete channel-levee-overbank system defined acoustically (see Mississippi Fan, Chapters 36 and 39).

Major COMFAN Conclusions

The intent of the COMFAN review is not to discourage turbidite studies, but to emphasize the need for thorough, careful field investigations using consistent terminology. The observational restrictions and difficulties in comparison just outlined should only serve to make the reader aware of some of the pitfalls and help avoid reporting meaningless conclusions or poorly received fan models. The COMFAN participants themselves had difficulties with scale effects during the many discussions, demonstrating that it is easier to discuss the pitfalls than avoid them. The major conclusion reached during the 1982 COMFAN meeting was that the participants did not feel ready to reach any major conclusions for re-casting submarine fan research.

At this time, it is extremely difficult and dangerous to make direct comparisons between modern submarine fans and ancient turbidite sequences, basically because of differences in data types and scale relations.

Slope failure of local and possibly regional extent is more important and common than heretofore realized. More detailed studies are required and serious geotechnical studies may provide more definitive answers.

A general consensus was that major fan accumulation takes place during periods of relative low sea level. This may be important for both active and passive margin settings, and may be the prime reason for fans that develop in a passive margin.

References

[1] Normark, W. R., Mutti, E., and Bouma, A. H., 1983/84. Problems in turbidite research: a need for COMFAN. Geo-Marine Letters, v. 3, pp. 53–56.

[2] Normark, W. R., Piper, D. J. W., and Hess, G. R., 1979. Distributary channels, sand lobes, and mesotopography of Navy Submarine Fan, California Borderland, with applications to ancient fan sediments. Sedimentology, v. 26, pp. 749–774.

[3] Mutti, E., and Ricci-Lucchi, F., 1972. Le torbiditi dell'Apennino Settentrionale: introduzione all'analisi de facies. Memorie della Societa Geologica Italiana, v. 11, pp. 161–199.

[4] Nelson, C. H., and others, 1978. Thin-bedded turbidites in modern submarine canyons and fans. In: Stanley, D. J., and Kelling, G. (eds.). Modern and Ancient Submarine Canyons and Fans. Dowden, Hutchinson & Ross, Stroudsburg, PA, pp. 177–189.

[5] Mutti, E., and Sorrino, M., 1981. Compensation cycles: a diagnostic feature of turbidite sandstone lobes. In: International Association of Sedimentologists; 2nd European Regional Meeting, Bologna, 1981, Abstracts, pp. 120–123.

CHAPTER 3

Diagnostic Parameters for Comparing Modern Submarine Fans and Ancient Turbidite Systems

Neal E. Barnes and William R. Normark

Abstract

The comparison of modern and ancient submarine fans and related turbidite deposits has been hindered by the lack of common descriptive parameters and a failure to understand the effects of the extreme size range of the deposits. To address these problems, a summary of diagnostic parameters was prepared that includes: 1) morphometric maps that present a variety of physiographic characters and sediment parameters, 2) a table of key descriptive features that allows a quick comparison of the major physical parameters, and 3) fan outlines presented at the same scale (1:5 × 10^6) that make it possible to visually compare sizes of 21 major "submarine fans."

Introduction

Both formal and informal discussions during the COMFAN (COMmittee on FANs) meeting focused on the problems in reconciling the discrepancies among published submarine fan models. One of the major problems was found to be that of scale. It became apparent in many of the discussions that we are not only dealing with the problem of comparing extremely varied types of turbidite deposits ("apples and oranges"), but also of comparing "watermelons and grapes."

Too little attention has been given to the direct comparison of the variety of fan sizes, shapes, and morphologic features. To alleviate this shortcoming, it was suggested that a "wall chart" presentation of various descriptive parameters on modern submarine fans and ancient turbidite systems be added to the special issue of *Geo-Marine Letters* [1] proposed for publication by COMFAN.

Using data provided by scientists who had studied the various depositional systems, we assembled three major units for comparison: 1) morphometric maps, 2) diagnostic parameters, and 3) fan outlines at an equal scale. All COMFAN authors were asked to provide the descriptive data in a standardized format for the preparation of morphometric maps and a data sheet outlining the diagnostic parameters requested for the table (in back of book).

Morphometric Maps

The front of the wall chart presents a schematic representation of key morphologic features of 13 modern fans, including selected data on surface gradients, channel dimensions, and sediment lithology (grain size). Only those fans that are described in the special issue of *Geo-Marine Letters* [1] are included. All maps were drafted by one drafting group (Gulf Research and Development Company) to ensure uniformity in the style of presentation. The various degrees of "completeness" or detail in the maps represent the extent of research that has been conducted relative to the size of a particular fan as well as the interests of the particular researchers (i.e., sedimentological versus geophysical).

The morphometric maps clearly demonstrate that few submarine fans display a typical "fan shape." Of the 13 fans presented, only the Amazon, Magdalena, and possibly the Indus show a "classic" fan shape; in fact, of the 21 modern fans listed in the descriptive features table, only the Indus, Magdalena, and the Nile Fans are described as fan-shaped. The Astoria and Rhone Fans appear to have started with a "true" fan shape, but are now restricted by sediments coming from adjacent turbidite systems. The Crati, Delgada, Laurentian, Mississippi, Monterey, and Navy Fans show varying degrees of basin control on their outlines. Controls on fan shape are discussed in detail elsewhere (Chapters 4 to 6). In addition, several fans cannot be described with the standard upper, middle, and lower fan divisions (e.g., the Crati, Ebro, and Laurentian Fans); this is not surprising when one considers that there is no agreed upon usage for these terms.

Table of Descriptive Features

The table is generally self explanatory, but a few comments about its compilation are appropriate. Each contributing COMFAN author received a data sheet requesting information for this table. In addition, other researchers were solicited to provide the data for turbidite systems not covered by one of the chapters. These data were then compiled into a conformable format for ease of presentation. It should be noted that for several entries, especially the dimensions, some of the data are only best guesses and for some fans no significant data are available for presentation at this time. The table is viewed as the most ephemeral part of the wall chart, constantly evolving as more data are collected or updated. The table's power lies in its allowing the reader to compare several different parameters and key descriptive features of one or more submarine fans.

The fans are listed in two major groups, modern (8–23) and ancient (24–35), and their locations are given on the world map presented on the reverse side of the chart. Modern fans are considered to be those that are still active or that were only recently cut off from their sediment source. The Zodiac Fan, which has been inactive since Miocene, can thus be considered as transitional between modern and ancient.

The table further subdivides the fans by margin type as either active (accretionary, transform, and subduction) or passive. Using this classification, the modern fans can be divided about equally, with 11 on active margins and 10 on passive margins. Among the ancient turbidite systems, however, eight are considered to be on active margins and two on passive margins. From this, one conclusion is that fan deposits have a better chance of being preserved if they are initially deposited on continental crust along an active margin. This is further supported by the two ancient systems on passive margins, the Brae and the Torok-Fortress Mountain systems, which are defined using well data (see column on instrumentation) rather than outcrop data.

One should exercise caution in making generalizations based on a single parameter from the table (e.g., dominant grain size). With the exception of the La Jolla and San Lucas Fans, all modern fans list the dominant grain size in the mud size classification (i.e., mud, silt, or clay). The ancient systems all list sand or sand and mud as the dominant grain size. Whether this range in grain size for modern fans is truly representative or just a manifestation of a lack of long enough core samples is unknown. Another explanation could be that only the sandy sections of ancient systems are adequately exposed and easy to map, thus resulting in a biased appraisal of the "dominant" grain size. In general, the value of the entries in the table must be judged by, at the very least, checking the column showing type and extent of data.

The table does not include all known fans and turbidite systems. A more exhaustive list can be found on a recent map of productive and nonproductive sediment provinces [2].

Comparison of Fan Outlines at a Scale of 1:5,000,00

The comparison of fan outlines at an equal scale is the most striking feature of the wall chart and is the central theme around which the chart evolved. The outlines have been arranged for aesthetics and the efficient use of space, but it is noteworthy that all 23 fans presented, except for the Indus and Chugach Fans, could be crowded inside the outline of the Bengal Fan. As a point of reference, the area of the Indus Fan is about one-third greater than the state of Texas, while the Navy Fan would easily fit inside the submarine canyon feeding the Bengal Fan. The Cengio turbidite system and Crati Fan both had to be drafted slightly oversized so that they would be visible after reproduction.

Conclusion

Every effort has been made to make the chart as accurate as possible with the data available. However, the reader should feel free to inform us of any errors, especially in the table, and update his or her own copy as new data become available.

References

[1] Normark, W. R., Mutti, E., and Bouma, A. H. (eds.), 1983/84. Submarine Clastic Systems: Deep Sea Fans and Related Turbidite Facies. Geo-marine Letters, v. 3, no. 2–4, 172 p.

[2] St. John, B., Bally, A. W., and Klemme, H. D., 1984. Sedimentary Provinces of the World—Hydrocarbon Productive and Nonproductive. American Association of Petroleum Geologists. Map and text, 35 pp.

CHAPTER 4

Sedimentary, Tectonic, and Sea-Level Controls

Dorrik A.V. Stow, David G. Howell, and C. Hans Nelson

Abstract

To help understand factors that influence submarine fan deposition, we outline some of the principal sedimentary, tectonic, and sea-level controls involved in deep-water sedimentation, give some data on the rates at which they operate, and evaluate their probable effects. Three depositional end-member systems, two submarine fan types (elongate and radial), and a third nonfan, slope-apron system result primarily from variations in sediment type and supply. Tectonic setting and local and global sea-level changes further modify the nature of fan growth, the distribution of facies, and the resulting vertical stratigraphic sequences.

Introduction

Numerous modern fans have been studied over the past 15 years, and many examples of ancient turbidite sequences have been interpreted as fans or parts of fans. Several different descriptive models have been developed from both modern and ancient examples to characterize morphological features and the pattern of facies distribution. Clearly, several different fan types exist, but it is equally clear that many turbidites and associated sediments are deposited in other settings, including slope aprons in small and large basin systems, submarine canyons, and trenches.

We believe it is possible to gain further insight into where, why, and what types of fan systems are formed by considering the primary and secondary controls on their development [1–4]. In this paper, we first outline these main controls, giving some data on the rates at which they operate, and then discuss their probable effects.

Controls and Rates

Three primary controls on fan development and deep-sea sedimentation can be identified (Figure 1): (1) sediment-type and supply, (2) tectonic setting and activity, and (3) sea-level variations. These controls are by no means independent; for example, tectonic factors play an important part in determining sediment supply or local sea-level changes.

Sediment Type and Supply

Various types of sediment are available for redeposition. Terrigenous material is the most abundant worldwide, with muds being between two and ten times as important volumetrically as sands and gravels. Biogenic debris from carbonate reefs and platforms is common at low latitudes, and calcareous and siliceous oozes may be locally redeposited from areas of high pelagic accumulation. Evaporites, volcaniclastics and organic-carbon-rich sediments can all occur as turbidites and associated facies but rarely form a complete submarine fan complex. The sediment grain size affects the process and distance of transport and hence the geometry of the deposit. Biogenic particles behave differently from terrigenous grains during transport, so that carbonate facies and fans differ from their clastic counterparts [5].

The volume and rate at which sediments are supplied to an area and therefore made available for redeposition are other important variables. Major river-delta systems, such as the Ganges, Indus, and Mississippi, can provide a large and rapid supply of sediment to the shelf, although the availability of

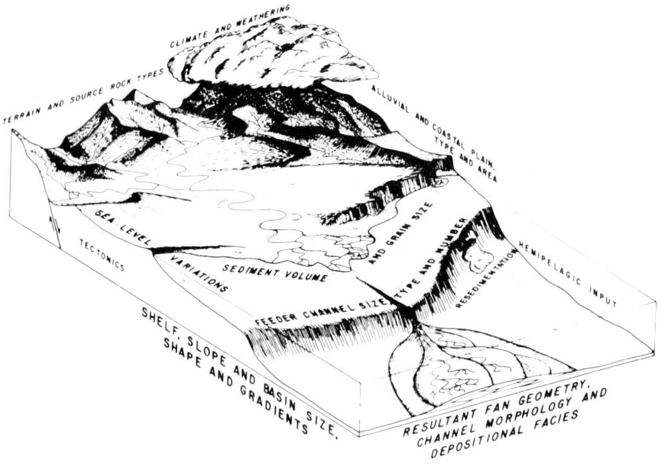

Figure 1. Schematic diagram highlighting the variety of physical processes that control submarine fan development.

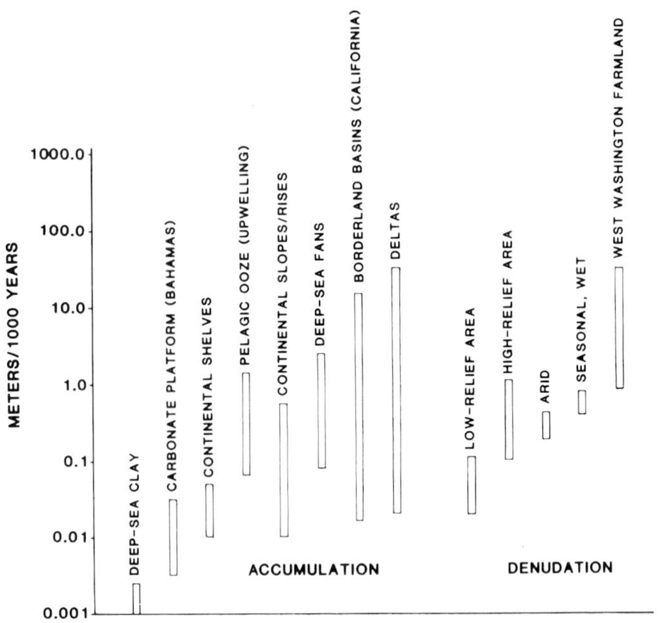

Figure 2. Log-plot comparing span of rates for sedimentary accumulation and denudation. Based on data in Howell and Von Huene [6] and Blatt and others [7] (1980).

this material for downslope resedimentation will depend on sea-level and shelf width. Wave-stirred, canyon-indented shelves will generally provide less sediment to the outer margin. In high latitudes, glaciers and floating ice shelves may greatly increase the supply of terrigenous material to the shelf margin. Low-latitude carbonate platforms and topographic highs covered with pelagic material commonly provide lower rates of sediment supply.

The number and spacing of input points along a given margin will determine whether single or isolated fans are developed, or whether an overlapping-fan/slope-apron system is produced.

Secondary factors that influence sediment type and supply are also illustrated schematically in Figure 1. They include: (1) the original source-rock type that affects composition, size, and erodability of detritus; (2) climate and vegetation that affect the nature and degree of weathering and the mode of supply (fluvial, glacial, wind, etc.); (3) the relief and tectonic activity in the hinterland that affect the rate of denudation and of supply to the transitional source; (4) the distance between the original and transitional sources and the mode of transport that affect the compositional and textural maturity of sediments; (5) the topography, tectonic activity, and sediment residence time in the transitional source area that affect sediment compaction, erodability, and type; and (6) local marine conditions (currents, Coriolis force, water temperature, upwelling, etc.) that affect the supply of biogenic detritus and organic carbon, bioturbation, or physical reworking and suspension of sediment, as well as the final sediment distribution.

The rate at which land areas are eroded (Figure 2) will have an effect on sediment supply [6,7]. On average, denudation of a low-relief terrain is at a rate of 0.01 to 0.1 m/1000 yr, whereas for high-relief areas the rates are as much or an order of magnitude higher, 0.1 to 1 m/1000 yr. Areas of high seasonal precipitation tend to erode at twice the rate as those that are semiarid. The rates of sediment accumulation (Figure 2), particularly in the transitional source area, are equally important. In the long term, these rates rarely exceed 50 mm/1000 yr. Local rates can be as high as 1 m/yr, such as in front of active delta distributaries. Typical long-term accumulation rates on carbonate or clastic shelves are from 10 to 40 mm/yr. Pelagic ooze sedimentation will normally not exceed 30 mm/1000 yr, although under upwelling areas it may reach 0.1 m/1000 yr.

Resedimentation of material to deeper water results in accumulation rates on modern deep-sea fans from 0.1 to 2 m/1000 yr, and up to 10 m/1000 yr in small tectonically active basins. Turbidity currents are one process by which resedimentation occurs, and estimates of their frequency mostly range from 1 per 500 to 10,000 years for deep-sea clastics and from 1 per 20,000 to 100,000 years for carbonates [2,6]. Muddy turbidity currents off some active river-delta systems may occur with a frequency as high as one per year.

Tectonics

The major tectonic settings in which submarine fans and associated deep-water systems can develop include mature passive margins (eastern North American and Gulf of Mexico), active rifting margins (Red Sea, Gulf of California), convergent margins with arc or trench systems (Aleutian Trench, Nan Trough), transform margins (California Continental

Borderland, offshore Venezuela), marginal seas and back-arc basins (Lau Basin, Phillipine Sea), oceanic basins flanked by ridges and seamounts (Indus Cone), and intracratonic basins on continental shelves and within continents (Bering Sea Shelf, Cretaceous seaway of North America).

These tectonic settings exert a first-order control on the types of fans developed by affecting the rates of uplift and denudation, drainage patterns, coastal plain and shelf widths, continental margin gradients, gross sediment budgets, the morphology of receiving basins, and local sea-level changes. The specific tectonic parameters that immediately determine fan type are the size and internal geometry of the basin, including gradients of the basin margin and floor [8]. The style and frequency of seismic activity and faulting, both in the original and transitional source areas, are also of primary significance since these factors influence the frequency and volume of sediment gravity flows feeding the basin, e.g., mature passive margins experience infrequent, but commonly large, earthquakes, which may trigger very large slumps that develop into debris flows and turbidity currents. Frequent earthquakes along active margins do not permit a large build-up of sediment in transitional settings.

Secondary factors involved in tectonic activity include the rates of horizontal and vertical motion, the maturity of the margin, and the relationship of a particular setting to neighboring plates. The rates of motion can be critical. If deposition rates are slower than tectonic rates, fan growth will be controlled by tectonism rather than by sedimentary factors such as fluctuating gradients and migrating channels, distributaries, and terminal lobes.

The rates at which various tectonic processes operate are illustrated in Figure 3 [6,7]. Rates of uplift in mountains, mainly along convergent and transform margins, are relatively fast, ranging from 1 m to as much as 75 m/1000 yr, although they are generally between 3 and 10 m/1000 yr. Such uplift is generally of short duration (10^5 to 10^6 yr). More long-term epeirogenic uplift in nonmountainous areas is commonly from 0.1 to 3.7 m/1000 yr, although isostatic uplift (and subsidence) can be an order of magnitude greater, from 4 to 40 m/1000 yr. All uplift is probably episodic with short periods of rapid vertical movement alternating with longer periods of stability or only very slow movement.

Subsidence is equally variable and also episodic in nature. The greatest rates, up to 12 m/1000 yr, occur locally in small basins within transform margins and in arc and trench basins (up to 5 m/1000 yr) along convergent margins. Average rates outside of these tectonically active regions rarely exceed 1 m/1000 yr and are significantly lower (10 to 40 mm/1000 yr) along passive margins and adjacent oceanic basins.

By comparison with vertical motion, horizontal plate movements may be an order of magnitude greater, from 10 to 100 m/1000 yr. Spreading rates across oceanic ridges and displacement rates within transform zones are often of these magnitudes.

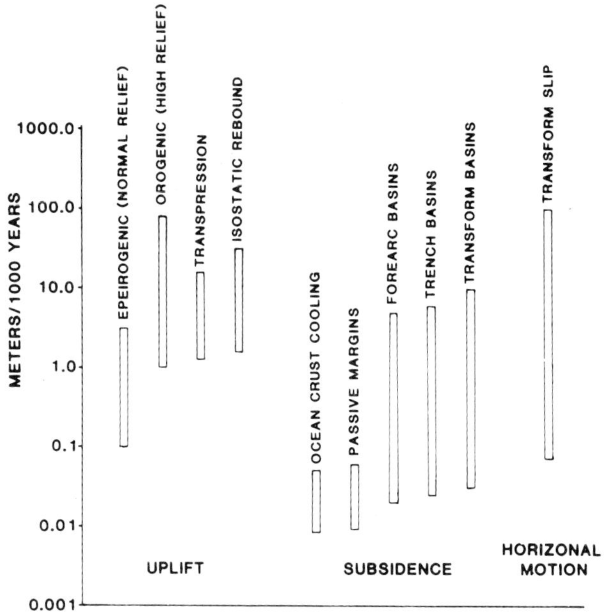

Figure 3. Log-plot comparing span of rates for tectonic uplift, subsidence, and horizontal plate motion. Based on data in Howell and Von Huene [6] and Blatt and others [7].

Sea-Level

Fluctuation in sea-level not only affects the nearshore realm of sedimentation, but also profoundly influences deep-sea depositional and resedimentation patterns. Shoreline sources such as rivers or littoral drift cells either may have direct access to basin slopes during periods of low sea-level or indirect access through paralic and continental shelf environments during periods of high sea-level [8].

Sea-level changes may be global (eustatic) or regional in nature. Eustatic fluctuations occur as a result of a change in the total volume of ocean basins or a change in the volume of sea water. The volume of ocean basins is affected by four secondary factors, for which Pitman [9] has estimated the maximum sea-level change that can result: (1) Variation in sediment input can cause sea-level fluctuation of up to 2.0 mm/1000 yr; (2) continental collision and subduction has been calculated to change sea-level by 1.6 mm/1000 yr in the case of the collision of India into the Asian plate; (3) growth of seamount chains causes minimal sea-level change of only 0.2 mm/1000 yr; whereas (4) swelling and shrinking of midocean ridge systems can cause changes of up to 6.7 mm/1000 yr.

The variation in the total volume of sea water as a result of subduction, volcanism, and sea-floor spreading processes seems unlikely to produce sea-level changes in excess of 2 mm/1000 yr [9]. However, the locking-up of very large amounts of water in expanded polar ice sheets during glacial periods and its release during warm climatic epochs can cause

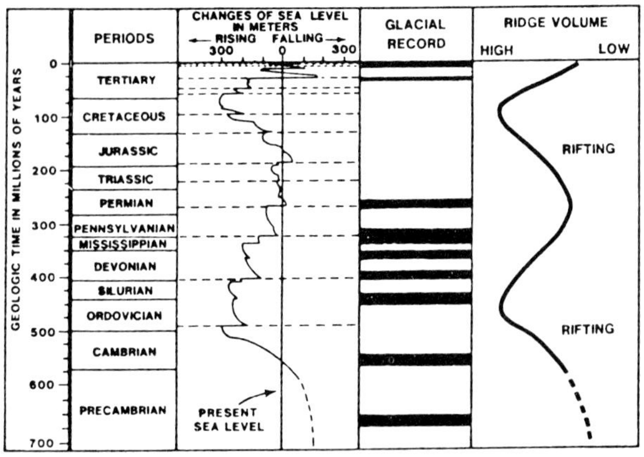

Figure 4. Comparison of global changes of sea level, glacial record, and midoceanic ridge volume. After Shanmugam and Moiola [32] 1982.

enormous sea-level fluctuations of up to 10 m/1000 yr.

These rates, apart from those induced by climate, appear too small in comparison with tectonic rates of uplift and subsidence to have any appreciable effect on fan growth along active margins (Figure 3). However, prolonged eustatic changes, due to changes in spreading rate and volume of midocean ridges, will have a significant effect on sea-level along the more stable coastlines.

Regional fluctuations in sea-level that result from local tectonic and isostatic factors can be much greater than eustatic changes, as evidenced by the rates of tectonic processes discussed in the preceding section. They are not always readily distinguished from and may completely mask the eustatic change. Nonetheless, efforts have been made to chronicle the eustatic fluctuations and to construct an average worldwide sea-level curve [10] (Figure 4).

Effects and Examples

We are able to identify the likely effects on fan sedimentation of the major controls by assessing the features and controls of numerous modern and ancient fans. Clearly, as our examples show, many of the controlling influences interact to produce "hybrid" fan types that combine features of several end-member models. In this relatively short summary, we have not attempted to be exhaustive, but point out basic relationships that exist between fans and the controls that produce these patterns.

Sediment Type and Supply

We identify two end-member submarine fan types and a third end-member nonfan system (slope-apron) (Figure 5) that result primarily from variations in sediment type and in the nature and rate of sediment supply.

The *elongate fan* develops in response to a medium to high sediment input of mixed size grades, but with mud and very fine sand as the dominant sediment type. These fans have a single major primary source (e.g., large river, delta, ice channel) and commonly have one active channel at any given time and one or more abandoned or periodically active feeder channels across the slope. These channels may have elaborate tributary systems, and the major fan valleys may have extensive distributary networks. An elongate shape with an irregular to smooth downfan morphology is developed by the effective funnelling of sands and coarser material to terminal lobes at the ends of channels on the distal fan.

Numerous authors have described fans of this type and various synonyms exist: large deep-water or open-basin fan [8,11], high-efficiency fan [12], muddy fan (D. J. W. Piper, personal communication, 1982), delta-fed fan [3], and morphologically poorly-developed fan [13]. Modern examples include the Astoria, Bengal, Indus, Mississippi, Amazon, Laurentian, and Rhone Fans among others, and the facies distribution is best characterized by Stow [11] and Nelson [8]. Recognition of the very large elongate fans in the ancient record is almost impossible, but some examples from the Italian flysch [14,15] and from the North Sea Tertiary [16] may be similar types on a smaller scale. Mutti and Ricci Lucchi's fan model [17] best describes these ancient sequences and their facies distribution.

The *radial fan* results from a smaller sediment input than elongate fans, commonly with the sand grade sediment of equal or greater abundance than mud. There is a single general source area, although not necessarily a single river or delta system, and a single feeder canyon or channel across the slope. A main upper fan valley generally divides into a limited distributary network in the midfan. The fan shape is typically radial, and the classical [18] tripartite morphological divisions are developed.

This type of fan has been variously termed a restricted-basin fan [8], low efficiency fan [12], sandy fan (D. J. W. Piper, personal communication, 1982), canyon-fed fan [3], small fan, and morphologically well-developed fan [13]. Among many modern examples are La Jolla, Navy, San Lucas, and Redondo Fans on the western margin of North America from which Normark has derived his model [3,18]. A number of ancient turbidite sequences have been related to the Normark model, although, strictly speaking, we do not generally have the morphological evidence to ascertain that they are indeed radial fans. Subsurface examples such as the Magnus [19] and Frigg [20] Fans from the North Sea Tertiary, or the Stevens and related fans [21] of the Great Valley, California, may fit into this category. Walker [22] has developed a composite model to describe the facies types and distributions on this type of fan.

The *slope-apron* (also debris-apron) system, which includes both the slope and base-of-slope associations, has been placed at the third apex of our ternary diagram (Figure 5). It is *not* a submarine fan because of the absence of channels,

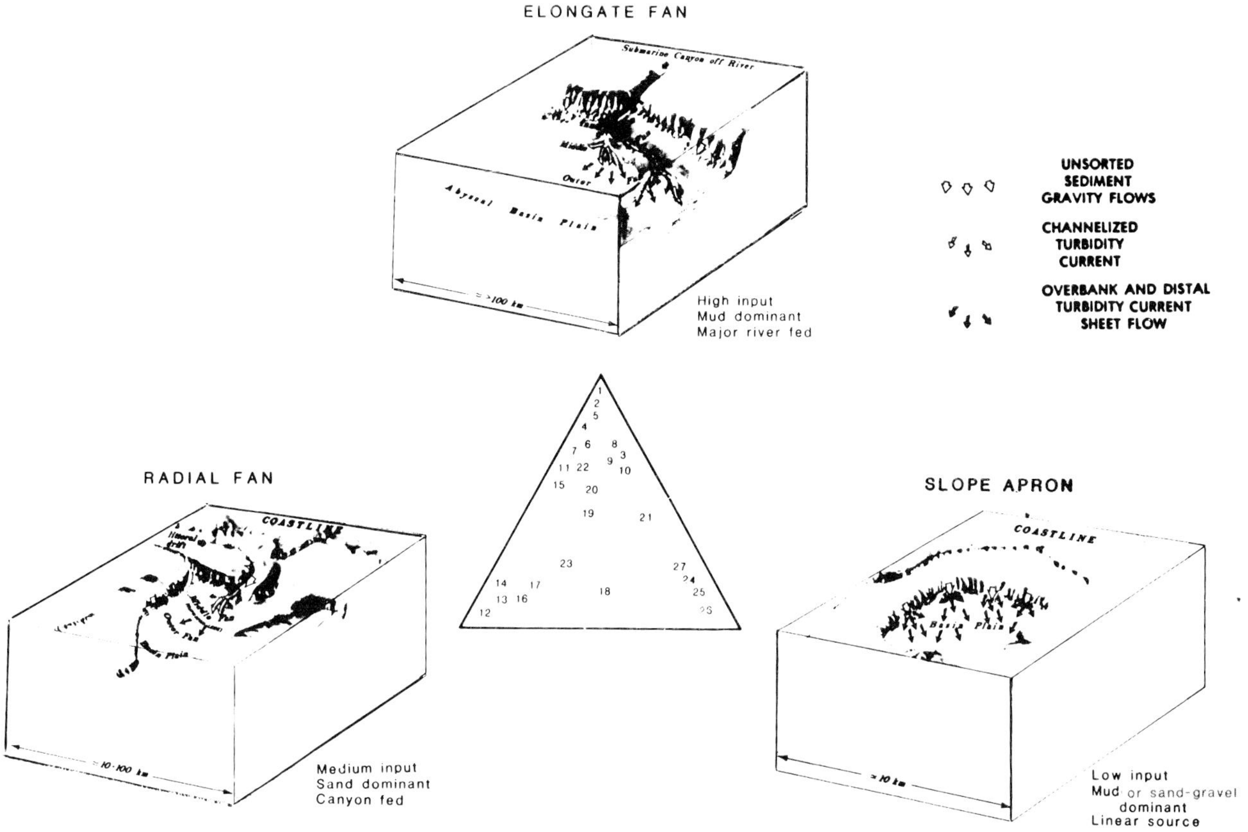

Figure 5. Triangular diagram showing estimated positions of various modern fans with respect to the three end-member modes: elongate fan, radial fan, and slope-apron system (fan block diagrams modified from Nelson [33] © Kendall-Hunt Publishing Co.).

1. Bengal
2. Indus
3. Mississippi
4. Zaire
5. Amazon
6. Reserve
7. Rhone
8. Laurentian
9. Monterey
10. Delgada
11. Astoria
12. La Jolla
13. Redondo
14. Navy
15. Nitinat
16. Coronado
17. San Lucas
18. Ebro
19. Hudson
20. Orange
21. Crati
22. Nile (Rosetta)
23. Menorca
24. Normal slopes and rises
25. California Continental Borderland basin slopes

but is a closely related turbidite system. It is characterized by a low to medium sediment supply rate, and a mixed sediment type that is commonly sand and even gravel-rich in very small tectonically active basins, but may be mud dominated along muddy continental slopes. There is a multiple or linear sediment source feeding directly across a nonchannelized or "straight"-gullied slope. Most present-day oceanic slopes and rises as well as small slope basins fit into this general category, of which several subdivisions can be made. Recently, Stow [23] has attempted to synthesize a slope-apron morphological and facies model similar to that developed for carbonate slopes [24]. Several examples of probable slope-apron deposits have also been described from the ancient record [22,26,30].

Hybrid fan types that are gradational between our end-member models are probably the norm rather than the exception (Figure 5). Even the examples listed above may not be true end-members; for example, the Laurentian Fan is considered a typical elongate fan type, but it has many elements of the normal slope-apron development as shown by the adjacent Scotian and Grand Banks margins. The Ebro "Fan" in the western Mediterranean combines elements of a slope-apron system and a radial fan system because tectonic subsidence has disrupted fan development [27]. The Monterey, Delgada, and Astoria Fans appear to lie somewhere between the radial and elongate types. The Astoria system clearly has elongate growth and facies patterns because it fills a trench parallel to the margin. Channels have migrated to parallel the margin and funnel sand to outer fan depocenters. In the absence of competing tectonic or sea-level controls, normal sedimentary processes (channel erosion and filling, lobe switching, and so on) will produce characteristic horizontal facies distributions and vertical facies sequences (Figure 6), [11,17,22]. These will differ somewhat for each end-member system; the sandy facies, for example, will be distributed concentrically on the radial fan, in more linear channels and terminal lobes on the elongate fan, and in sheets, stringers and isolated lobes in the slope-apron system. The small-scale compensation cycles recently described by Mutti and Sonnino [28] are also the result of normal sedimentary controls.

Tectonic Setting and Activity

First-order tectonic factors clearly exert a primary control on both the sediment type and supply by their influence on relief, rock types, resedimenting processes, and eustatic or local sea-level changes. More specifically, tectonic setting determines the basin size and shape and the slope gradients that confine and control depositional patterns, and the rate of tectonic processes that my disrupt normal fan growth.

On stable and *mature passive margins,* rates of fan deposition exceed rates of tectonic motion (Figures 2 and 3). In these relatively quiescent settings large, mature fans of either type may develop, depending on sediment type and supply factors; very thick slope-apron systems may form between fans, such as the Atlantic rise prisms [11].

On *transform margins* and *convergent margins* (arc basins, trenches), rapid uplift, subsidence, or horizontal movement along fault zones may disrupt normal growth patterns in fans that have typical sedimentation rates (Figures 2 and 3). Short-term fans usually develop along these margins since supply areas, feeder channels, and receiving basins tend to be transient and subject to varied tectonics. The exception is in trench floor areas where, if a large sediment source exists, "elongate" fans can form as a result of structural confinement. For the most part, relatively small radial fans develop in at least two different configurations [6]: In the first case (e.g., Baranof Fan, off southwest Alaska), the primary slope feeder channel is landward of the transform fault so that fans are continuously transported away from the sources by slip on the fault; in the second case (e.g., Delgada and Monterey Fans, off California), the major slope feeder channel is seaward of the fault and allows a single larger fan to develop.

The slope-apron systems between fans accumulate relatively thin deposits and are subject to considerable slumping and other mass movements.

Immature *rifted passive margins* and portions of *marginal seas* and *transform margins* will all evolve through periods of predominantly vertical tectonics. In this case, considerable quantities of sediment may be shed across a steepened and fault-scarp slope into the adjacent rapidly subsiding basin. No well-defined fan morphology will develop, but in-

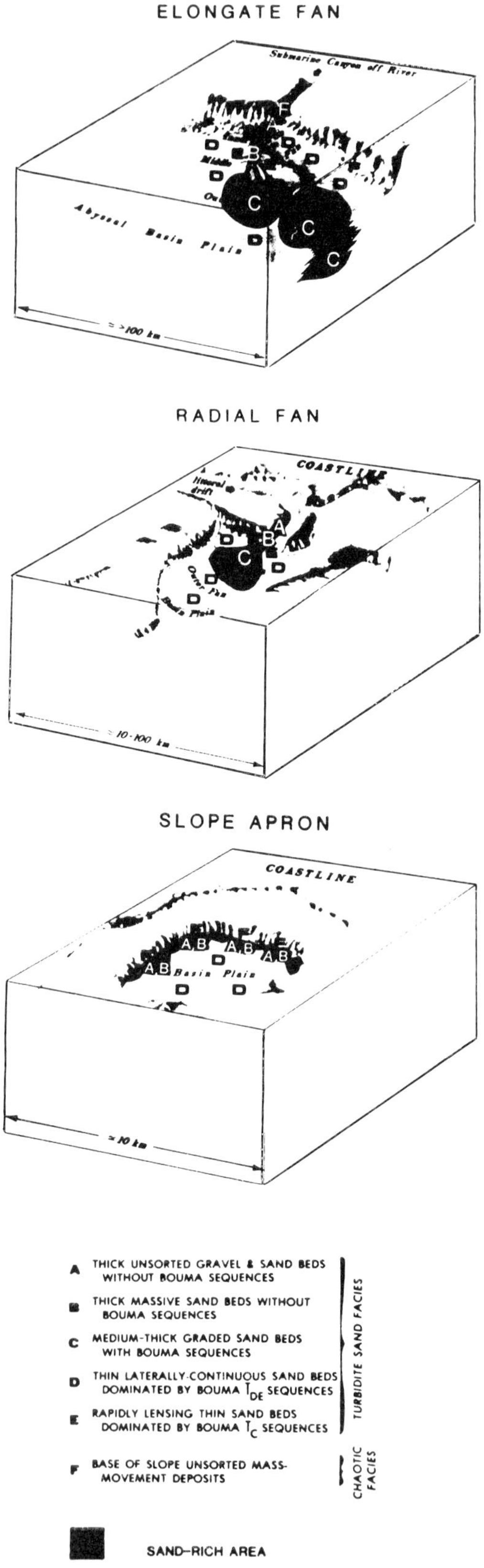

Figure 6. General organization of facies in elongate fan, radial fan, and slope-apron systems (modified from Nelson [33] © Kendall-Hunt Publishing Co.).

stead a very thick and somewhat irregular sediment fringe will accumulate in a narrow band in the base-of-slope area that forms a fault-scarp slope-apron system analogous to that seen at the base of the Crater Lake Caldera walls. Modern examples are found along parts of the Mediterranean Sea margins [27,29], south of the Arabian peninsula in the Gulf of Aden (J. C. Faugeres, personal communication, 1983), and elsewhere. Ancient examples include some of the Mesozoic-Cenozoic basin margins off California [6], and the Jurassic rifted margins of eastern Greenland [30], and the North Sea [1].

Along active convergent or transform margins, small morphologically constricted basins are developed with variable shapes, sediment supplies, and duration. There is a strong morpho-tectonic control exerted on the style of sediment accumulation in these basins. The complex combinations of slope aprons and fans may be better termed "basin-fill systems," rather than submarine fans or slope aprons because they have their own characteristic facies distribution, sequences, and morphological evolution. These are not discussed further here, but a preliminary synthesis of basin models has been made by Stow [23].

In addition to the general tectonic setting, the magnitude, location, and periodicity of tectonic activity are very significant controlling factors on fan development. Channel erosion, filling, abandonment, or rejuvenation can all be influenced by local tectonic activity. Vertical facies sequences will evolve that mirror the coarsening- or fining-upward sedimentary sequences, but that have tectonic rather than sedimentary causes.

Diapiric activity in thick slope successions of either active or passive margins will exert further specific controls on the distribution of facies and on the development of different fan, slope, and basin types [31].

Sea-Level Fluctuations

Changes in global and local sea-level clearly have profound effects on sedimentation throughout the marine realm [6] as dramatically displayed by the late Cenozoic fluctuations of fan depositional regimes in response to glacial cycles [8]. Local changes in sea-level are a major factor in controlling turbidite sedimentation and fan growth within a specific basin, but these sea-level fluctuations do not always reflect a worldwide eustatic variation and may be only a local event.

High stands of sea-level coincide with relatively inactive phases in many fans. Sediment transport across the shelf and into canyons will be at reduced levels since most detritus is trapped in estuaries, lagoons, and other nearshore environments (e.g., Astoria [8], Laurentian [11], Cap Ferret [4], Ebro [27], and Monterey Fans [3]). Various types of bottom currents other than turbidity currents may remain or become active over the fan and help to mold fluvial-like features (braided and meandering channels, point bars, etc.) and leave traction-current deposits (thick cross-bedded sets, lag deposits) [11,20]. Off major rivers and deltas, large elongate fans or slope-aprons may continue to grow, but at reduced rates, even during high stands of sea-level, e.g., Mississippi (J. M. Coleman, personal communication 1982), Ebro [27], and Amazon Fans (J. E. Damuth, personal communication 1982), because thick Holocene deposition of river mud on the slope results in deposition of major debris sheets over the fan surface.

Low stands of sea-level, on the other hand, will lead to narrower shelves with more rapid sedimentation as well as the direct funnelling of sediments through canyons and fan valleys to deeper basins. Several authors have noted the probable correlation of low sea-level with more frequent turbidity current activity and greater supply of sands and gravels to the deep sea [6,8,32].

Fluctuation in sea-level will result in different styles of fan growth: progradational sequences, rejuvenation, and channel incisement during lowering of sea-level, and regradational sequences with channel and lobe abandonment during raising of sea-level. Cyclicity in vertical sequences may thus be related to repeated sea-level fluctuations as well as to normal channel abandonment and compensation cycles [28].

Summary

We have attempted, in this brief survey, to show that there may be some coherence in the plethora of fan types and models that have been described over the past 15 years. We do this by addressing not the models themselves, but the primary and secondary controls that influence fan development and the rates at which these controls operate.

Two end-member fan types and the third nonfan slope-apron system result primarily from variations in sediment type and supply. Elongate fans are mud-dominated with high input rates and are fed by major rivers, deltas, and one or more delta-front troughs or canyons; radial fans are sand-dominated with medium input rates and are canyon-fed from local rivers and littoral drift cells; and slope-apron systems are mostly low-input, mud- or sand-gravel dominated with a linear source. Hybrid fan types are the rule rather than the exception.

The tectonic setting exerts a first-order control of fan development. Large mature fans of either type and thick slope-apron sequences develop along parts of transform and convergent margins. Where there is significant vertical movement on these margins, a thick narrow belt of sediment accumulates as fault-scarp slope-apron systems.

High stands of sea-level are often periods of relative fan dormancy, except perhaps for large "delta-fed" elongate fans where the delta has prograded to the slope edge, or fans where the canyon head incises into estuaries or near littoral drift cells. Low stands of sea-level commonly result in active fan growth. The complex variation of sea-level fluctuation as well as tectonic activity and normal sedimentary factors will all affect the style of fan growth, facies distribution, and vertical sequences.

Acknowledgments

We are grateful to many colleagues for helpful discussion in the evolution of our ideas for this paper. Particularly stimulating was the COMFAN meeting in September 1982 hosted by GULF Research and Development Company. DAVS acknowledges personal support from the Royal Society of Edinburgh and secretarial and drafting assistance at the Grant Institute of Geology. CHN thanks Lee Baily and Tau Rho Alpha for assistance with drafting. We thank Arnold Bouma, Michel Cremer, and William Normark for helpful comments on an earlier version of the manuscript.

References

[1] Stow, D. A. V., Bishop, C. D., and Mills, S. J., 1982. Sedimentology of the Brae Oil Field, North Sea: fan models and controls. Journal of Petroleum Geology, v. 5, pp. 129–148.

[2] Howell, D. G., and Normark, W. R., 1982. Sedimentology of submarine fans. American Association of Petroleum Geologists Memoir 31, pp. 365–404.

[3] Normark, W. R., 1978. Fan valleys, channels and depositional lobes on modern submarine fans: characters for recognition of sandy turbidite environments. Bulletin of the American Association of Petroleum Geologists, v. 62, pp. 912–931.

[4] Cremer, M., 1983. Approches sédimentologiques et géophysiques des accumulations turbiditiques. Unpublished thesis, Bordeaux University, France, 344 pp.

[5] Stow, D. A. V., and others, 1984. Depositional model for calcilutites: Scaglia Rossa limestones, Umbro-Marchean Apennines. In: D. A. V. Stow and D. J. W. Piper (eds.), Fine-grained sediments: deep-water processes and facies. Geological Society of London Special Publication, v. 14, pp. 223–240.

[6] Howell, D. G., and von Huene, R., 1980. Tectonics and sediment along active continental margins. Society of Economic Paleontologists and Mineralogists Short Course, San Francisco, 1980.

[7] Blatt, H., Middleton, G., and Murray, R., 1980. Origin of Sedimentary Rocks, 2nd ed., Prentice-Hall, New Jersey, 782 pp.

[8] Nelson, C. H., and Kulm, L. D., 1973. Submarine fans and channels. Society of Economic Paleontologists and Mineralogists Short Course, Anaheim, 1973, pp. 39–78.

[9] Pitman, W. C., 1979. The effect of eustatic sea level changes on stratigraphic sequences at Atlantic margins. American Association of Petroleum Geologists Memoir 29, pp. 453–460.

[10] Vail, A. R., and Mitchum, R. M., 1979. Global cycles of relative changes of sea level from seismic stratigraphy. American Association of Petroleum Geologists Memoir 29, pp. 469–472.

[11] Stow, D. A. V., 1981. Laurentian Fan: morphology, sediments, processes and growth pattern. Bulletin of the American Association of Petroleum Geologists v. 65, pp. 375–398.

[12] Mutti, E., 1979. Turbidites et cones sous-marins profonds. In: P. Homewood (ed.), Sedimentation Détritiques (Fluviatile, Littoral et Marine). Institut Géologique Université de Switzerland, Fribourg, v. 1, pp. 353–419.

[13] Pickering, K. T. In press. The shape of deep-water siliciclastic systems—a discussion. Geo-Marine Letters.

[14] Mutti, E., 1974. Examples of ancient deep-sea fan deposits from circum-Mediterranean geosynclines. In: R. H. Dott and R. H. Shaver (eds.), Modern and Ancient Geosynclinal Sedimentation, pp. 92–105.

[15] Ricci Lucchi, R., 1981. The Miocene Marnoso–Arenacea turbidites, International Association of Sedimentologists 2nd European Meeting, Bologna, Excursion Guidebook.

[16] Carman, G. J., and Young, R., 1980. Reservoir geology of the Forties oilfield. In: L. V. Illing and G. D. Hobson, (eds.), Petroleum Geology of the Continental Shelf of Northwest Europe. Heyden; London, pp. 371–379.

[17] Mutti, E., and Ricci Lucchi, F., 1972. Le torbiditi dell'Appenino settentrionale—introduzione all'analisi di facies. Societa Geologica Italiana Memorie, v. 11, pp. 161–199.

[18] Normark, W. R., 1970. Growth patterns of deep-sea fans. Bulletin of the American Association of Petroleum Geologists, v. 54, pp. 2170–2195.

[19] De'Ath, N. G., and Schuyleman, S. F., 1981. The geology of the Magnus oilfield. In: L. V. Illing and G. D. Hobson, (eds.), Petroleum Geology of the Continental Shelf of Northwest Europe. Heyden, London, pp. 342–351.

[20] Heritier, F. E., Lossell, P., and Wathne, E., 1979. Frigg Field—large submarine fan trap in lower Eocene rocks of the North Sea Viking Graben. Bulletin of the American Association of Petroleum Geologists, v. 63, pp. 1999–2020.

[21] Macpherson, B. A., 1978. Sedimentation and trapping mechanism in Upper Miocene Stevens and older turbidite fans of southeastern San Joaquin Valley. Bulletin of the American Association of Petroleum Geologists, v. 62, pp. 2243–2274.

[22] Walker, R. D., 1978. Deep water sandstone facies and ancient submarine fans: models for exploration for stratigraphic traps. Bulletin of the American Association of Petroleum Geologists, v. 62, pp. 932–966.

[23] Stow, D. A. V., 1985. Deep-sea clastics review—where are we and where are we going? Geological Society of London Special Publication.

[24] McIlreath, I. A., and James, N. P., 1978. Facies models 13. Carbonate slopes. Geoscience Canada, v. 5, pp. 189–199.

[25] Cook, H. E., 1979. Ancient continental slopes and their value in understanding modern slope development. Society of Economic Paleontologists and Mineralogists Special Publication, v. 27, pp. 287–306.

[26] Piper, D. J. W., Normark, W. R., and Ingle, J. C., 1976. The Rio Dell Formation: a Plio–Pleistocene basin slope deposit in Northern California. Sedimentology, v. 23, pp. 309–328.

[27] Nelson, C. H., Maldonado, A., and Coumes, F., in press. The Ebro deep-sea fan: a channelized, restricted basin type fan. Geo-Marine Letters.

[28] Mutti, E., and Sonnino, M., 1981. Compensation cycles: a diagnostic feature of turbidite sandstone lobes. International Association of Sedimentologists 2nd European Meeting, Bologna, Abstracts, pp. 120–132.

[29] Wezel, F. C., and others, 1981. Plio–Quaternary depositional style of sedimentary basins along insular Tyrrhenian margins. In: F. C. Wezel (ed.), Sedimentary Basins of Mediterranean Margins, CNR Italian Project of Oceanography, pp. 239–269.

[30] Surlyk, 1978. Submarine fan sedimentation along fault-scarps on tilted fault blocks (Jurassic/Cretaceous boundary, East Greenland). Bull. Gronlands Geologiske Undersogelse, v. 128, p. 108.

[31] Bouma, A. H., 1981. Depositional sequences in clastic continental slope deposits, Gulf of Mexico. Geo-Marine Letters, v. 1, pp. 115–121.

[32] Shanmugam, G., and Moiola, R. J., 1982. Eustatic control of turbidites and winnowed turbidites. Geology, v. 10, pp. 231–235.

[33] Nelson, C. H., 1983. Modern submarine fans and debris aprons: an update of the first half century. In S. J. Boardman (ed.), Revolution in the Earth Sciences, Advances in the Past Half-Century. Kendall/Hunt, Dubuque, Iowa, pp.148–166.

CHAPTER 5

Eustatic Control of Submarine Fan Development

G. Shanmugam, R. J. Moiola, and J. E. Damuth

Abstract

Global changes in sea-level control both siliciclastic and calciclastic turbidity-current deposition in the deep sea. On modern active and passive margins, growth of submarine fans occurred mainly during the Pleistocene glacials (low sea-level). During interglacials (high sea-level), most fans were dormant. In the rock record, the occurrence of most turbidites and winnowed turbidites closely correlates with lowstands of sea-level. Sea-level fluctuations appear to have been the primary control on fan growth throughout the Phanerozoic. In most cases, tectonism appears to be of secondary importance.

Introduction

This chapter reviews studies that document global lowering of sea-level as the primary factor in the generation of both siliciclastic and calciclastic turbidity currents as well as the cause of vigorous contour currents in the deep sea [1–3]. We used the global sea-level curve of Vail, Mitchum, and Thompson [4], now referred to as the coastal onlap curve [5], as our standard reference because of its coverage of the entire Phanerozoic. Also, the "Vail curve" is in good agreement with most previously published curves (Fig. 1). Our studies show that the duration of maximum sea-level lowering is more important to our thesis than the rate of rise or fall [12]. Global changes in sea-level are primarily controlled by tectonism and glaciation; however, glaciation is considered to be the only mechanism capable of causing relatively rapid (> 1 cm/1000 yr) fluctuations in sea-level [13]. The relative effects of tectonism and glaciation on sea-level fluctuations are illustrated in Figure 2. Long-term gradual fluctuations in global sea-level appear to be controlled primarily by changes in mid-oceanic ridge volume (spreading rate) as well as by subsidence of continental margin and sediment compaction. Short-term rapid fluctuations in sea-level appear to be related to glaciation.

Glacio-Eustatic Control of Submarine Fan Development

Sedimentation and growth of most modern submarine fans have been controlled during the last few million years by Plio-Pleistocene glacio-eustatic sea-level fluctuations. During the relatively short (5,000 to 20,000 years) interglacial phases such as the Holocene, recession of continental glaciers caused the sea-level to rise to or above its present level. Such high stands move the locus of river sedimentation from the vicinity of the shelf break to as much as several hundred kilometers inland across the continental shelf. The great width and low gradient of the shelf generally restrict river deposition to deltas on the innermost shelf, and large amounts of terrigenous sediment that are needed to build submarine fans cannot reach the continental slope or rise. Hence, fan development is temporarily halted during such high sea-level stands. In contrast, during glacial phases such as the Wisconsin Glacial, sea-level is lowered 40 to 150 m below the present level. Most continental shelves become emergent, and rivers discharge their sediment loads directly into the heads of submarine canyons at or near the shelf break. Thus, large quantities of terrigenous sediment are continuously transported to the deep sea via turbidity currents and related mass-gravity flows, and submarine fan development is rapid.

Such glacio-eustatic control of fan sedimentation has been documented in detail for the Amazon Deep-Sea Fan and the adjacent continental margin off northeast Brazil

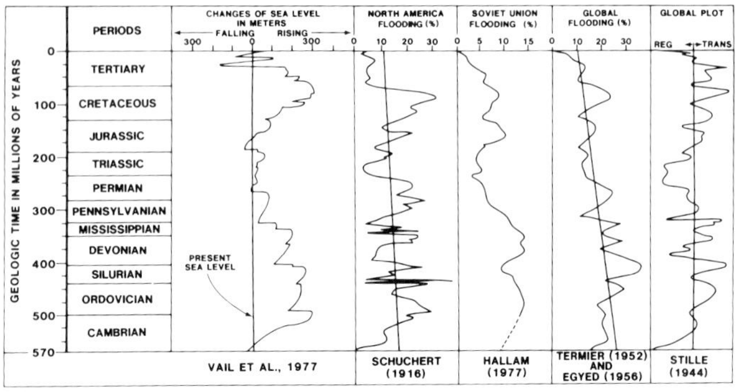

Figure 1. Comparison of sea-level curve by Vail and others [4] with curves by Schuchert [6], Hallam [7], Termier and Termier [8], Egyed [9], and Stille [10]. See Wise [11] for details.

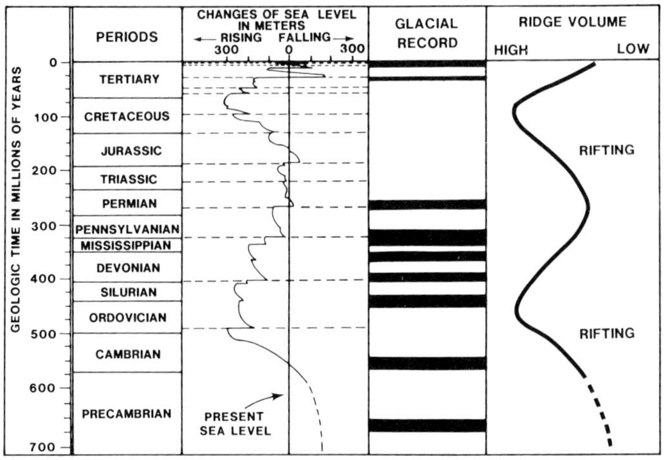

Figure 2. Comparison of global changes in sea-level [4], glacial record, and mid-oceanic ridge volume. From Shanmugam and Moiola [1].

[14–16; Damuth and Flood, Chapter 15 this volume]. Sediment cores show that although terrigenous sediments accumulated rapidly and continuously throughout the Wisconsin Glacial (Y Zone), this sedimentation was abruptly halted at the beginning of the Holocene (~ 11,000 yr B.P.) when sea-level quickly rose in response to rapid deglaciation (Fig. 3). Sea-level rise moved the locus of Amazon River sedimentation from the shelf break inland by as much as 350 km to its present location. The great shelf width and low gradient plus strong longshore currents have confined Amazon River sediments to the estuary and innermost shelf during the Holocene. The Amazon Fan has been inactive during this time and received only a thin (< 1 m) veneer of pelagic sediment (Fig. 3).

Studies of late Quaternary climatic fluctuations (see references in [16]) have revealed that warm interglacial intervals (and accompanying high sea-level stands), such as the Holocene, tend to be relatively short (10,000 to 20,000 years) and occur with a periodicity of about 100,000 years (see sea-level curve in Fig. 3). During the remaining 80,000 to 90,000 years of each cycle, the glacial mode is predominant as ice gradually builds up on the continents and sea-level fluctuates between 40 and 150-m below the present level. The generalized sea-level curve (based on published oxygen-isotope curves) shows sea-level fluctuations during the last complete glacial/interglacial cycle. During this period, sea level was at or near its present level only during the Holocene (0 to 11,000 yr B.P.) and at the beginning of the last interglaciation (~ 120,000 to 127,000 yr B.P.).

The generalized core log in the center of Figure 3 summarizes the sediment lithology observed on the Amazon Fan and adjacent continental margin during the last glacial/interglacial cycle. During the period from 120,000 to 11,000 yr B.P., sea-level was low and turbidity flows continuously deposited sandy turbidites and silty hemipelagic clays across the fan at rates of 25 to greater than 150 cm/10^3 yr. It was during this and previous glacial phases that the fan was actively growing. The core log shows that during the high sea-level stands (1 to 11,000 and 120,000 to 127,000 yr B.P.), only pelagic biogenic (foram-rich marl) sediments accumulated on the fan at rates of less than 5 cm/10^3 yr. Hence, during these interglacial high sea-level stands, the Amazon Fan was temporarily inactive.

Glacio-eustatic sea-level fluctuations have also controlled sedimentation on most other modern deep-sea fans during the Plio-Pleistocene, especially fans on passive margins with wide shelves. The Bengal and Indus Fans, the two largest modern fans, both show an abrupt cessation of terrigenous deposition during the Holocene ([17]; V. Kolla and G. Griep, personal communication). Recent D.S.D.P. results from the Mississippi Fan ([18]; Bouma and Coleman, Chapter 36 of this volume) show a sedimentation regime similar to that observed for the Amazon Fan; i.e., slow pelagic deposition (3 to 30 cm/10^3 yr) during the Holocene (Z Zone) and Last Interglacial (X Zone), and rapid (up to 1200 cm/10^3 yr) turbidity-current deposition during the Wisconsin (Y Zone) Glacial. In general, sediment cores from deep-sea fans, continental rises, and abyssal plains throughout the Atlantic and Indian Oceans show lithostratigraphic relationships that are similar to those shown in Figure 3 and which demonstrate that downslope deposition of terrigenous sediment is largely controlled by glacio-eustatic sea-level fluctuations [17]. The right side of Figure 3 shows examples of sedimentation rates on four modern fans during the last glacial/interglacial cycle.

Two notable exceptions to this sedimentation pattern are the Congo and Magdalena deep-sea fans, which have continued to receive turbidity flows and mass-transport deposits throughout the Holocene [19,20]. In these cases, the continental shelf is exceptionally narrow and the submarine canyon that feeds the fans extends across the entire shelf directly into the river mouth. Hence, sediments are discharged directly into the canyon even during the present high sea-level.

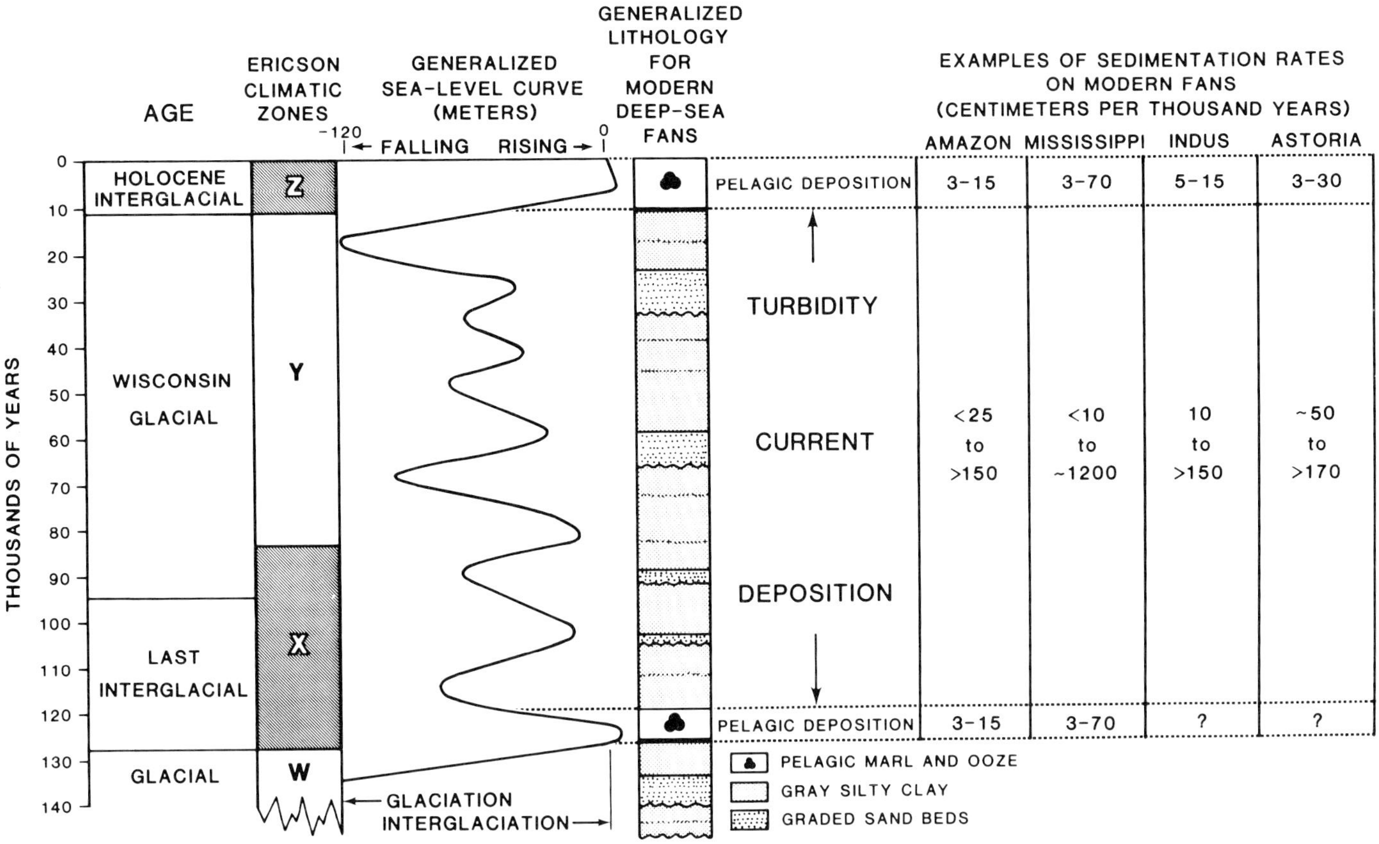

Figure 3. Generalized diagram showing the relationship between glacio-eustatic sea-level fluctuations and sediment lithology on Amazon and other deep-sea fans and adjacent areas during the last 130,000 years. Age boundaries, Ericson climatic zones, sea-level curve, and lithologic log are modified from Damuth [16]. Examples of sedimentation rates on modern fans and adjacent regions are shown on the right. Amazon [15] and [16]; Mississippi [18] and [28]; Indus (G. Griep, personal communication); and Astoria [29].

In summary, glacio-eustatic sea-level fluctuations are apparently the most important factor controlling sedimentation and growth of most modern deep-sea fans on passive margins, and at least some fans on active margins as well. It is difficult to evaluate the effects of glacio-eustatic fluctuations on fan development prior to the Pliocene because the occurrence, timing, and magnitude of glacial cycles are largely unknown. However, when major periods of glaciation did occur (Fig. 2), they probably had a profound influence on deep-sea fan development.

Eustatic Control of Ancient Submarine Canyons and Fans

Global changes in sea-level appear to control the origin of submarine canyons and fans. The following hydrocarbon-bearing submarine canyon and fan deposits appear to have originated during periods of low sea-level: 1) the Pennsylvanian (Atokan) Red Oak Sandstone in Oklahoma; 2) the early Permian Cook Channel of the Jameson Field in Texas; 3) the Upper Cretaceous Woodbine-Eagle Ford Interval in Texas; 4) the Paleocene sequence of Forties and Montrose Fields in the U.K., North Sea; 5) the Paleocene Balder Field in the Norwegian North Sea; 6) the Paleocene Cod Fan in the Norwegian North Sea; 7) the early Eocene Yoakum Channel in Texas; 8) the early Eocene sequence of Frigg Field in the border of U.K. and Norwegian North Sea; 9) the late Oligocene Lower Hackberry Sandstone in Texas; 10) the late Oligocene Puchkirchen Formation in Austria; 11) the late Miocene Stevens Sandstone of southeastern San Joaquin Valley in California; 12) the late Miocene Puente Formation of Wilmington Field in California; 13) the early Pliocene Repetto Formation of Ventura Field in California; and 14) the Pleistocene Mississippi Canyon in Louisiana. The relationship of Tertiary and Quaternary hydrocarbon-bearing submarine canyon and fan deposits to global sea-level stands are shown in Figure 4 (see references in [2]).

Eustatic Control of Winnowed Turbidites

In modern oceans, thermohaline-induced bottom currents are commonly known as contour currents because of their tendency to flow parallel to bathymetric contours [21,22]. Contour currents are capable of erosion, transportation,

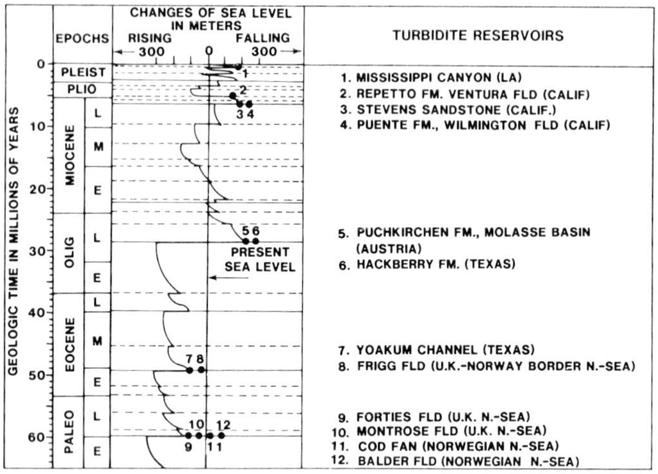

Figure 4. Hydrocarbon-bearing submarine canyon and fan deposits that closely correspond with Vail and Hardenbol's [13] lowstands of sea-level. From Shanmugam and Moiola [2].

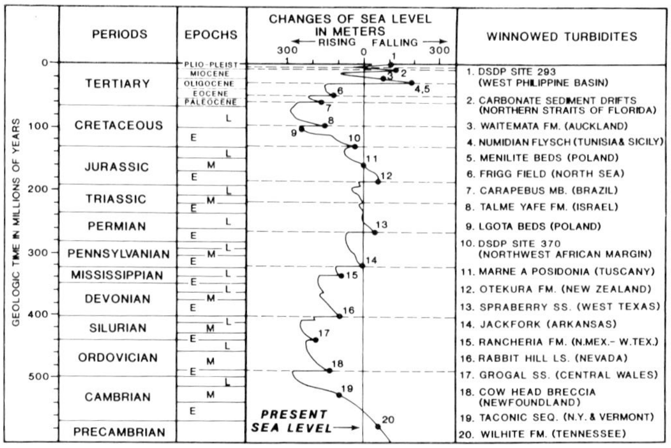

Figure 5. Correlation of winnowed turbidites with lowstands of sea-level by Vail and others' [4] curve. From Shanmugam and Moiola [1].

dissolution, and redeposition of vast quantities of sediment, and some portions of the continental rise appear to be largely constructed of such deposits (e.g., Western North Atlantic). Theoretically, contour currents should be more vigorous during global lowstands of sea-level. This should cause major winnowing of turbidites. To test this hypothesis, Shanmugam and Moiola [1] plotted ages of winnowed turbidites on the global sea-level curve of Vail, Mitchum, and Thompson (Fig. 5). This plot (Fig. 5) shows a strong correlation between winnowed turbidites and relative lowstands of sea-level.

Eustatic Control of Calciclastic Turbidites

Reworked neritic fossils in central Pacific pelagic sediments (Fig. 6) have been attributed to erosion of shallow-water carbonates during periods of low sea-level [23]. The presence of shallow-water-derived limestone clasts in these deep-sea pelagic sediments can only be explained by downslope transportation from shallow-water carbonate platforms by debris flows and associated turbidity currents. Furthermore, these data (Fig. 6) imply a possible lowstand of sea-level in the Late Cretaceous (Campanian-Maestrichtian) that has not been previously recognized. Similar to these Cenozoic and Mesozoic examples, Shanmugam and Moiola [3] showed that the following calciclastic turbidites of the Paleozoic age also correlate with low sea-level: 1) the late Permian (Guadalupian) Pinery and Rader Limestones in west Texas; 2) the Pennsylvanian (Atokan) Dimple Limestone in Texas; 3) the late Mississippian (Meramecian) Rancheria Formation in New Mexico and west Texas; 4) the early Devonian Rabbit Hill Limestone in Nevada; 5) the early Middle Ordovician Cow Head Breccia in Newfoundland; 6) the early Ordovician Hales Limestone in Nevada; and 7) the early Cambrian Sekwi Formation in Northwest Territories. These data suggest that deep-sea carbonate sediments of turbidity-current origin also tend to form during relative low sea-level stands.

Effects of Tectonic Setting on Fan Development

Effects of active and passive margins are generally reflected in associated submarine fans. Active margins tend to produce small sand-rich fans, whereas passive margins develop large mud-rich fans (Table 1). Tectonic activity may disrupt normal fan growth on active margins (Stow and others, Chapter 4 this volume), but fans may continue to grow without tectonic interruptions on passive margins. The relationship between tectonic setting and type of fan is less clear for ancient fans when compared with modern fans. For example, the Eocene Hecho Group in Spain and its mud-rich, highly efficient submarine fan is associated with an active margin setting. An explanation for this apparent discrepancy may be found in the modern fans and the timing of their development.

Modern fans of both active (e.g., Astoria, Navy, Coronado, and Monterey) and passive (e.g., Bengal, Indus, Amazon, and Mississippi) margins were all active during the Pleistocene Glacials (low sea-level), but most were either dormant or relatively inactive during the Holocene and previous interglacials (high sea-level). The development of a submarine fan, therefore, is controlled primarily by fluctuations in sea-level, and not by tectonic setting.

The conventional wisdom that tectonic uplift usually increases sediment supply and, thus, submarine fan growth may not be valid. In order for a fan to develop, a large supply of unconsolidated sediment is required to generate the major turbidity currents that form the fan. Such volumes of sediment are seldom produced by initial tectonic uplift

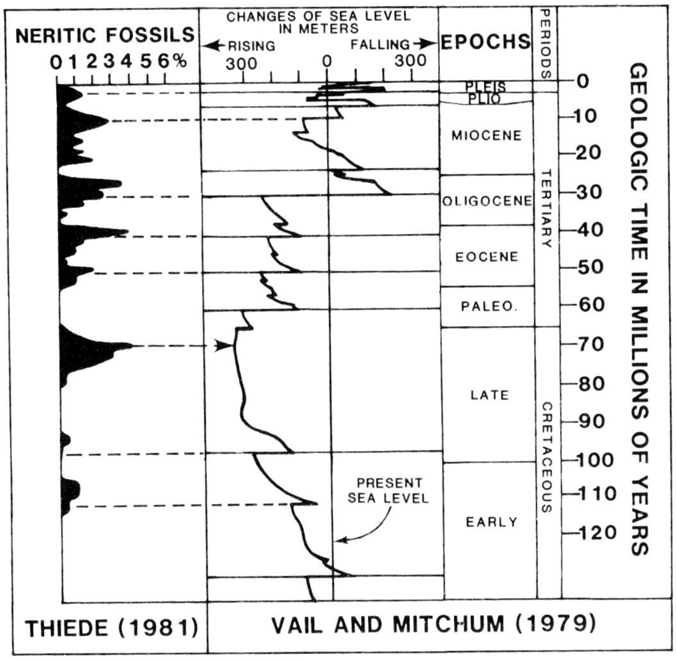

Figure 6. Correlation of calciclastic turbidites [23] and lowstands of sea-level [24]. Arrow points to a possible low sea-level during Campanian-Maestrichtian time not shown on Vail and Mitchum's curve [24]. From Shanmugam and Moiola [3].

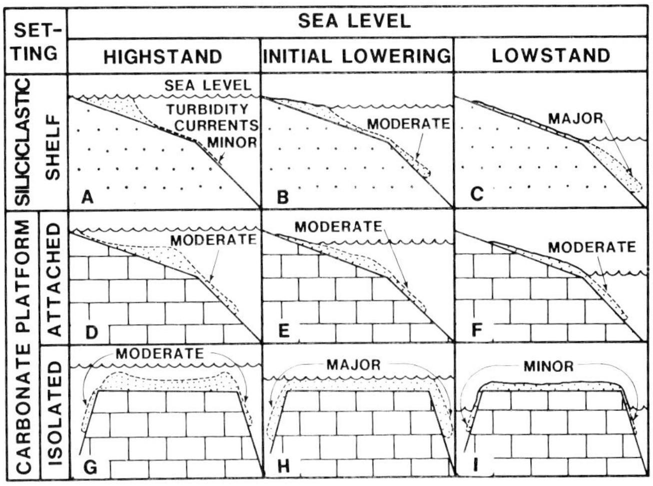

Figure 7. Eustatic model for development of siliciclastic and calciclastic turbidites.

because the uplifted land is a lithified mass. Initial erosion of the uplifted land mass does not produce large quantities of sand- to clay-sized sediment. Because of the lack of sand- to clay-sized sediment, the major turbidity currents needed to develop submarine fans cannot be generated. The coarse debris from initial erosion must first be broken into a finer size by fluvial and shallow-marine processes before it can be transported by turbidity currents to deep-sea fans. This important aspect is often overlooked in advocating tectonics as a causal factor for submarine fan growth.

Shelf and related environments are major areas of sediment accumulation where delta, offshore bars, storm deposits, sand ridges, and sand waves develop. The shelf is a "transit lounge" for sand- to clay-sized sediments that are bound for deep-sea fans during lowstands of sea-level. As a generality, passive margins with wide shelves should develop large-scale fans, and active margins with narrow shelf should produce small-scale fans. In summary, sea-level changes exert primary control on fan development as evidenced by the preferential growth of fans during lowstands of sea-level on both active and passive margin settings. However, tectonics play an important secondary role on fan growth along active margins by limiting shelf width and sediment supply.

Table 1. Factors Affecting Fan Development on Active and Passive Margins

Factor	Active Margin	Passive Margin
1. Eustatic influence	Low to high	High
2. Tectonic influence	High	Low
3. Shelf width	Narrow	Wide
4. Shelf exposure/ sediment availability	Small	Large
5. Sediment transport	Short distance	Long distance
6. Fan sediment	Sand-rich	Mud-rich
7. Fan size	Small (tens of km)	Large (hundreds of km)
8. Modern example	Navy Fan	Amazon Fan

Summary and Conclusions

Our studies [1,2] show that major packages of siliciclastic turbidites throughout the geologic record are associated with lowstands of sea-level. This is presumably because of increased exposure and erosion of the shelf plus the discharge of sediment by rivers directly into the heads of submarine canyons (Fig. 7C). As subareal exposure of carbonate platforms eventually results in meteoric cementation [25], most major turbidity currents derived from isolated carbonate platforms should occur during the initial stage of sea-level lowering (Fig. 7H), just prior to prolonged exposure of the platform. At this time, subaqueous carbonate sediments would be unaffected by meteoric cementation, and lowered wave base would result in slope instabilities that would produce slumps, debris flows, and turbidity currents. Such downslope transport apparently correlates with initial lowering of sea-level from the Sangamon highstand in Exuma Sound [26]. Thick deep-water carbonates of the Roncal Unit of the Hecho Group (Cuisian and Lutetian) in northern Spain, which are considered to be analogous to carbonate debris sheets in Exuma Sound, also correlate with a major sea-level low during the Eocene [27]. Combined effects of high carbonate productivity during highstands of sea-level and erosion of the platform during lowstands would result in moderate turbidity-current activity from attached carbonate platforms during all phases of sea-level changes (Fig. 7D-F), with perhaps major flows developing during initial lowering of sea-level (Fig. 7E). Development of submarine fans in the deep sea are, therefore, expected to correspond to the timing of major turbidity flows from both clastic shelves and carbonate platforms during lowstands of sea-level.

Acknowledgments

We thank Mobil Research and Development Corporation for permission to publish this review and S. L. Dunham for typing the manuscript.

References

[1] Shanmugam, G., and Moiola, R. J., 1982. Eustatic control of turbidites and winnowed turbidites. Geology, v. 10, pp. 231–235.

[2] Shanmugam, G., and Moiola, R. J., 1982. Prediction of deep-sea reservoir facies. Transactions Gulf Coast Associations of Geological Societies, v. 32, pp. 275–281.

[3] Shanmugam, G., and Moiola, R. J., 1984. Eustatic control of calciclastic turbidites. Marine Geology, v. 56, pp. 273–278.

[4] Vail, P. R., Michum, R. M., Jr., and Thompson, S., III, 1977. Seismic stratigraphy and global changes of sea level, part 4: global cycles of relative changes of sea level. In: C. E. Payton (ed.), Seismic Stratigraphy—Applications to Hydrocarbon Exploration. American Association of Petroleum Geologists Memoir 26, pp. 83–97.

[5] Vail, P. R., and Todd, R. G., 1981. Northern North Sea Jurassic unconformities, chronostratigraphy and sea-level changes from seismic stratigraphy. In: L. V. Illing and G. D. Hobson (eds.), Petroleum Geology of the Continental Shelf of North-West Europe. Heyden and Son, Ltd., London, pp. 216–235.

[6] Schuchert, C., 1916. Correlation and chronology on the basis of paleogeography. Geological Society of America Bulletin, v. 27, pp. 491–513.

[7] Hallam, A., 1977. Secular changes in marine inundation of USSR and North America through the Phanerozoic. Nature, v. 269, pp. 769–772.

[8] Termier, H., and Termier, G., 1952. Histoire Geologique de la Biosphere. Masson, Paris, 312 p.

[9] Egyed, L., 1956. Change of earth dimensions as determined from paleogeographical data. Geofisica Pura e Applicata, v. 33, pp. 42–48.

[10] Stille, H., 1944. Geotektonische Gliederung der Erdgeschichte. Abh Preuss Akad Wiss Math Nat Klasse, v. 3, 80 p.

[11] Wise, D. U., 1974. Continental margins, freeboard and the volumes of continents and oceans through time. In: C. A. Burk and C. L. Drake (eds.), The Geology of Continental Margins. Springer-Verlag, New York, pp. 45–58.

[12] Hallam, A., 1981. A revised sea-level curve for the early Jurassic. Journal of the Geological Society of London, v. 138, pp. 735–743.

[13] Vail, P. R., and Hardenbol, J., 1979. Sea-level changes during the Tertiary. Oceanus, v. 22, pp. 71–79.

[14] Damuth, J. E., and Fairbridge, R. W., 1970. Equatorial Atlantic deep-sea arkosic sands and ice-age aridity in tropical South America. Geological Society of America Bulletin, v. 81, pp. 189–206.

[15] Damuth, J. E., and Kumar, N., 1975. Amazon cone: morphology, sediments, age and growth pattern. Geological Society of America Bulletin, v. 86, pp. 863–878.

[16] Damuth, J. E., 1977. Late Quaternary sedimentation in the western equatorial Atlantic. Geological Society of America Bulletin, v. 88, pp. 695–710.

[17] McGeary, D. F. R., and Damuth, J. E., 1973. Postglacial iron-rich crusts in hemipelagic deep-sea sediment. Geological Society of America Bulletin, v. 84, pp. 1201–1212.

[18] Leg 96 Scientific Party, 1984. Challenger drills Mississippi Fan. Geotimes, v. 29, no. 7, pp. 15–18.

[19] Heezen, B. C., and others, 1964. Congo submarine canyon. American Association of Petroleum Geologists Bulletin, v. 48, pp. 1126–1149.

[20] Heezen B. C., 1956. Corrientes de turbidez del Rio Magdalena. Bol Soc Geog Colombia, v. 51–52, pp. 135–143.

[21] Heezen, B. C., Hollister, C. D., and Ruddiman, W. F., 1966. Shaping of the continental rise by deep geostrophic contour currents. Science, v. 152, pp. 502–508.

[22] Hollister, C. D., and Heezen, B. C., 1972. Geologic effects of ocean bottom currents: western north Atlantic. In: A. L. Gordon (ed.), Studies in Physical Oceanography, Volume 2. Gordon and Breach Science Publishers, New York, pp. 37–66.

[23] Thiede, J., 1981. Reworked neritic fossils in Upper Mesozoic and Cenozoic central Pacific deep-sea sediments monitor sea-level changes. Science, v. 211, pp. 1422–1424.

[24] Vail, P. R., and Mitchum, R. M., Jr., 1979. Global cycles of relative changes of sea level from seismic stratigraphy. American Association of Petroleum Geologists Memoir 29, pp. 469–472.

[25] Kendall, C. G. St. C., and Schlager, W., 1981. Carbonates and relative changes in sea level. Marine Geology, v. 44, pp. 181–212.

[26] Crevello, P. D., and Schlager, W., 1980. Carbonate debris sheets and turbidites, Exuma Sound, Bahamas. Journal of Sedimentary Petrology, v. 50, pp. 1121–1148.

[27] Johns, D. R., and others, 1981. Origin of a thick, redeposited carbonate bed in Eocene turbidites of the Hecho Group, south-central Pyrenees, Spain. Geology, v. 9, pp. 161–164.

[28] Ewing, M., Ericson, D. B., and Heezen, B. C., 1958. Sediments and Topography of the Gulf of Mexico. In: L. G. Weeks (ed.), Habitat of Oil. American Association Petroleum Geologists, pp. 995–1053.

[29] Nelson, C. H., 1968. Marine Geology of Astoria Deep-Sea Fan. Ph.D. Thesis, Oregon State University, Corvalis, Oregon, 278 p.

CHAPTER 6

Submarine Fan Models: Problems and Solutions

G. Shanmugam and R. J. Moiola

Abstract

Submarine fan models have created considerable confusion in the literature because of misuses of terminology and concepts. We propose that ancient submarine fans should be considered only as systems in which lobes are attached to feeder channels. The popular detached lobe model is based solely on the Eocene Hecho Group in Spain. This tectonically controlled fan system should not be used as a general sedimentologic model for fans. Classification of fans into "highly efficient" and "poorly efficient" systems is not practical because individual fans often possess properties of both systems. Fans are most active during periods of low sea-level.

Introduction

Submarine fans constitute major hydrocarbon reservoirs in various parts of the world. Consequently, a clear understanding of their geometry and facies relationships is critical for exploring and exploiting these deposits effectively. In this regard, submarine fans have become one of the most thoroughly studied depositional systems in the rock record [1]. However, our understanding of submarine fans is quite controversial, and dispute exists in the usage of terminology, facies models, and concepts [2–5]. The purpose of this report is to address selected problems related to submarine fan models and suggest possible solutions.

Ancient Versus Modern Fans

Jacka and others [6] were the first to propose a three-part submarine fan model that they applied to a Permian sequence in the Delaware basin of New Mexico and Texas. Although, numerous fan models have been proposed since the late 1960's, two models that are often cited and debated in the literature are Mutti and Ricci Lucchi's for ancient fans [1] and Normark's for modern fans [7].

Differences in terminology between modern [7,8] and ancient [1] fan models are shown in Figure 1. In reality, the two models are identical except that the boundary between channelized and non-channelized sequences (dashed line in Fig. 1) occurs in the middle part (i.e., suprafan) of modern fans and in the lower part of ancient fans. When they prograde, both systems produce similar vertical sequences. In a lateral sense, both suprafan lobes (modern) and lower fan lobes (ancient) represent the non-channelized outer part of a submarine fan. Therefore, assigning lobes to either suprafan or lower fan is inconsequential because this difference exists only in terminology and not in terms of depositional processes or facies [3].

Significance of Modern Fans

Normark [8] reported that the study of modern fans is lacking in "ground-truth" samples. In addition, sediment cores from the upper 1 to 10 m of modern fans (e.g., Navy fan [9]) may not be representative of the entire fan package that may reach a maximal thickness of up to several thousand meters (e.g., Bengal fan).

Most major submarine fans develop during periods of low sea-level [10]. Because the upper veneer of most modern fans is apparently intact as a result of the recent rise in sea level, cores from the upper few meters should represent sediment deposited by turbidity currents during the waning stage of fan development. Unlike modern fans, upper veneers of ancient fans probably underwent several

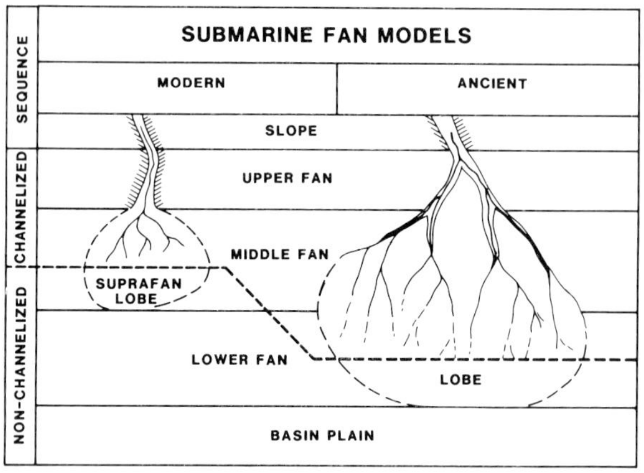

Figure 1. Comparison of modern [7,8] and ancient [1] fan models.

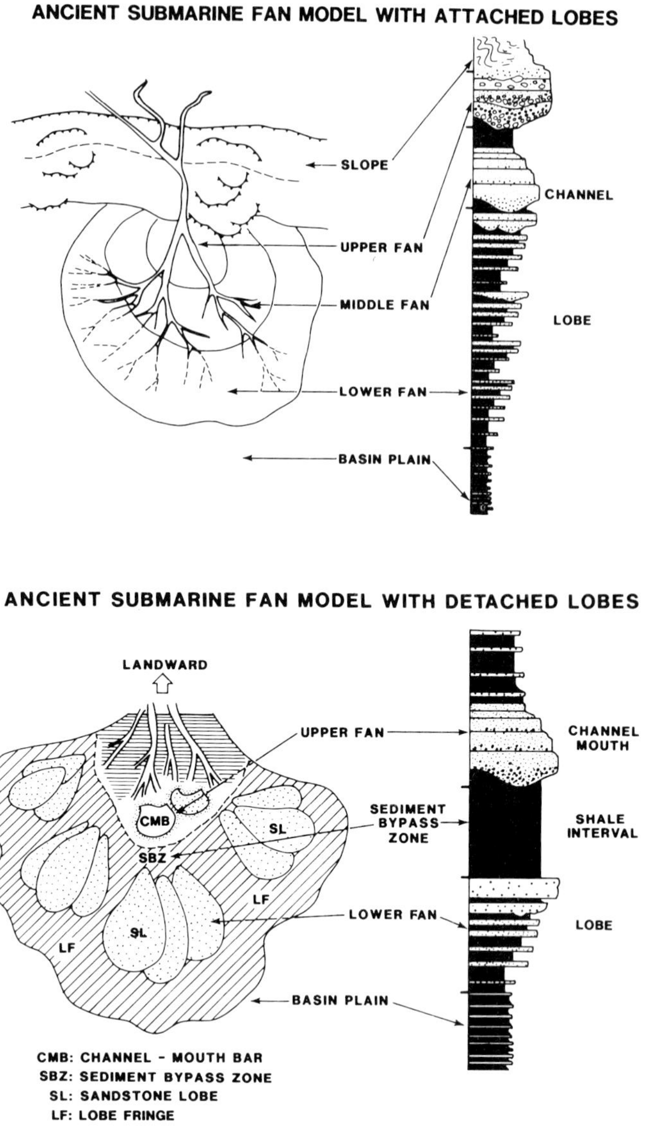

Figure 2. Top: Ancient submarine fan model with attached lobes [1]; vertical progradational sequence [11] showing thinning-upward channel and thickening-upward lobe cycles. Bottom: Ancient submarine fan model with detached lobes [12].

cycles of erosion during subsequent periods of low sea-level, a condition not yet encountered by modern fans. Thus, comparison of surface sediments deposited during the waning stage of modern fan development, with the bulk of ancient sequences representing the prime stage of fan development, is questionable.

Seismic reflection methods are effective in understanding the morphologic features of modern fans [8]. In studying ancient fans, however, outcrops and cores are more useful than seismic sections because of a general lack of time-correlative units and the imprints of diagenetic and structural effects on seismic response. As a result, morphologic relationships between ancient fans are difficult to establish. Until modern fans are extensively cored and their facies are understood, the direct application of modern fan concepts and terminology to ancient fan systems [11] should be done with caution.

Attached Versus Detached Lobes in Ancient Fans

Mutti and Ricci Lucchi [1] initially proposed a submarine fan model in which lobes are attached to feeder channels (Fig. 2). A subsequent model [12], however, advocated detachment of lobes from their feeder channels as a result of sediment bypassing (Fig. 2). If bypassing persists for a considerable length of time, the only recognizable feature in outcrop would be a zone of "thick" hemipelagic shale separating channel mouth deposits from lobe deposits. Otherwise, progradational events in both attached and detached lobe systems would result in identical vertical sequences. Even the presence of a "thick" shale interval between channel mouth and lobe deposits does not necessarily indicate bypassing because channel avulsion can also result in a shale interval between genetically unrelated but adjacent channel mouth and lobe deposits. The significance of this shale interval in hydrocarbon exploration is that it could act as a permeability barrier between lower fan lobes and other potential reservoir facies in the middle and upper fan.

In the Eocene Hecho Group in Spain, the type locality for the detached lobe model, the distribution of channels and lobes (Fig. 3A) with respect to the growing Boltana anticline suggests a tectonic control (Fig. 3C) for sediment bypassing. Mutti [13], however, claims that a hydrodynamic readjustment of turbidity flows in front of the channel mouth results in nondeposition and a related zone of bypassing (Fig. 3B). Although Mutti [14] recognized that the anticline was actively growing during the deposition of the Hecho Group, he did not perceive the role of the anticline in developing the zone of bypassing.

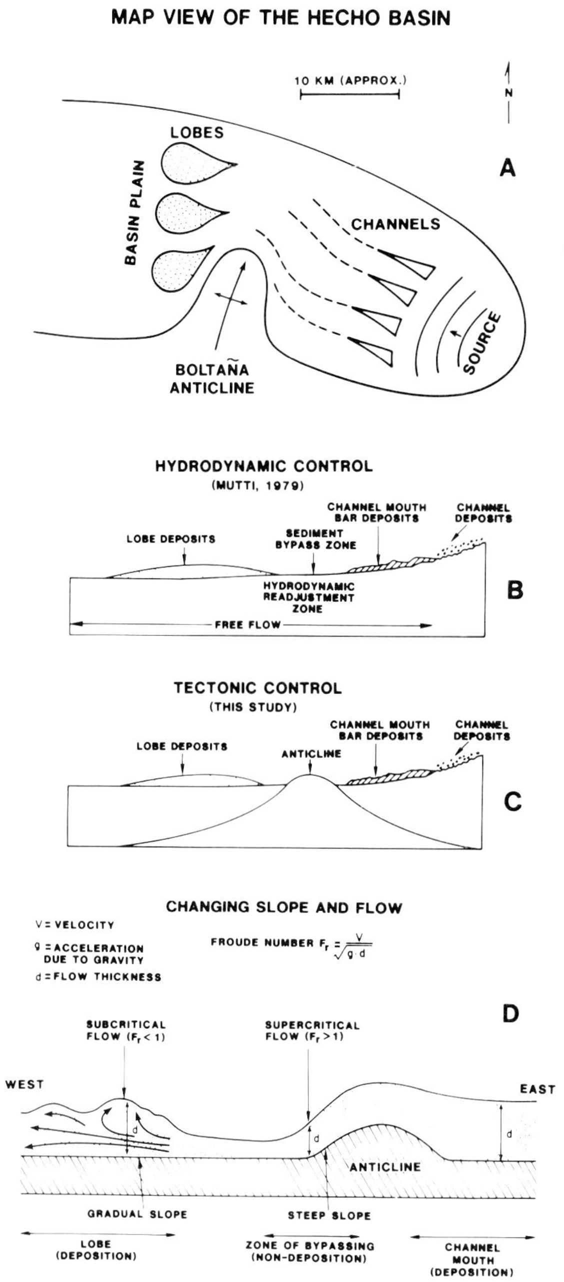

Figure 3. A: Schematic map view of the Hecho Basin in Spain showing channels and detached lobes. B: Longitudinal section showing hydrodynamic control for the development of sediment bypass zone. C: Tectonic control for the development of sediment bypass zone. D: Details of tectonic influence on bypassing; hydraulic model is modified after Walker [16] and Komar [17].

On theoretical grounds, when a flow spreads out from a channel, it will loose its velocity and results in deposition rather than bypassing [15]. Bypassing, however, is possible when a flow encounters a sudden increase in slope. If a flow had been thick enough to overflow the Boltana anticline, supercritical flow might have occurred because of an increase in slope and a related reduction in flow height

Table 1. Characteristics of Highly Efficient and Poorly Efficient Fan Systems (13,19,20)

Characteristics*	Highly Efficient	Poorly Efficient
1. Sediment	Mud-rich	Sand-rich
2. Source area	Large	Restricted
3. Sediment feeding system	River-delta	Beach-canyon
4. Size of fan	Large (hundreds of km)	Small (tens of km)
5. Gradient	Low	High
6. Distance of transport	Long	Short
7. Amount of fines in suspension	High	Small or absent
8. Channels	Detached from lobes	Attached to lobes
9. Sandstone lobes	Large	Small
10. Lobe cycles	Well developed thickening-upward trends	Poorly developed thickening-upward trends
11. Zone of bypassing	Present	Absent
12. Fan fringe deposits	Well developed	Poorly developed or absent
13. Basin plain deposits	Well developed	Poorly developed or absent
14. Ancient example	Eocene Hecho Group, Spain [14]	Eocene Rock Sandstone, CA [21]
15. Modern example	Bengal Fan, India [22]	Navy Fan, off Southern CA [9]

*See text for problems in applying this classification to fans such as the Bengal Fan.

(Fig. 3D). Such an increase in flow energy was apparently responsible for developing a zone of bypassing along the western limb of the anticline.

In order for a facies model to be effective, it should act as a norm, a framework, a guide, a predictor, and a basis for hydrodynamic interpretation [18]. The Hecho Group fails to act as a norm, a framework, or a guide because no other well documented detached lobe fan sequences have been described. The distribution of facies cannot be predicted using the bypass model because of the presumed detachment of lobes from channels. Finally, tectonic control of bypassing cannot be generally used as a basis for hydrodynamic interpretation.

Highly Efficient Versus Poorly Efficient Fan Systems

Mutti [13] proposed two types of fan systems on the basis of their efficiency to transport sand. He suggested that turbidity currents of a mud-rich system transport sand efficiently over a long distance, whereas the transport efficiency of a sand-rich system is relatively poor (Table 1).

The detached lobe fan model and the highly efficient fan system are synonymous according to Johns and Mutti [23]. Thus, there are many problems in applying this classification scheme and its terminology:

1. Mutti [13] used the term "efficiency" only with re-

spect to sediment transport, but some workers have misused this concept. For example, Ricci Lucchi [24] referred to highly efficient mud-rich systems as "sand-efficient." The term "sand-efficient" could easily be misinterpreted as indicating a sand-rich, poorly efficient system. The real meaning of the term "sand-efficient" is that the system is rich in mud and, therefore, it is efficient in transporting sand.

2. The world's largest fan, the Bengal, which is fed by the Ganges River, should theoretically be classified as a "highly efficient" system. The continuous presence of channels throughout the entire 2500-km length of the fan [22] defies such a classification because a zone of sediment bypassing is apparently absent.
3. Ricci Lucchi [24,25] classified the modern Crati Fan in the Ionian Sea as a highly efficient system, but the characteristics of the Crati Fan, such as the presence of attached lobes and its small size (about 15-km long), favor its classification as a poorly efficient system.
4. According to Mutti [13], the zone of bypassing has a definite sedimentologic connotation. It refers only to the area between channel mouths and lobes of a submarine fan. Without realizing this, some authors such as Labude [26] have referred to the slope as a zone of bypassing.
5. In a highly efficient system, suprafan lobes do not occur [13]. Scott and Tillman [27], however, misused Mutti's concept by proposing a detached lobe model (i.e., highly efficient system) with suprafan lobes for the Miocene Stevens Sandstone in California.

In summary, it is not practical to classify fans on the basis of their transport efficiency because a single fan can and often does possess properties of both "highly efficient" and "poorly efficient" systems.

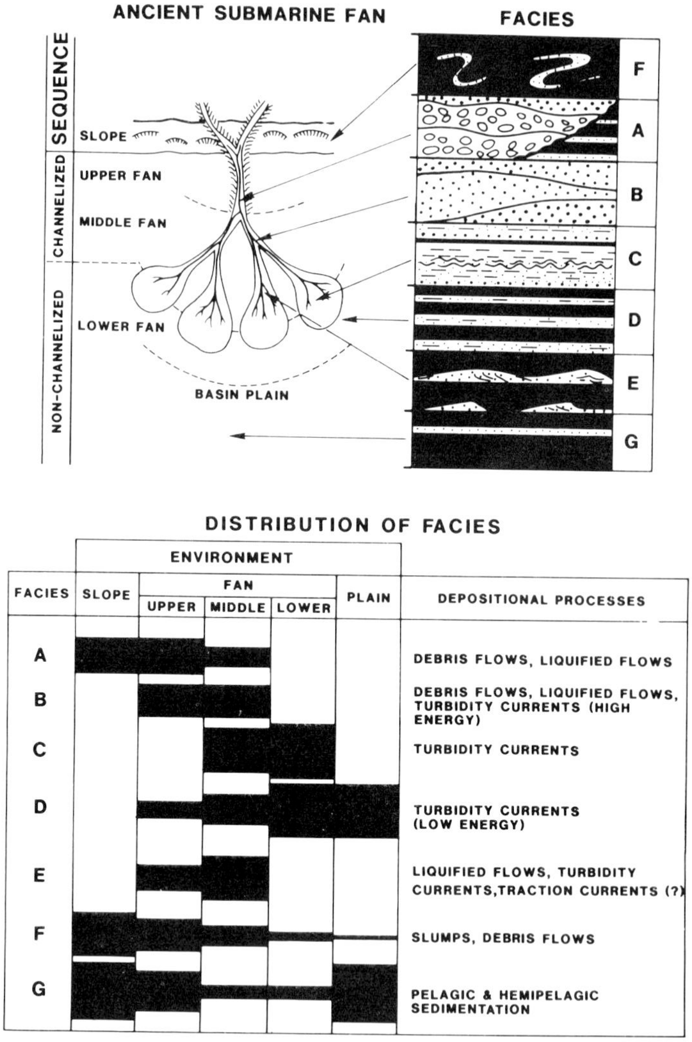

Figure 4. Top: Submarine fan model showing facies associations for channelized and non-channelized sequences; vertical sequence does not represent progradation. Bottom: Distribution of facies in submarine fan and associated environments; facies nomenclature is after Mutti and Ricci Lucchi [1].

Ancient Submarine Fan Cycles and Facies

Depositional cycles in submarine fan systems are interpreted in terms of fluvial and deltaic analogs [1,28]. Thinning- and fining-upward cycles have been related to channel deposition because an upward-widening channel section results in the emplacement of successively thinner beds. In some cases, however, well developed levees may preclude the development of thinning- and fining-upward cycles. Thickening- and coarsening-upward cycles suggest non-channelized lobe deposition (analogous to the delta front) because of basinward progradation and resultant deposition of thicker beds caused by increased sediment supply from subsequent flows. Minor thickening-upward cycles, characteristically associated with lobe sequences, are considered to represent compensation features produced by progressive smoothing out of the depositional relief caused by lobe-upbuilding [29]. Our study of Carboniferous longitudinal fan systems in the Ouachita Mountains of Arkansas and Oklahoma reveals that depositional cycles are of complex and often variable nature [30]. Therefore, depositional cycles should be interpreted in the context of their overall facies association.

Ancient submarine fan facies have been discussed by Mutti and Ricci Lucchi [1]. Based on our observations, the distribution of these facies in a submarine fan complex is shown in Figure 4.

In the rock record, it is easier to describe submarine fans

according to the presence or absence of channelization because this can be recognized not only in outcrops, but also in seismic sections. In general, a channelized sequence (upper and middle fan) with its thinning-upward cycles is composed of facies A, B, C, and E, whereas a non-channelized sequence (lower fan) with its thickening-upward cycles is represented by facies C, D, and G (Fig. 4). Although facies F and G occur in all environments, facies F is characteristic of the slope and facies G is common in the basin plain as well as in the slope.

The facies concept can be applied effectively only to major turbidite systems with well developed fan components. Such fans usually develop during lowstands of sea level [10]. During highstands of sea level, major fans generally do not develop because most land-derived sediment is trapped on the continental shelf (Shanmugam et al., Chapter 5, this volume). Only infrequently do turbidity flows reach the deep sea where they tend to develop a non-fan sequence. A non-fan sequence may superficially resemble the non-channelized portion of a fan system because both consist of facies C, D, and G. The main difference, however, is that a non-fan sequence is characteristically non-cyclic, whereas a non-channelized fan sequence will exhibit thickening- and coarsening-upward cycles. In the modern Atlantic, turbidite sand layers of non-fan affinity have been described by Pilkey and others [31]. These and other present-day turbidites appear to be controlled by the recent rise in global sea-level.

Summary

1. Comparison between modern and ancient submarine fans should be done with caution until modern fans have been more extensively cored and studied.
2. Ancient fans should be considered as systems in which depositional lobes are attached to feeder channels.
3. The detached lobe fan model proposed by Mutti and Ricci Lucchi [12] is based on the Eocene Hecho Group whose deposition was tectonically influenced to produce the observed "bypassing" or detachment of fan lobes. Therefore, this model has limited application to the geologic record.
4. Classification of fans based on their transport efficiency as either "highly efficient" or "poorly efficient" systems is not practical because individual fans often possess properties of both systems.
5. Major submarine fan deposits usually develop during periods of low sea-level. They are characterized by the association of channelized (thinning-upward) and non-channelized (thickening-upward) sequences. Minor turbidite packages appear to develop during periods of high sea-level; they are characteristically non-cyclic.

Acknowledgments

We thank Mobil Research and Development Corporation for permission to publish this report. A. H. Bouma, J. M. Coleman, J. E. Damuth, A. Lowrie, and W. R. Normark made valuable comments. We are thankful to E. Mutti for guiding us through the Hecho Basin in Spain and for discussions on submarine fans. D. Magill typed the manuscript.

References

[1] Mutti, E., and Ricci Lucchi, F., 1972. Turbidites of the northern Apennines: introduction to facies analysis (English translation by Nilsen TH 1978). International Geology Review, v. 20, pp. 125–166.

[2] Nilsen, T. H., 1980. Modern and ancient submarine fans: discussion of papers by R. G. Walker and W. R. Normark. American Association of Petroleum Geologists Bulletin, v. 64, pp. 1094–1101.

[3] Walker, R. G., 1980. Modern and ancient submarine fans. American Association of Petroleum Geologists Bulletin, v. 64, pp. 1101–1108.

[4] Hiscott, R. N., 1981. Deep-sea fan deposits in the Macigno Formation (Middle-Upper Oligocene) of the Gordana Valley, northern Apennines, Italy. Journal of Sedimentary Petrology, v. 51, pp. 1015–1021.

[5] Ghibaudo, G., 1981. Deep-sea fan deposits in the Macigno Formation (Middle-Upper Oligocene) of the Gordana Valley, northern Apennines, Italy. Journal of Sedimentary Petrology, v. 51, pp. 1021–1026.

[6] Jacka, A. D., Beck, R. H., St. Germain, L. C., and Harrison, S. C., 1968. Permian deep-sea fans of the Delaware Mountain Group (Guadalupian), Delaware basin. Society of Economic Paleontologists and Mineralogists Permian Basin Special Publication 68-11, pp. 49–90.

[7] Normark, W. R., 1970. Growth patterns of deep sea fans. American Association of Petroleum Geologists Bulletin, v. 54, pp. 2170–2195.

[8] Normark, W. R., 1978. Fan valleys, channels, and depositional lobes on modern submarine fans: characters for recognition of sandy turbidite environments. American Association of Petroleum Geologists Bulletin, v. 62, pp. 912–931.

[9] Normark, W. R., and Piper, D. J. W., 1972. Sediments and growth pattern of Navy deep-sea fan, San Clemente basin, California Borderland. Journal of Geology, v. 80, pp. 192–223.

[10] Shanmugam, G., and Moiola, R. J., 1982. Eustatic control of turbidites and winnowed turbidites. Geology, v. 10, pp. 231–235.

[11] Walker, R. G., 1978. Deep-water sandstone facies and ancient submarine fans: models for exploration for stratigraphic traps. American Association of Petroleum Geologists Bulletin, v. 62, pp. 932–966.

[12] Mutti, E., and Ricci Lucchi, F., 1975. Turbidite facies and facies associations. In: Examples of Turbidite Facies and Facies Associations From Selected Formations of the Northern Apennines, Field Trip Guidebook A-11. International Sedimentological Congress IX, Nice, pp. 31–36.

[13] Mutti, E., 1979. Turbidites et cones sous-marins profonds. In: P. Homewood (ed.) Sedimentation Detrique (Fluviatile, Littorale et Marine). Institute of Geology, Universite de Fribourg, pp. 353–419.

[14] Mutti, E., 1977. Distinctive thin-bedded turbidite facies and related depositional environments in the Eocene Hecho Group (south-central Pyrenees, Spain). Sedimentology, v. 24, pp. 107–131.

[15] Walker, R. G., 1980. Exploration of turbidites and other deep water sandstones. American Association of Petroleum Geologists Fall Education Conference, Houston, Texas, pp. 1–72.

[16] Walker, J., 1981. The charm of hydraulic jumps, starting with those observed in the kitchen sink. Scientific American, v. 244, pp. 176–184.

[17] Komar, P. D., 1983. Shapes of streamlined islands on Earth and Mars: Experiments and analyses of the minimum-drag form. Geology, v. 11, pp. 651–654.

[18] Walker, R. G., 1979. Facies and facies models, general introduction. In: R. G. Walker (ed.) Facies Models. Geoscience Canada Reprint Series, v. 1, pp. 1–7.

[19] Mutti, E., and Johns, D. R., 1978. The role of sedimentary by-passing in the genesis of fan fringe and basin plain turbidites in the Hecho Group system (south-central Pyrenees). Memoir Societa Geologica Italiana, v. 18, pp. 15–22.

[20] Mutti, E., and Ricci Lucchi, F., 1981. Introduction to the excursions on siliciclastic turbidites. In: Excursion Guidebook. International Association of Sedimentologists 2nd European Regional Meeting, Bologna, Italy, pp. 1–3.

[21] Link, M. H., and Nilsen, T. H., 1980. The Rock Sandstone, an Eocene sand-rich deep-sea fan deposit, northern Santa Lucia range, California. Journal of Sedimentary Petrology, v. 50, pp. 583–601.

[22] Curray, J. R., and Moore, D. G., 1974. Sedimentary and tectonic processes in the Bengal deep-sea fan and geosyncline. In: C. A. Burk and C. L. Drake (eds.) The Geology of Continental Margins. Springer-Verlag, New York, pp. 617–627.

[23] Johns, D. R., and Mutti, E., 1981. Facies and geometry of turbidite sandstone bodies and their relationship to deep sea fan systems. Abstracts, International Association of Sedimentologists 2nd European Regional Meeting, Bologna, Italy, p. 89.

[24] Ricci Lucchi, F., 1981. Contrasting the Crati submarine fan with California fans and models. Abstracts, International Association of Sedimentologists 2nd European Regional Meeting, Bologna, Italy, pp. 157–160.

[25] Ricci Lucchi, F., 1981. The Marnoso-arenacea: a migrating turbidite basin (over-supplied) by a highly efficient dispersal system. In: Excursion Guidebook. International Association of Sedimentologists 2nd European Regional Meeting, Bologna, Italy, pp. 232–275.

[26] Labude, C., 1981. Slope to basin sedimentation and paleobathymetry in Maastrichtian of the central Apennines (Italy). Abstracts, International Association of Sedimentologists 2nd European Regional Meeting, Bologna, Italy, pp. 90–93.

[27] Scott, R. M., and Tillman, R. W., 1981. Stevens sandstone (Miocene), San Joaquin basin, California. In: C. T. Siemers, R. W. Tillman, and C. R. Williamson (eds.) Deep-Water Clastic Sediments, A Core Workshop. Society of Economic Paleontologists and Mineralogists Core Workshop No. 2, San Francisco, pp. 116–248.

[28] Ricci Lucchi, F., 1975. Depositional cycles in two turbidite formations of northern Apennines (Italy). Journal of Sedimentary Petrology, v. 45, pp. 3–43.

[29] Mutti, E., 1983. Facies cycles and related main depositional environments in ancient turbidites systems (abstract). American Association of Petroleum Geologists Bulletin, v. 67, p. 523.

[30] Moiola, R. J., and Shanmugam, G., 1984. Submarine fan sedimentation, Ouachita Mountains, Arkansas and Oklahoma. Transactions Gulf Coast Association Geological Societies, v. 34, pp. 175–182.

[31] Pilkey, O. H., Locker, S. D., and Cleary, W. J., 1980. Comparison of sand-layer geometry on flat floors of 10 modern depositional basins. American Association of Petroleum Geologists Bulletin, v. 64, pp. 841–856.

CHAPTER 7

Potential Petroleum Reservoirs on Deep-Sea Fans off Central California

Pat Wilde, William R. Normark, T. E. Chase, and Christina E. Gutmacher

Abstract

Evaluation of the petroleum potential of the Monterey and Delgada submarine fans based on single-channel seismic reflection profiles and surficial sediment data indicates two high, six good, and four moderate prospects. Sediment thicknesses in a northwest-southeast basement trough at the base of the continental slope generally exceed 1 km; within internal depressions thicknesses can reach 2 to 3 km. Organic carbon contents up to 1.3 weight percent occur on the upper fan. An average heat flow of 2 HFU suggests the occurrence of a thermal gradient sufficient to exceed catagenic conditions within the trough if potential source beds are old enough.

Introduction

There is increasing interest in the possibility of petroleum accumulation along continental margins and, in particular, in submarine fans [1–7]. The hydrocarbon potential of specific deep-sea areas has been investigated in the Gulf of Mexico [1,8,9], the Nile Cone [10], and the U.S. Atlantic continental rise [11].

The pioneer work on central California submarine fans [12–15] was extended to incorporate localized detailed survey data [16–18]. The bathymetry presented in this report (Fig. 1) is our revision of earlier charts [19–20]. In a previous study [21], we discussed the petroleum potential of the deep continental margin off central California, primarily on the Monterey submarine fan. Since 1976, geophysical and sample data from six U.S. Geological Survey cruises (Fig. 1) have been made available to augment and modify the earlier discussion of the petroleum potential in the central California margin. These new data give a more detailed view of the Monterey Fan and provide a reconnaissance seismic survey on the similarly sized Delgada Fan to the north. The study area (Fig. 1) extends from latitude 34° to 40° N. and from the base of the continental slope westward to longitude 127° W., and encompasses approximately 1.8×10^5 km^2. Even with the seismic coverage from these new cruises, large areas of the fans are crossed by only a few single-channel seismic profiles taken with sparker and air-gun sound sources.

We do not imply that direct evidence exists for significant petroleum reserves on these or other modern submarine fans. This can be determined only by drilling. High energy costs, however, accompanied by the extension of drilling and production capabilities for deeper water areas indicate that deep-water environments of continental margins may be exploratory targets in the near future. Particularly attractive are areas off coasts with both high marine biogenic productivity and extensive sedimentary wedges or basins.

Conditions for Petroleum Accumulation

The basic requirements for the accumulation of petroleum in submarine fans are the same as for nearshore continental sedimentary basins. These are: 1) source beds brought to the appropriate thermal-maturation conditions for petroleum generation; 2) these source beds are adjacent or linked to reservoir rocks that have sufficient porosity and permeability to allow the migration and collection of hydrocarbons; and 3) stratigraphic or structural traps involving reservoir rocks favorable for the accumulation of hydrocarbons [22].

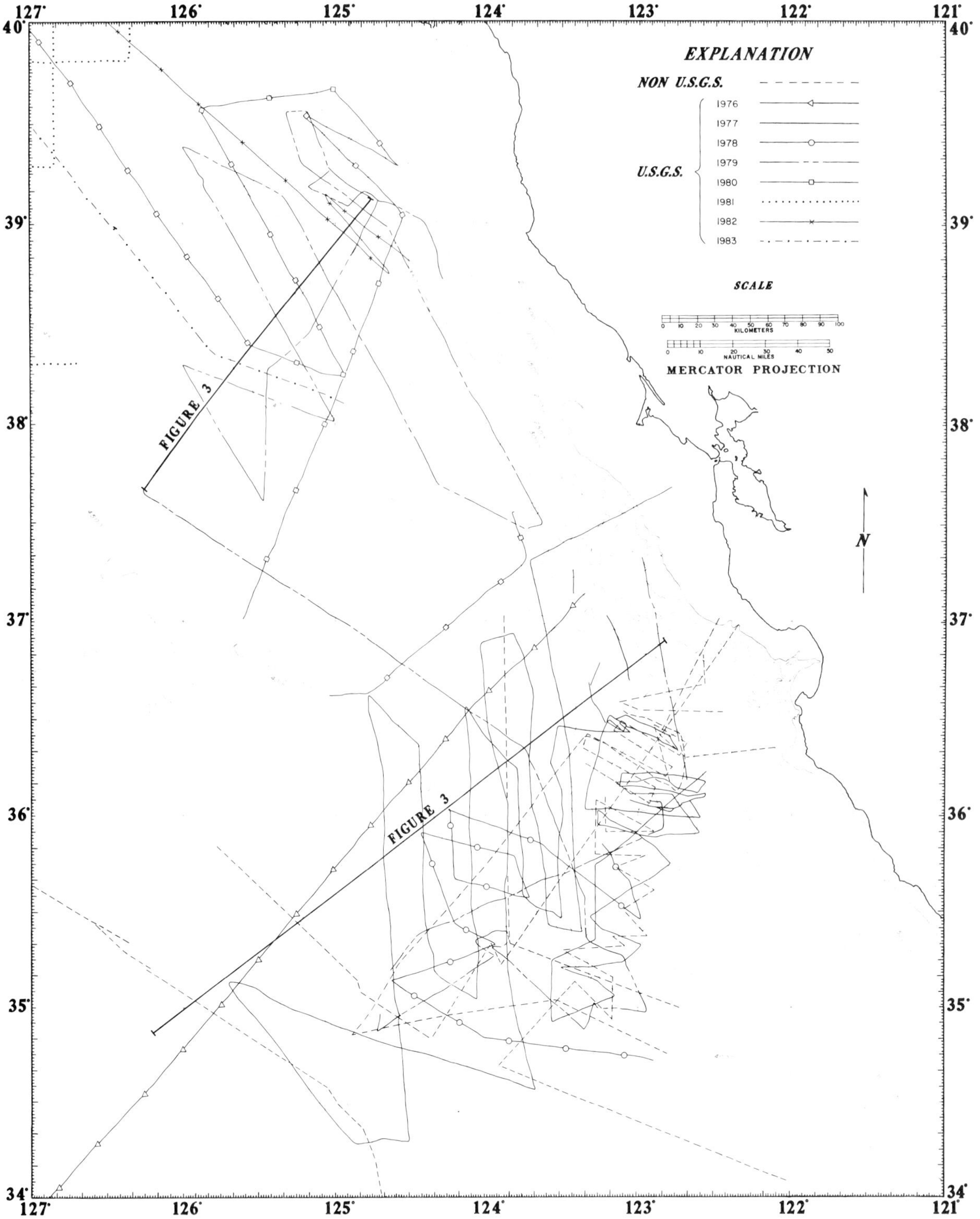

Figure 1. Trackline chart of available seismic reflection profiles of Monterey and Delgada Fans.

Source Beds

The central California continental rise is potentially a favorable area for the formation of hydrocarbon source beds as a result of the seasonal upwelling and high organic productivity that occurs there [22,23]. Analyses for organic carbon in sediments from recently collected piston cores from the middle and upper parts of the Monterey and Delgada Fans shows values about double those reported for Deep Sea Drilling Project (DSDP) Sites 32, 33, and 34 on the more distal part of the Delgada Fan [24]. The total organic carbon content in the DSDP samples resembles that

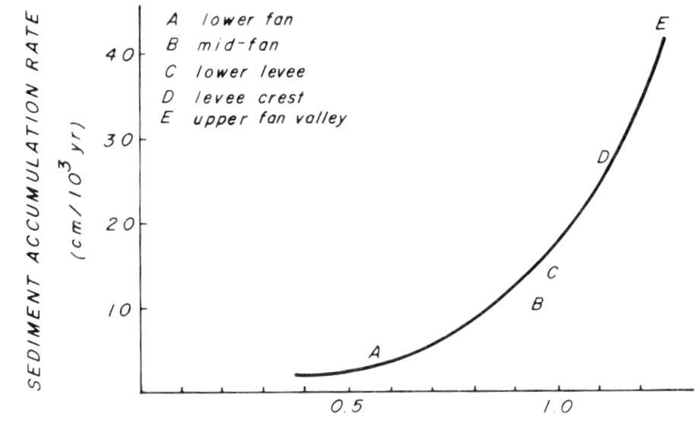

Figure 2. Organic carbon content versus sedimentary accumulation rate for selected depositional environments on Monterey and Delgada Fans. Approximate values for sedimentary accumulation rates are based on the time transgressive change in radiolarian/foraminifer ratio between the Holocene and Pleistocene. The organic carbon contents were measured in more recently collected cores than those used for determining sediment accumulation rate.

determined in piston cores from the distal parts of the Monterey Fan. The higher contents reflect better preservation of organic material, probably resulting from higher depositional rates on the mid-fan and upper fan as the average organic carbon content within the cores increases with higher sedimentary accumulation rates (Fig. 2). This relation suggests that the total organic matter locally may approach that measured in deposits in California Continental Borderland basins [25].

An 8-m core obtained in 1979 from the lower Monterey Valley levee contained sufficient gas to blow off the end caps on core liners. Eventually, we flared the gas from four of the 1.5-m long sections. Analyses of the gas content of the core showed significant amounts of residual hydrocarbons. The methane concentration exceeded 1 ml/L of wet sediment, even after a storage period of four weeks (K. A. Kvenvolden and T. M. Vogel, written communication, 1980). The ratio of methane to other gaseous components indicates a biological origin for the gas. The total organic carbon content measured in four samples from the core was only mid-range for our Monterey Fan samples (average 1.0 weight percent). Thus, it is unlikely that such a high biogenic gas content exists just in the lower part of only this core. The evidence from one core 8-m long from a fan whose thickness is measured in kilometers only can suggest the existence of source beds on these fans. That similar processes were or are operative in the deeper and older parts of the fans seems a valid hypothesis, in lieu of deeper analyzed samples, by analogy with the depositional history of the fans. Unquestionably, the existence, location, and extent of source beds are the biggest unknowns in evaluating the petroleum potential of these fans.

The actual alteration of organic matter in marine sediment to hydrocarbons is a function of the cumulative effects of

Table 1. Time-Temperature-Depth Data for Petroleum Generation

	A	B	C
Age (m.y.)	Temperature[1] of Catagenesis	Depth[2,3] Using 70°C/km Gradient (°C)	Depth[2,4] Using 35°C/km Gradient (km)
2: Base of Pleistocene	182	2.6	5.2
5: Base of Pliocene	152	2.2	4.3
22: Base of Miocene	111	1.6	3.2
38: Base of Oligocene	97	1.4	2.8
55: Base of Eocene	89	1.3	2.5

[1]ln t = (5,879 × 1/T − 12,224; where t = the time in m.y., and T = the temperature in K [27].
[2]Depth of deposits below sea floor needed to attain temperature in column B; catagenic conditions exist below this depth for age given in column A.
[3]Calculated from measured heat flow of 2 HFU on the upper fan.
[4]Normal geothermal gradient in the oceans corresponds to approximately 1 HFU [21].

temperature (depth) and age [26,27]. Table 1 illustrates the estimated time-temperature-depth data for central California submarine fans, using Connan's [27] formulation for catagenic conditions and the limited heat-flow data available for this region [19,28]. The importance of reliable heat-flow data in estimating the petroleum-generating potential in deep-water deposits has been illustrated by Simoneit and others [29], who reported petrogenic (C_{2-8}) hydrocarbons in gravity-core material from the Guaymas Basin, Gulf of California, in a region of high heat flow (6.2 HFU) [30]. Thus, the time required to achieve catagenic conditions in thick sedimentary sections locally may be substantially decreased if regions of higher than average heat flow are nearby.

Sediment Thickness

Radial sections of the Monterey and Delgada Fans display a wedge shape typical of submarine fans (Fig. 3). The thickest sections occur at or near the base of the slope, as expected, although local basement relief results in deep and relatively isolated basins well away from the base of the slope. The bathymetric chart of the fans (Fig. 1), however, shows a sharp contrast in the shape of the two fans (see also the chapters on Monterey and Delgada Fans, this volume). The Delgada Fan includes two separate elongate lobes formed by prograding leveed-valley systems; these lobes are adjacent to and perhaps were deposited on a relatively smooth ramp of continental rise deposits [31]. In contrast, the morphology of the Monterey Fan reflects a series of changes in the channel systems that results in switching of active depositional (lobe?) areas. Extensive basement relief, recognizable on the bathymetric chart (Fig. 1) as ridges and seamounts rising above the fan surface, has greatly influenced the growth pattern of the fan (see Fig. 5 in Chapter 13).

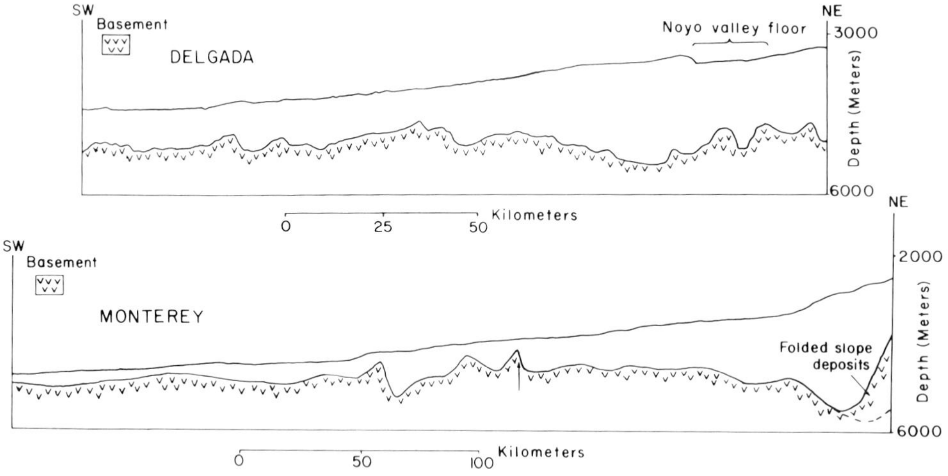

Figure 3. Radial profiles of Delgada and Monterey Fans.

Our isopach maps (Figs. 4 and 5) show that many seamounts and basement ridges have been completely buried by fan deposits. The trough adjacent to the continental slope on the upper part of the Monterey Fan, and the elongate basins associated with the Chumash fracture zone on the lower and middle parts of the Monterey Fan, are the principal potential hydrocarbon source areas based on sedimentary thicknesses and an assumed uniform heat flow of 2 HFU. Table 1 presents the time-temperature-depth data as a function of the minimal age of the deposits. The depth-time calculations [27] can be used to outline potential source areas by plotting the catagenic depths on Figures 4 and 5. The applicability of these time boundaries depends on the actual age of the deposits under the thicker parts of the fan. Furthermore, the target areas may expand or contract locally if the heat flow differs from the few values available. Higher heat flows might occur along active tectonic features. For example, residual volcanism along the Chumash fracture zone during deposition of turbidites in the nearby troughs might contribute to petroleum generation. The Monterey Fan, with numerous seamounts and ridges under fan deposits, may contain more localized high heat-flow regions and, thus, have a higher petroleum generation potential with depth than the Delgada Fan with an acoustic basement of low relief.

Reservoir Beds

Sand beds of valley and channel systems and depositional lobes around the distributary channels on the mid-fan and lower fans are the best potential reservoir sections in modern fans. On the upper fan area, the relatively clean sand of the floor and levee crests of the large depositional valley on Monterey Fan [32] provides broad (possibly as wide as 20 km) areas of potential reservoir beds. The deeply incised valley of the main Monterey Fan valley appears to be floored by the coarsest deposits observed on the fan [16,32], and filling of this valley with similar sediment could create a thick section of massive sand and gravel beds extending across two-thirds of the fan's radius. The major valley systems on the Monterey Fan, both active and abandoned, appear to end in unleveed distributary-channel systems that feed a broad sandy lobe as much as 100-km across (see Fig. 5 in Chapter 13).

In our earlier report [21], we emphasized linear bodies of channel/valley sands as the primary reservoir sections. Sand lobes adjacent to distributary-channel systems also would provide good reservoirs if the adjacent source beds were buried sufficiently to allow catagenesis of organic material. Sand is deposited most extensively on the mid-fan or lower fan where the sedimentary thickness rarely exceeds 500 m. Thus on the Monterey Fan, the heat flow would have to be higher to expect petroleum generation in those areas. Sand beds in the active depositional lobes on the Monterey Fan are more poorly sorted and have a smaller average grain size than those in the upper fan channels and valleys. However, the larger areal extent of these lobes increases their attractiveness as reservoir beds. The most attractive targets would be the older, and perhaps coarser, sand lobes deposited close to the continental margin during the early stage of fan growth, or those lobes formed in the deeper basins on the mid-fan.

On the Delgada Fan, only the prominent depositional-valley systems or the underlying continental rise deposits are likely to contain potential reservoir beds. Apparently only a few sand beds occur in the upper 10 m of the valley-floor and levee deposits. No major channel distributary system nor active sandy depositional lobe was recognized at the termination of the valley. Thus, the Delgada Fan seems to lack the thick sand sections in the levee and valley sequences that are seen on the Monterey Fan. The continental rise around the depositional valleys, however, is substantially sandier according to limited core data on near-

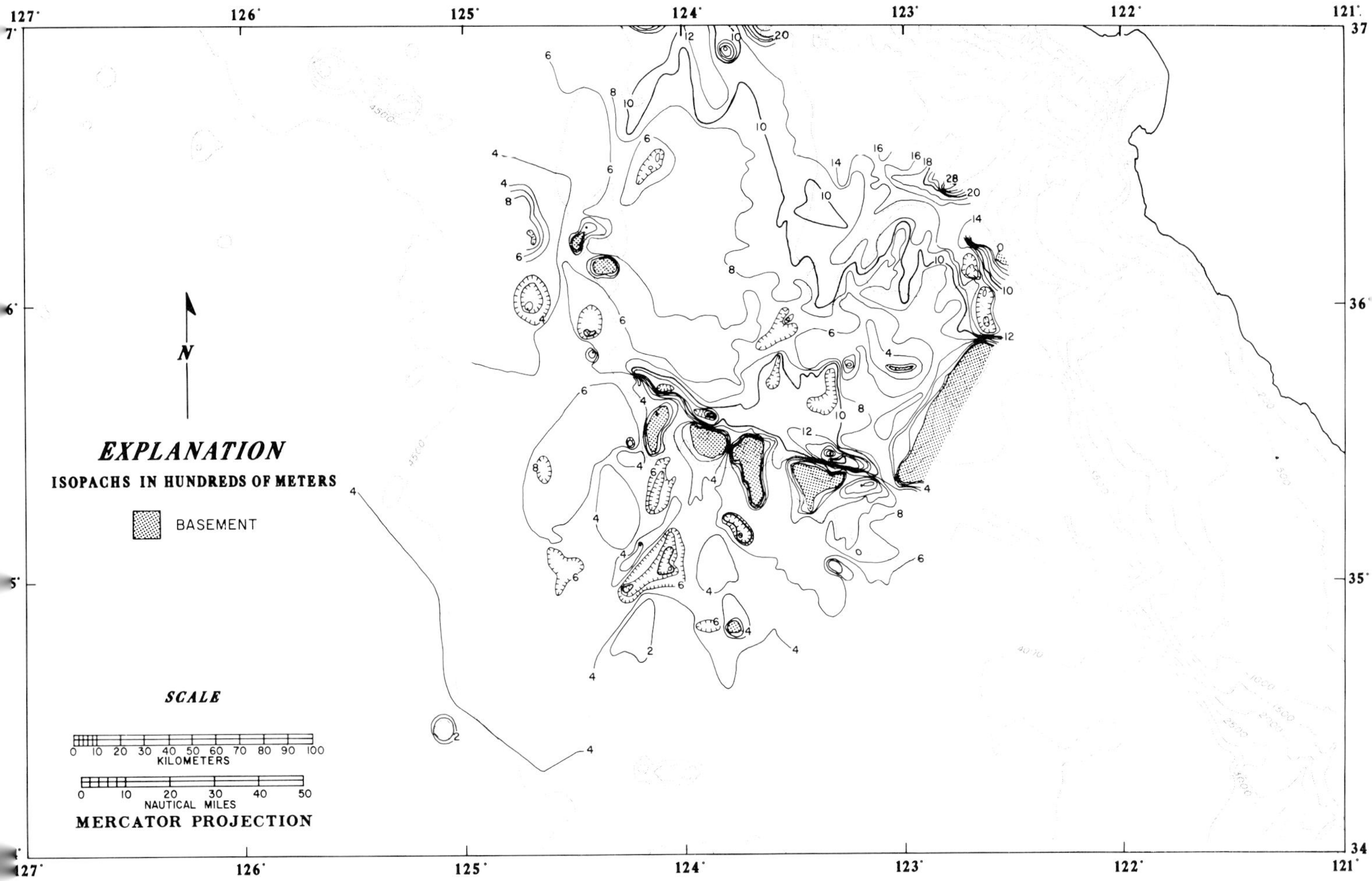

Figure 4. Isopachous map for Monterey Fan, corrected for sound velocity from multi-channel profile on Monterey Fan and unpublished CDP profiles from Delgada Fan (D. McCulloch, written communication).

surface sediments (see Chapter 10). The most likely reservoir section on the Delgada Fan may lie in these rise sections now buried under the thick depositional-valley sequences (Fig. 5).

Traps

The single-channel reflection profiles show several distinct types of stratigraphic traps on modern fans; these types include a lateral transition from valley-floor sand to the muddier levee facies and the broad anticlinal shape of beds within the levees. Isolated basins filled by turbidites on the mid-fan represent sites of onlap onto the basin flanks that would provide potential hydrocarbon traps. Onlap of upper fan sections onto the lower slope deposits can create other potential traps; specifically, onlap of the thick and coarse Ascension valley-floor deposits onto continental slope muds may be one of the best traps on the Monterey Fan. Differential compaction between silt and clay sections and adjacent sand and gravel units of channels and lobes can provide additional stratigraphic traps similar to those on the Frigg Fans in the North Sea [33].

Such structural traps as anticlinal folds or areas along faults have not been detected on either of these central California submarine fans. Although some deposits may be folded in the deeper parts of basement troughs, these flexures may be an artifact of the acoustic record because of the contrast in travel time between the deposits and the adjacent higher velocity basement materials. Multi-channel seismic reflection lines, especially 3-D surveys, are needed to resolve these geometric features.

Traps with the best petroleum potential would be those areas subjacent to zones of source beds exceeding catagenic conditions or those sections connected by permeable beds with catagenic sections downdip. Accordingly, potential target areas on the Monterey Fan probably exist beyond the limits of the catagenic zones shown in Figures 4 and 5.

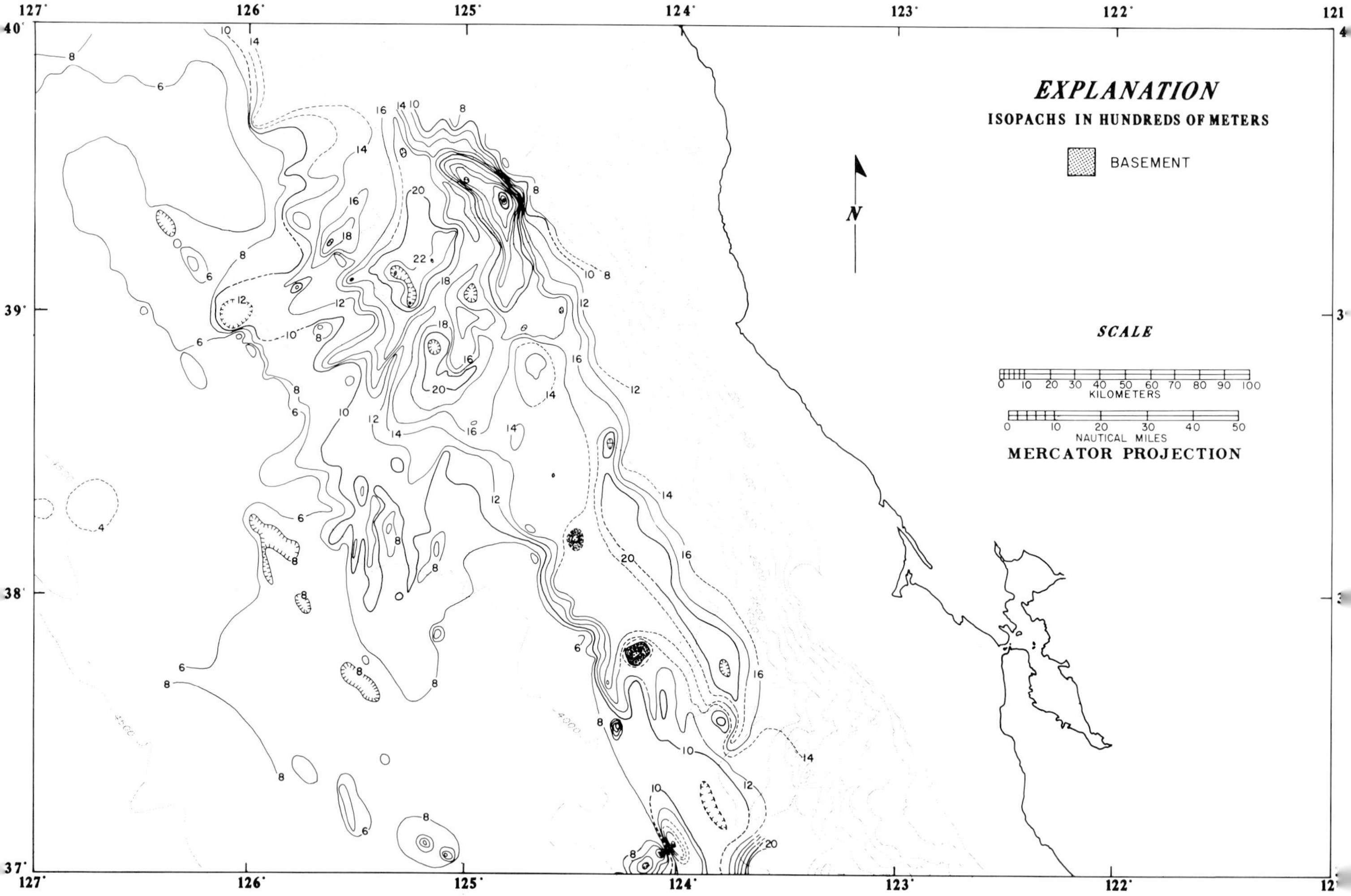

Figure 5. Isopachous map for Delgada Fan, corrected for sound velocity from multi-channel profile on Monterey Fan and unpublished CDP profiles from Delgada Fan (D. McCulloch, written communication).

Prospects and Conclusions

The best hydrocarbon prospects on deep-sea fans off central California are stratigraphic traps in the older parts of the relatively thick sections near the base of the continental slope. Most mapped channels are in the upper 500 m of these sections. Buried (but as yet unmapped) channel sections may be common in the deeper sections of the upper fans. The upper part of the Delgada Fan contains an extensive area and volume of deposits that might exceed catagenic conditions. Prospects on the Monterey Fan are less attractive because of the limited areal extent of deep basins or thick sections along the base of the slope. At present, prospects on the mid-fan to outer fan are rare. Only a few basins are recognized on the better studied Monterey Fan, particularly in the Chumash fracture zone trough. The more extensive areas of thick deposits on the Delgada Fan may indicate greater potential petroleum resources because the highest heat-flow (3.45 HFU) [28] are found there. Table 2 lists the potential prospects and a relative ranking of their attractiveness.

Technology represented by the DSDP drilling vessel *Glomar Challenger* and its ODP successor, the drilling vessel *SEDCO 471*, already exists to drill test wells in the water depths of deep-sea fans. The proximity of prospects in the central California deep-sea fans to existing refineries and the ample local markets in California indicate that the relatively well studied Monterey and Delgada Fans are possible areas to test the hypothesis of significant petroleum accumulation in the deeper parts of the continental margin.

Table 2. Prospects for Central California Fans Trap: Stratigraphic Updip Pinch-out

Rank[1]	Reservoir: Valley/Channel Sands	Source Beds[2]
	Location	
1	Apex, Monterey Fan	A
2	Apex, Monterey Fan	B
1	Apex, Delgada Fan	A
3	Apex, Delgada Fan	B
2	Middle part of fan	B
2	Chumash fracture-zone trough	A
2	Chumash fracture-zone trough	B
	Reservoir: Suprafan and Lobes	
	Location	
2	Apex, Monterey Fan	A
2	Apex, Delgada Fan	A
3	Apex, Delgada Fan	B
3	Middle part of fan	B
2	Chumash fracture-zone trough	A
3	Chumash fracture-zone trough	B

[1]Rank of relative attractiveness: 1 = high; 2 = good; 3 = moderate.
[2]Relation to source beds: A, reservoir within catagenic zone; B, reservoir above catagenic zone.

Acknowledgments

We thank G. Hess for the use of some of his unpublished data on sedimentation rates and D. McCulloch for unpublished sound velocity data. A. Stevenson was chiefly responsible for the success of the coring operations. The manuscript benefited from critical reviews by D. G. Howell and L. Cohen. The word-processing and editing were done by J. Edgar.

References

[1] Moore, G. T., and others, 1977. Mississippi fan, Gulf of Mexico—physiography, stratigraphy, and sedimentational patterns. In: A. H. Bouma, G. T. Moore, and J. M. Coleman (eds.) Framework, Facies, and Oil-Trapping Characteristics of the Upper Continental Margin. American Association of Petroleum Geologists, Studies in Geology, v. 7, pp. 155–191.

[2] Moore, G. T., 1973. Oil on the shelf. Oceanus, v. 17, pp. 11–17.

[3] Moore, G. T., and Fullam, T. J., 1973. Deep water channels and their potential value in petroleum localization. Gulf Coast Association Geological Society Transactions, v. 23, pp. 256–258.

[4] Beck, R. H., and P. Lehner, 1974. Oceans: new frontier in exploration. American Association of Petroleum Geologists Bulletin, v. 58, pp. 376–395.

[5] Thompson, T. L., 1976. Plate tectonics in oil and gas exploration of continental margins. American Association of Petroleum Geologists Bulletin, v. 60, pp. 1463–1501.

[6] Sangree, J. B., and J. M. Widmier, 1977. Seismic interpretation of clastic depositional facies. In: Seismic Stratigraphy—Applications to Hydrocarbon Exploration. American Association of Petroleum Geologists Memoir, v. 26, pp. 165–184.

[7] Walker, R. G., 1978. Deep-water sandstone facies and ancient submarine fans: Models for exploration for stratigraphic traps. American Association of Petroleum Geologists Bulletin, v. 62, pp. 932–966.

[8] Caughey, C. A., and C. J. Stuart, 1976. Where the potential is in the deep Gulf of Mexico. World Oil, v. 183, no. 1, pp. 67–72.

[9] Moore, G. T., and others, 1979. Investigation of Mississippi Fan, Gulf of Mexico. In: J. S. Watkins, L. Montadert, and P. W. Dickerson (eds.) Geological and Geophysical Investigations of Continental Margins. American Association of Petroleum Geologists Memoir, v. 29, pp. 423–442.

[10] Summerhayes, D., Ross, D. A., and Stoffers, P., 1977. Nile submarine fan: Sedimentation, deformation, and oil potential. 9th Offshore Technology Conference Proceedings, v. 1, pp. 35–40.

[11] Mattick, R. E., and others, 1978. Petroleum potential of U.S. Atlantic slope, rise, and abyssal plain. American Association of Petroleum Geologists Bulletin, v. 63, pp. 592–608.

[12] Dill, R. F., Dietz, R. S., and Stewart, H. B., 1954. Deep-sea channels and delta of the Monterey submarine canyon. Geological Society of America Bulletin, v. 65, pp. 191–194.

[13] Menard, H. W., 1955. Deep-sea channels, topography, and sedimentation. American Association of Petroleum Geologists Bulletin, v. 39, pp. 236–255.

[14] Menard, H. W., 1960. Possible pre-Pleistocene deep-sea fans off central California. Geological Society of America Bulletin, v. 71, pp. 1271–1278.

[15] Heezen, B. C., and H. W. Menard, 1963. Topography of the deep-sea floor. In: M. H. Hill (ed.) The Sea: Ideas and Observations on Progress in the Study of the Seas, Volume 3, The Earth Beneath the Sea. History. John Wiley & Sons, New York, pp. 233–280.

[16] Wilde, P., 1965. Recent sediments of the Monterey deep-sea fan. University of California, Berkeley, Hydraulics Engineering Laboratory Report HEL 2-13, 153 p.

[17] Shepard, F. P., 1966. Meander in valley crossing a deep-sea fan. Science, v. 154, pp. 385–386.

[18] Normark, W. R., 1970. Channel piracy on Monterey deep-sea fan. Deep-Sea Research, v. 17, pp. 837–846.

[19] Chase, T. E., Normark, W. R., and Wilde, P., 1975. Oceanographic data of the Monterey deep-sea fan, 34°-37°N 120°-127°W. University of California Institute of Marine Resources Technical Report TR-58.

[20] Wilde, P., Normark, W. R., and Chase, T. E., 1976. Oceanographic data off central California 37° to 40° North. U.S. Lawrence Berkeley Laboratory Publication 92.

[21] Wilde, P., Normark, W. R., and Chase, T. E., 1978. Channel sands and petroleum potential of Monterey deep-sea fan, California. American Association of Petroleum Geologists Bulletin, v. 62, pp. 967–983.

[22] Dow, W. G., 1977. Petroleum source beds on continental slopes and rises. American Association of Petroleum Geologists Course Note Series, No. 5, pp. D1–D37.

[23] Dow, W. G., 1979. Petroleum source beds on continental slopes and rises. In: J. S. Watkins, L. Montadert, and P. W. Dickerson (eds.) Geological and Geophysical Investigations of Continental Margins. American Association of Petroleum Geologists Memoir, v. 29, pp. 423–442.

[24] Weser, O. E., 1970. Lithologic summary. In: Initial Reports of the Deep-Sea Drilling Project. U.S. Government Printing Office, Washington, D.C., pp. 569–620.

[25] Emery, K. O., 1970. The Sea Off Southern California–A Modern Habitat of Petroleum. John Wiley & Sons, New York.

[26] Lopatin, N. V., 1971. Tremperatura i geologicheskoe vremya kak faktory uglefikatsii. Ivestia Akadamii Nauk Kazakstan Sovetskik Sotsialisticheskikh Respublik Series in Geology, no. 3, pp. 95–106 (in Russian).

[27] Connan, J., 1974. Time-temperature relation in oil genesis. American Association of Petroleum Geologists Bulletin, v. 58, pp. 2516–2521.

[28] von Herszen, R. P., 1964. Ocean-floor heat-flow measurements west of the United States and Baja, California. Marine Geology, v. 1, pp. 225–239.

[29] Simoneit, B. R. T., and others, 1979. Organic geochemistry of recent sediments from Guaymas Basin, Gulf of California. Deep-Sea Research, v. 26, pp. 879–891.

[30] Lawver, L. A., and others, 1973. Heat flow measurements in the southern portion of the Gulf of California. Earth and Planetary Science Letters, v. 12, pp. 198–202.

[31] Winterer, E. L., Curray, J. R., and Peterson, M. N. A., 1968. Geologic history of the intersection of the pioneer fracture zone and the Delgada deep-sea fan, northeast Pacific Ocean. Deep-Sea Research, v. 15, pp. 509–520.

[32] Hess, G. R., and W. R. Normark, 1976. Holocene sedimentation history of the major fan valleys of Monterey Fan. Marine Geology, v. 22, pp. 233–251.

[33] Heritier, F. E., Lossel, P., and Wathne, E., 1979. Frigg Field-Large submarine-fan trap in lower Eocene rocks of North Sea Viking Graben. American Association of Petroleum Geologists Bulletin, v. 63, pp. 1999–2020.

II

Modern Submarine Fans

Active Margin Setting

CHAPTER 8

Astoria Fan, Pacific Ocean

C. Hans Nelson

Abstract

The Astoria Fan, a modern system, is located on a subducting oceanic crust and fills a north-south-trending trench along the Oregon continental margin. Well-developed channels cross the entire fan length; they display classic inner-fan leveed profiles but evolve into distributaries in the midfan area where the gradient decreases sharply. During periods of low sea level, inner- and middle-fan channels funnel sand to distal depositional sites in the outer-fan area where the sand/shale ratios are highest. This pattern of sand displacement and efficiency of transport appears to be characteristic of elongate fans fed by a major river and submarine canyon.

Introduction

The modern Astoria Fan is situated on the continental rise off the coast of Oregon (Fig. 1). The wedge of fan sediment radiates asymmetrically southward from the mouth of the Astoria Canyon, which heads off the Columbia River. From the canyon mouth, the fan extends about 100 km to its western boundary, the Cascadia Channel. The fan proper ends 160 km south of the canyon mouth, although its depositional basin extends southward another 150 km to the Blanco Trough [1].

The Astoria Fan is an excellent example of an elongate fan developed in an unconfined basin and fed by a large river and submarine canyon point source [2]. It has unusually well-preserved topography (Fig. 1) [1] and tuffaceous turbidite marker beds containing Mazama ash [3]. Both of these characteristics permit good definition of typical fan physiography, sedimentary processes, and depositional facies.

Good bathymetric and stratigraphic information exists for the Astoria Fan, although few seismic and no deep-tow side-scan sonar profiles are available. Thorough bathymetric soundings were taken in the canyon mouth and the slope-base fan valleys extending from it (Fig. 2). Transverse sounding lines, 10 km or less apart, cover these linear features, from which smooth sheets with closely interpolated depths were constructed.

A total of 40 piston cores provides a means of establishing Holocene and Late Pleistocene stratigraphy for the Astoria Fan [4]. These cores include unique, correlative sand layers containing ash from the Mount Mazama eruption 6600 yr B.P. [3]. The only subsurface information available, except for fan piston cores, are the data from Site 174 of the Deep Sea Drilling Project (DSDP) and a few site-survey reflection profiles [5]; these data and profiles suggest that the facies relationships seen in the Pleistocene section of the piston cores persist at depth.

Geologic Setting

The continental shelf in the active continental margin area off the Columbia River is relatively narrow and steep with a shelf edge at 160 m (Fig. 1). Near Astoria Canyon, the continental slope is 63-km wide and extends to a depth of 2013 m, but southward toward the outer fan, it widens to 96 km and deepens to 2740 m (Fig. 3). The slope has an average gradient of 33/1000, lower than those of most continental slopes.

The Astoria Canyon heads 100 m below sea level and follows a mildly sinuous path west-southwestward for approximately 120 km to a depth of 2085 m (Fig. 3). The mouth opens onto the Astoria Fan and connects with the main fan valley, the Astoria Channel (Fig. 1).

The Astoria Fan is an asymmetric wedge of sediment (2000 m thick) that radiates from the Astoria Canyon and forms the continental rise. From the mouth of the Astoria Canyon, the fan extends about 102 km to its western boundary (Cascadia Channel) but only about 28 km to its northern boundary

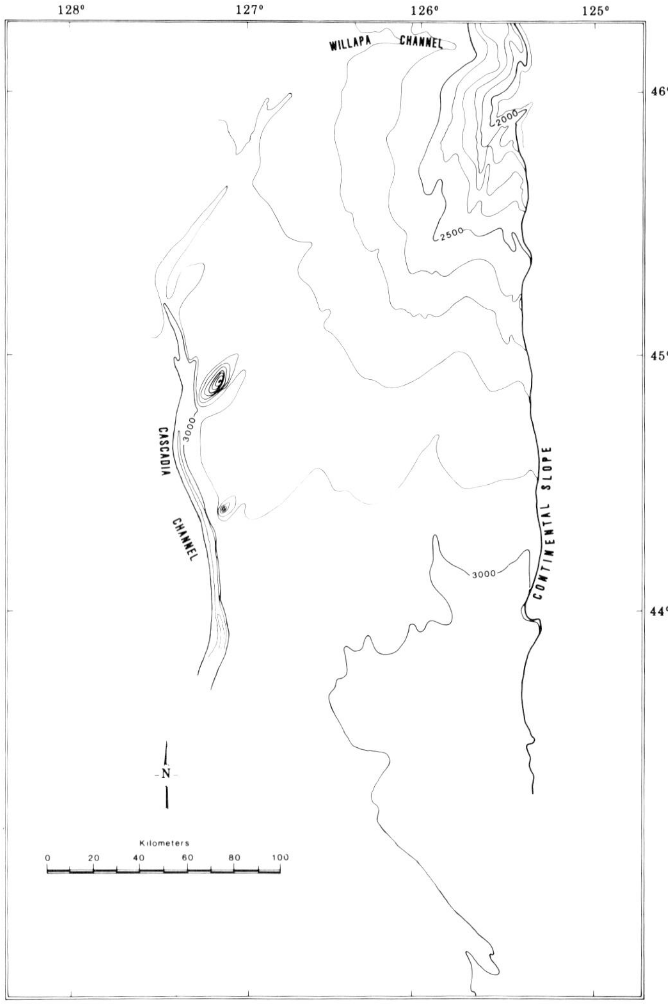

Figure 1. Bathymetric chart of Astoria Canyon, Astoria Fan, and adjacent continental margin off northern Oregon coast after [1]. Dashed line denotes fan boundary.

(Willapa Channel) (Figs. 1, 2). The 2800-m contour 139 km south of the canyon mouth is the southernmost contour to outline the conical shape of the fan, which in this area extends westward 140 km to the Cascadia Channel. The continental slope forms the eastern border of the fan.

The fan constitutes a westward-thinning north-south-trending wedge of sediment along the margin of the continental slope, rather than a simple conical-shaped body oriented perpendicular to the slope. It fills and covers an apparent trench at the eastern edge of the eastward-subducting Gorda-Juan de Fuca plate [5]. The Astoria Fan is 284-m thick at its southwestern distal end, 1000-m thick along its eastern margin at the base of the continental slope, and more than 1000-m thick at its apex. Pleistocene sediment of the Astoria Fan rests unconformably on a thin sequence of Pliocene and Pleistocene abyssal plain turbidites [5].

The Cascadia Channel is a major deep-sea channel situated along the northern and western margins of the Astoria Fan (Figs. 1, 2). The channel appears to have restricted the seaward and radial growth of the fan by limiting and capturing sediment gravity flows westward from the Astoria Canyon [6]. The elongate growth of the Astoria Fan to the south also was encouraged by progressive lefward shift of the Astoria Fan valley systems over time (Fig. 2) [1].

Various sediment types have been transported to the canyon by the Columbia River; this is the third largest river of North America, and has an average discharge of 10 to 12 × 10^6 t of suspended sediment per year [1]. Sediment source terranes include the Coast Range, the Cascade Mountains, the Columbia Plateau, and the Rocky Mountains.

Physiography

The head of the Astoria Canyon now lies 17-km west of the Columbia River mouth, but buried Pleistocene channels formerly connected it to the river mouth [1]. The canyon has

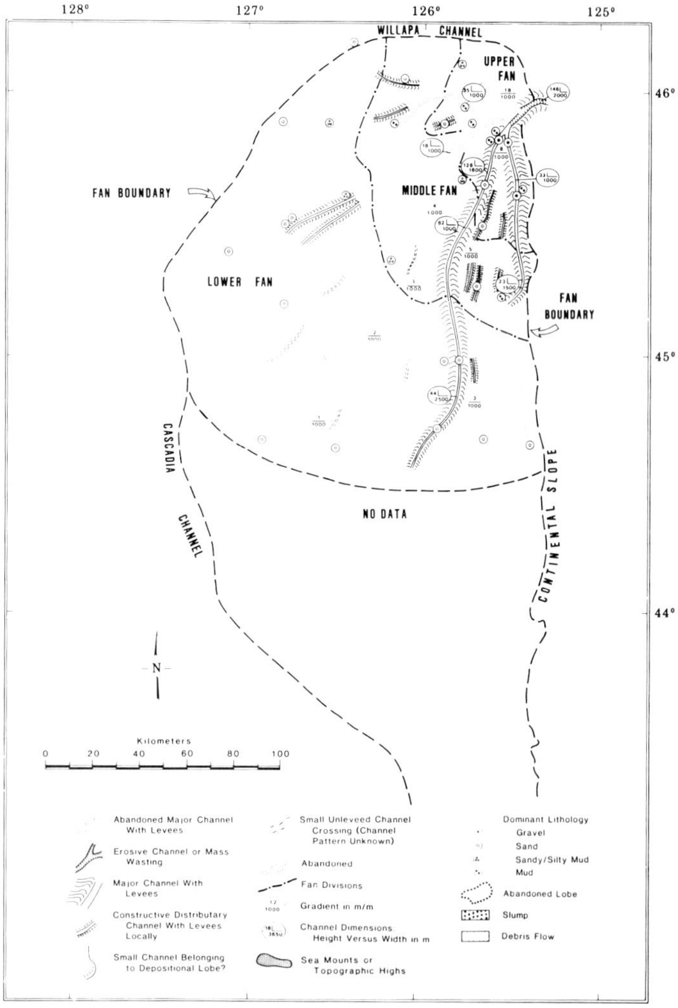

Figure 2. Astoria Fan morphology and generalized lithology, modified from [1]. Sharp breaks in slope define inner, middle, and outer fan (See Fig. 3A).

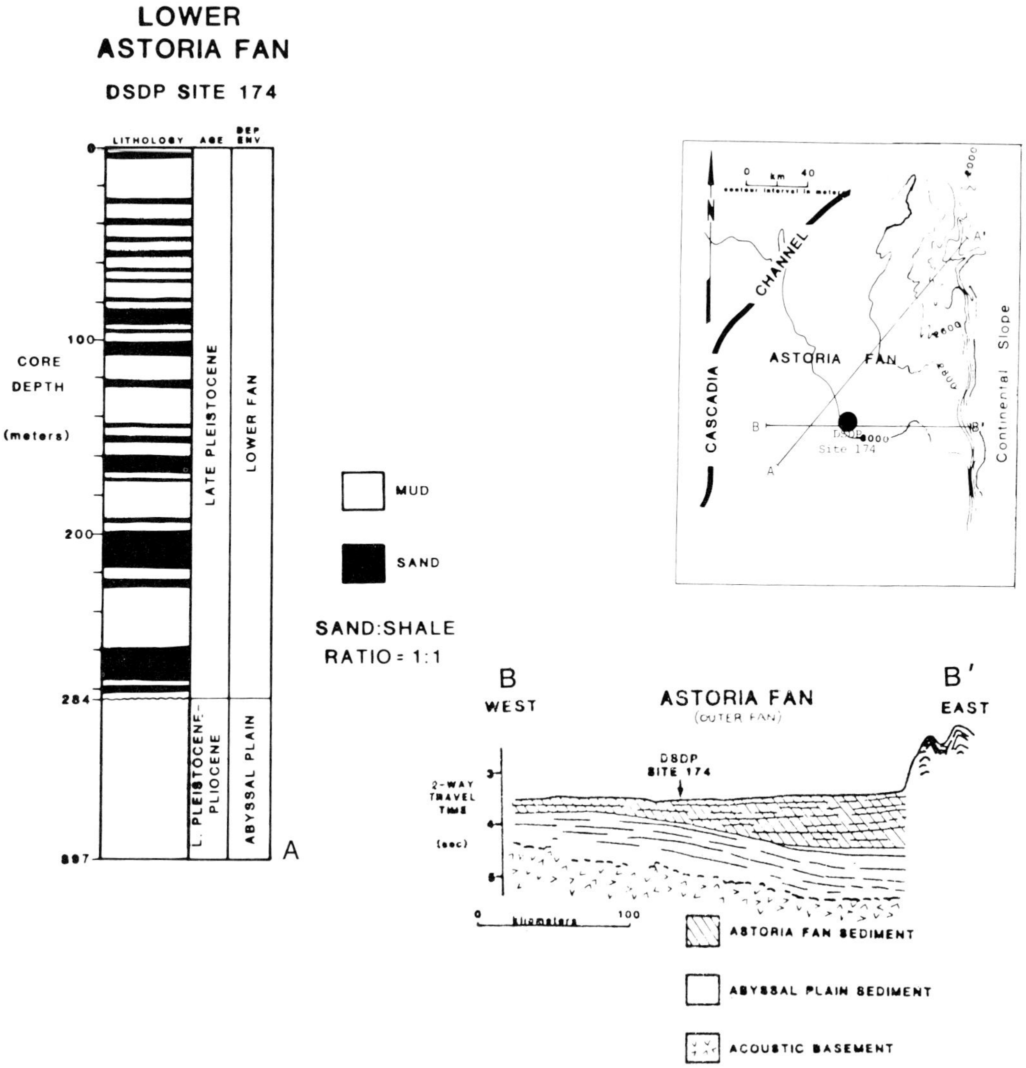

Figure 3. (A) Seismic profile across Astoria Fan, modified from [5]. (B) Simplified stratigraphic section of DSDP Site 174, modified from [5].

several hundred meters of relief, V-shaped profiles, numerous tributaries, and a pathway coincident with major structural trends across the continental margin. The canyon mouth merges smoothly into the Astoria Channel and the slope-base fan valley. These fan valleys are characterized by the youngest Mazama ash turbidites, U-shaped profiles, up to 200 m of relief, and levee development (Fig. 2). Singularly deep and narrow inner-fan valleys are present on the concave-upward steep and rough (surface relief 18 to 55 m) inner-fan surface. Fan valleys break into distributaries on the middle fan, where there is convex-upward morphology and the sharpest change in fan surface gradient (18.5 to 5.3/1000 m) (Figs. 1, 2, 3). The main valleys broaden and become shallow in the outer fan, whereas the generally concave-upward fan surface grades to nearly a flat sea floor with less than 10 m relief (Fig. 2). The gradients of fan-valley floors vary from 8/1000 in the inner fan to 5/1000 in the middle fan and 3/1000 in the outer fan [1].

Progressively, toward the northwestern part of the fan, more extensively modified and older channel systems occur (Figs. 1, 2) [1]. The most easterly channel segments, Astoria Channel and the slope-base fan valley, connect as continuous channel systems extending far across the fan, whereas the more northwesterly segments appear to be discontinuous fragments of channels. Because channel floor gradients toward the northwest become increasingly steeper and out of adjustment with gradients of the most recently active Astoria Channel, the more northwesterly fragments appear to be progressively older [1].

Stratigraphy

DSDP site-survey seismic profiles (20-in^3 airgun and 160-kJ sparker) crossing DSDP Site 174 show that (1) thick Pleistocene Astoria Fan turbidite sand layers overlie Pliocene and Pleistocene abyssal plain pelagic mud and thin-bedded turbidites, and (2) these fan and abyssal plain sedimentary deposits cover subducting oceanic crust (Fig. 3b) [5]. DSDP

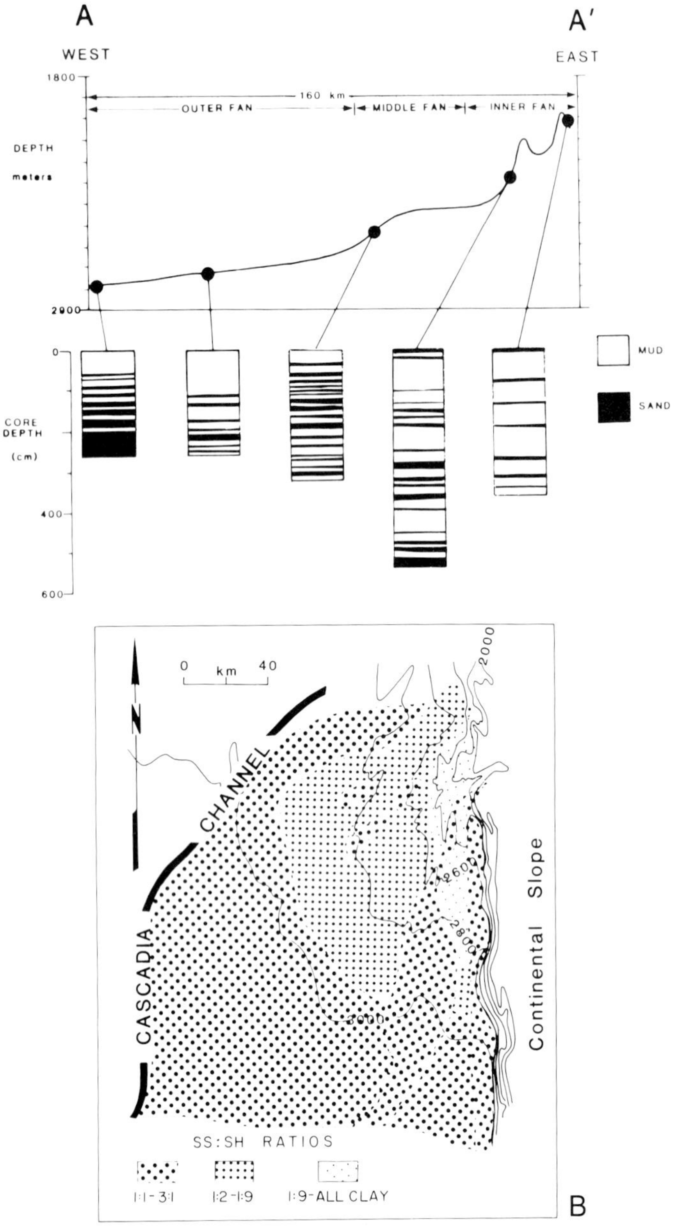

Figure 4. (A) Stratigraphic sections of selected cores from Astoria Fan, modified from [6]. See Fig. 3 for locations of profiles. (B) Areal distribution of sand/shale ratios, based on 60 6-m piston cores from Astoria Fan (modified from [6]). Mud thickness is reduced to one-third for sand/shale ratio calculations.

Site 174, drilled through this section in the outer fan, indicates that the high sand/shale ratio (1:1) of 6-m piston cores in this distal area persist throughout the entire fan section (Figs. 3, 4).

Dating of the fan sediment at DSDP Site 174 suggests a Late Pleistocene age for this part of the fan [5]. This age appears to be a minimum because the site is located on the youngest and most recently active part of the fan, according to the physiographic and sedimentary history of the channels (Figs. 1, 4). On the basis of sediment-budget history, the fan northwest of DSDP Site 174 must be older, and deposition probably began with the onset of major Pleistocene glaciation [1]. Late Pleistocene piston-core stratigraphy suggests that mainly, (1) glacial marine sedimentation (mud containing ice-rafted pebbles) occurred throughout times of maximum glaciation and low sea levels, (2) major turbidite sedimentation occurred during maximum glacial-melt phases and initial rise of sea level [4], and (3) hemipelagic clay deposition occurred within interglacial periods of high sealevel [4].

Sedimentary History and Processes

Both Late Pleistocene and Early Holocene hemipelagic clay is interrupted by sediment gravity-flow deposits. The proximal-fan-valley floors contain thick, muddy, and very poorly sorted sand and gravel beds with poorly developed internal sedimentary structures. In contrast, fan-valley floors of the middle to outer fan and interchannel areas of the outer fan contain thick, clean, moderately sorted fine to medium sand layers that are vertically graded in texture, composition, and sedimentary structures (Fig. 4). Tuffaceous turbidites (containing Mazama ash, 6600 years old) can be traced as thick deposits (~30 to 40 cm) throughout the Astoria Channel system and as thin correlative interbeds (~1 to 2 cm) in interchannel areas [3]. Similarly, sand/shale ratios are high throughout the fan valleys and the middle-fan and outer-fan areas of distributaries, but low in the inner-fan and middle-fan interchannel areas (Fig. 4).

These depositional trends indicate that high-density turbidity currents carrying coarse traction loads remain confined to inner-fan but not to outer-fan valleys (Fig. 5). Fine debris is selectively sorted out from channelized flows into overbank suspension flows that spread over the fan and deposit clayey-silt turbidites. High contents of mica, plant fragments, and glass shards (in tuffaceous turbidites) characterize deposits of these overbank flows, which are a major process in the building of inner-fan levees and interchannel areas [4].

During the late Pleistocene, turbidity currents funneled most coarse debris through inner-fan channels to depositional sites in the middle-fan and outer-fan distributaries. These distributaries periodically shifted, anastomosed, and braided to spread sand layers throughout the outer-fan area. At this time, depositional rates were much higher (>50 cm/1000 years) than during the Holocene (8 cm/1000 years) [4] (Fig. 5).

During the Holocene rise of sea level, (1) the shoreline shifted and Columbia River sediment was trapped in the estuary; (2) high-density, turbidity-current activity slackened from one major event per 6 years during the late Pleistocene to one event per 1000 years during the Early Holocene and none since the Mount Mazama eruption (6600 yr B.P.) [4];

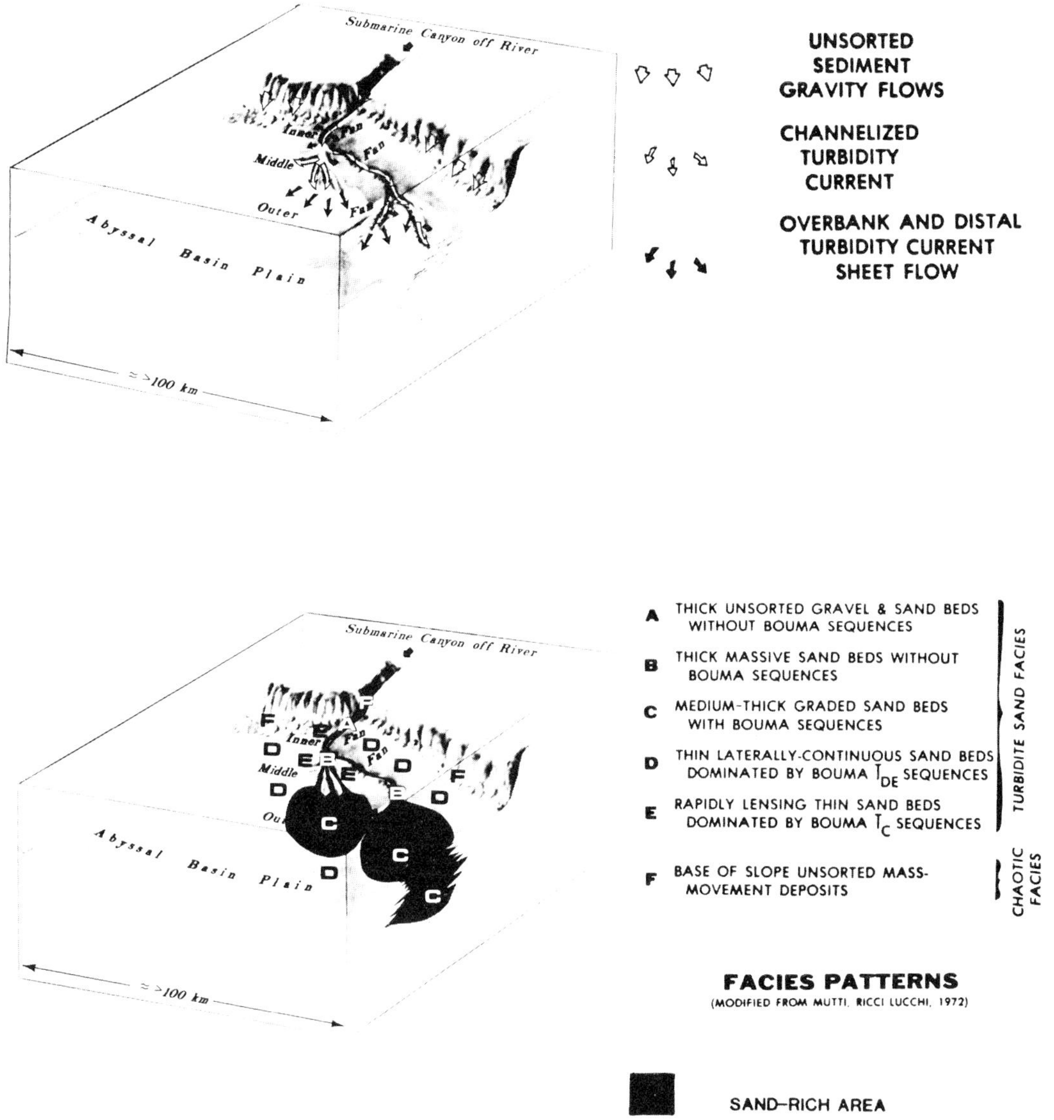

Figure 5. (A) Depositional processes in large, elongate submarine fans. (B) Facies patterns in large, elongate submarine fans. Facies designations modified from [11].

and (3) turbidites became muddier and were deposited as thick beds within main channels that contributed to a Holocene deposition rate three times higher there (25 cm/1000 years) than in interchannel areas. Turbid-layer flows, moving westward from the continental terrace, also were trapped and funneled down channel systems, helping to augment rapid sedimentation on the fan-valley floors.

During the Late Holocene, continuous particle-by-particle deposition of hemipelagic clay has dominated and is characterized by a biogenous coarse fraction of radiolarian tests; this contrasts with the foraminifer-rich Pleistocene clay. These hemipelagites contain progressively more clay size and biogenous debris offshore. The hemipelagic sedimentation during high-sea-level epochs, however, is insignificant in comparison with the turbidite deposition during low-sea-level periods.

Geologic Significance

Increasing sand/shale ratios downfan (Figs. 3b, 4) indicate that turbidity currents feeding the Astoria Fan funnel coarse debris to outer-fan sites via a distributary-channel system extending throughout the entire fan system (Fig. 5). The high silt and clay content of the major Columbia River point source results in large sediment gravity flows that carry sand efficiently and develop an extensive distributary-channel system to funnel most of the sand to distal outer-fan and abyssal-plain depositional sites [6]. The same pattern of sand depocenters in the outer fan also is noted in other large river-fed abyssal sea-floor fans such as the Amazon [8] and the Indus [9] deep-sea fans.

In contrast, in smaller (<100-km diameter) radial fans [7] such as the Navy [10], channelized turbidity–current flow ceases in middle fan, and major sand depocenters occur there. Thus, facies *C* thick sand turbidites occur in the outer-fan lobes of larger systems, whereas they occur more proximally in the midfan or suprafan areas of smaller systems in restricted basins [10]. Basin-plain facies *D* turbidites are deposited distally on the abyssal plains of large fans with unrestricted growth across the unconfined deep-sea floor. In contrast, suspension flows pond in outer-fan areas of smaller basins enclosed [2] or semienclosed [10] by basin margins.

The key factors controlling depositional patterns, other than basin size and setting, appear to be the number, volume, and grain size of sediment sources feeding the fans. The greater number of smaller sources with coarser material typically feeding radial fans [10] or debris aprons [2] in restricted basins inhibits both the length of sand transport and the development of extensive distributary-channel systems to carry sand to distal depocenters. A single fine-grained point source of large river-fed systems results in efficient distal transport and outer-fan depocenters of sand like those found in Astoria Fan.

References

[1] Nelson, C. H., and others. Development of the Astoria Canyon—Fan physiography and comparison with similar systems. Marine Geology, v. 8, pp. 259–291.

[2] Nelson, C. H., 1983. Modern submarine fans and debris aprons: an update of the first half century, In: S. J. Boardman (ed.), Revolution in the Earth Sciences: Advances in the Past Half Century. Kendall Hunt, Dubuque, IA.

[3] Nelson, C. H., and others, 1968. Mazama ash in the northeastern Pacific. Science, v. 161, pp. 47–49.

[4] Nelson, C. H., 1976. Late Pleistocene and Holocene depositional trends, processes, and history of Astoria deep-sea Fan, Northeast Pacific. Marine Geology, v. 20, pp. 129–173.

[5] Kulm, L. D., von Huene, R. E., and others, 1973. Initial Reports of the Deep Sea Drilling Project, v. 18. U.S. Government Printing Office, Washington, DC.

[6] Nelson, C. H., and Nilsen, T. H., 1974. Depositional trends of modern and ancient deep-sea fans. In: R. H. Dott, Jr., and R. H. Shaver (eds.), Modern and Ancient Geosynclinal Sedimentation: Society of Economic Paleontologists and Mineralogists Special Publication 19, pp. 69–91.

[7] Nelson, C. H., and Kulm, L. D., 1973. Submarine fans and deep-sea channels: In: G. V. Middleton and A. H. Bouma, (eds.), Turbidites and Deep Water Sedimentation. Society of Economic Paleontologists and Mineralogists, Pacific Section, Short Course, Anaheim, CA, pp. 39–78.

[8] Damuth, J. E., and Kumar, N., 1975. Amazon Cone—morphology, sediments, age, and growth pattern. Geological Society of America Bulletin, v. 86, pp. 863–878.

[9] Kolla, V., and others, 1979. Morpho-acoustic and sedimentologic characteristics of the Indus Fan. Program 1979 Annual Meetings Geological Society of America, San Diego, CA, pp. 459–460.

[10] Normark, W., and Piper, D. J. W., 1983. Navy Fan, California Borderland: Growth pattern and depositional processes. Geo-Marine Letters, v. 3, in this issue.

[11] Mutti, E., and Ricci Lucchi, F., 1978. Turbidites. Reprinted from International Geology Review, v. 20, no. 2, pp. 125–166.

CHAPTER 9

Crati Fan, Mediterranean

Franco Ricci Lucchi, Albina Colella, Gianni Gabbianelli, Sergio Rossi, and William R. Normark

Abstract

The Crati Fan is located in the tectonically active submerged extension of the Apennines chain and foretrough. The small fan system is growing in a relatively shallow (200 to 450 m), elongate nearshore basin receiving abundant input from the Crati River. The fan is characterized by a short, steep, channelized section (inner or upper fan) and a smooth, slightly bulging distal section (outer or lower fan). The numerous subparallel channels head in the shelf or littoral zone and do not form branching distributary patterns. Sand and mud depositional lobes of the outer fan stretch over more than 60% of fan length.

Setting

The Crati Fan is growing in the Corigliano Basin (Fig. 1), a nearshore embayment of the Ionian Sea bordered by a narrow and steep (1° to 3°) shelf. The basin is located on the southwestern (inner) tectonically active flank of the southern Apennines Foredeep, made by piled nappes or thrust sheets. The Taranto Valley, crossing the Gulf of Taranto from northwest to southeast, marks the thrust front [1], and the Apulian foreland (with a mostly carbonate sedimentary cover) forms the northeast side of this valley and the nearby land. The rugged topography of the rigid-plastic thrust complex is further complicated by recent (still active) extensional tectonics; one of the downthrown blocks is the Corigliano Basin, which is partly silled by an alignment of submarine highs. Within the basin, and along its margin, a considerable topographic smoothing has resulted from recent sedimentation (rates as high as 6 mm/yr); an aggradational mud blanket, for example, masks former shelf and base-of-slope breaks.

The Crati Fan is a small (70 km^2), steep (0.5° to 3°), and relatively shallow-water (200 to 450 m) system. It is remarkable for being young (probably not older than 6000 yr), active, and connected with a coarse-grained, torrential-type delta on the shelf. Fan thickness and volume are quite small (least values in this COMFAN comparison; see Barnes and Normark, this volume). The fan is characterized by multiple, prominent leveed channels that cross the delta slope and merge downdip into long depositional lobes (Fig. 2) The shape of the fan is distorted and elongated east-west (16-km length, 4 to 5-km width), because of its confinement in a narrow depression that is probably of tectonic origin. The fan gradually merges updip with slope sediment at about 190- to 200-m depth and thus has no defined apex (Fig. 3A). Its distal end coincides with the basin margin; a true basin plain does not exist because the undulating basin floor has not yet been covered by ponded turbidite sediment. Turbidite-free areas flanking the fan are basinal *l.s.* The longitudinal profile of the fan is concave upward and the transversal profile slightly convex-up especially in the lobe area (Fig. 3C).

Data

The Crati Fan surveys [1,2] that provided the data used in this report obtained echo-sounding and seismic-reflection profiles spaced 100- to 500-m apart in the shelf-slope area (50 to 200-m water depth). The trackline spacing is wider on the outer fan (0.5 to 1.5 km). The total length of profile data is 520 km.

Side-scanning sonar data are less extensive and are basically restricted to the main fan valley. A total of approximately 60 km of side-scanning-sonar tracklines (with 200-m total swath width) has been used to help map the sinuous trends of the main valley system and adjacent features [3]. A limited number of sediment samples were obtained to

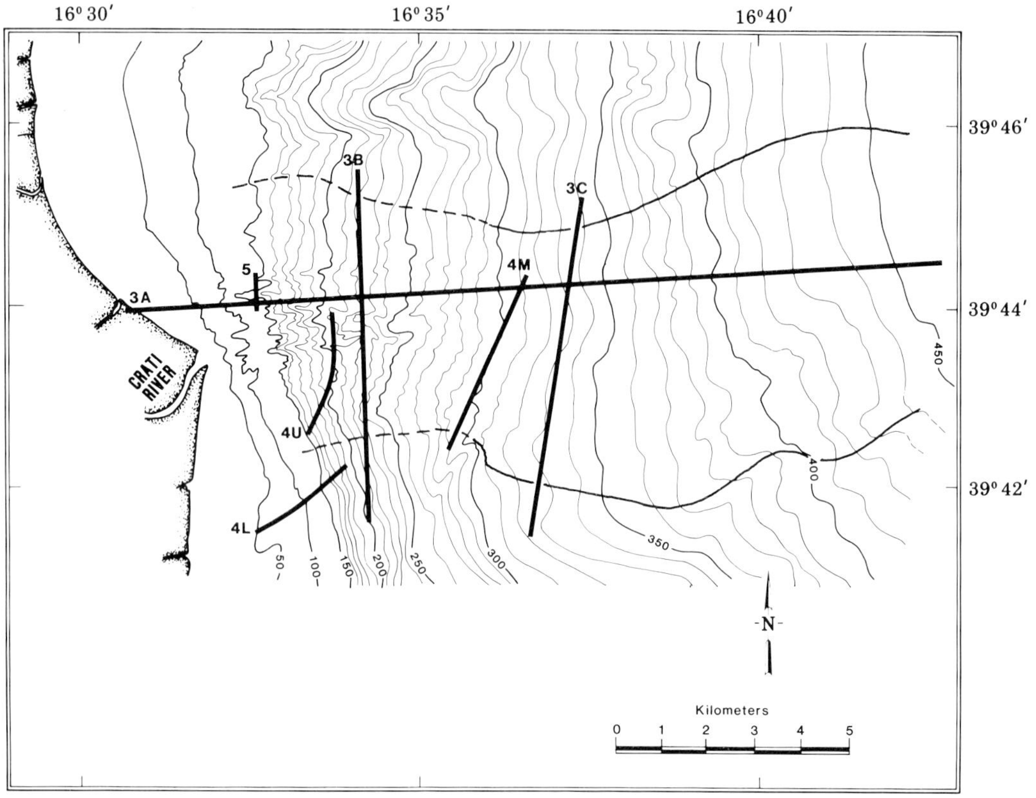

Figure 1. Preliminary bathymetric map modified from [2]. Profile locations for representative sections (Fig. 3) and seismic-reflection profiles (Figs. 4, 5) are shown.

characterize the various morphologically distinct features on the fan.

Morphology and Fan Division

The longitudinal topographic profile can be split into three segments of different gradients (Fig. 3A). The upper and intermediate segments, however, are not clearly differentiated in terms of local morphologic relief or sedimentology. For example, the erosional character of the channels is more pronounced in the upper segment, and a depositional character dominates in the intermediate segment; the changes are gradual, however, and apparently not influenced by the topographic break at 230- to- 240-m depth. Furthermore, levees are present in the upper part of the channels even though they are erosional in character. Thus, a twofold zonation is more appropriate (Fig. 2), with an *inner or channelized fan* (Fig. 3B) and an *outer nonchannelized fan* (Fig. 3C). A midfan region is not clearly definable.

The channelized fan is characterized by: (1) a major (although short) entrenched and leveed valley with a well-developed and complex tributary system but no distributaries at the distal end; the valley width reaches 0.5 to 0.8 km, the depth 30 m, and the topography is fairly complex (thalweg, bottlenecks, terraces, etc.; see Figs. 2, 4 upper). The tributary pattern is probably controlled by slumping. Bathymetric and morphometric maps portray a tentative schematization (Figs. 1, 2); major tributaries reach a depth of 40 m. (2) Minor, more or less parallel, channels with few or no tributaries head independently in the shelf or nearshore zone; the largest of these, south of the main valley, has a dominantly depositional character (high and narrow levees, flat and muddy bottom). (3) Well-developed, undulating, and hummocky levees start along the upper slope at about 130-m depth; the right hand, or southern, levee of the main valley is wider than the left one and complicated by minor channels (in part of crevasse type?), slides, and sediment waves (Fig. 5).

The channel system thus consists of a main northern trunk and a subordinate southern trunk. This subdivision is matched in the outer fan by two parallel, slightly upbulging, depositional lobes separated by a shallow interlobe depression (Figs. 3C, 4, middle). The northern lobe is larger and shows a rougher (meso)topography [4] with small channels or depressions less than 2 m in depth. The proximal parts of the lobes are thickest (up to 12 to 15 m), similar to the lobes found on the Navy Fan [4]. A distinct suprafan bulge or distinctive midfan area does not stand out, however, on the Crati Fan, and the lobes are much more extensive relative to the channels than is seen on Navy Fan [4,5].

Sediments and Acoustic Facies

The sediment distribution on the fan is interpreted from core, dredge, and grab samples (Fig. 6), bottom photography

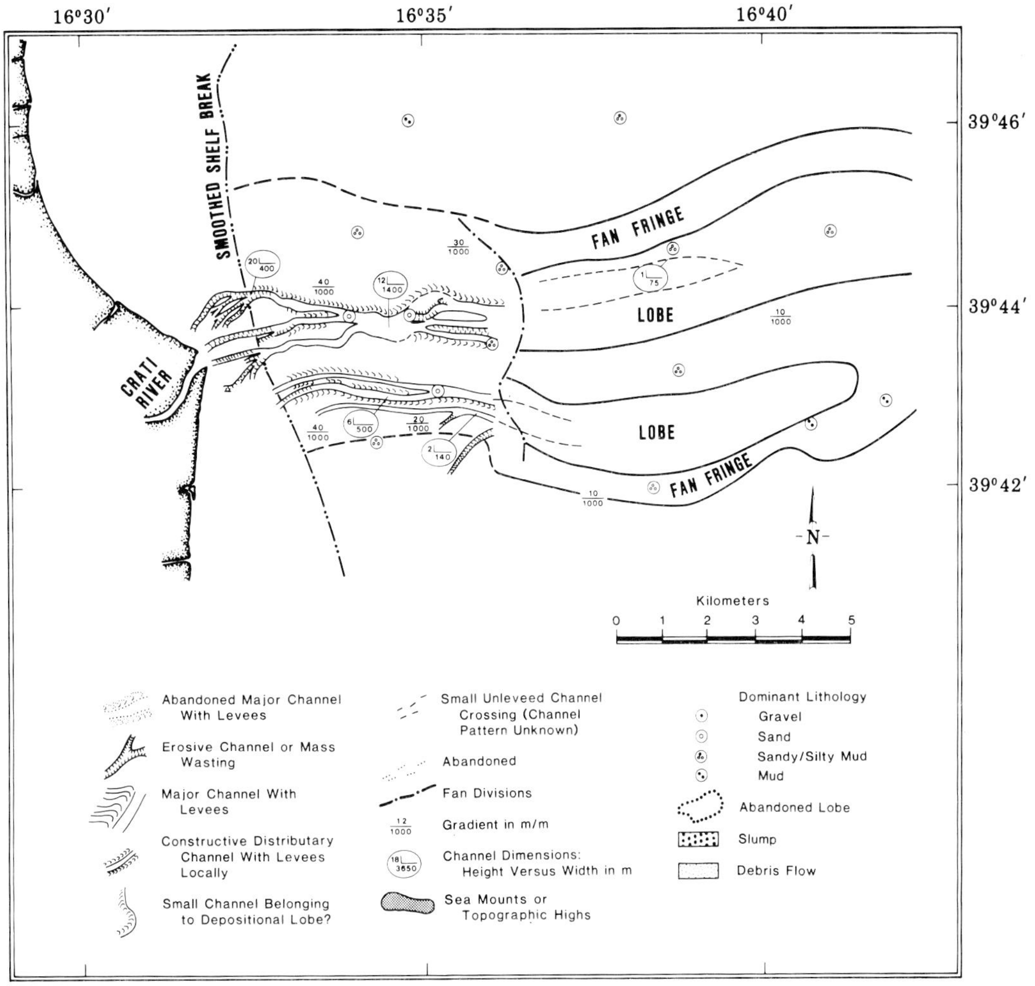

Figure 2. Morphometric map of Crati Fan.

(Benthos photocamera), and high-resolution reflection profiling. The ship could only be equipped with a light gravity corer (0.2 *t*) that resulted in a maximum penetration of 2.7 m and core recovery of 1.6 m. The following types of bottom–subbottom reflection characters can be distinguished (similar to the schemes of Damuth and Normark and others [4,6,7]):

(1) Continuous, distinct bottom echoes with several continuous, parallel subbottom reflections (see [6], type IB); occurence: shelf, slope, basin, margin, and interchannel areas in the upper part of the fan;
(2) Distinct to prolonged bottom echoes without subbottom reflections, locally with subbottom overlapping hyperbolae (intermediate character between types IA-1 and IIB in [6]); occurence: main (northern) lobe, axial portion;
(3) Prolonged bottom echoes with discontinuous subbottom reflectors (see [6], type IIA); occurence: levees, southern lobe, margins or fringes of northern lobe, base-of-slope areas outside the fan. A vertical threefold subdivision is usually recognizable: surficial discontinuous reflectors, an intermediate transparent lens, and deeper downbent reflectors.
(4) Irregular, overlapping or single, hyperbolic reflectors with widely varying elevations above sea floor (see [6]); occurrence: deeper portions of major channels;
(5) Regular, overlapping hyperbolic reflectors with varying elevations above sea floor ([6]); occurrence: slope near fan channels, levees on slope and upper fan; and
(6) Broad, single, gently rolling hyperbolic reflectors with distinct bottom echoes and several conformable or disconformable subbottom reflectors (see [6]); occurrence: levees, slope, and base-of-slope areas.

Photographs of the sea floor show that the whole fan surface is veneered by intensely bioturbated mud; sand is exposed only along channel bottoms, mainly in the upper and intermediate reaches of the main valley. In general, the uppermost meter of sediment is completely muddy or contains a variable number of thin (0.2 to 9 cm) turbidite sands [8,9] (Fig. 6).

The muddy sediment of the fan is wholly terrigenous (planktonic remains are scarce) with a predominant brown-

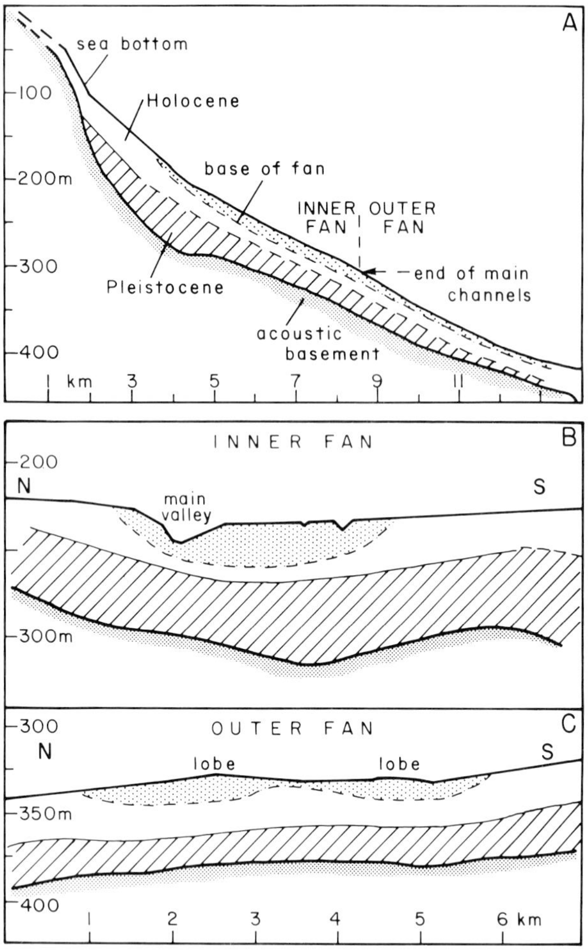

Figure 3. Longitudinal (A) and transverse (B and C) sections based on interpretation of seismic-reflection profiles (modified from [2]). Locations in Fig. 1.

gray color; lighter and softer interbeds that occur with decreasing frequency seaward are regarded as prodelta muds flocculated from turbid plumes; the darker muds are inferred to be turbiditic. The coarser turbidite beds have a sand or silt basal unit and grade to a structureless unit with a sharp contact with overlying mud (at places with a thin Bouma T_c division). Muddy turbidites are thicker (max. 60 cm), particularly in the outer fan, and consist of E_1, E_2 and E_3 sequences [10], not always complete.

Combining information from bottom samples and echo character, the following sediment distribution can be derived (Fig. 2): mud and mud–sand interbeds in overbank areas (with a downfan increase of sand content) and in the bottom of several channels and distal parts of the main valley; sand in deeper channel bottoms, on levees near channel termini, and on the proximal part of the main lobe; and sand interbedded with mud in the smaller lobe and lobe-fringe areas. Thus, the length and mud content of lobes might be positively correlated and taken as evidence of efficiency of sand transport (*sensu* Mutti, [11]).

The main sediment supply, the Crati River, carries to the sea a tremendous load (1,730,000 t/yr), two-thirds of which is suspended sediment. Most coarse-grained detritus should be trapped on the shelf in the Crati delta and adjacent beaches and beach–ridge complexes.

Slump and Mass-Flow Features

The best and most extensive examples of slump and related mass-flow deposits are displayed by the gullied slope updip from the fan; slumping or sliding could be related to the origin of the channels themselves (prodelta failures resulting from oversteepening, underconsolidation of sediment, etc). Both open-slope deposits and levee sediments are affected. The average slope gradient is 4° to 4.5°.

Bottom undulations, generally interpreted as sediment waves [6,12], might also represent, at least in part, mud drapes on top of slides especially in base-of-slope and basinal areas far from channels [13,14].

Growth Pattern and Sediment Source

Longitudinal and transversal cross-sections of the fan based on 1-kj seismic profiles show maximum fan thickness in the channelized, inner-fan section seaward of a break in slope seen in the acoustic basement that is preserved by deposition of recent prefan sediments (Fig. 3A). The observed sediment thickness on the shelf is consistent with data reported north of the study area and attributed to the Holocene (Flandrian) [15]. It reaches a maximum of 25 m (assuming a velocity of 1.5 km/sec). The basement is therefore regarded as an erosional Pleistocene surface marking a glacial low stand of sea level. Seaward of the shelf break observed on the Pleistocene surface, the sedimentary sequence can be split into a lower part with numerous, relatively strong reflectors and an upper part that is more acoustically transparent and, therefore, probably muddier. The lower unit reaches a thickness of 25 m. It pinches out against the "basement" slope (probably a buried fault scarp), thus marking a regressive phase of unconfined turbidite deposition during a Late Pleistocene (Wurm?) low stand of sea level (Figs. 3A, 4 lower). The upper unit has the same thickness and acoustic character as the sediments that mantle the shelf. We attribute this upper unit, which includes the lensoid fan body (confused echo returns in Fig. 4 upper), to the Holocene transgression. The tributary channels of the fan system cut into the upper sequence and locally reach the acoustic basement.

These observations suggest that channel cutting and fan growth started when sea level became stabilized at the transgressive acme (climatic optimum, 5000 to 6000 years ago), and the Crati delta prograded across the shelf providing an adequate sediment supply to submarine channels. The Crati River drains tectonically unstable reliefs (Sila Massif and

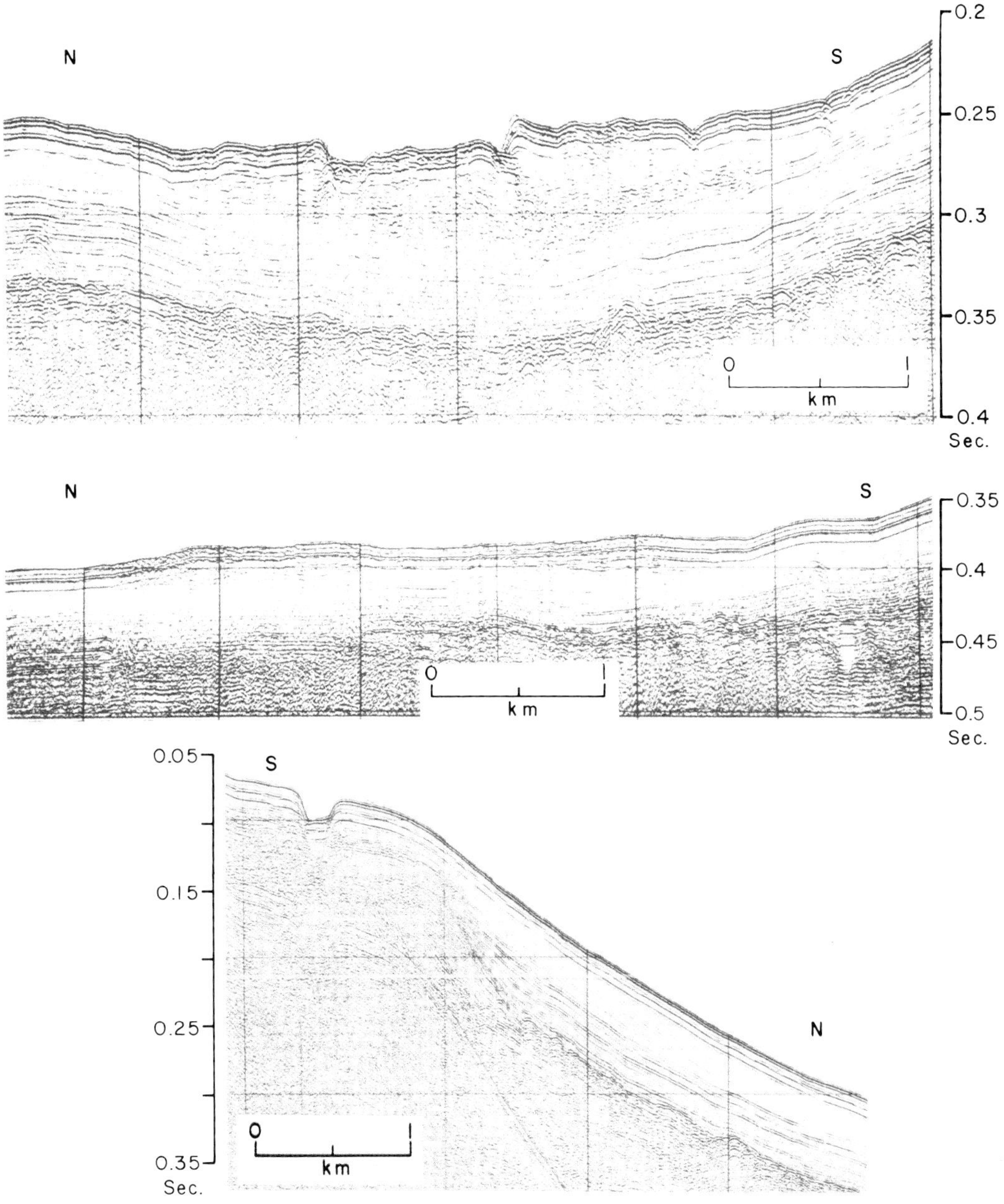

Figure 4. One-kilojoule sparker profiles; vertical scales in seconds of roundtrip-travel time. Upper: transverse profile of inner, channelized fan. Middle: transverse section across lobe area. Lower: basin-slope section from shelf area (left). Locations in Fig. 1.

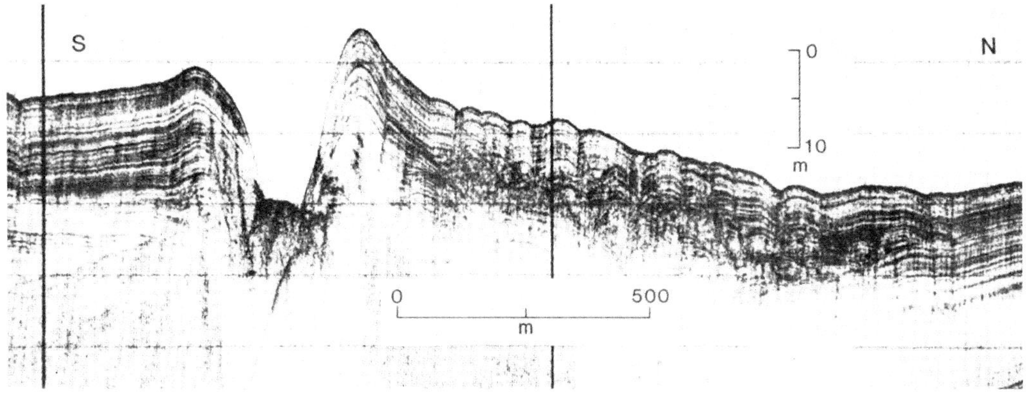

Figure 5. High-resolution (3.5 kHz) profile across main valley on inner fan showing sediment waves on levee and limited subbottom reflectors in channel axis. Location in Fig. 1.

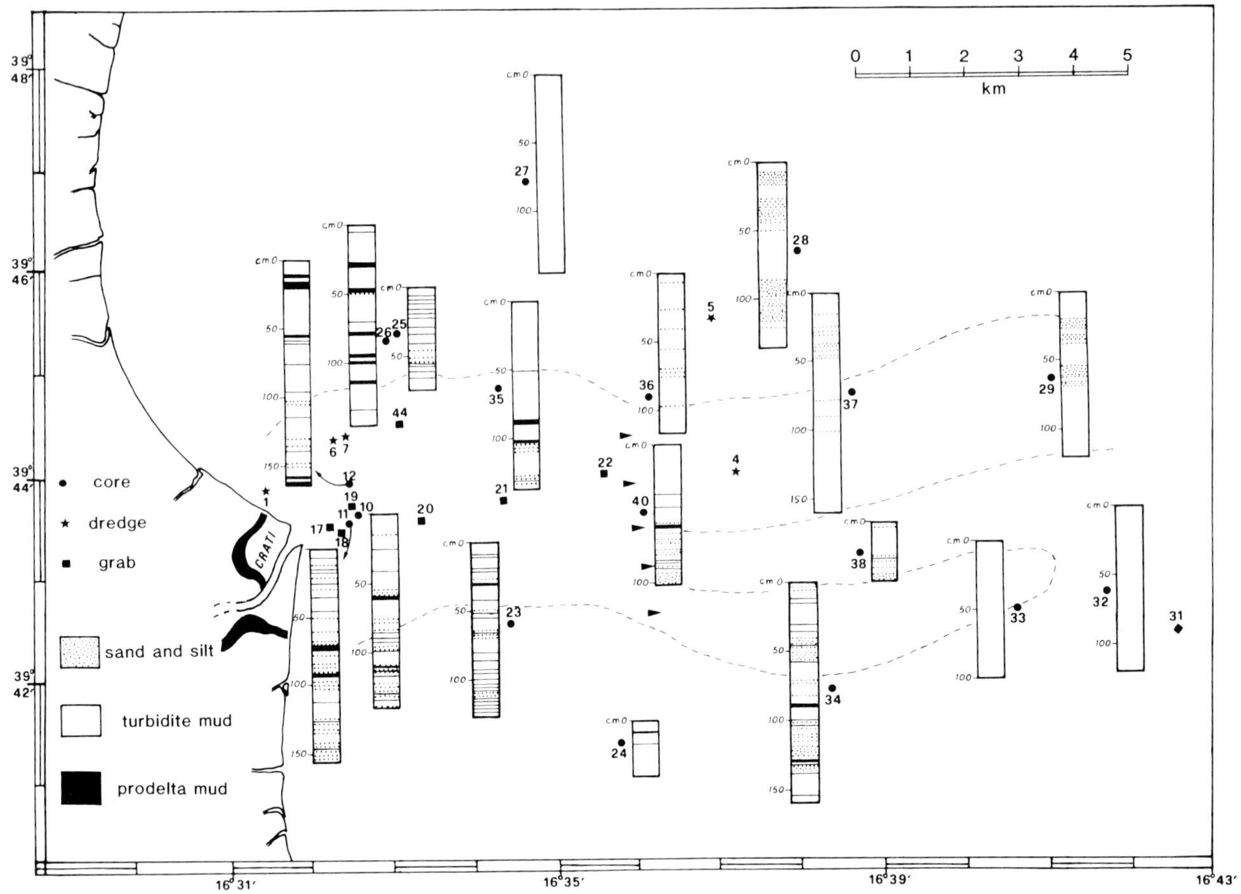

Figure 6. Sediment sample locations and stratigraphic sections, modified from [8].

Coastal Calabrian Range) of 1332-km^2 area that have been intensely uplifted since the Late Miocene. The fan delta is arcuate, wave-dominated, and occupies an area of 70 km^2 at the end of an alluvial plain that fills a graben (Sibari Trough) with a rate of subsidence as high as 3 to 4 mm/y.

Conclusions

The most significant aspect of the Crati Fan is that its mud cover does not reflect Holocene abandonment or semiactivity, but a mud-rich, active system fed by a river. The fan seems to have been active since the end of the Holocene transgression, with feeding channels heading into the mouth of a single-channel (at present) delta. The delta plain and littoral drift trap most of the coarse stream load, but a part of it escapes seaward and finds its way through the tributary gullies. The whole mud load, on the contrary, settles out on the fan and adjoining basin. The mud is carried either by surface plumes (northward) or by turbidity currents (eastward, *i.e.*, downslope). This river-fed submarine fan, in essence, occupies the position of a prodelta area with respect to deltaic systems and results in a fan deposit very similar to the sublacustrine fan of the Rhone delta in Lake Geneva [16]. The Crati Fan is, consequently, a trap for mud with additional, sporadic sand supply via gravity flows.

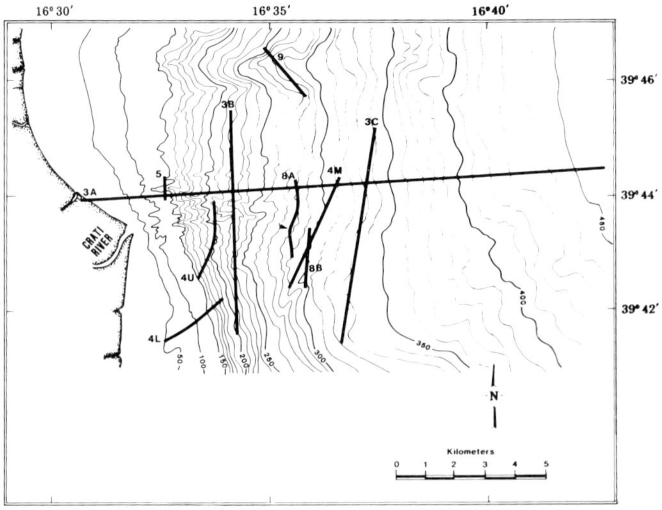

Figure 7. Locations for 3.5-kHz seismic-reflection profiles from Crati Fan; modified from Figure 1. Profile numbers indicate figure numbers.

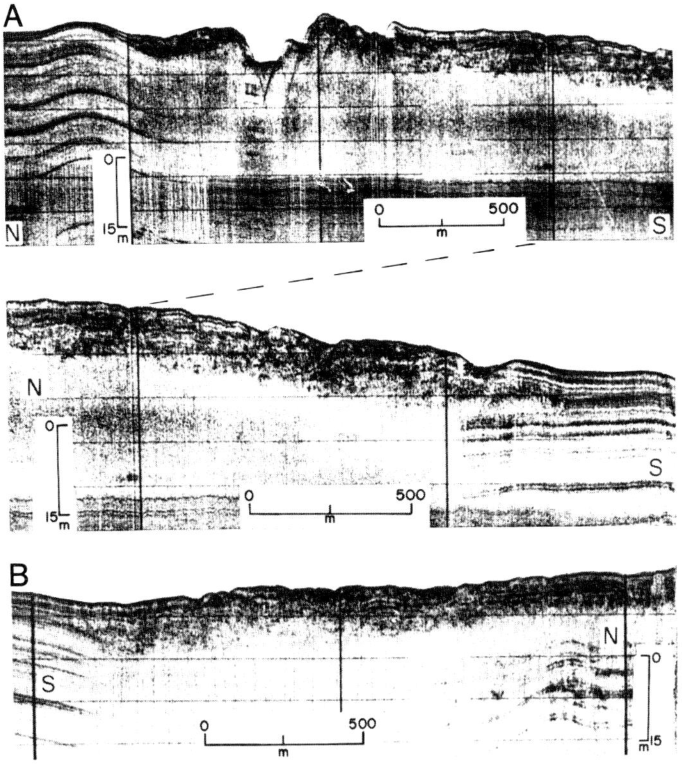

Figure 8. (A) High-resolution reflection profile (3.5 kHz) across inner-fan channels near the transition with the depositional lobes. The subbottom reflectors underlying the channel floor probably indicate a lack of sand fill within the channel. Dashed line indicates the degree of overlap for the two segments of this profile. (B) High-resolution reflection profile (3.5 kHz) across the channel-lobe transition showing small-scale channelling. Profile locations in Figure 7.

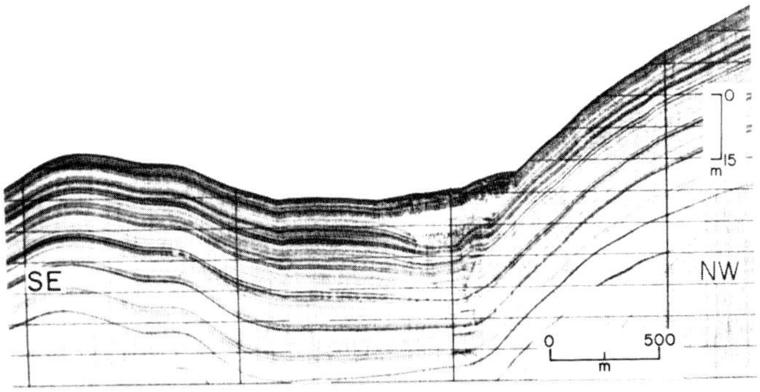

Figure 9. High-resolution reflection profile (3.5 kHz) showing acoustically transparent lens at the base of the slope at north edge of Crati Fan. Transparent lens is interpreted as a debris flow. Profile location in Figure 7.

Addendum

The bathymetry of Crati Fan as presented in Figure 1 is repeated in Figure 7 to show the locations of the additional seismic-reflection profiles of Figures 8 and 9. The 3.5-kHz seismic reflection profiles from near the transition between the inner-fan channels and outer-fan lobes (Figs. 8A and B) clearly illustrate that the entire Crati Fan is rather thin (< 30 m). Seismic reflectors from pre-fan basin sediments can be observed both under the margins of the fan sediments and below the valley floor (Fig. 8A). Debris-flow deposits fill a channel-like re-entrant area on the basin slope immediately north of the fan (Fig. 9).

References

[1] Rossi, S., and Gabbianelli, G. N., 1978. Geomorfologia del Golfo di Taranto. Bolletin Societa Geologica Italiana, v. 97, pp. 423–437.

[2] CRATI Group, 1981. The Crati Submarine Fan, Ionian Sea. A Preliminary Report. International Association of Sedimentologists 2nd European Meeting, Bologna, Abstract Volume, pp. 34–39.

[3] Colella, A., and Normark, W. R., 1984. Sinuous delta slope and inner fan channels of the late Holocene Crati fan system, Southern Italy. International Association of Sedimentologists 5th European Meeting, Marseille, Abstract Volume, pp. 114, 115.

[4] Normark, W. R., Piper, D. J. W., and Hess, G. R., 1979. Distributary channels, sand lobes, and mesotopography of Navy Submarine Fan, California Borderland, with applications to ancient fan sediments. Sedimentology, v. 26, pp. 749–774.

[5] Ricci Lucchi, F., 1981. Contrasting the Crati Submarine Fan with California fans and models. International Association of Sedimentologists 2nd European Meeting, Bologna, Abstract Volume, pp. 157–160.

[6] Damuth, J. E., 1978. Echo character of the Norwegian–Greenland Sea: Relationship to Quarternary sedimentation. Marine Geology, v. 28, pp. 1–36.

[7] Damuth, J. E., 1980. Use of high-frequency (3.5–12 kHz) echograms in the study of near-bottom sedimentation processes in the deep sea: a review. Marine Geology, v. 38, pp. 51–76.

[8] Colella, A., 1981. Preliminary core analysis of Crati Submarine Fan deposits. International Association of Sedimentologists 2nd European Meeting, Bologna, Abstract Volume, pp. 26–28.

[9] Colella, A., and others, in press. I Depositi Attuali della Conoide Sottomarina del Crati (Golfo di Taranto). Atti del 5° Congresso della Associazione Italiana di Oceanologia e Limnologia.

[10] Piper, D. J. W., 1978. Turbidite muds and silts on deep-sea fans and abyssal plains. In: D. J. Stanley and G. Kelling (eds.), Sedimentation in Submarine Canyons, Fans, and Trenches. Dowden, Hutchinson & Ross, Stroudsburg, PA, pp. 163–176.

[11] Mutti, E., 1979. Turbidites et cones sous-marins profonds. In: P. Homewood (ed.), Sedimentation Detritique (Fluviatile, Littoral et Marine) Institut Géologique Université de Fribourg, Switzerland, v. 1, pp. 353–419.

[12] Normark, W. R., Hess, G. R., Stow, D. A. V., and Bowen, A. J., 1980. Sediment waves on the Monterey Fan levee: a preliminary physical interpretation. Marine Geology, v. 37, pp. 1–18.

[13] Gabbianelli, G., and Ricci Lucchi, F., 1981. Soft sediment deformation in the Crati Fan and Basin, Ionian Sea. International Association of Sedimentologists 2nd European Meeting, Bologna, Abstract Volume, pp. 63–64.

[14] Embley, R. W., 1980. The role of mass transport in the distribution and character of deep-ocean sediments with special reference to the North Atlantic. Marine Geology, v. 38, pp. 23–50.

[15] De Maio, A., and others, 1979. Atti Convegno Scientifica Nazionali Programmazione Finale Oceanografia, v. 3, pp. 1333–1348.

[16] Houbolt, J. J. H. C., and J. B. M. Jonker, 1968. Recent sediments in the eastern part of Lake Geneva (Lac Leman). Geologie en Mijnbouw, v. 47, pp. 131–148.

CHAPTER 10

Delgada Fan, Pacific Ocean

William R. Normark and Christina E. Gutmacher

Abstract

The Delgada Fan, an irregularly shaped turbidite deposit extending more than 350 km offshore from northern California, consists of two large leveed-valley units each fed by a separate complex of coalescing submarine canyons and slope gullies. Although the leveed-valley units head within 25 km of each other, both appear to have developed independently during fan growth. The larger southern leveed-valley system has not developed middle-fan distributary channels and appears to illustrate a period of progressive valley abandonment. Although the lower-fan area is underlain by sandy sediments, little sand has been recovered in piston cores from the leveed-valley unit.

Introduction

This paper, which presents a brief summary of the growth pattern of Delgada Fan, is based on limited seismic-reflection profile data and sediment-core data obtained over the last 4 years by the U.S. Geological Survey. These new data include approximately 4200 km of 3.5-kHz high-resolution profiles and 3000 km of low-frequency single-channel seismic-reflection profiles. In addition, six core samples, including two piston cores, were obtained in 1979. The southern part of the fan is the main focus of the paper because most of the data was collected from this area (Fig. 1; inset); however, the reconnaissance nature of the data collection together with local unresolved problems in the available bathymetry require caution in accepting the deduced sedimentation history even in this part of the fan.

Fan Setting and Morphology

The Delgada Fan lies in deep water (2600 to 4500 m) adjacent to the California coast north of San Francisco Bay (Fig. 1). Cores from DSDP sites 32–34 suggest that the fan began to develop in the Late Miocene [1]. Throughout the late Neogene, the Pacific plate and, thus, Delgada Fan have moved northwest with respect to the North American plate along the San Andreas and related transform faults [2]. The relatively narrow apex of the fan (Fig. 2) suggests either that the coalescing canyon systems on the continental slope have remained with the fan and adjacent sea floor during the Neogene transform offset or that the two main fan lobes (see below), which include the bulk of the upper-fan deposits (Fig. 3), are relatively recent features.

The areal limits of the Delgada Fan shown on the bathymetric map (Fig. 1) are based on morphology and internal structure as interpreted from both single-channel and 3.5-kHz seismic profiles. The fan consists of two main parts: (1) a northern sector that lies mostly north of latitude 39°N and is apparently fed by the northern canyon complex; one of the tributary canyons to this complex is the Delgada Canyon, for which the fan was named [3,4]; (2) a much larger southern sector that includes a complex leveed-valley system extending about 200 km from the mouth of the southern canyon (Figs. 1 and 2). The basement relief of the oceanic crust has locally controlled the growth pattern of the fan, especially in the southern sector, which feeds sediment a long distance westward along a trough on the south side of the Pioneer Fracture Zone at 38°40′N (Fig. 2). Fracture zone relief on the oceanic crust is not observed in seismic-reflection profiles, however, east of longitude 125°40′W.

The inner half of the southern sector is characterized by a pronounced topographic rise that is recognizable in the bathymetry between water depths of 2700 and 4200 m (Fig. 1). A prominent channel feature is found near the crest of this rise for much of its length and, in most reflection pro-

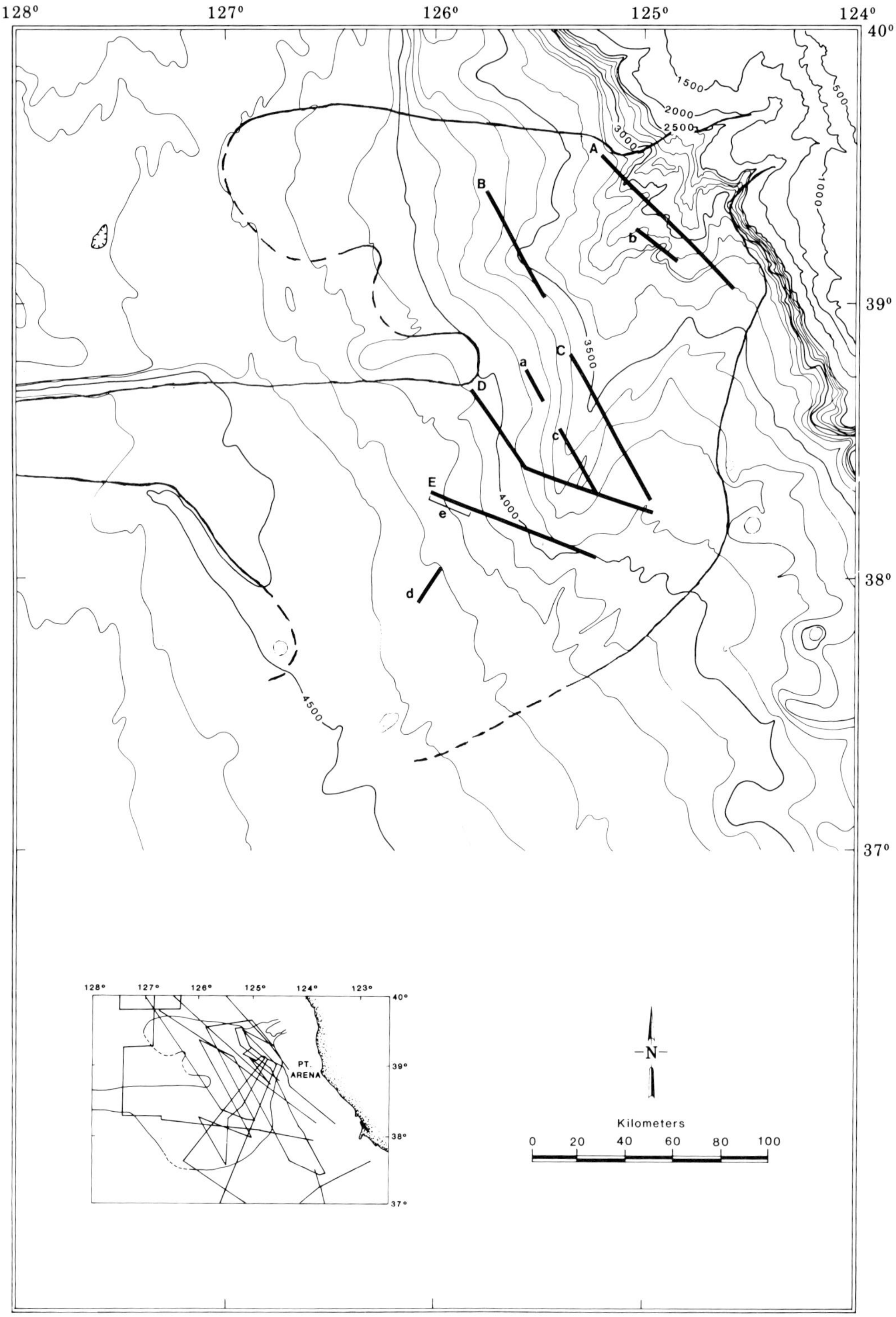

Figure 1. Bathymetric map of Delgada Fan and adjacent areas (modified from [4]). Outline of the fan and subdivisions are based on interpretation of 3.5-kHz and single-channel seismic-reflection profiles and sea-floor morphology. Tracklines for single-channel profiles (A to E, Fig. 3) and 3.5-kHz profiles (a to e, Fig. 4) are shown. Inset shows tracklines for all seismic profiles available for this study.

files, it appears to be a typical leveed-valley system common to large deep-sea fans, e.g. Figure 3C and D [5,6,7]. There is no evidence for structural control on the trend or length of this leveed-valley system, which attains a thickness of nearly 3 km on the upper fan. The easternmost channel in the southern sector, which terminates about 20-km south of latitude 39°N (Fig. 2), has little levee-type relief and may be a recently formed pathway for sediment to bypass the large leveed-valley system.

The northern sector of the fan is not only smaller, but it lacks channel relief over much of its length and appears as a large prograding "lobe" (Fig. 3B). At this point, there is

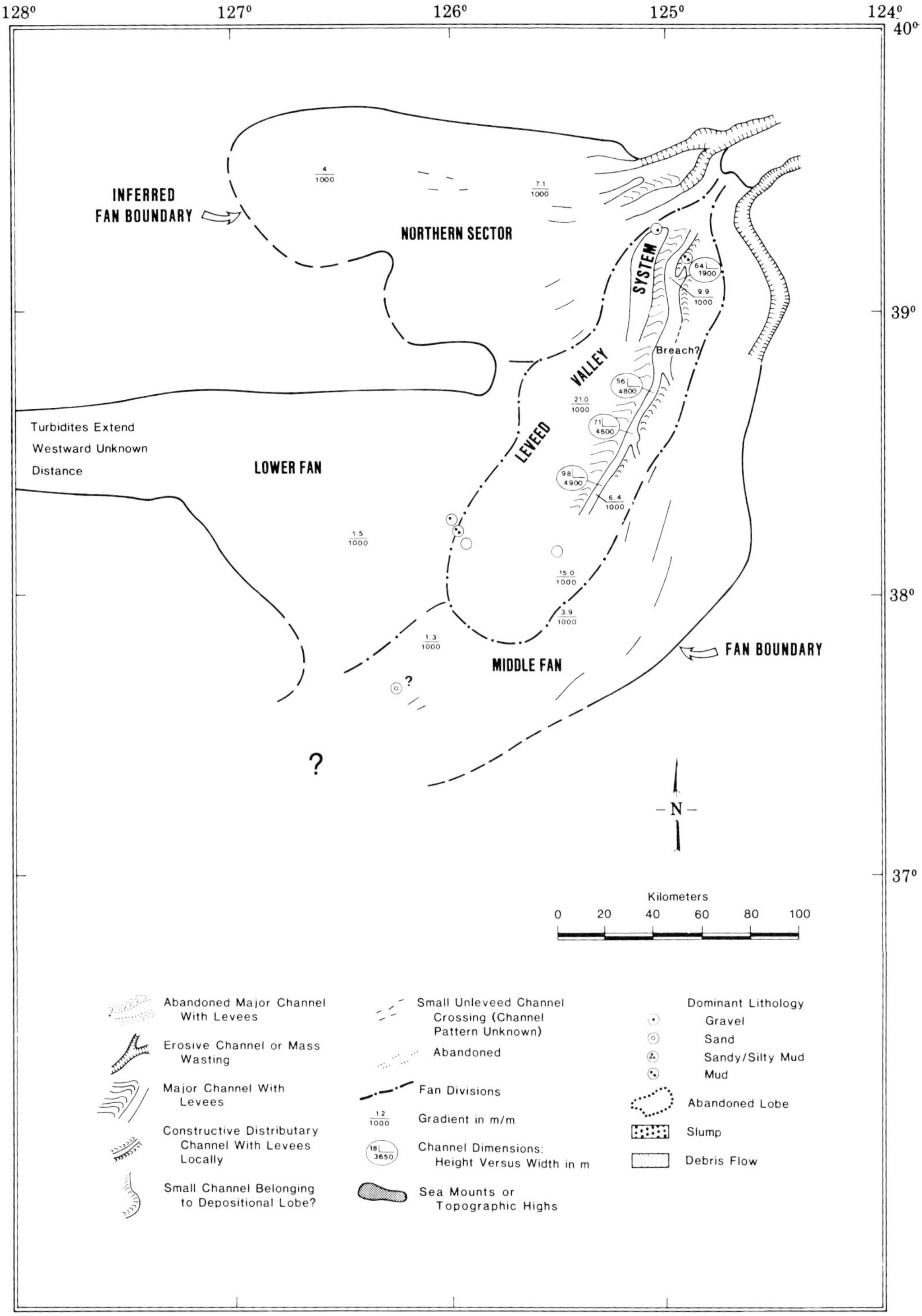

Figure 2. Schematic map of Delgada Fan showing channel pattern, channel dimensions, fan divisions, gradients, and dominant lithology.

insufficient information about the coalescing canyon complexes to relate canyon development to fan growth.

Structure and Near-Surface Acoustic Facies

The surface relief and internal structure of the southern lobe is highly variable (Fig. 3). Profiles C and D from the outer part of this feature show a close resemblance to broad depositional fan valleys. These two profiles show relatively wide (3 to 5 km), smooth, elevated valley floors bounded by broad levees 20- to 40-km wide. As is common in large depositional valley systems [5,7], the larger righthand (northern) levee has a smooth levee crest, although sediment waves are common on the backside (away from the channel floor). The position of the valley floor does not appear to be controlled

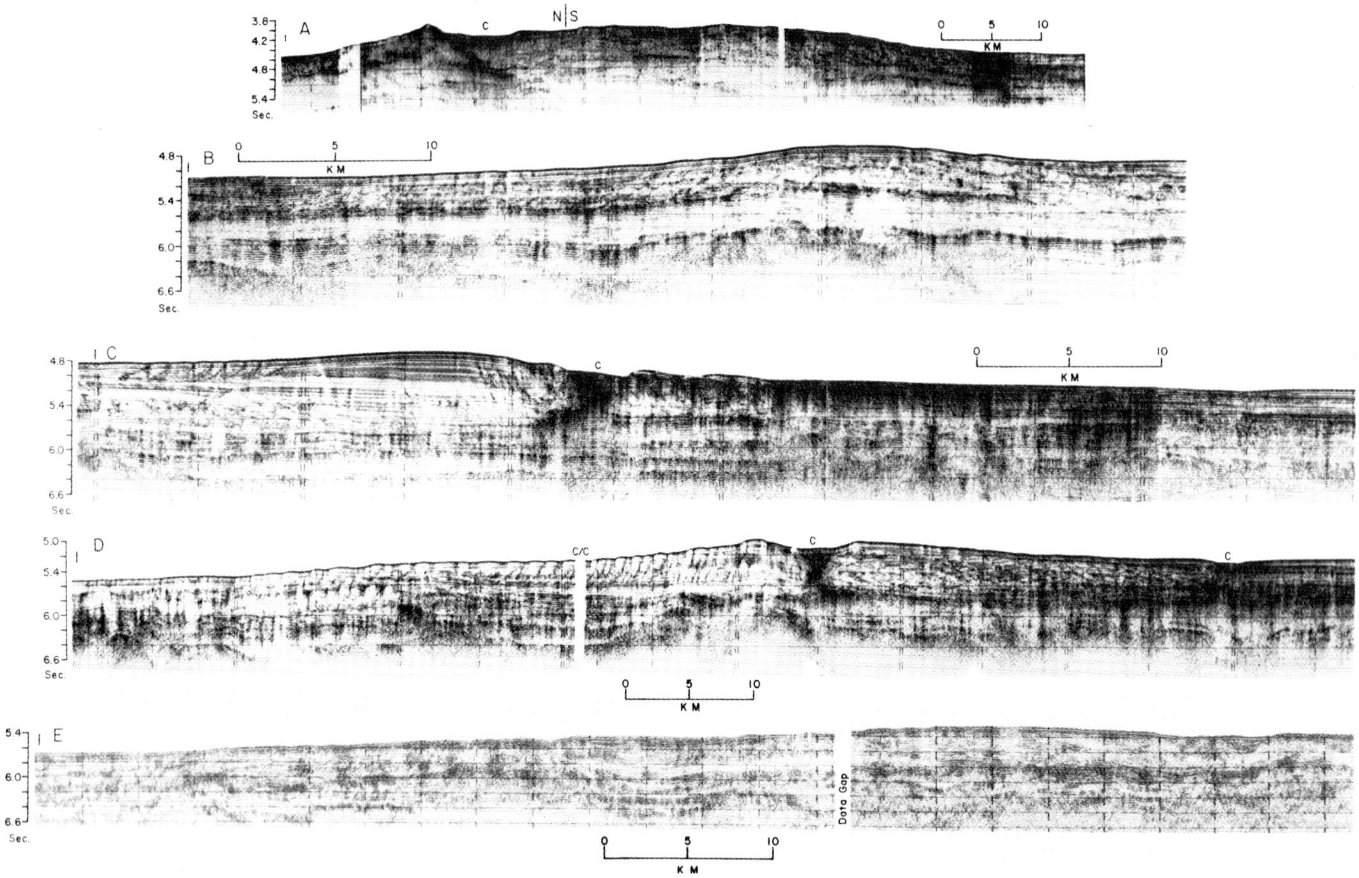

Figure 3. Single-channel seismic-reflection profiles from Delgada Fan. Vertical scales in seconds of roundtrip travel time. Small scale bar (left end of each profile) shows 100 m of sea-floor relief. Profiles are ordered from inner fan seaward, and north or west is to the left. See Fig. 1 for location. (A) Proximal sections of Delgada Fan, both northern and southern lobes. Vertical line above profile shows approximate division between the lobes. Note channel feature (c) on northern lobe. (B) Northern lobe without obvious channel features but with limited areas of sediment waves on north flank. (C) Southern lobe with channel features (c) near crest and along southern edge. (D) Southern lobe farther seaward than profile C and showing well-developed sediment waves in addition to channel features (c) at crest and at southern edge. (E) Southern lobe below termination of large channel features.

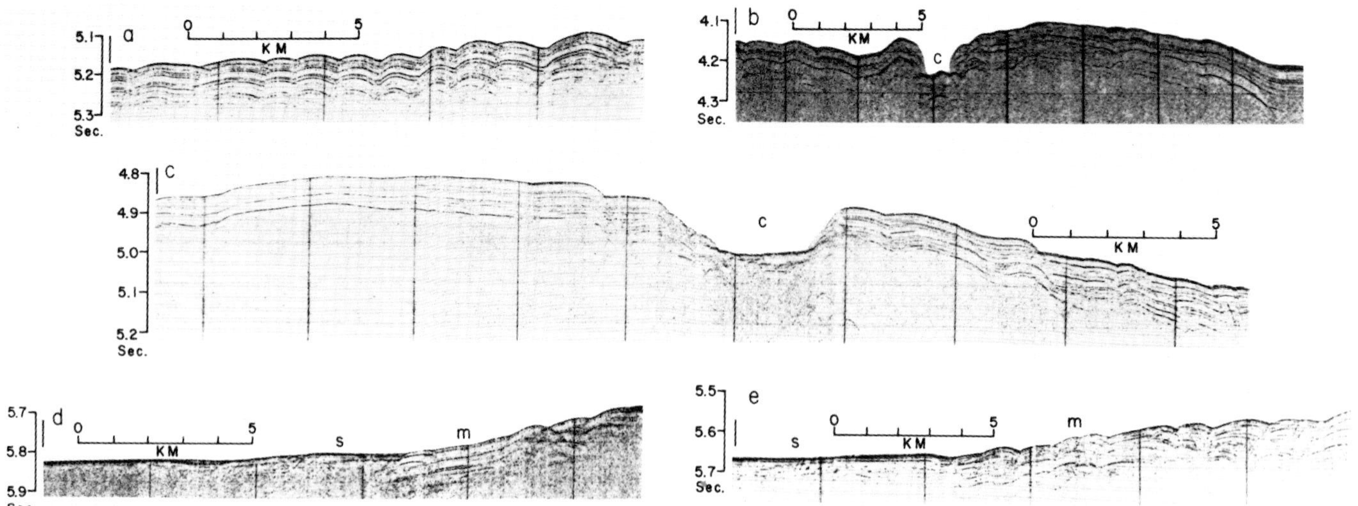

Figure 4. High-resolution 3.5-kHz reflection profiles from Delgada Fan. Vertical scales in seconds of roundtrip travel time. Small scale bar shows 50 m of sea floor relief. North or west is to the left. See Fig. 1 for location. (a) Well-developed sediment waves on western flank of southern lobe. (b) Proximal section of southern lobe with channel (c). (c) Main channel feature (c) of southern lobe (between seismic profiles 3c and 3d). (d,e) Transition from well-bedded, muddier sediments of lobe (m) to adjacent areas of poor acoustic penetration, limited continuity of reflectors, and sandier sediment(s).

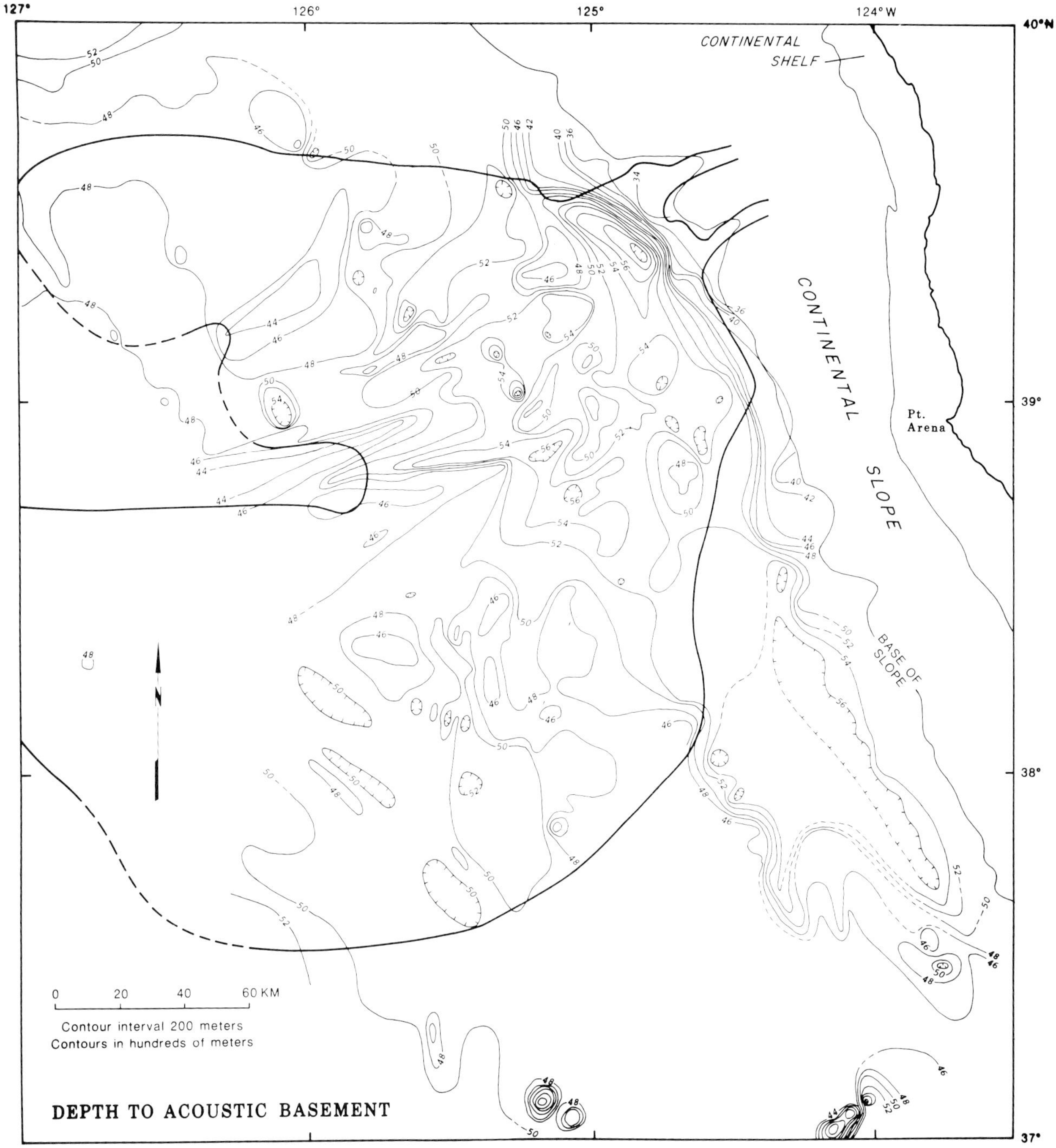

Figure 5. Basement topography under Delgada Fan. Depth to basement is corrected for both sound velocity in sea water (using Matthews' tables) and sound velocity within the sediments using a velocity profile derived from interval velocities for multichannel seismic profiles from the easternmost part of the fan (unpublished data, D. McCulloch, U.S. Geological Survey). Fan outline as shown in Figure 1.

by prefan basement topography. Near the southwestern termination of this feature (Fig. 3E), most of the typical fan-valley characteristics are absent. There is no evidence for a distributary channel system that would be fed by the major valley seen in profile D.

The interpretation of events at the proximal end of the southern leveed-valley feature is also puzzling. Figure 3A shows that the leveed-valley lobe is narrow and has no large-channel features. The northern lobe (Fig. 2) generally lacks large-channel relief (Fig. 3B) although the sediment-wave relief on the north side may be related to levee deposition.

The high-resolution 3.5-kHz reflection profiles show that the upper 30 to 40 m of sediment on the southern lobe on Delgada Fan are relatively muddy. Profiles a, b, and c (Fig.

4) are characterized by numerous subparallel reflectors that persist over distances of 10 or more km. This type of 3.5-kHz record is indicative of mud-dominated areas on modern submarine fans [8,9] and is best expressed in the area of sediment waves on the northern levee (Fig. 4a). The most proximal crossing of a channel feature on top of the lobe (Fig. 4b) shows that even the channel floor lacks an acoustic character that would indicate much sandy sediment. Farther southwest (Fig. 4c), the valley floor shows more discontinuous reflectors and poorer depth of acoustic penetration, which might indicate somewhat coarser sediment.

In contrast, the sea floor adjacent to the leveed-valley lobe shows an acoustic character suggesting more sandy sediments (Fig. 4d and e). The depth of acoustic penetration is limited and reflector continuity is poor even though the adjacent depositional lobe shows little evidence of sandy sediment. The limited core data from Delgada Fan support the inferences based on the 3.5-kHz reflection profiles; the number of sand or silt beds per meter of core is distinctly higher on the sea floor around the depositional lobe.

Discussion and Fan History

The morphometric map of the Delgada Fan (Fig. 2) summarizes our current interpretation of the history of part of the southern sector of the Delgada Fan and also serves to indicate the major unresolved problems. The overall morphology and structure of the southern lobe suggests that it is, or was, a major depositional valley system. Unlike most large fan valleys [5,6,7], the channel relief on the southern lobe does not show a uniform decrease in size (width of channel floor and levee-channel relief) along its length. Channel shape also varies in an unsystematic fashion (Figs. 3 and 4). Our seismic profiles and available bathymetric data suggest that there are at least two breaches of the lefthand levee (see Fig. 2). The whole leveed-valley complex may be cut off at present, and this could account for the lack of coarse sediment in our cores from this elongate lobe. The channel feature along the continental slope east of the apex of the southern lobe may be the bypass channel.

Although we have not observed a distributary channel system extending from the southern leveed-valley system, the flatter sea floor around the lobe exhibits shallow channels on many of the profiles. Some of these channels are not part of a distributary system extending from the continental slope; they first appear on the gently sloping rise south of the main leveed valley. The most prominent of these "headless" channels appears just northeast of profile C and is best developed where crossed on the eastern end of profile D (Fig. 3). The coarser nature of sediments off the lobe together with the independent channel features not connected to the lobe valley also could be consistent with a general abandonment of the lobe. At present, there is neither a detailed enough bathymetric map nor sufficient core data to provide a more specific history.

With no core data and limited seismic-reflection profiles from the northern lobe, we cannot determine how it developed. Deposition on the northern lobe appears to coincide with growth of the main leveed-valley sequence on the southern lobe (Fig. 3B and C). Thus, at least for the upper part of the turbidite sequences on Delgada Fan, the leveed-valley systems maintained separate feeding canyon systems.

Addendum

A map of the depth to acoustic basement (Fig. 5) shows that the morphology of the ocean crust under Delgada Fan does not indicate any obvious topographic control on the development of the elongate leveed-valley complex that dominates the southern sector of the fan. Several east-northeasterly trends just south of 39° N latitude might be the easternmost extension of the Pioneer Fracture Zone. In general, the crustal relief under the Delgada Fan is less extreme, with fewer seamounts, than exists under the Monterey Fan further south (Chapter 13, this volume).

References

[1] McManus, D. A., and others, 1970. Site 32. In: D. A. McManus and others, Initial Reports of the Deep Sea Drilling Project, v. 5. U.S. Government Printing Office, Washington, DC, pp. 15–56.

[2] Atwater, T., 1970. Implications of plate tectonics for the Cenozoic tectonic evolution of western North America. Geological Society of America Bulletin, v. 81, pp. 3513–3536.

[3] Menard, H. W., 1960. Possible pre-Pleistocene deep-sea fans off central California. Geological Society of America Bulletin, v. 71, pp. 1271–1278.

[4] Wilde, P., Normark, W. R., and Chase, T. E., 1976. Oceanographic data off Central California, 37° to 40° North including the Delgada deep-sea fan. Lawrence Berkeley Laboratory Publication No. 92, Berkeley, CA.

[5] Hamilton, E. L., 1967. Marine geology of abyssal plains in the Gulf of Alaska. Journal of Geophysical Research, v. 72, pp. 4189–4213.

[6] Normark, W. R., 1970. Growth patterns of deep-sea fans. American Association of Petroleum Geologists Bulletin, v. 54, pp. 2170–2195.

[7] Normark, W. R., 1978. Fan valleys, channels, and depositional lobes on modern submarine fans: characters for recognition of sandy turbidite environments. American Association of Petroleum Geologists Bulletin, v. 62, pp. 912–931.

[8] Damuth, J. E., 1975. Echo character of the western equatorial Atlantic floor and its relationship to the dispersal and distribution of terrigenous sediments. Marine Geology, v. 18, pp. 17–45.

[9] Normark, W. R., Piper, D. J. W., and Hess, G. R., 1979. Distributary channels, sand lobes, and mesotopography of Navy Submarine Fan, California Borderland, with applications to ancient fan sediments. Sedimentology, v. 26, pp. 749–774.

CHAPTER 11

La Jolla Fan, Pacific Ocean

Steven B. Bachman and Stephan A. Graham

Abstract

Development of the upper and middle La Jolla Fan system is controlled by the regional California Borderland structural grain. This tectonic fabric produces three upper fan distributary paths. The position of La Jolla Canyon is controlled by the geometry of a buried hard-rock structure. The northern Newport Canyon channels contain seismically detectable vertical stacks of channel sand bodies. Sediment from northern feeders and the central La Jolla Canyon-Fan form three-dimensional wedges that interfinger at their distal ends. La Jolla Fan system contains an interleaved set of sediment wedges derived from multiple sources, woven around the regional structural fabric of uplifts and basins.

Introduction

La Jolla Fan, offshore of San Diego, California, is one of the best studied of modern submarine fans. Most studies have used shallow-penetration seismic or coring methods and are essentially morphologic studies. We used seismic data with a greater depth of penetration to reveal aspects of the underlying architecture of the fan and the controls on its development.

La Jolla Fan (Fig. 1) is a typical borderland fan lying in 470 to 1100 m of water [1]. Even at a first-order level, the morphology of the fan departs markedly from the radial cone of sediment comprising the ideal model of a submarine fan [2,3]. Instead of growing onto an unrestricted basin floor, La Jolla Fan has filled local structural basins and wrapped around fault-block uplifts that form the tectonic fabric of the southern California Borderland. The faults bounding these borderland uplifts and basins are probably a combination of strike-slip faults of the San Andreas transform system [4,5] and thrust/reverse faults with compression directed normal to the San Andreas fault [6]. Because of this structural complexity, the limits of the La Jolla Fan are vaguely defined. Most authors confine the term to mean the morphologic mounding in the vicinity of the "La Jolla Fan" label in Figure 1. We also include the proximal (Newport Canyon and Loma Sea Valley) and the distal (San Diego Trough) portions of the "La Jolla Fan system" to describe the larger deep-water sediment dispersal patterns offshore of San Diego [7].

The age of the La Jolla Fan system is not well defined. Fan sedimentation has been ongoing for most of the Cenozoic, although the "undeformed" [8] sediments that apparently reflect the latest periods of fan deposition are upper Pliocene and younger [7]. Major sediment bypassing of the upper fan may have occurred during the Holocene lowstand [9], resulting in the growth of a new suprafan far downfan in the San Diego Trough (Fig. 1) [10].

Our interpretations are based on a colinear grid of CDP reflection-seismic, conventional sparker, and high-resolution uniboom data recorded in the mid-late 1970's. The 1.6 × 3.3-km grid of 18 to 24 kJ sparker lines shows that the upper and middle La Jolla Fan system is comprised of a complex meshing of sediment wedges.

Upper and Middle La Jolla Fan

The most prominent feature of the upper portion of La Jolla Fan is the main fan valley. The La Jolla Fan system, however, has three major and several minor structurally controlled upper fan valleys that funnel sediment into the fan. The morphologically prominent main feeder, the La Jolla Fan Valley, has been studied extensively (see references in [7]). The southern feeder, the Loma Sea Valley, and the

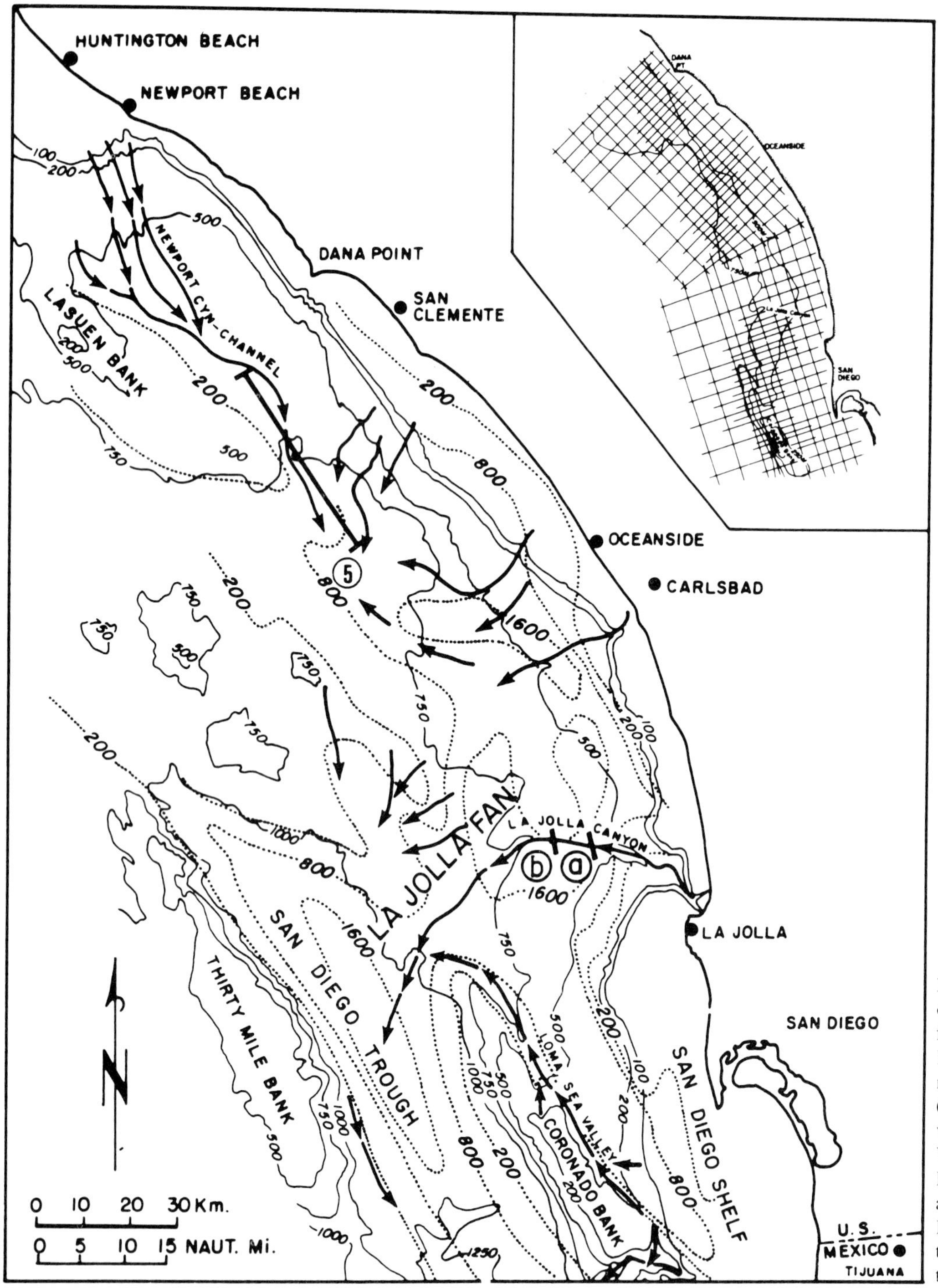

Figure 1. Bathymetric and generalized map of La Jolla submarine fan and feeder canyons/channels. Bathymetry (solid lines, in m) and modern fan sediment thickness (dotted lines, in m) are from previous studies of the La Jolla Fan Valley [7 and references therein]. Drainages are shown with heavy arrows. Circled numbers locate Figures 5, 6a, and 6b. Inset shows tracklines of seismic data used in this study.

northern feeders, Newport Canyon-Channel and smaller tributaries, have been described in less detail. Most published studies of the La Jolla Fan tend to emphasize the La Jolla Fan Valley as the main sediment conduit to the present fan system. However, the deep-penetration seismic data reveal that the other feeders have contributed major amounts of sediment to the fan system.

Newport and Carlsbad Canyon-Channels

Northerly feeders to the La Jolla Fan system include canyons and fan channels heading near Newport, Oceanside, and Carlsbad (Fig. 1). At the present time, only Newport Canyon-Channel appears to be an important conduit of sediment [11]. It heads about 0.4 km from shore as a series of slope rills which tap longshore drift sources of sandy sediment [11]. As the rills coalesce downslope into a single channel, the sediment dispersal pattern becomes constrained between the fault-controlled Lasuen Bank on the southwest and the fault-controlled continental slope on the northeast (Fig. 2) [7]. Although currently flat-bottomed and aggraded, the channel clearly is incised into an apron of sediment between the topographic highs (Figs. 2 and 3). Flanking levees apparent on seismic data (Fig. 4) are made up of relatively continuous reflectors bundled in lenticular packages [7] that consist of thin-bedded silty turbidites in shallow-penetration cores [11]. Channel areas (Fig. 3) include relatively chaotic seismic reflectors [7] and contain sand and gravel [11].

A buried predecessor channel can also be detected below Newport Channel (Fig. 3) and can be mapped from line-to-line (Fig. 2). Continued structural relief along the faults in this area has thus contained the channel system over a period of several million years [7] and has produced a series of stacked linear sand and gravel bodies encased in finer-grained overbank facies (Figs. 2 and 3). Farther to the southeast, Lasuen Bank plunges into the subsurface and no longer provides a westerly barrier to Newport Channel (Fig. 1). A series of anastomosing small low-relief channels [7,11] form below this point. Reflector continuity increases greatly in this area with the reduction of restricted channel deposits and the development of more continuous, non-channelized midfan-like deposits. An isopach map of sediment associated with the modern channel shows that the break into anastomosing channels coincides with a local depocenter that fills a structural depression southwest of San Clemente (Fig. 2).

La Jolla Fan sediment is also derived from the canyon heading near Oceanside (Fig. 1) [11]. Our seismic data show that over time channel-mouth aprons from both Newport and Oceanside Canyons form three-dimensional sediment wedges that interfinger at their distal ends. The wedges follow a radial growth pattern, unless constrained by structural controls, and form the basic building blocks of the larger fan system. A sparker line recorded longitudinally down Newport Channel (Fig. 5) illustrates these relationships. The line shows four main seismic stratigraphic packages above acoustic basement: deformed Miocene-Pliocene deep-water sediments; onlapping sediments associated with the ancestral Newport Channel of Figure 2; a southerly interleaving wedge of sediment from the Oceanside-Carlsbad area; and sediment associated with the active Newport Channel. Thus, the upper La Jolla Fan system is a complex sediment accumulation derived from multiple point sources.

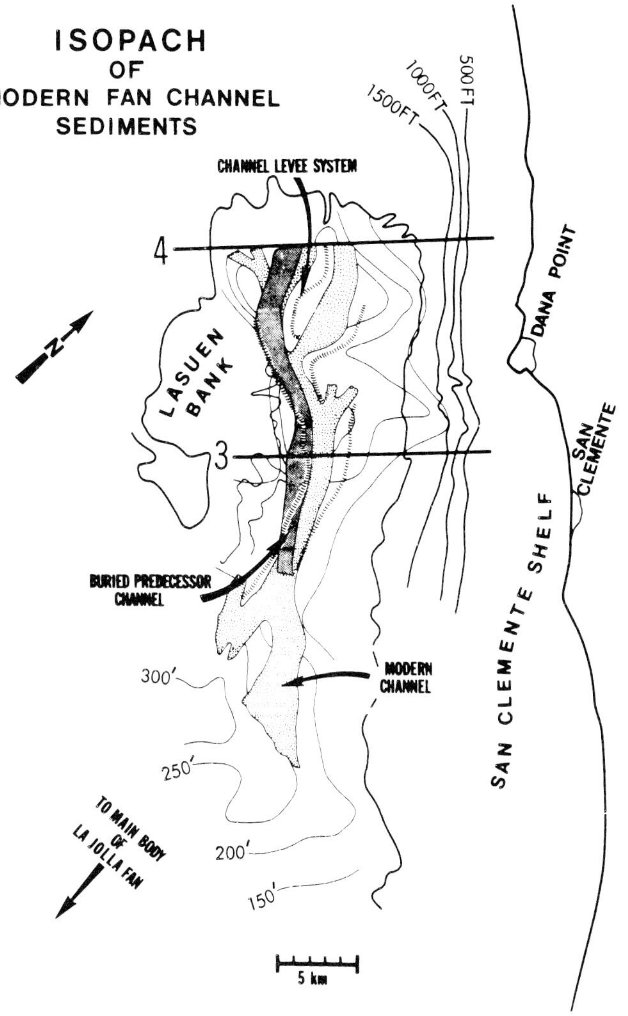

Figure 2. Map of Newport Channel with modern and buried predecessor channels. Bathymetry shown with heavy lines. Isopachs of sediments associated with the modern Newport channel are shown with finer lines. Straight lines locate positions of Figures 3 and 4.

La Jolla Canyon-Fan Valley

The main upper fan channel, La Jolla Fan Valley, has an incised, erosional character that may have resulted from eustatic sea-level changes [3]. Although most coarse-grained sediment is apparently bypassing the upper fan on its way to the suprafan [9,10], our data indicate that depositional levees are well developed (Fig. 6), particularly on the north bank. Thus, sediment apparently is deposited in overbank locations during sediment-bypassing in the channel.

Although the left hook of the La Jolla Canyon and Channel (Fig. 1) may be partly related to Coriolis effects [12], local structure plays an important role in the location of the system [7]. The fan valley is banked on its southern side against Miocene-Pliocene deformed sediments that apparently act as a buried hard-rock barrier to fan valley migration (Fig. 6a). This structural barrier, the northern extension of the San Diego shelf, deflects sediment to the north and results in the curvilinear shape of La Jolla Canyon.

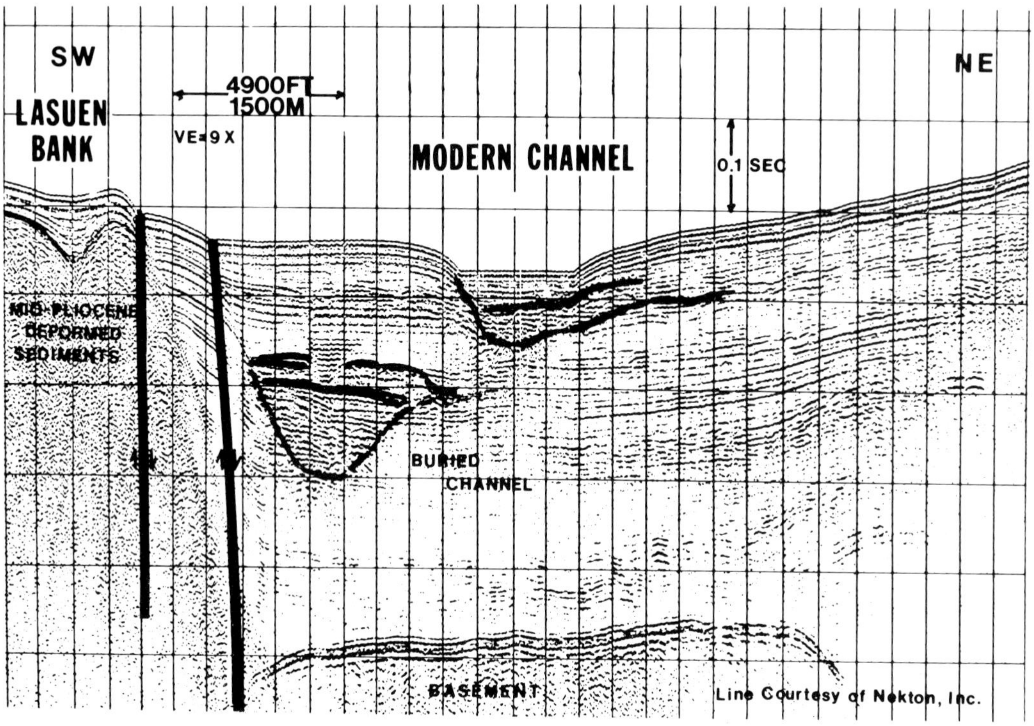

Figure 3. A 24 kJ sparker line across Newport Channel showing modern and buried channel complexes. Location indicated on Figure 2.

The constructional leeves that flank La Jolla Fan Valley are built in part on older levee deposits (Fig. 6) that can be mapped in three dimensions. These levees can be recognized on seismic lines by their convex upper surfaces, lenticular shapes with long-dimension subparallel to the valley axis, and uneven-to-contorted internal reflectors. Individual levee packets are generally about 50 ms thick (approximately 40 m) and extend outward from (normal to) the fan valley for 5 to 10 km.

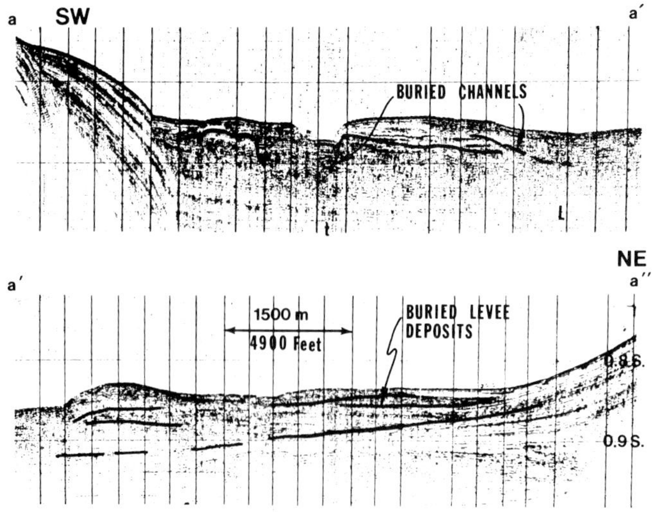

Figure 4. Uniboom line across Newport Channel showing details of channel incision, filling, and levee development. Location indicated on Figure 2.

Loma Sea Valley

Loma Sea Valley (Fig. 1) is the most important southern feeder to the La Jolla Fan system [2]. Its position is structurally controlled between Coronado Bank on the west and the San Diego shelf on the east. The strike of the valley follows the regional structural grain, and both the east flank of Coronado Bank and the San Diego shelf are fault-bounded. The thickness of channel-fill in the valley varies from 200 to 500 m. An unusual sediment dispersal mechanism dumps sediment at the mouths of tributary channels in the valley, thus clogging the valley floor. This sediment tapers in a wedge down the valley and resembles the tributary mouth bars of major fluvial systems [7].

Conclusions

The upper portions of the La Jolla submarine fan are a complex departure from a simple pattern of radial growth because of structural constraints on sediment dispersal and fan growth. A major tectonic effect is the structural control on the positions of multiple upper fan channels. Several of these distributary systems parallel the continental margin, as does the regional structural grain. These feeder channels contribute interleaving sediment wedges—many individ-

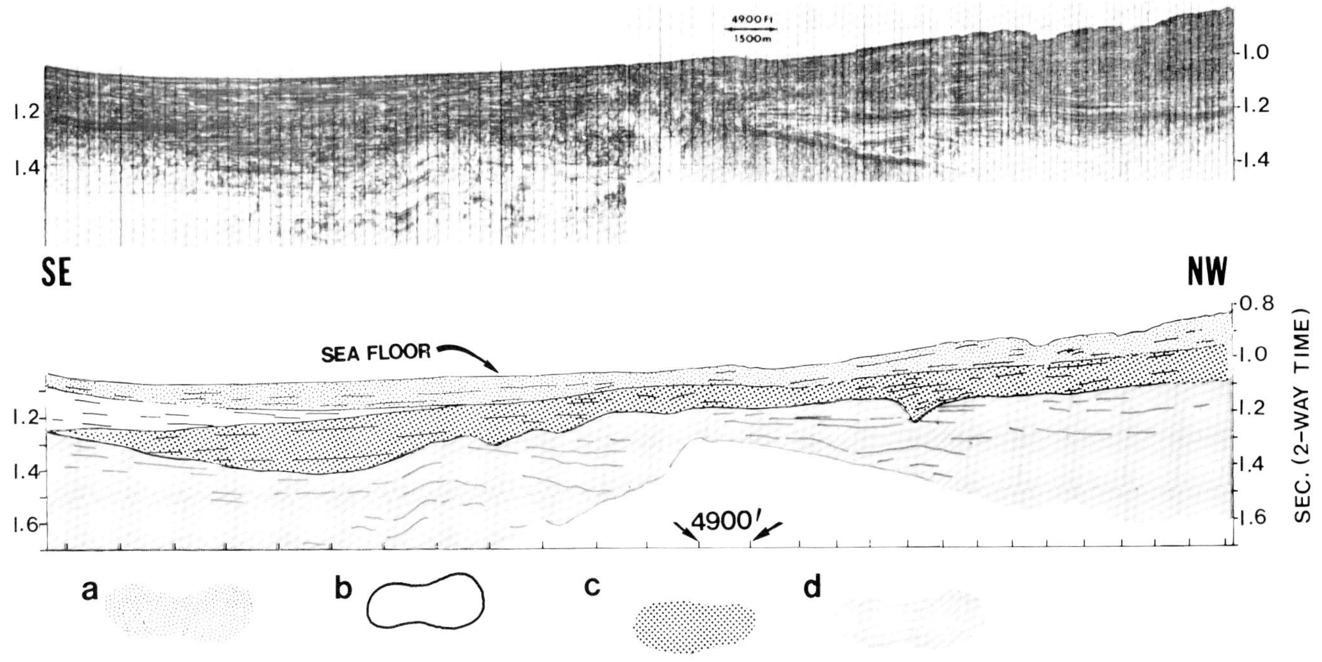

Figure 5. A 24 kJ sparker record and line-drawing recorded down the axis of Newport Channel showing seismic stratigraphic units: (a) sediments of modern Newport Channel, (b) sediments derived from Oceanside Canyon, (c) sediments associated with the buried channel in the Newport Channel area, and (d) deformed Miocene-Pliocene sediments. Basement rocks are likely "Catalina Schist." Location indicated on Figure 1.

ually displaying radial growth patterns—which, as a system, are woven around borderland uplifts.

Seismic facies in the upper part of La Jolla Fan are well defined, particularly in the channel/levee setting. Within fault-controlled conduits, long-lived channels stack vertically and can be mapped for long distances. These multistory sand bodies are encased in finer-grained facies, and both the channels and levees can be recognized within the upper one second of seismic penetration.

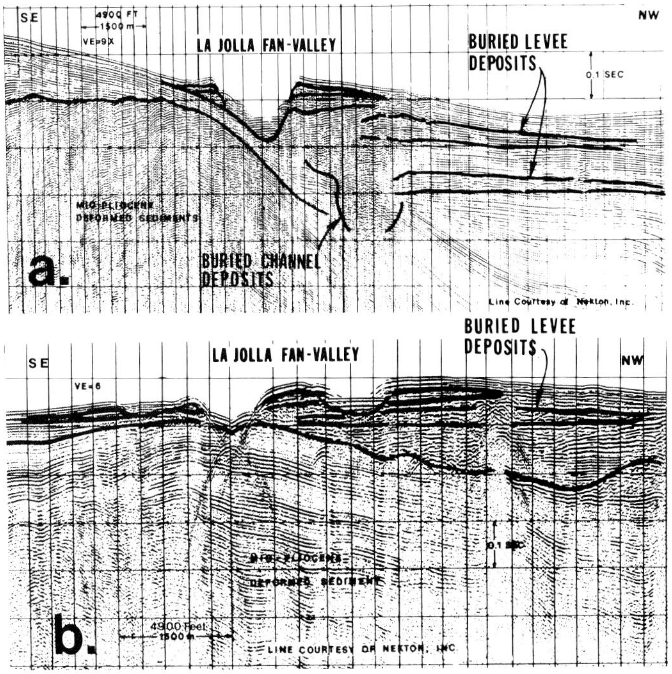

Figure 6. Two 24 kJ sparker lines across La Jolla Fan Valley illustrating the erosional nature of the canyon and stacked-buried levee deposits. Structural control on fan valley is indicated by banking against older deformed sediments on the south (6a). Sections located on Figure 1.

References

[1] Normark, W. R., 1974. Submarine canyons and fan valleys: factors affecting growth patterns of deep-sea fans. Society of Economic Paleontologists and Mineralogists Special Publication 19, pp. 56–68.

[2] Shepard, F. P., Dill, R. F., and von Rad, U., 1969. Physiography and sedimentary processes of La Jolla submarine fan and fan-valley, California. American Association of Petroleum Geologists Bulletin, v. 53, pp. 390–420.

[3] Normark, W. R., 1970. Growth patterns of deep-sea fans. American Association of Petroleum Geologists Bulletin, v. 54, pp. 2170–2195.

[4] Crowell, J. C., 1974. Origin of the late Cenozoic basins in southern California. In: Tectonics and Sedimentation. Society of Economic Paleontologists and Mineralogists Special Publication 22, pp. 190–204.

[5] Smith, D. L., and Normark, W. R., 1976. Deformation and patterns of sedimentation, south San Clemente basin, California Borderland. Marine Geology, v. 22, pp. 175–188.

[6] Crouch, J. K., Bachman, S. B., and Shay, J. T., 1984. Post-Miocene compressional tectonics along the central California margin. In: J.

K. Crouch and S. B. Bachman (eds.) Tectonics and Sedimentation Along the California Margin. Pacific Section-Society of Economic Paleontologists and Mineralogists, v. 38, pp. 37–54.
[7] Graham, S. A., and Bachman, S. B., 1983. Structural controls on submarine-fan geometry and internal architecture: upper La Jolla fan system, offshore southern California. American Association of Petroleum Geologists Bulletin, v. 67, pp. 83–96.
[8] Moore, D. G., 1969. Reflection profiling studies of the California Continental Borderland–structure and Quaternary turbidite basins. Geological Society of America Special Paper 107, 142 p.
[9] Piper, D. J. W., 1970. Transport and deposition of Holocene sediment on La Jolla deep-sea fan, California. Marine Geology, v. 8, pp. 211–227.
[10] Normark, W. R., 1978. Fan valleys, channels, and depositional lobes on modern submarine fans: characters for recognition of sandy turbidite environments. American Association of Petroleum Geologists Bulletin, v. 62, pp. 912–931.
[11] Hand, B. M., and Emery, K. O., 1964. Turbidites and topography of north end of San Diego trough, California. Journal of Geology, v. 72, pp. 526–542.
[12] Shepard, F. P., and Buffington, E. C., 1968. La Jolla submarine fan valley. Marine Geology, v. 6, pp. 107–143.

CHAPTER 12

Magdalena Fan, Caribbean

V. Kolla and R. T. Buffler

Abstract

The Magdalena Fan can be divided into: upper fan—1:60–1:110 gradients, channels with well-developed levees, generally several subbottom reflectors on 3.5-kHz records, and fine-grained sediments; middle fan—1:110–1:200 gradients, channels with very subdued levees, several to few subbottom reflectors on 3.5-kHz records, and chaotic and discontinuous reflections on multichannel seismic (MCS) records; lower fan—<1:250 gradients, small channels and relatively smooth seafloor, generally coarse-grained sediments, few or no subbottom reflectors on 3.5-kHz records, and flat continuous reflections on MCS records. In addition to the turbidity currents, slumping along the continental slope and elsewhere also influenced sedimentation in the fan.

Introduction

The Magdalena Fan forms a broad arcuate bathymetric feature in the southern Colombian Basin, Caribbean Sea, and is about 230 km long (Figs. 1 and 2). The fan grades downdip to the northwest into the deep abyssal plain of the basin and terminates updip against the deformed northwestern continental margin of Colombia. According to Duque-Caro [1], the northwestern Colombia margin may be considered an accretionary margin.

Terrigenous sediments have been supplied to the Magdalena Fan primarily by the Magdalena River and to some extent by the Sinu River. Both these rivers drain the Andean Cordillera, which forms three branches in northern Colombia: the western, central, and eastern Cordillera [2]. Because of out-building of the Magdalena River delta, the continental shelf in this area is virtually absent (Fig. 1). Several slope valleys and gullies, as well as mud diapirs, exist off the Magdalena River [3]. Occurrence of cable breaks presumably resulting from slumping and turbidity currents off the Magdalena River were reported by Heezen [4], and by Heezen and Munôz [5]. On the continental slope west of the Magdalena River, valleys and channels occur, which apparently served as conduits for Magdalena River sediments in the past (Figs. 1 and 2).

The crust underlying the Colombia Basin is oceanic and may have formed in Late Cretaceous [6]. Although sediments in this basin have been deposited since that time, the Magdalena Fan sequence was not deposited until much later time. Kolla and others [7] distinguished six seismic sequences in the sedimentary section of the Colombia Basin. The upper three seismic sequences (units) were probably influenced significantly by terrigenous sedimentation since Miocene coinciding with episodes of diastrophism and orogeny on land and may make up the fan sediments. The uppermost seismic unit, designated as *F* by Kolla and others [7], is definitely a fan sequence and has been deposited since about middle Pliocene, coinciding with the most important of northern Andean orogenies. The thickness of this unit is 0.8 secs two-way travel time or less.

Data Base

From the 3.5-kHz records obtained during various cruises by Lamont-Doherty Geological Observatory (LDGO) and the

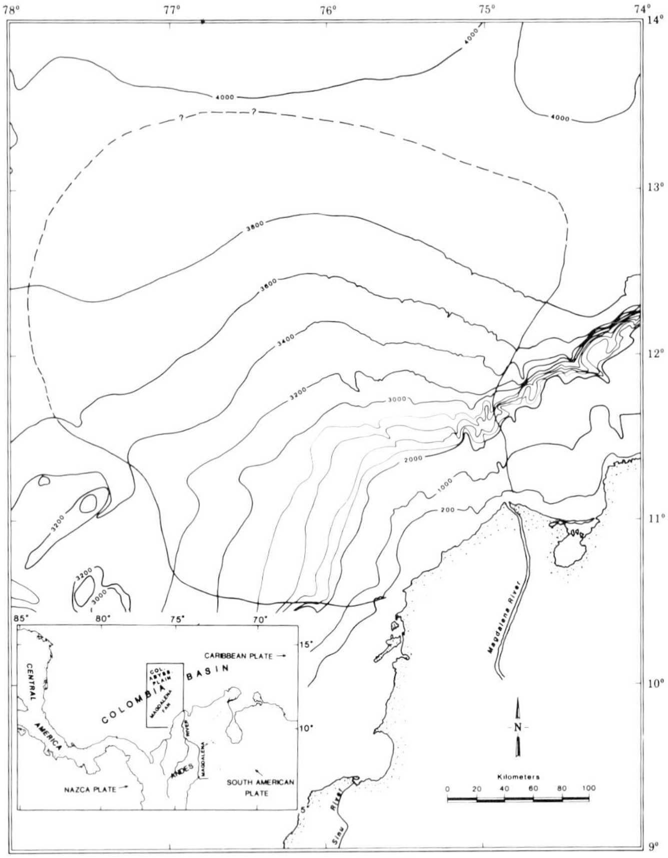

Figure 1. Bathymetry of the Magdalena Fan region with simplified courses of the Magdalena and Sinu rivers.

University of Texas Institute for Geophysics (UTIG), details of the morphology of the sea floor, types of echoes, subbottom penetration, and subbottom reflector continuity were noted (Figs. 2 and 3). The lithology of Quaternary sediments in the piston cores available at LDGO's core library was described (Fig. 4). We utilized primarily multichannel seismic (MCS) data obtained by LDGO and UTIG in studying the characteristics of deeper sections of the sediment column (Fig. 5).

Morphologic, Acoustic, and Sedimentologic Characteristics of the Fan

Although many of the characteristics change gradually downfan, the combined morphologic, acoustic, and sedimentologic properties of the upper part of the sediment column based on 3.5-kHz records, seismic data, and cores allow the Magdalena Fan to be divided into upper, middle, and lower fan.

Upper Fan

The gradients of the continental slope are steeper than 1:60 whereas those of the upper fan are 1:60 to 1:110 (Table 1). The continental slope is characterized by an irregular sea floor, canyons, and slump features whereas the upper fan has very well-developed channel-levee complexes [7]. The channel relief in the upper fan extends to >100 m (Table 1). The shallow (3.5 kHz) acoustics of the upper fan are generally characterized by distinct echoes with several continuous subbottom reflectors and good penetration (Fig. 3A). Bedforms in the form of regular hyperbolae or sediment wave-like features, as observed on 3.5-kHz records, are present on the backsides of the levees as well as far away from the levees [7]. Irregular bedforms are present on the valley floors. The upper fan has fine-grained muds except in valleys where Bouma turbidite sequences T_{a-e} with coarse to fine-grained sands are present. The generally good subbottom penetration of the upper fan on 3.5 kHz results from the fine-grained nature of sediments, similar to other fans (Fig. 4) [8,9]. On MCS records the levees in the upper fan show up as overlapping and coalescing wedge-shaped reflection packages (Fig. 5). The channel floors themselves are characterized by onlap fill reflection configurations. In Fig. 5, channels (CH) flanked by levees are identified within the topmost seismic unit *F* of Kolla and others [7]. Several periods or episodes of channel activity (CH-I, CH-II, CH-IIA, CH-III) in the upper parts of the upper fan region within unit *F* can be recognized on this seismic line. More than one episode of sedimentation can also be recognized on other seismic lines. We believe that these episodes reflect several uplifts in the source region (Andes), lowered sea levels, and/or the shifting of the Magdalena Delta in space and time, subsequent to the major orogeny in middle Pliocene.

Middle Fan

The middle fan has gradients between 1:110 and 1:200. This region has good 3.5-kHz subbottom penetration in its upper part (Fig. 3B) and poor penetration with discontinuous subbottom reflectors or fuzzy subbottoms in its lower part (Fig. 3C). The decreasing subbottom penetration downfan generally reflects the increasing sand content in the sediments compared with the upper fan (Fig. 4). Occurrence of numerous channels appear to be a distinctive characteristic of the middle fan on 3.5-kHz and seismic records (Fig. 2). The channels of the upper part of the middle fan have levees, but they have relatively low heights compared to those on the upper fan (Table 1). In the lower part of the middle fan, levees are either poorly developed or absent. Regular hyperbolic and wave-like features are common in the middle fan, and their heights gradually decrease downfan [7]. Although the general morphology of the whole middle fan is very subdued compared with the upper fan, portions of the middle fan are elevated somewhat above the sea floor owing to up-building by numerous channel-levee complexes. However, we do not believe that any clear-cut suprafan morphology, with dominance of sandy sediments as defined by

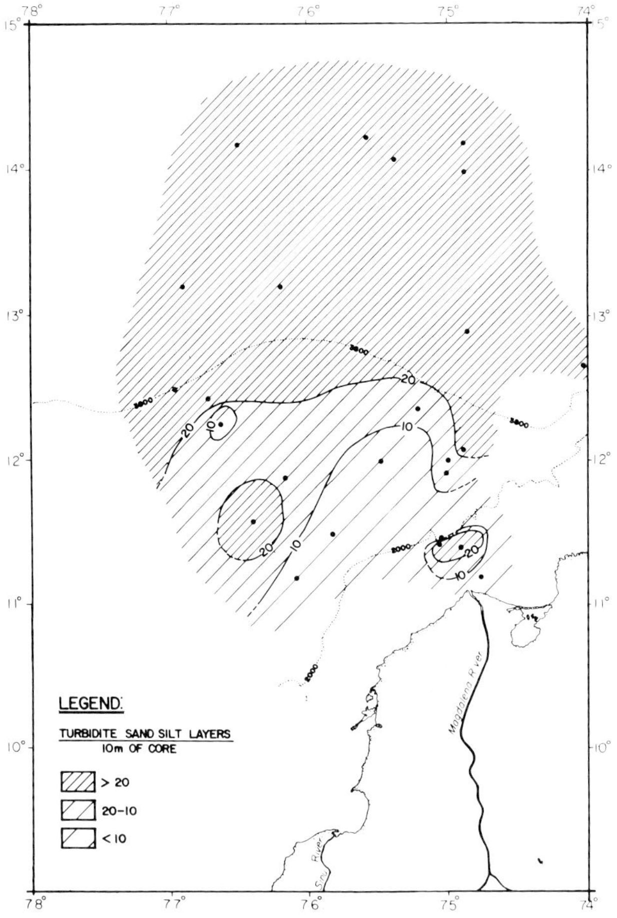

Figure 4. Number of sand-silt layers per 10 m of core. Note area of high sand content in 1000 to 2000-m water depths in a fan valley opposite the present Magdalena River mouth.

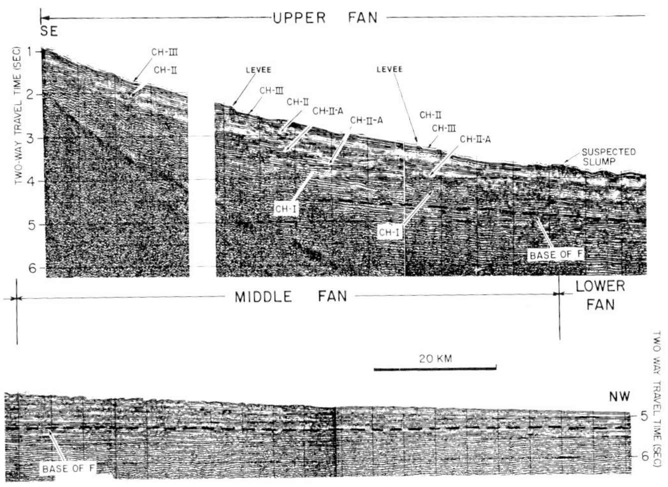

Figure 5. Portions of MCS line showing detailed characteristics of the upper, middle, and lower fan regions. CH indicates the location of channels flanked by levees. Within the uppermost seismic unit *F*, several periods of channel activity (CH-I, CH-II . . .) in the upper fan can be identified.

The bedforms indicated by regular hyperbolae and wave-like features on the levees of channels in the upper and middle fan regions may be related to overbank spilling of turbidity currents [12]. However, these bedforms seem to be common even in areas far away from the channel (or valley) levees. Several indications of slump scars and typical slump features (irregular bedforms) have been observed from the continental slope off Colombia. Locally in the upper fan, there are patches of sea floor with no 3.5-kHz penetration [7], which may be indicative of slump or erosional scars. These features combined with the gradual decrease of relief of hyperbolic and wave-like bedforms downfan suggest that

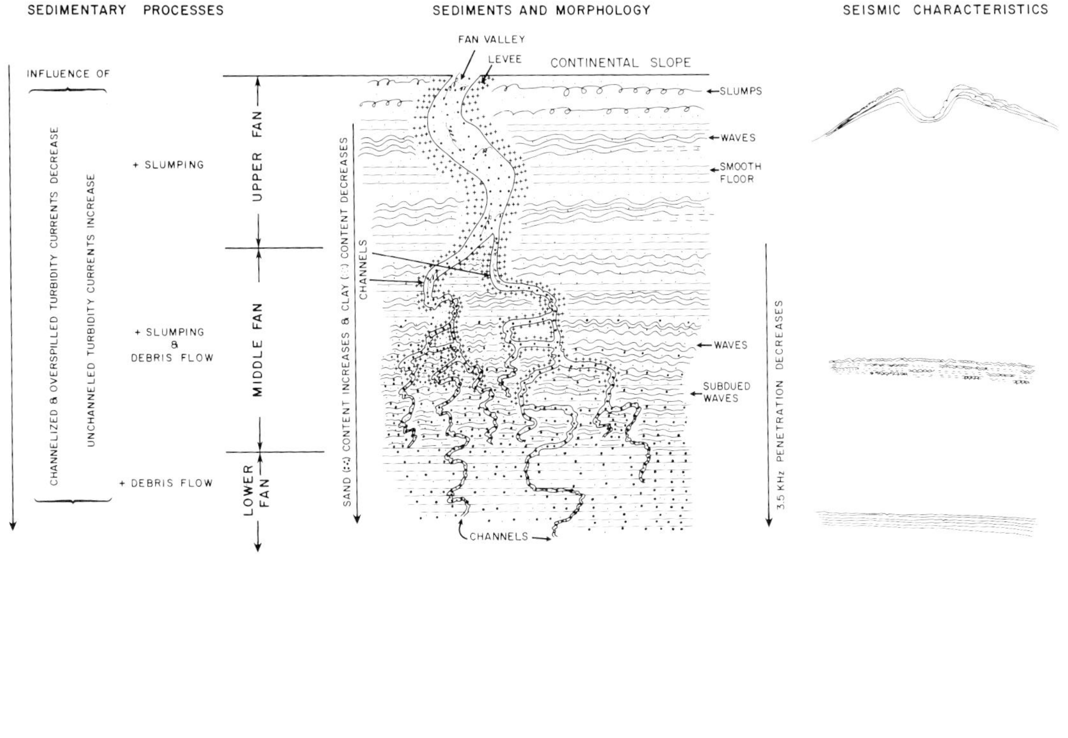

Figure 6. A schematic representation of characteristics and model of processes and evolution of Magdalena fan during one episode of sedimentation.

Table 1. Summary of Characteristics of Magdalena Fan

	Characteristics	Upper Fan	Middle Fan	Lower Fan
	Gradient	1:60 to 1:110	1:110 to 1:200	1:250
	Longitudinal profile and relief	Irregular; usually >25 m	Irregular to smooth; 6–20 m	Smooth; <5 m
	Channels and levees	Channel depth up to >100 m, levee heights up to >100 m. Limited number of channel-(valley) levee complexes	Channel depth 30–40 m or less, levee height 20 m or less; many channels	Channel depth <20 m, levees absent; infrequent occurrence of channels
Acoustics	3.5 kHz	Distinct echoes with continuous subbottom reflectors; regular hyperbolae and sediment waves common	Distinct to indistinct echoes, with continuous to fuzzy subbottom reflections; regular hyperbolae and sediment waves common	Indistinct echoes with few or no subbottom reflectors; poor to no penetration; smooth, flat sea floor
Acoustics	Multichannel seismics	Overlapping or coalescing wedge-shaped levee sequences, channel floors may contain high amplitude, discontinuous reflections and onlap fill sequences	Discontinuous, hyperbolic, and chaotic reflections common: coalescence of small wedge-shaped reflections in the upper middle fan; mounding owing to many small channel-levee complexes	Relatively continuous, flat reflections
Sedimentology	Lithology	Coarse-grained sediments in channels; dominance of fine-grained sediments in other areas	Type of sediment intermediate between the upper and lower fan	Coarse-grained (sand) sediments dominate
Sedimentology	Processes	Channelized and overbank spilled turbidity currents Slumping: from continental slope within channels back sides of levees	Channelized and overbank spilled turbidity currents, unchannelized turbidity currents, slumping and debris flows	Unchannelized turbidity currents; debris flows

slumping processes along broad fronts on the continental slope or in the upper fan regions outside the valley, and the debris flows and turbidity currents that might evolve from such processes, could have also caused the bedforms on the Magdalena Fan. Slumping could also occur on the banks of channels as well as on back sides of oversteepened levees that might generate the bedforms. Mud diapirism, active folding and faulting [3], undercompaction of sediments, and relatively steep gradients are probably responsible for slumping on the Magdalena Fan.

The Magdalena Fan, although located along an active margin, is essentially similar in characteristics and evolution to river-fed fans along passive margins. Although nothing is known about the vertical sedimentary sequences in the Magdalena Fan, in view of the generally increasing sand content in the lower fan, this fan may be classified as a "high-efficiency" deep-sea fan [13]. However, unlike the high-efficiency deep-sea fans of Mutti [13], channels are present, to varying degrees, in the lower Magdalena Fan as well as in several other lower deep-sea fan regions (e.g., Amazon, Indus, and Bengal fans) of the modern ocean fed by river systems. The channels in the lower Magdalena and Amazon Fans lack levees, whereas the ones in the Indus and Bengal Fans have levees.

In seismic studies of deep-sea fans, the middle and lower fan regions have been shown to have mound-shaped morphologies with dip reversals, and hyperbolic, chaotic, and discontinuous internal reflection configurations [14,15]. These fan characteristics have been interpreted to be indicative of sand-prone areas. In the case of Magdalena Fan, the middle-fan region typically has the above seismic characteristics. However, although the sand content is generally more in the middle than in the upper fan, it is in the lower Magdalena Fan that sands dominate.

Acknowledgments

The work was done when the senior author was at Lamont-Doherty Geological Observatory. Support for the work was provided by Office of Naval Research Contract (N0014-80-C-0098) to Lamont-Doherty Geological Observatory and by Industrial Sponsors to Institute for Geophysics, University of Texas. A. Bouma, W. Schweller, and D. Ford critically read the manuscript and made helpful suggestions. Lamont-Doherty Geological Observatory Contribution No. 3645 Institute for Geophysics, University of Texas at Austin Contribution No. 594.

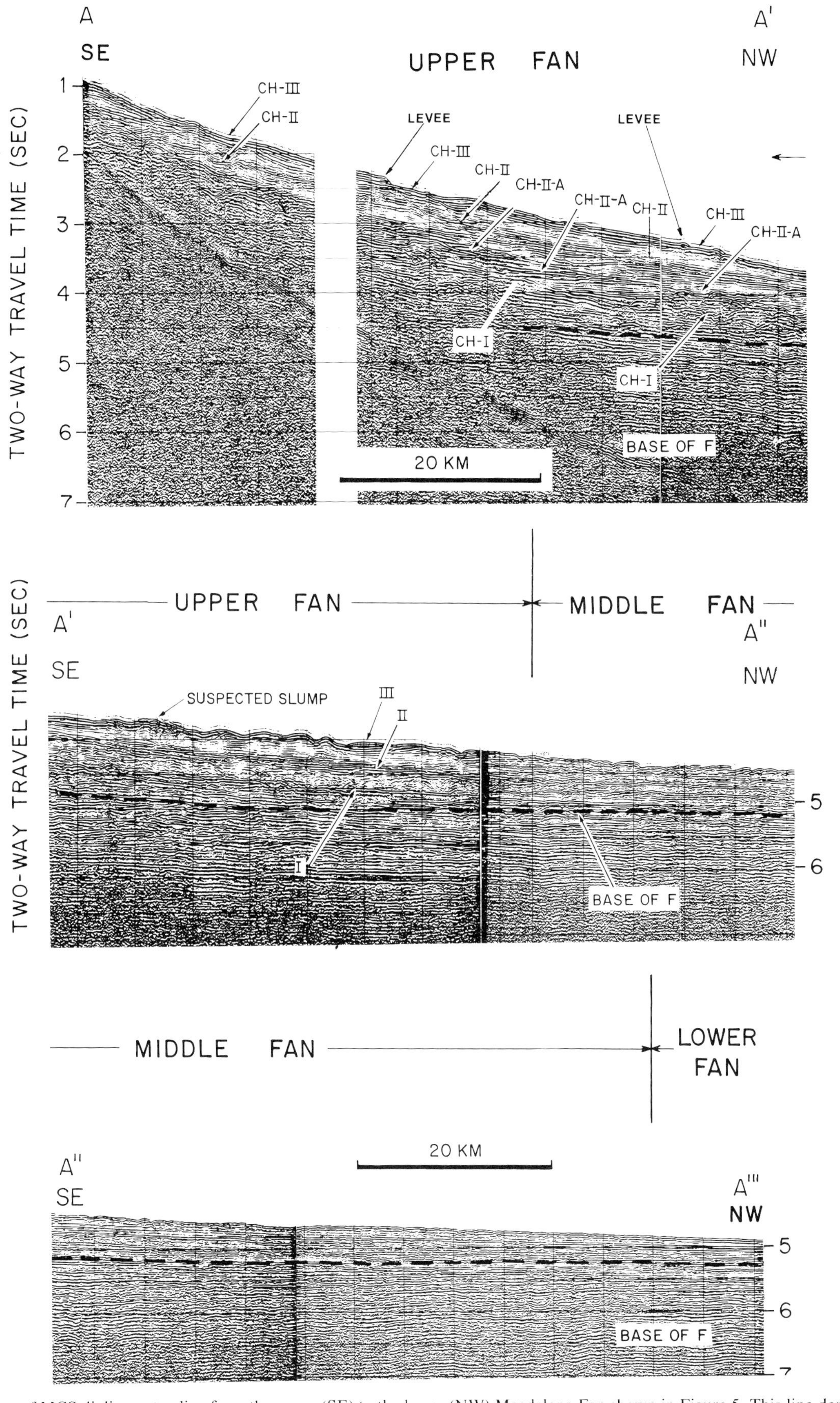

Figure 7. An enlarged version of MCS dipline extending from the upper (SE) to the lower (NW) Magdalena Fan shown in Figure 5. This line depicts seismic characteristics of different regions of the fan, in particular, several periods of channel activity (CH-I, CH-II, CH-IIA, CH-III) corresponding to different episodes of sedimentation (I, II, and III).

Addendum

An enlarged version of the multichannel seismic (MCS) line, presented in Figure 5, is shown in Figure 7 to show more clearly the fan characteristics. This line is split into three segments—the first segment AA′, starting in the southeast, and the two following segments, A′A″ and A″A‴ extending to northwest, are all given on the previous page.

References

[1] Duque-Caro, H., 1979. Major structural elements and evolution of northwestern Columbia. In: J. S. Watkins, L. Montadert, and P. W. Dickenson (eds.), Geological and Geophysical Investigations of Continental Margins. American Association of Petroleum Geologists, Memoir 29, pp. 330–351.

[2] Van Houten, F. B., and Travis, R. B., 1968. Cenozoic deposits, upper Magdalena Valley, Colombia. American Association of Petroleum Geologists Bulletin, v. 52, pp. 675–702.

[3] Shepard, F. P., 1973. Sea floor off Magdalena delta and Santa Marta area, Colombia. Geological Society of America Bulletin, v. 84, pp. 1955–1972.

[4] Heezen, B. C., 1956. Corrientes de turbidez del Rio Magdalena. Boletin de La Sociedad Geografica de Colombia, nomeros 51 y 52, pp. 135–143.

[5] Heezen, B. C., and Munôz, J., 1965. Magdalena turbidites in deep sea sediments. Fourth Caribbean Geological Conference, Trinidad, 1965, p. 342.

[6] Christofferson, E., 1976. Colombian Basin magnetism and Caribbean plate tectonics. Geological Society of America Bulletin, v. 87, pp. 1255–1258.

[7] Kolla, V., and others, 1984. Seismic stratigraphy and sedimentation of the Magdalena Fan, southern Colombian Basin, Caribbean Sea. American Association of Petroleum Geologists Bulletin, v. 68, pp. 316–332.

[8] Normark, W. R., 1978. Fan valleys, channels and depositional lobes on modern submarine fans: characters for recognition of sandy turbidite environments. American Association of Petroleum Geologists Bulletin, v. 62, pp. 912–931.

[9] Damuth, J. E., and Kumar, N., 1975. Amazon cone: morphology, sediments, age and growth pattern. Geological Society of America Bulletin, v. 86, pp. 863–878.

[10] Damuth, J. E., Kolla, V., and others, 1983. Distributary channel meandering and bifurcation patterns on the Amazon deep-sea fan as revealed by long-range side-scan sonar (GLORIA). Geology, v. 11, pp. 94–98.

[11] Prell, W. L., 1978. Upper Quaternary sediments of the Colombia Basin. Geological Society of America Bulletin, v. 89, pp. 1241–1255.

[12] Normark, W. R., and others, 1980. Sediment waves on the Monterey Fan Levee: a preliminary physical interpretation. Marine Geology, v. 37, pp. 1–8.

[13] Mutti, E., 1979. Turbidites et cones sous-marins profonds. In: P. Homewood (ed.), Sédimentation Détritique (Fluviatile, Littorale et Marine). Institut Géologique Université de Fribourg, Switzerland, pp. 353–349.

[14] Sarg, J. F., and Skjold, L. J., 1982. Stratigraphic traps in Paleocene sands in the Balder Area, North Sea. In: M. T. Halbouty (ed.), The Deliberate Search for the Subtle Trap. American Association of Petroleum Geology, Memoir 32, pp. 197–206.

[15] Berg, O. R., 1982. Seismic detection and evaluation of delta and turbidite sequences: their application to exploration for the subtle trap. American Association of Petroleum Geologist Bulletin, v. 66, pp. 1271–1289.

CHAPTER 13

Monterey Fan, Pacific Ocean

William R. Normark, Christina E. Gutmacher, T. E. Chase, and Pat Wilde

Abstract

Monterey Fan is the largest modern fan off the California shore. Two main submarine canyon systems feed it via a complex pattern of fan valleys and channels. The northern Ascension Canyon system is relatively inactive during high sea-level periods. In contrast, Monterey Canyon and its tributaries to the south cut across the shelf and remain active during high sea level. Deposition on the upper fan is controlled primarily by the relative activity within these two canyon systems. Deposition over the rest of the fan is controlled by the oceanic crust topography, resulting in an irregular fan shape and periodic major shifts in the locus of deposition.

Fan Setting and Morphology

The Monterey Fan lies in deep water (3000 to 4700 m) adjacent to the California coast between Monterey Bay and Pt. Arguello, California [1] (Fig. 1). The fan is built on oceanic crust of Oligocene age (based on the magnetic anomaly time scale) that formed along the Pacific–Farallon spreading center, which died off central California by the Late Oligocene [2]. Thus, Monterey Fan probably did not begin to develop until Latest Oligocene or Early Miocene time when turbidity currents from the California margin could first reach the fan area. Based on their character in the single-channel seismic profiles, most of the sediments of the fan appear to be turbidites, but there is no direct method to date onset of fan growth.

Throughout the Neogene, the fan area has been moving northwest relative to the North American plate along faults of the San Andreas system [2]. All of the large fan–valley systems on the fan head within the general area of Monterey Bay, and seismic-reflection profiles from these valleys show that these sediment pathways have persisted throughout most of the fan growth [3] despite the proposed transform fault displacements of the canyon-head areas during the Neogene [4]. This suggests that the slope segments of the canyons have moved with the fans during the Neogene transform motion.

The limits of the fan shown in Figure 1 are based on seafloor morphology and the acoustic character of the sediments interpreted from 3.5-kHz and single-channel seismic-reflection profiles. Available seismic-reflection tracklines are shown in the inset to Figure 1. The solid tracklines indicate data obtained on three cruises between 1977 and 1979 for which we also have 3.5-kHz high-resolution profiles. In addition, 27 cores taken during these cruises provide the basis for the sediment distribution shown in Figure 2.

Upper Fan

The upper-fan division (Fig. 2) is marked by a complicated system of leveed fan valleys from two canyon systems, the Monterey and Ascension. The levee slopes, both away from the channel axis and along the levee crest, are among the steepest anywhere on the fan (generally >5/1000; Fig. 2). In addition, much of the levee surfaces is marked by large fields of sediment waves [5] (Fig. 3A and C).

The leveed valleys on the upper fan indicate a complicated history of development. The only active valley without abrupt changes in axial gradient is the one fed by the modern Monterey Canyon; its axial gradient is relatively constant (3.5 to 4.0/1000) over the 150-km reach below the canyon mouth (Fig. 2). The Ascension fan valley is a hanging tributary with its floor nearly 200-m higher than the Monterey Fan valley at their juncture. The Monterey East valley(s) south of the

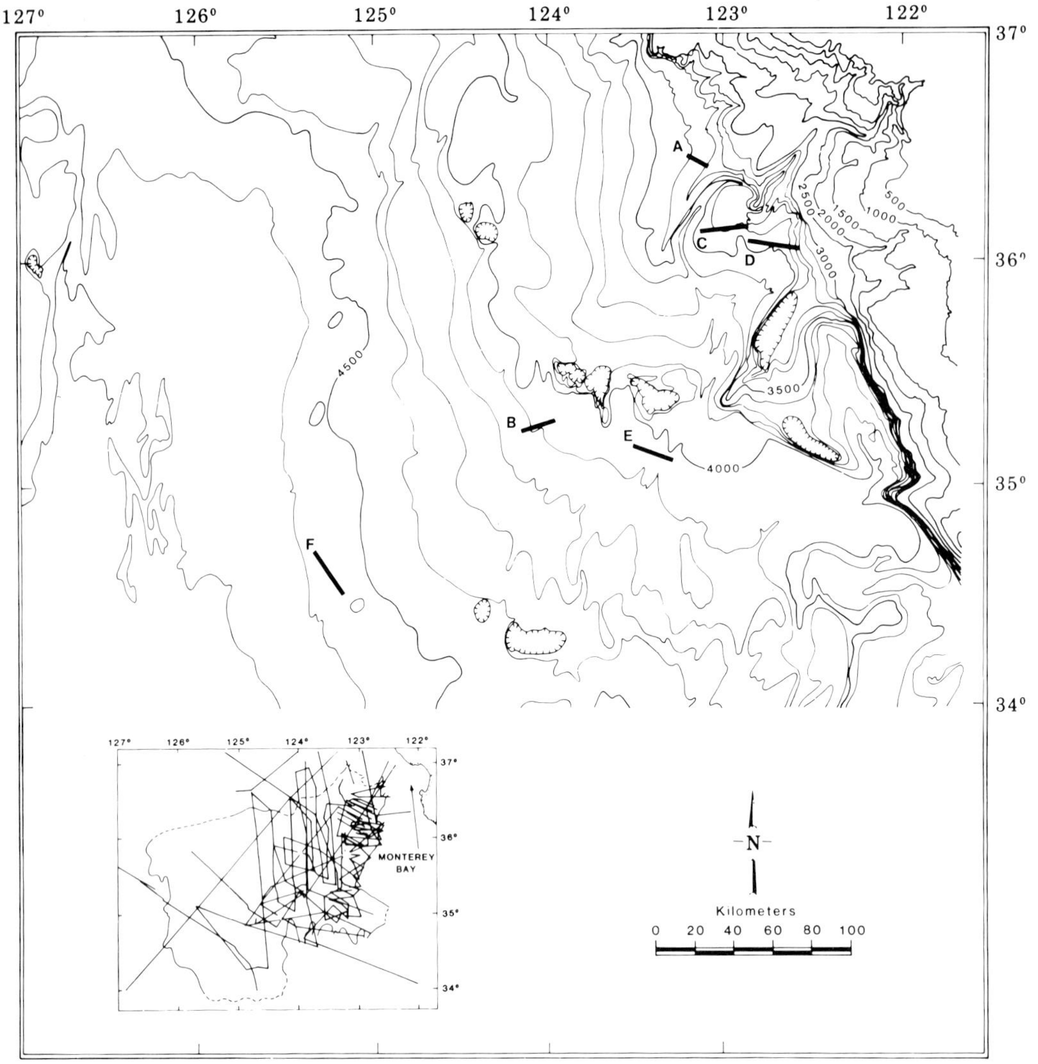

Figure 1. Bathymetric map of Monterey Fan and Canyon systems (from [1]). The outer limit of the fan is determined from interpretation of seismic-reflection profiles and sea-floor morphology. Profile locations for Fig. 3 are shown. Inset shows tracklines for single-channel and 3.5-kHz profiles.

tight meander in the Monterey valley are a hanging distributary system with about the same relief above the Monterey valley (Fig. 2 and 3). Normark [3] and Hess and Normark [6] discuss the evidence for channel piracy and erosional deepening involving the present Monterey Fan valley.

An abandoned and partially buried leveed valley on the western upper fan trends southwest through the present wide levee of the main Monterey Fan valley near latitude 36°N (Fig. 2). The age of this system is not known. In addition, a smaller unnamed valley appears to emerge from under a large submarine slide on the eastern side of the upper fan that lies adjacent to the base of the continental slope between around 35°30′ to 36°N latitude (Fig. 2). This small valley appears to join the Monterey valley immediately north of the Chumash Fracture Zone. The 3.5-kHz profile of Figure 3D shows the fan surface dipping eastward under the slide debris. This dipping surface may be the western margin of the buried channel. Farther north, a deep-tow profile crosses this channel where it lies immediately west of the slide (Fig. 2 in [6]).

The submarine slide covers more than 900 km^2 and forms a major olistostrome on the upper-fan surface. A series of gravity (2- to 3-m long) and piston (6- to 8-m long) cores show that this slide unit consists in part of mud clasts from sediment originally deposited on the upper slope (J. C. Ingle, oral communication, 1979) and of contorted turbidite units presumably from the fan surface. High-resolution profiles across the slide unit show a characteristic hummocky surface with no subbottom structure. On profile D (Fig. 3), both the lower continental slope surface and the fan surface on the west can be traced under the edges of the slide.

Middle Fan

The middle-fan segment on Monterey Fan appears to consist of at least three distinct parts: (1) the western middle fan,

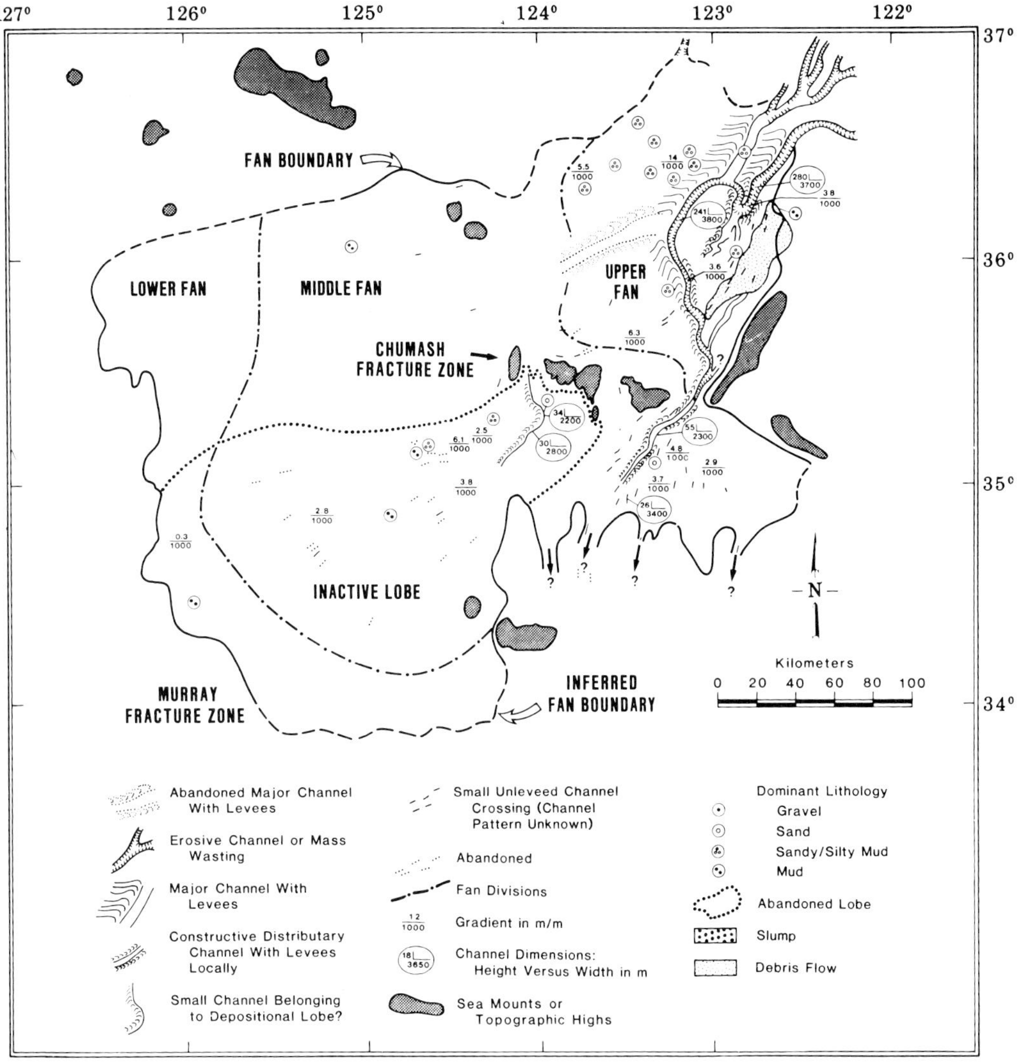

Figure 2. Schematic map of the Monterey Fan showing fan divisions, gradients, channel pattern and dimensions, and dominant lithology at every core site where core is longer than 1 m.

which lies north of latitude 35°N and is not fed by any main valley; (2) the abandoned southwestern part of the middle fan, which is lobate in shape; and (3) the active southern part of the middle fan, which is, at present, the smallest part, yet the most extensive area of sand deposition on the fan.

The southern part of the middle fan is fed directly by the main Monterey Fan valley, which may continue with small levees south of latitude 35°N. Many small, unleveed channels are seen in profiles across this part of the middle fan. The 3.5-kHz reflection profiles across most of this part of the fan record no subbottoms and show a strong, diffuse surface reflection (Fig. 3E) suggesting a sand-rich environment [7,8]. Limited core samples are sandy in this area [6]. Thus, this sector of the middle fan is the active depositional lobe fed by the main valley.

A southwestern lobe area on the middle fan immediately west of the presently active lobe has many of these same acoustic characters on 3.5-kHz profiles except that much of this area is blanketed by a 6- to 10-m thick acoustically transparent layer (Fig. 3F). Channels, and other relief seen on a highly reflective surface with few subbottoms, have been preserved during deposition of this transparent layer. As is the case for the presently active lobe to the east, no channel pattern has been recognized yet and this area appears to be fed by one small, leveed valley that extends from the pass through the Chumash Fracture Zone (Figs. 2 and 3B). The acoustically transparent layer covering most of the lobe thickens eastward into the levee of this small valley, and several internal reflectors are present (Fig. 3B), suggesting that limited turbidite deposition continues on the apex of this lobe. In general, however, this area of the middle fan appears to be the previously active sandy depositional lobe on Monterey Fan.

The short leveed valley at the head of this lobe does not connect (either on the surface or in the subbottom) with any valley on the upper fan. It appears to be a valley formed by

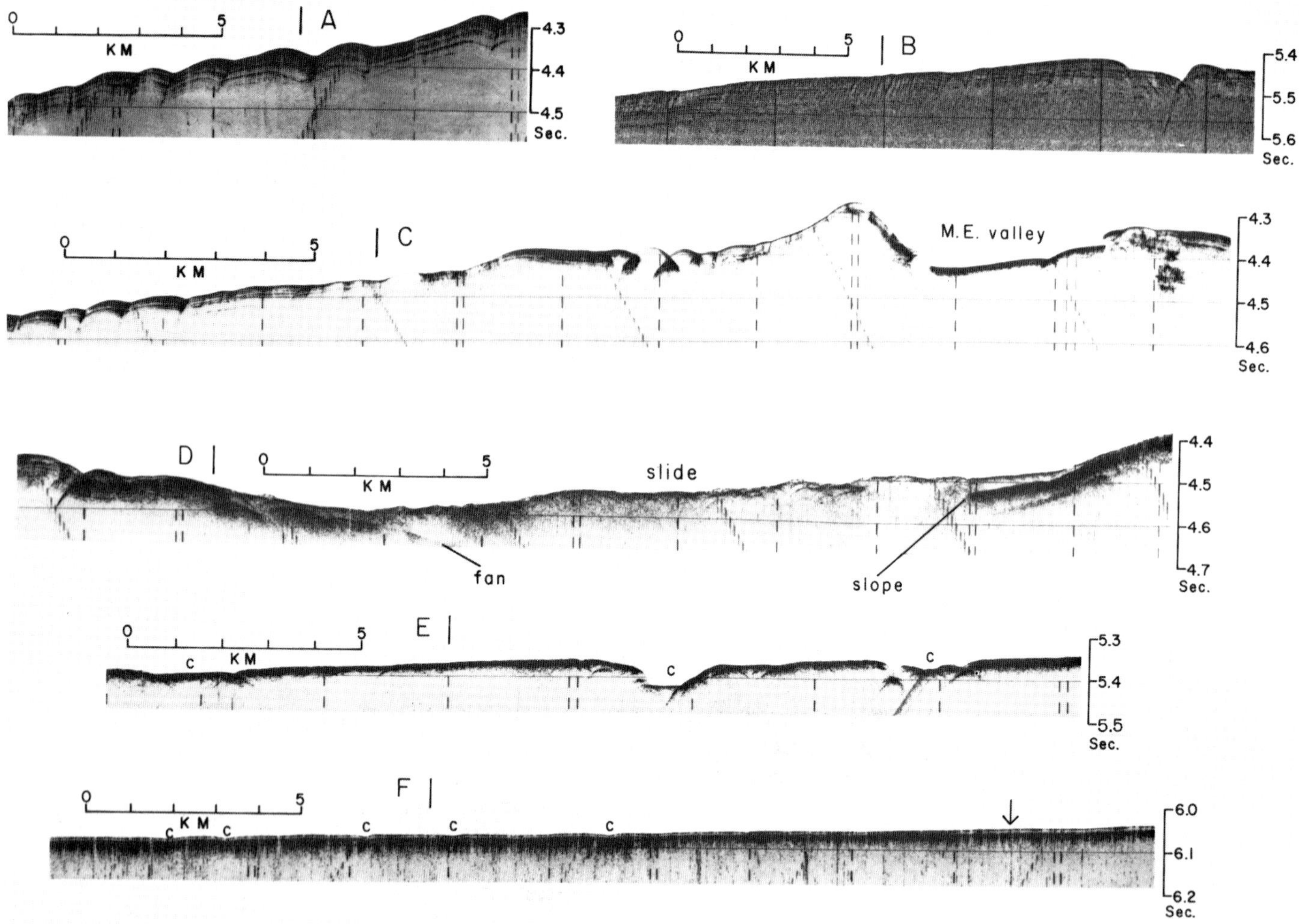

Figure 3. High-resolution 3.5-kHz profiles from selected depositional environments on Monterey Fan. Locations given in Fig 1. Small scale bar shows 50 m of sea-floor relief. (A) Sediment waves on righthand levee on Ascension valley (see [5]). (B) Leveed valley on middle fan formed by turbidity currents flowing through Chumash Fracture Zone. Righthand (western) levee with small sediment waves is much larger than narrow (2 km wide) eastern levee. (C) Monterey East Valley [3] with irregular western levee including sediment waves at left end of profile (under scale). (D) Profile across large submarine slide showing characteristic hummocky surface and lack of internal structure. At its edges, the slide onlaps the pre-existing fan surface (west) and lower slope sediments (east). (E) Channels (c) with small or no levees on presently active midfan lobe south of Chumash Fracture Zone. Lack of subbottom reflectors indicates sandy sediment. (F) Abandoned middle-fan area with acoustically transparent unit (best seen at arrow) overlying strong diffuse reflector; relief of small channel (c) features is preserved during deposition of transparent unit.

rechannelization of overbank turbidity current flow from the main Monterey valley as flow is constrained through the pass in the fracture zone ridge. Even in this small leveed system, nevertheless, sediment waves distinguish the larger righthand levee (Fig. 3B).

The western part of the middle fan is relatively free of channels (Fig. 2), and the reflection profiles suggest that this may be the least sandy of the middle-fan segments. Apparently, most of the sediment has come from overflow of turbidity currents moving through the main Monterey valley because there are few channel segments to act as conduits.

Lower Fan

The lower-fan segment is the smallest of the major subdivisions. It represents channel-free, flat-lying turbidites that extend from the sloping middle fan into the valleys between abyssal hills. Gradients show a distinct break in slope over a distance of several kilometers between the middle and lower fan, which is the equivalent of a ponded basin plain (Fig. 2). One of the more striking features of Monterey Fan is that the presently active lobe to the south does not merge with a lower fan or basin–plain unit. In this area, the turbidity current deposition has not established smooth base level, and the turbidity currents continue to flow south along inherited valleys between abyssal hills in areas shown by arrows (Figs. 1 and 2).

Basement Relief

There is a sufficient density of single-channel seismic-reflection profiles from Monterey Fan to construct a basement

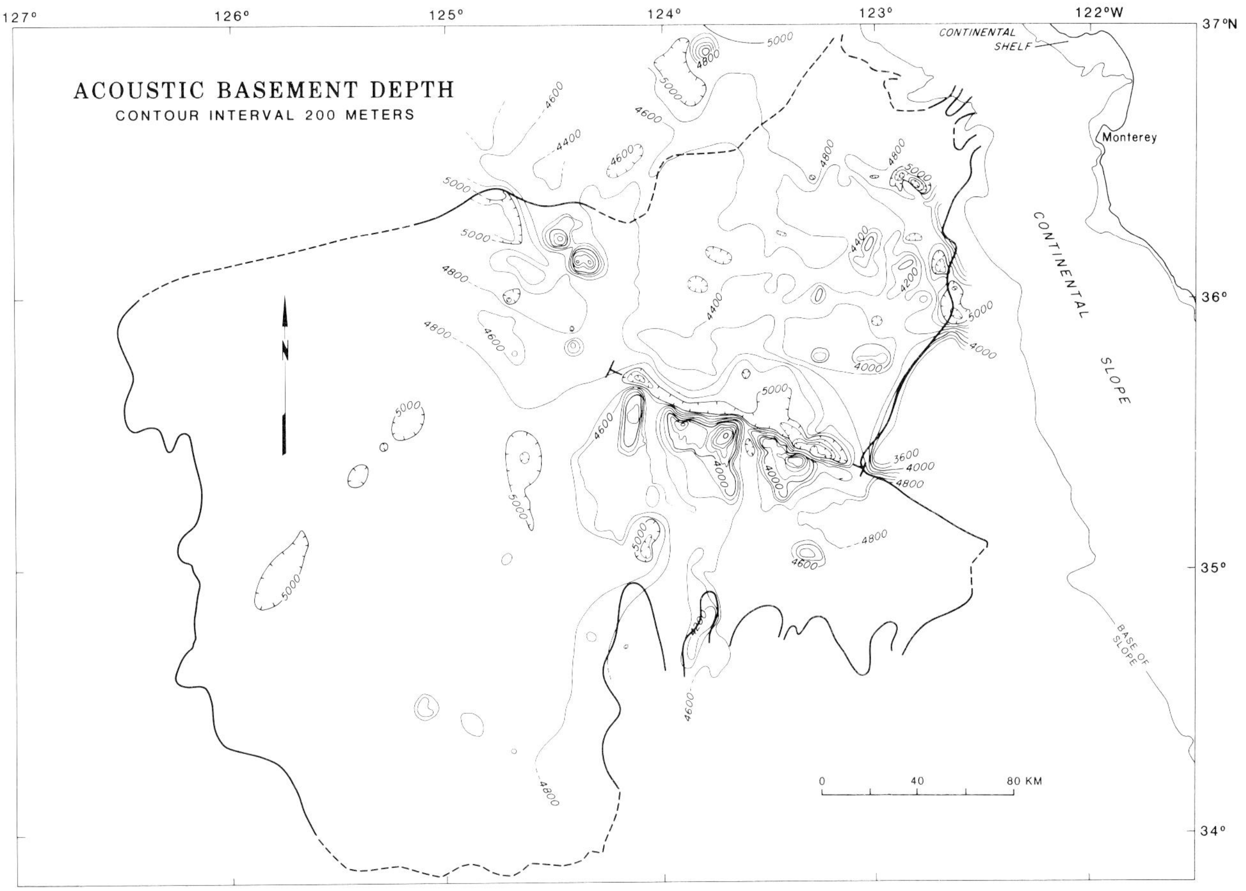

Figure 4. Basement topography under Monterey Fan. Depth to basement corrected for both sea-water sound velocity (using Matthews' tables, 1939) and sound velocity within sediments using velocity profile based on interval velocities from multichannel seismic profile across upper fan (D. McCulloch, written personal communication). Fan outline and divisions as shown in Figure 1. T-brackets delimit Chumash Fracture Zone.

contour map for the middle and upper fan (Fig. 4). The sea floor under the fan shows many seamounts and volcanic ridges, some of which extend well above the fan surface (Fig. 1). The fan is far from being a simple cone- or wedge-shaped body of sediment.

The major basement feature that has affected fan growth is the Chumash Fracture Zone (Fig. 2). A deep trough on the north side of the ridge contains up to 2 km of sediment fill (Fig. 4). This large trough had to be filled with sediment before either the southwestern or southern, active, lobes on the fan could begin to form. Menard [9] suggested that the volcanic ridge would act as a dam to the turbidity currents, and this damming effect probably continued long after the trough was filled. Eventually, turbidity currents flowed around the west end of the ridge (west of 124°W longitude) and began to build the southwestern (but presently abandoned) lobe. As turbidites continued to pond behind the ridge itself, a pass 20 km to the east (the feeder for the headless leveed valley) became a more direct route for turbidity currents flowing to the southwest. Sometime later, a newer pass was breached near 123° longitude that brought fan sediments to the presently active eastern part of the fan. The southern lobe area is deeper than the other parts of the middle fan, and headward erosion from the Chumash Fracture Zone pass has resulted in extensive downcutting of the Monterey Fan valley. This downcutting helped cut off sediment supply to the southwestern part of the middle fan.

Fan History

The long levee feature formed by the western levee of the Ascension valley and its continuation along the main Monterey valley south of their confluence appears to be one coherent feature at least to latitude 36°15′N [3]. This original leveed valley fed by Ascension Canyon has been present for most of the growth of the upper fan [3]. The Monterey Canyon may have fed a smaller valley during much of this time (Monterey East, Fig. 2), but it does appear that initially the Ascension fan valley was the main conduit for sediment (Fig. 5A). In time, the Ascension system began to bring sediment to the southwestern sector of the fan, perhaps through the now partially buried leveed valley (Fig. 5B). The Monterey East became a more important conduit as well and brought sediment directly to the Chumash Fracture Zone trough (Fig. 5B). The presence of the large western (righthand) levee on the Monterey East valley (Fig. 3C) suggests, moreover, that

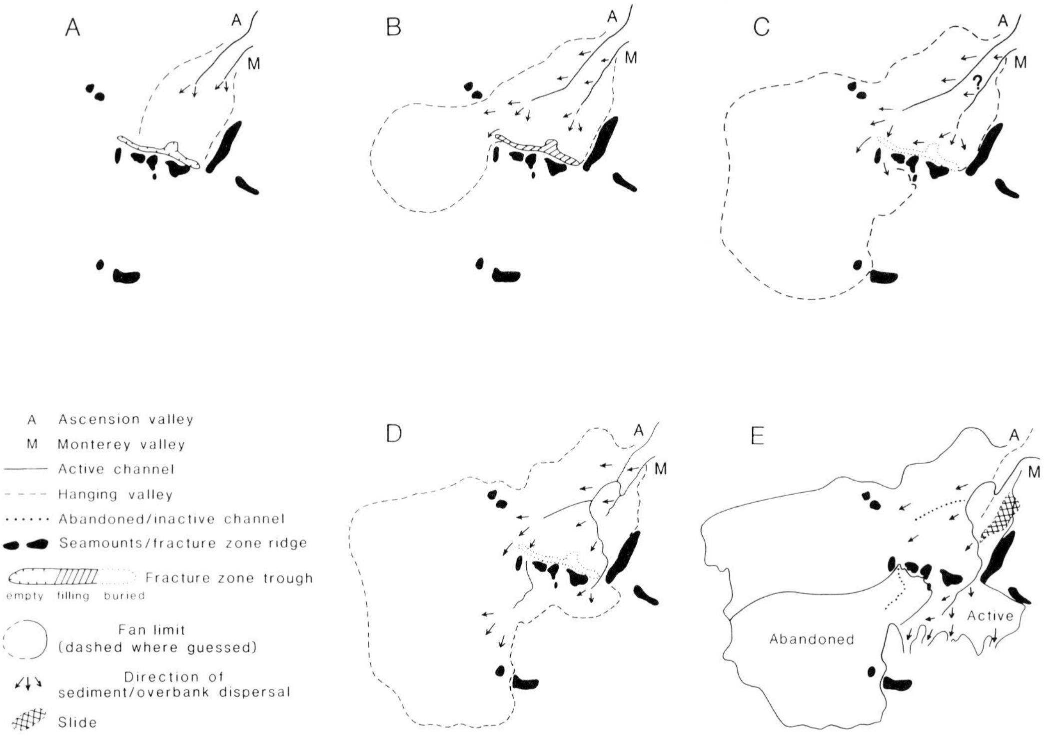

Figure 5. Growth pattern of Monterey Fan. Schematic geometry is not tied to time scale; only sequence can be determined. See text for explanation.

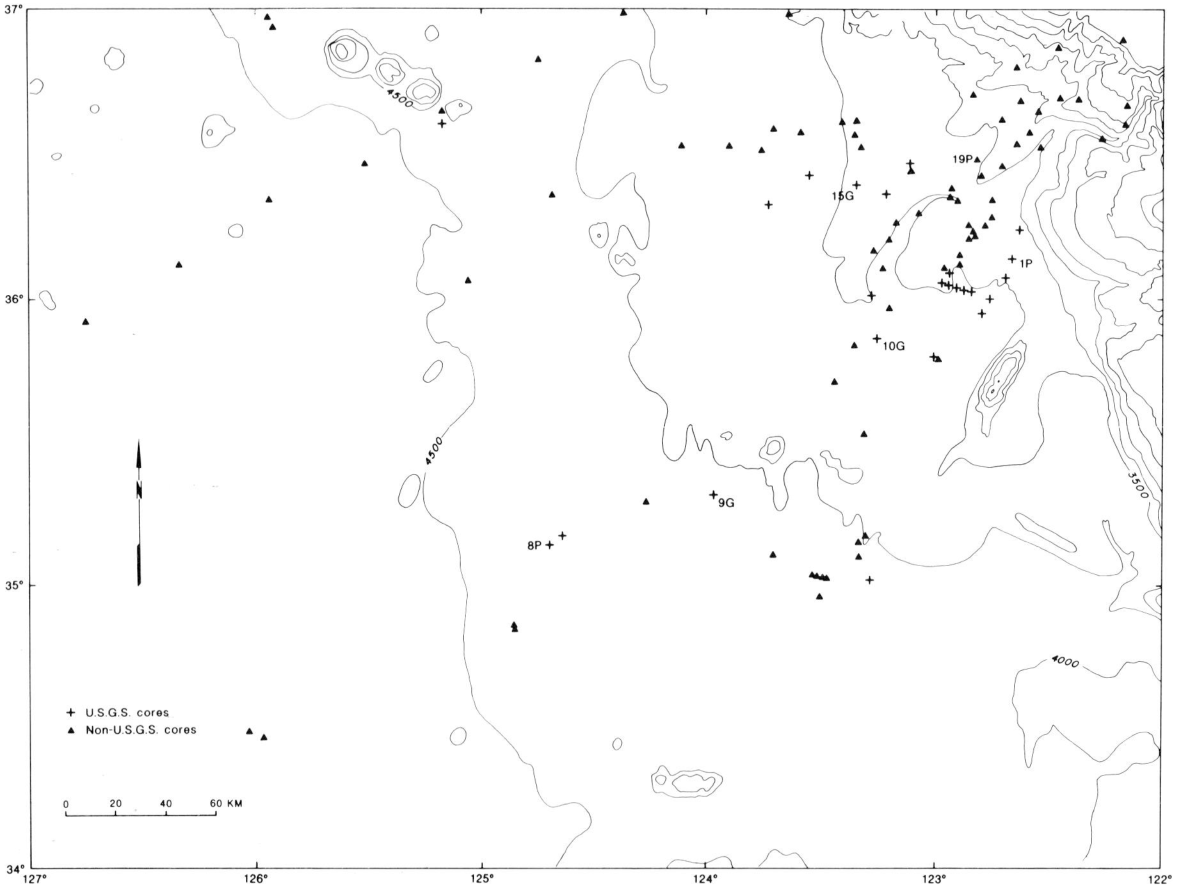

Figure 6. Core locations for Monterey Fan. Contour interval is 500 m. Numbered cores indicate those illustrated in Figure 7. G = gravity core; P = piston core.

much of the western overbank flow from the smaller Monterey valley would have been captured by the Ascension system, thus increasing the relative prominence of the latter leveed-valley complex.

The western part of the middle and lower fan (Fig. 2) has probably grown continuously by overbank deposition throughout the fan development. This area continued to receive substantial overbank sedimentation while the southwestern lobe was forming from turbidity currents moving around and through the west end of the fracture zone ridge after the Chumash trough was filled (Fig. 5C).

Slightly before or about the time that the latest (eastern) pass in the ridge was formed, the valley from Monterey Canyon had pirated the lower part of the Ascension system (Fig. 5D). We do not know the cause of this major change in the valley pattern, but failure of the common levee by slumping or erosion from a sinuous thalweg channel in either valley is possible. After the eastern pass became the main conduit, headward erosion began in the new valley system. This erosion probably accounts for 200 to 300 m of the valley relief in the area near the meander [3]. The present lack of a local base level for this southern lobe acts to maintain downcutting within the modern Monterey valley, and the lobe is growing slowly as sediment escapes to the south (Fig. 5E). The west-trending partially buried leveed valley may have been active until downcutting occurred. Both the western and southwestern parts of the fan now receive sediment only when large turbidity currents overflow the levees on the upper fan.

The main source of sediment for Monterey Fan now shifts between the Ascension and Monterey Canyons. The multiheaded Ascension Canyon system is, at least at present, confined to the upper slope and outer shelf and does not receive much coarse sediment during high stands of sea level. The Monterey Canyon, however, cuts across the shelf and remains active regardless of relative sea level. The prominence of the Ascension leveed-valley complex suggests that fluctuating sea levels did not influence fan growth initially but may have become more important in the Late Neogene. Shifting activity between the two canyon systems may have contributed to the channel piracy event.

One of the most recent events on the fan is the development of the large slide on the upper fan (Fig. 5E). This slide appears to bury an older valley, and it is overlain in our cores by only one turbidite unit. This turbidite may have been generated by the slide itself.

The brief fan history presented here is tentative because analysis of the seismic-reflection profiles and core data continues. The most critical areas for study include a better understanding of the facies changes between the middle and upper fan, determination of the channel pattern on the active middle fan, and placing time constraints on the major events described above.

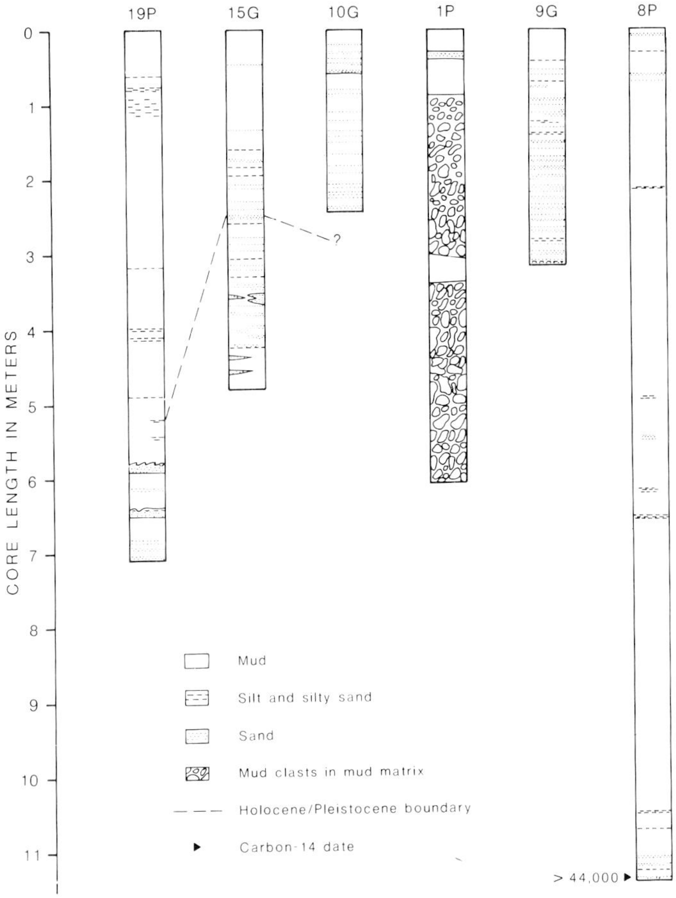

Figure 7. Lithologic sections for selected cores from the Monterey Fan. Holocene/Pleistocene boundary based on micropaleontological examination (C. Brunner, written personal communication, 1984). G = gravity core; P = piston core.

Addendum

The distribution of surficial sediment on Monterey Fan (Fig. 2) is based on 27 cores obtained during U.S. Geological Survey cruises between 1977 and 1979. These core locations, together with samples previously collected primarily on cruises from Scripps Institution of Oceanography [6], are shown in Figure 6. Representative core logs from selected depositional environments on Monterey Fan (Fig. 7) include one from the floor of Ascension fan valley (19P), two from the western levee of Monterey valley (15G, 10G), one from the large submarine slide on the eastern side of the fan (1P), and two from the channel system leading to the southwestern (inactive) lobe of the fan (9G, 8P). Analyses of the other cores from the fan support the general observations illustrated by these six examples. Muddy turbidity currents have deposited more than 5 m of sediment on the floor of Ascension Fan valley during the Holocene, while silty and sandy turbidity currents have deposited about half this amount on the levee of Monterey valley.

Overflow of turbidity currents is more frequent on the lower levee crest (10G) where the channel floor to levee relief is about 200 m less than in the area near core 15G.

The floor of the overflow valley south of the Chumash Fracture Zone is underlain by abundant thin-bedded units of sandy silts and fine sand (9G). Farther southwest on the lobe fed by this channel, only a few silt and sandy silt beds have been deposited during the last Pleistocene and Holocene (8P). Core 8P supports the interpretation of the high-resolution profiles that suggests relative inactivity on the southwestern part of Monterey Fan.

References

[1] Chase, T. E., Normark, W. R., and Wilde, P., 1975. Oceanographic data of the Monterey deep-sea fan. Institute of Marine Resources Technical Report Series TR-58.

[2] Atwater, T., 1970. Implications of plate tectonics for the Cenozoic tectonic evolution of western North America. Geological Society of America Bulletin, v. 81, pp. 3513–3536.

[3] Normark, W. R., 1970. Channel piracy on Monterey deep-sea fan. Deep-Sea Research, v. 17, pp. 837–846.

[4] Greene, H. G., 1977. Geology of the Monterey Bay region. U.S. Geological Survey Open-File Report 77-718.

[5] Normark, W. R., and others, 1980. Sediment waves on the Monterey Fan levee: a preliminary physical interpretation. Marine Geology, v. 37, pp. 1–18.

[6] Hess, G. R., and Normark, W. R., 1976. Holocene sedimentation history of the major fan valleys of Monterey fan. Marine Geology, v. 22, pp. 233–251.

[7] Normark, W. R., Piper, D. J. W., and Hess, G. R., 1979. Distributary channels, sand lobes, and mesotopography of Navy Submarine Fan, California Borderland, with applications to ancient fan sediments. Sedimentology, v. 26, pp. 749–774.

[8] Damuth, J. E., 1975. Echo character of the western equatorial Atlantic floor and its relationship to the dispersal and distribution of terrigenous sediments. Marine Geology, v. 18, pp. 17–45.

[9] Menard, H. W., 1955. Deep-sea channels, topography, and sedimentation. American Association of Petroleum Geologists Bulletin, v. 38, pp. 236–255.

CHAPTER 14

Navy Fan, Pacific Ocean

William R. Normark and David J. W. Piper

Abstract

Navy Fan is a Late Pleistocene sand-rich fan prograding into an irregularly shaped basin in the southern California Borderland. The middle fan, characterized by one active and two abandoned "distributary" channels and associated lobe deposits, at present onlaps part of the basin slope directly opposite from the upper-fan valley, thus dividing the lower-fan/basin-plain regions into two separate parts of different depths. Fine-scale mesotopographic relief on the fan surface and correlation of individual turbidite beds through nearly 40 cores on the middle and lower fan provide data for evaluating the Late Pleistocene and Holocene depositional processes.

Introduction

Navy Fan occupies approximately 560 km^2 within two offset parts of South San Clemente Basin, California Borderland. The basin has, at least in part, been formed by right-lateral movement during the Neogene on transform shear zones within the continental crust of the North American plate [1,2]. Navy Fan began to form in the Late Pleistocene and overlies slightly earlier Pleistocene(?) turbidite sediment that has been gently folded [2,3].

Navy Fan does not have a classic radial or cone shape because of the irregular form of the tectonically active basin (Fig. 1). Its upper fan is partially confined behind a northwest-trending, fault-bounded spur ridge that plunges beneath the middle-fan area. The fan sediments have prograded into an elongate northwest-trending trough and into an irregularly shaped part of South San Clemente Basin to the southwest; thus, the fan has two separate basin-plain areas differing in depth by about 30 m. Nevertheless, the relatively young fan does display the morphometric divisions that commonly develop on sand-rich submarine fans. Upper-, middle-, and lower-fan divisions, each with distinct morphologic characters, can be subdivided into two parts based on local relief and/or surface gradients. The source of sediments for the fan is the narrow, rock-bounded gorge that funnels sediment across the narrow sill from San Diego Trough. The coarse sediment (up to gravel grade) on Navy Fan is moved through Coronado Canyon from the Tijuana River, which drains volcanic and granitic terrane of the northern Baja California peninsula. The drowned Tijuana River bed extends to the head of the Coronado Canyon, which is a zone of active coarse sediment transport only during periods of low sea level [3,4].

Navy Fan is probably the most thoroughly studied submarine fan in the world (based on number of cores and line-kilometers of geophysical data per unit area of fan surface). We have surveyed Navy Fan during seven different cruises between 1968 and 1978. In addition to single-channel and high-resolution reflection profiles over most of the area, we have utilized near-bottom deep-tow instruments and have collected more than 100 cores. This review covers only a few selected topics dealing with morphology, acoustic facies, sediment distribution, and depositional processes. For more complete data presentations, several references are available [3,5,6,7].

Morphology

The morphometric divisions for Navy Fan presented in Figure 1 are based on distinct changes observed in gradients and local surface-relief features from the head of the fan to the basin plain. The middle-fan and immediately adjacent areas are known in great detail because this area was surveyed with

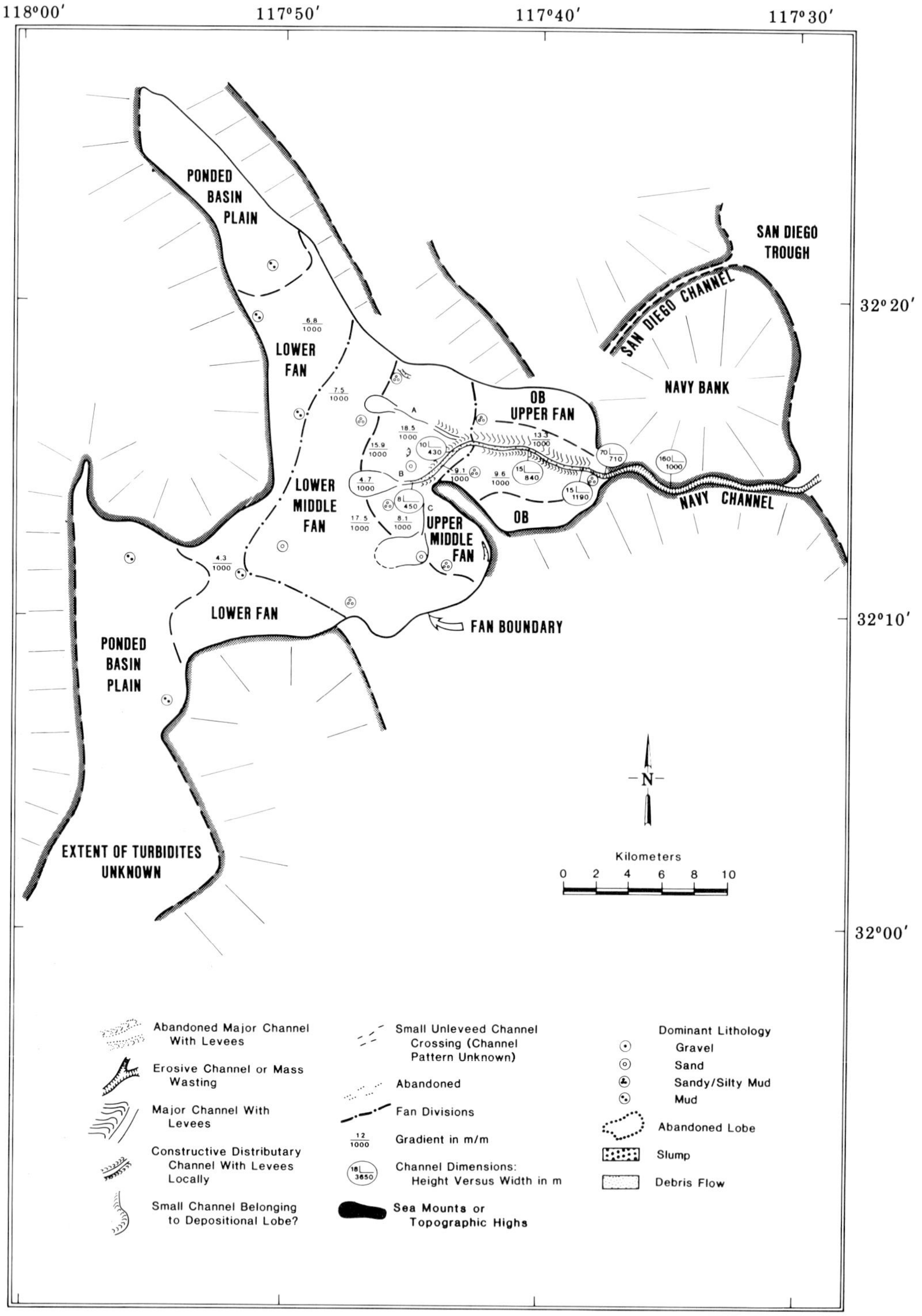

Figure 1. Schematic representation of Navy Fan showing morphologic divisions, channel pattern, channel dimensions, fan-surface gradients, dominant lithologies, and basin configuration.

the deep-tow instrument package [5,8]. In addition to narrow-beam echo-sounding profiles, high-resolution reflection profiles, and color photographs of the sea floor, the deep-tow survey provided nearly complete areal coverage with side-looking sonar images. The deep-tow instrument was also used to examine the transition from Navy Channel, which is the conduit for sediment moving from San Diego Trough to Navy Fan, to the valley on the upper fan [3]. Elsewhere on the fan, the morphology and acoustic facies interpretations are based on surface-ship 3.5-kHz profiles after correlation with the deep-tow profiles in the middle-fan area.

The bathymetric map (Fig. 2) combines available deep-tow and conventional surface-ship soundings and covers all of the fan except the outermost basin-plain regions.

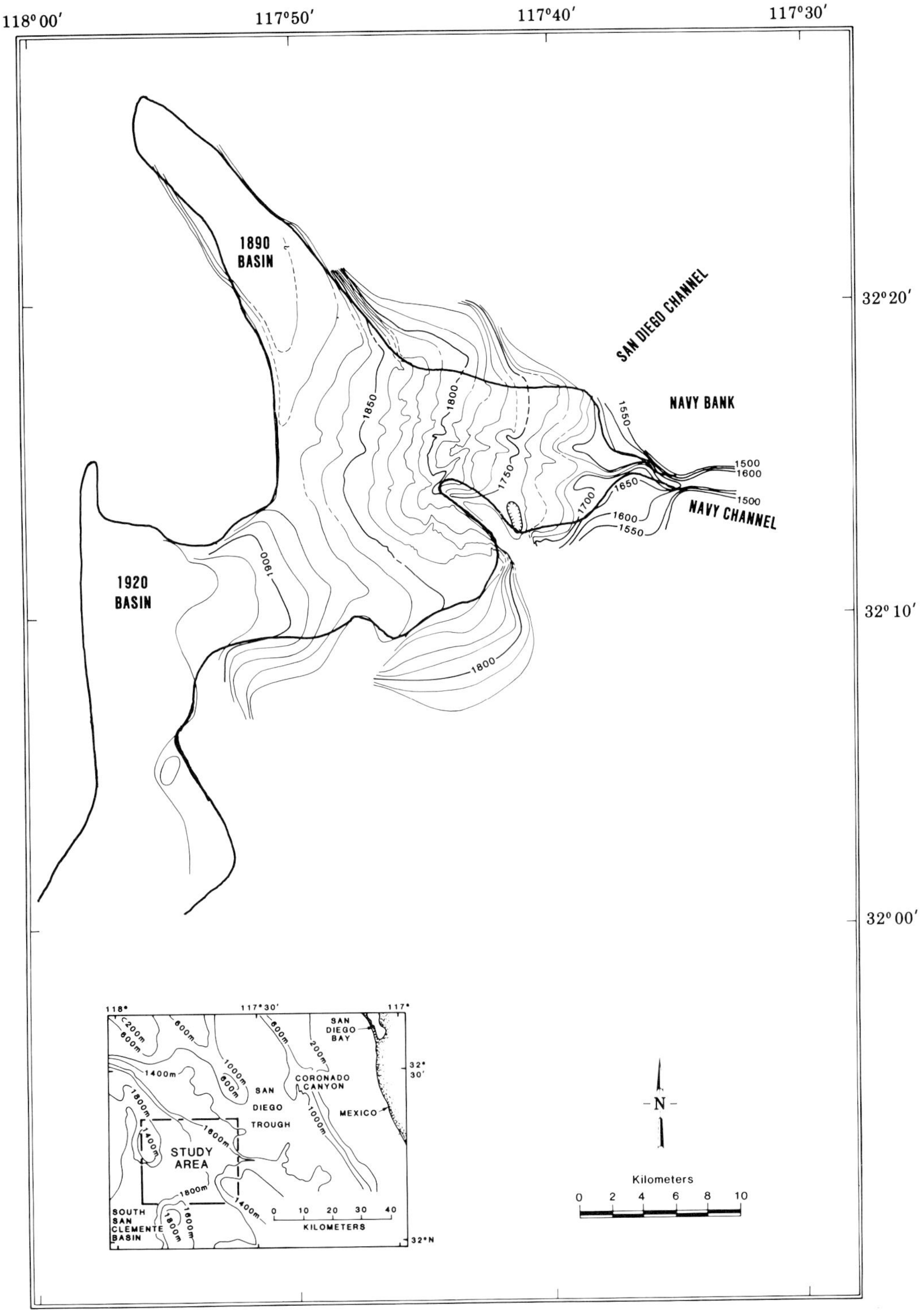

Figure 2. Bathymetric map of Navy Fan, combining deep-tow soundings (central area) and conventional shipboard 12-kHz soundings. Data from [3] and [5]. Inset shows location of Navy Fan study area.

Upper Fan

The valley on the upper fan begins as an erosional, unleveed, sinuous feature that changes to a wider, shallower, relatively straight, leveed valley within a few kilometers (Fig. 2). The leveed-valley complex is 4- to- 5-km wide and occupies about half of the upper-fan area, which is confined to a narrow part of the basin. The schematic map of the fan (Fig. 1) distinguishes the areas with recognizable levee relief from flat-lying areas closer to the basin walls that are termed overbank deposits. Turbidity currents overtopping the levee crests deposit sediment in both areas, but the morphologic term "levee" is restricted to the areas with topographic expression of the overbank deposition. Little is known about the mesotopo-

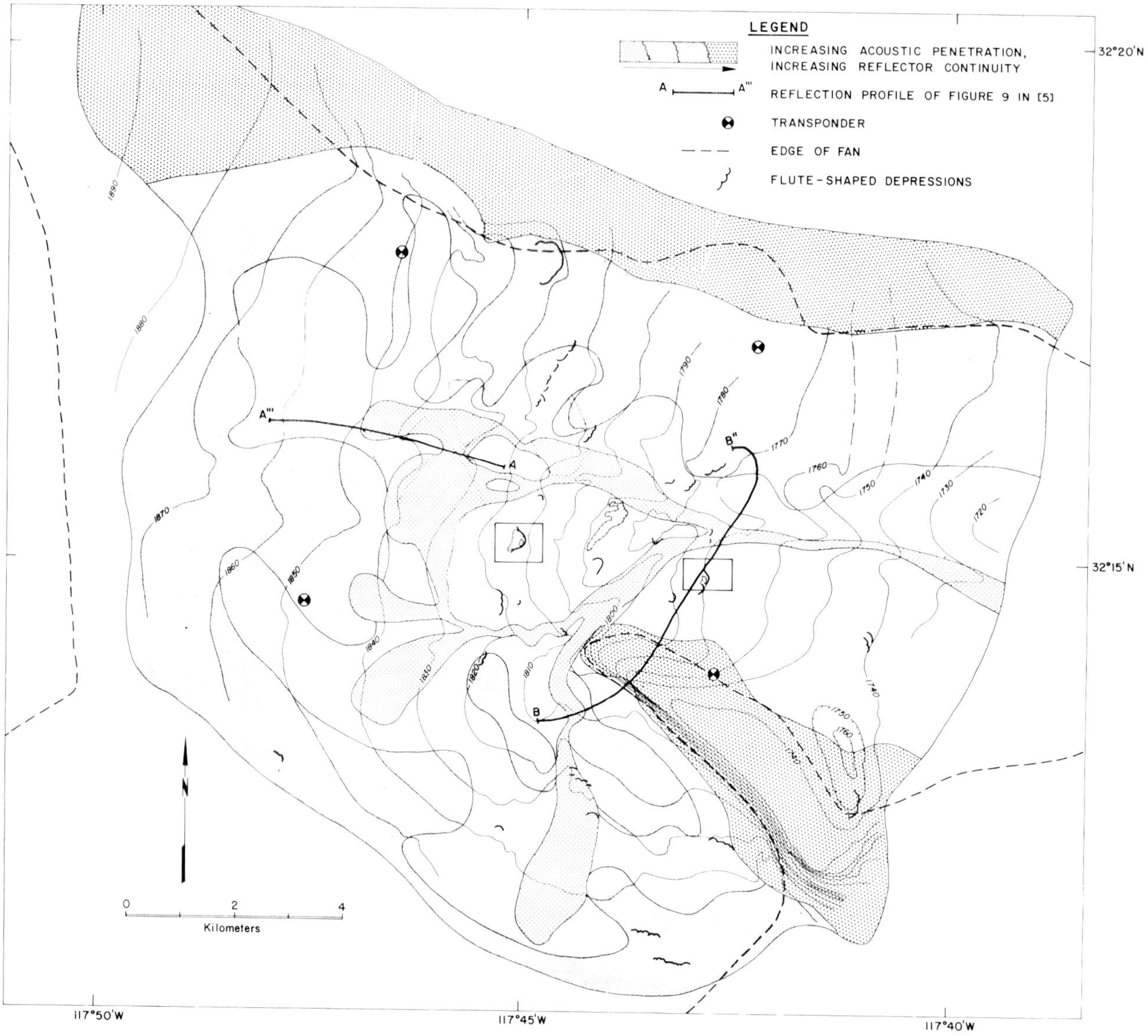

Figure 3. Acoustic facies distribution on Navy Fan based on character of 4-kHz high-resolution profiles taken with deep-tow instrument (from Fig. 7, [5]). Each pattern can be considered as a type of acoustic facies; see text for discussion.

graphic relief in this area because of limited surveying with deep-tow instrumentation. Mesotopography includes features larger than 2 to 3 m in vertical relief and 10 or 20 m across but too small to be adequately resolved with conventional surface-ship profiling systems. On Navy Fan, this means that most features less than 500 m across will probably not be correctly identified. Mesotopography is best mapped with deep-tow side-looking sonar and sounding profiles. The only deep-tow profile across the northern levee and valley floor shows numerous irregular, and commonly steep-sided depressions with depths of 4 to 8 m and 50 to 100 m in width. Surface-ship profiles cannot resolve these features (Fig. 4 in [9]).

Middle Fan

Because of the extensive deep-tow coverage, the middle fan can be characterized by several distinct morphologic characteristics. The fan valley from the upper fan appears to branch into three "distributary" channels with poorly developed levees only along their upper reaches (Figs. 1 and 2). These channels lead to smooth depositional lobes 2 to 3 km in length. No channels are observed on the lobes or even on the narrow-beam deep-tow sounding profiles. Only one of these channel/lobe systems, however, is joined to the upper-fan valley as an active distributary with a continuous downslope gradient (channel B, Fig. 1). The other two channels are sep-

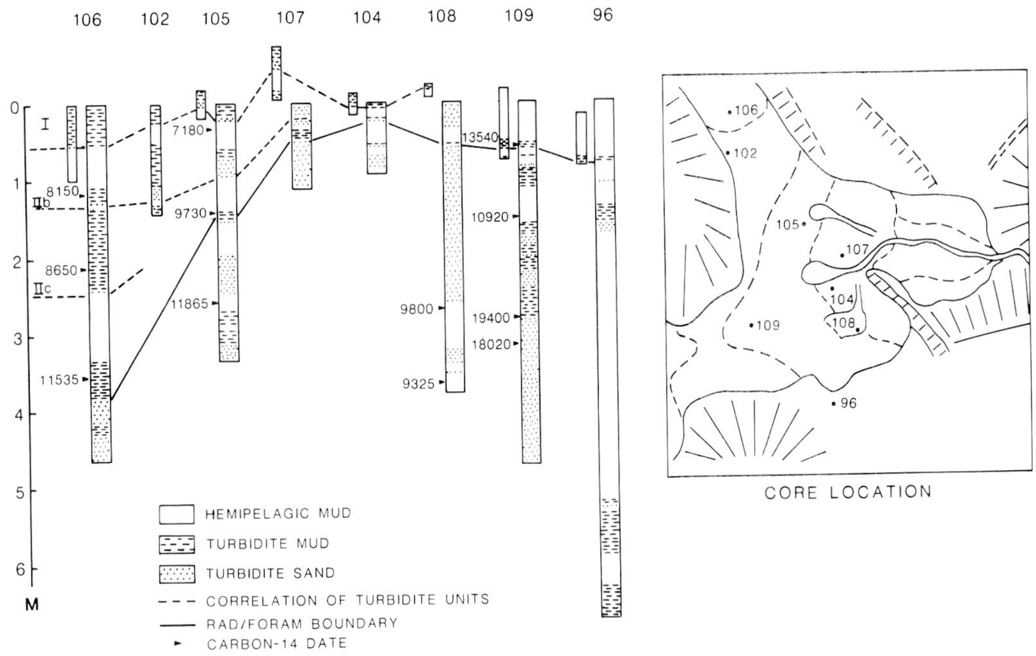

Figure 4. Sediment sequences and stratigraphy of piston cores from Navy Fan showing radiolarian marker (approximately the Holocene–Pleistocene boundary) and correlation of turbidite units. Core locations shown here on simplified version of schematic map (Fig. 1).

arated from the active distributary by low(?) levee deposits. These inactive channels can still receive overflowing turbidity currents, but they have been (or are being) cut off by progressive abandonment as each channel/lobe system aggrades and deposition is shifted to adjacent low areas [5].

The upper part of the middle fan, which includes the area between the upper-fan valley and the ends of the depositional lobes, is characterized by extensive mesotopographic surface relief. Little is known of the internal structure associated with these mesotopographic features because the coarseness of the sediment prevents acoustic penetration with high-resolution profiling systems. Thus, the flow conditions responsible for much of this mesotopography are not understood.

One common type of recognizable mesotopography is represented by large scour depressions that occur singly and in groups (Fig. 3). The largest of these flute-shaped depressions is 20 m in depth and 400 m across. The scours are most common near and in the channels and are oriented consistent with turbidity current flow spreading across the middle fan from the valley on the upper fan [5].

The widespread channel and mesotopographic relief on the upper middle fan produce the characteristic hummocky topography used to recognize suprafan morphology on surface echo-sounding profiles [3]. The lower middle fan generally appears free of channel relief or mesotopography even on the deep-tow narrow-beam profiles.

Lower Fan

The middle fan merges with the more gently sloping lower fan and ponded basin-plain segments (Fig. 1). Limited deep-tow data indicate no mesotopographic or channel relief extending to this part of the fan.

Acoustic Character

Surface-ship and deep-tow high-resolution (3.5 to 4.0 kHz) reflection profiles show distinct changes across Navy Fan. The depth of acoustic penetration and the continuity or lateral extent of reflecting surfaces both tend to increase with increasing distance from the channel/lobe areas (Fig. 3). This change can generally be related to the grain size of the surficial sediments, with little or no acoustic penetration in sand-rich section and the deepest penetration with extensive subbottom reflectors in muddy hemipelagic or pelagic sediment [3,5,10]. The character of the high-resolution profiles can thus be used to define acoustic facies for the fan sediments.

Navy Fan has two main acoustic facies: (1) parallel-bedded sediments with laterally continuous reflectors and good acoustic penetration (numerous subbottom reflectors) that mark the basin-plain areas and basin slopes and ridges standing above the fan surface; (2) fan areas underlain by no or short and laterally discontinuous reflectors; this facies has been qualitatively subdivided in Figure 3 to show lateral changes in acoustic character with respect to fan morphology. The channel floors and much of the depositional lobe surfaces are underlain by areas with no subbottom reflectors. The levees on the upper fan, interchannel areas, and the fan surface adjacent to the lobes generally show a few subbottom reflectors of very limited extent (50 to 100 m long). The overbank area on the upper-fan and lower-fan/basin-plain areas shows the deepest acoustic penetration on the fan (Fig. 3). The areas

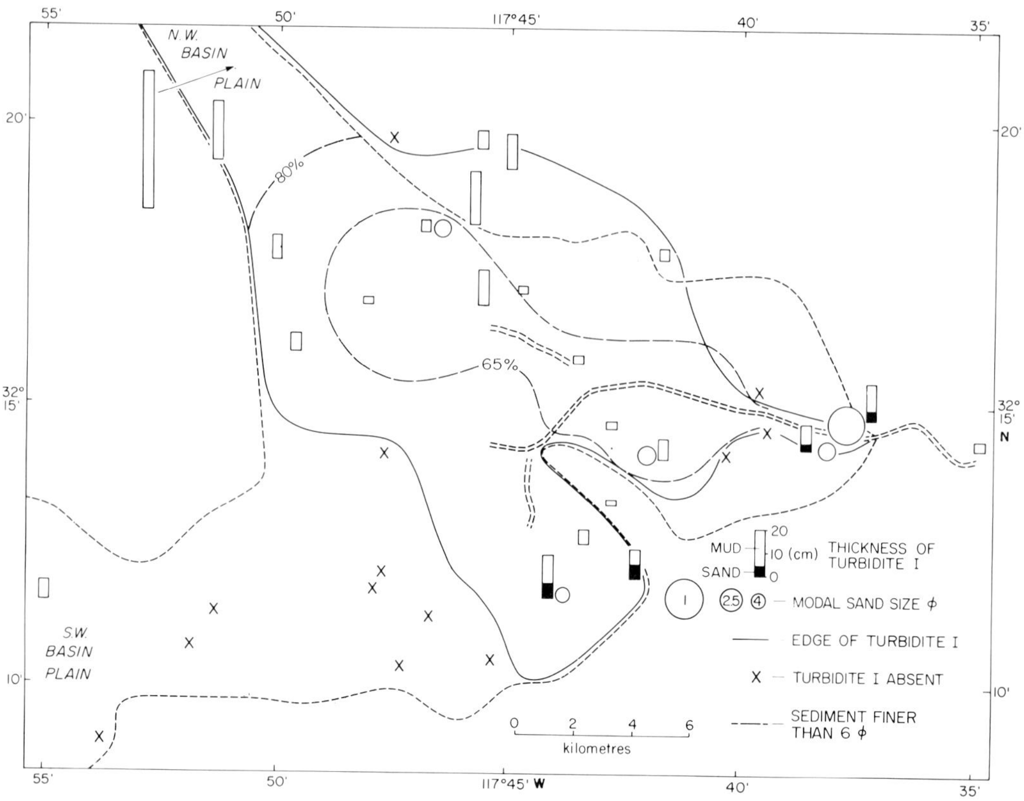

Figure 5. Proximal to distal changes in sediment characters for Turbidite I (summarized from [6]).

of poorest acoustic penetration generally correspond to the areas with greatest mesotopographic relief.

Sediments and Depositional Processes

The available piston-core samples support the interpreted general pattern of acoustic facies (Fig. 4). The thickest completely penetrated sand beds are on the depositional lobe and lower middle fan (cores 108, 109). The longest piston core from the fan area (106) is from the transition area from lower fan to basin plain and has the thinnest sand beds. Sand-bed thickness alone is not, however, a sufficiently reliable parameter to correlate with the acoustic facies distribution. For example, the shortest piston cores (104, 107) are from the interchannel areas and probably were stopped in coarse sediment; thus they are not fully representative of the sediment in this area. Core 96, from the gently sloping basin floor south of the middle fan, is mostly hemipelagic mud and has only a few thin turbidite units. High-resolution reflection profiles show the greatest acoustic penetration in this area (Fig. 8 in [3]). Most of the cored sand beds are Pleistocene in age based on carbon-14 dates and a change in radiolarian/foraminifera biostratigraphy in the interbedded muddy sediment (Fig. 4) [6].

In spite of the problems in obtaining representative cores from sandy areas of the fan, there are enough (30) short gravity cores to correlate a few Late Pleistocene–Holocene turbidites across most of the fan [6]. These correlations are supported by 12 radiocarbon dates from the piston core material and a distinctive stratigraphy within the muddy interbeds. Important results of this study [6] include the following: (1) there is a recognizable change in maximum grain size deposited as the turbidity current moves downfan; (2) at sharp bends in the channel system, that part of a turbidity current that rises above the levee crest will not follow the channelized portion of the flow through the bend but will continue in the prebend direction; this loss of the upper part of a current is referred to as flow stripping; (3) thick, but relatively slow, muddy turbidity currents can be several times the channel depth in thickness; thus, such currents do not follow the channel system and their movement is controlled by basin shape and Coriolis effects; (4) higher velocity, sand-transporting turbidity currents can remain mostly channelized; flow stripping of these currents will cause a marked decrease in velocity and thus an increase in rate of deposition. Some sand-carrying currents die within the channel/depositional lobe areas because of flow stripping.

The Holocene sediments of Navy Fan show many changes from proximal to distal environments of the type also found in ancient deep-sea fans [3]. There is a distal decrease in thickness and grain size of sand beds (as illustrated for turbidite 1 in Fig. 5), and sand beds also have a higher mica content distally. Mud beds increase in thickness and decrease in grain size distally. The few visible primary structures in sand beds indicate decreasing flow velocity down fan [3].

The presence of large scour features on the levees suggests

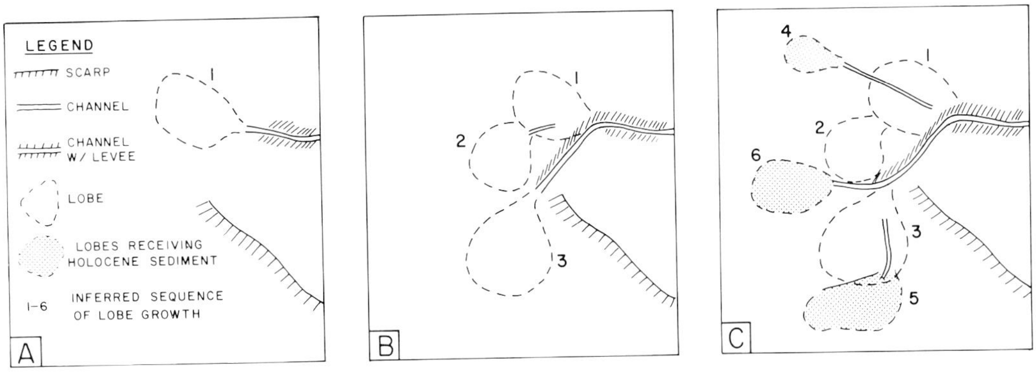

Figure 6. Sequence of lobe and distributary channel development on Navy Fan, modified from Normark and others, Fig. 12 [5].

that bankfull flow of a turbidity current does not approximate maximum size. Turbidity currents that just overtop the levees will aggrade the levees and adjacent areas while some may be large enough to deposit over most of the fan. Even larger currents, however, that are capable of eroding the levees on the upper fan may be those that provide most of the turbidite material in lower-fan/basin-plain areas.

Discussion

Navy Fan is a sand-rich fan that has almost stopped receiving turbidity currents after the Holocene sea-level rise. No true distributary channel system exists. Deep-tow side-looking sonar and narrow-beam profiling data show that two of the "distributary" channels are separated from the one active channel by low (several meters high) levee-like deposits. The individual channels consist of relatively straight segments with abrupt bends. There is no evidence for migration or braiding of these distributary channel features. Instead, it appears that aggradation of a channel, and the middle-fan lobe that it feeds, forces the channel to shift into the marginal low areas around the aggrading lobe. This produces a sharp bend in the channel near the apex of a now-abandoned lobe. Figure 6 shows the inferred sequence of lobe abandonment that has produced the present channel pattern on Navy Fan. This sequence is based on morphologic evidence only (see discussion in [5]).

The lobes on Navy Fan are without channels on their surface. Commonly, they are bordered by small channels that begin along the edges of the lobes, probably as a result of slightly accelerated turbidity current flow as the current moves off the lobe onto the adjacent lower surface of the fan. The middle fan may be composed of a succession of interfingered lobe deposits that are several kilometers wide but only 10-m thick or less; the thickness cannot be determined with existing seismic records, but it is inferred from the very low-relief bulge formed by the lobe in the fan surface (Fig. 2).

The deep-tow survey of Navy Fan showed many surface features (mesotopography) that are of small enough size to be recognized in large outcrops of ancient turbidites. Unfortunately, we know neither the origin of most of this relief nor how it might appear or be preserved in the rock record To correctly determine the range of depositional processes operating on submarine fans will require continued study of mesotopographic features on modern fans and recognition of these features (or the sediment sequences that compose them) in ancient turbidites.

Acknowledgments

We thank the crews, captains, and technicians of the seven cruises to Navy Fan on ships from Scripps Institution of Oceanography (SIO) and the U.S. Geological Survey. F. N. Spiess and D. E. Boegeman with the Marine Physical Laboratory of SIO were especially helpful during the two deep-tow operations. C. E. Gutmacher and G. R. Hess of the U.S. Geological Survey have spent many months helping us assemble the data and correcting our all too frequent mistakes on illustrations. The paper has benefited from critical reviews by G. R. Hess, H. A. Karl, and P. R. Hill.

References

[1] Junger, A., 1976. Tectonics of Southern California Borderland: In D. G. Howell, (ed.), Aspects of the Geologic History of the California Continental Borderland. Pacific Section of the American Association of Petroleum Geologists Miscellaneous Publication 24, pp. 486–498.

[2] Smith, D. L., and Normark, W. R., 1976. Deformation and patterns of sedimentation, South San Clemente Basin, California Borderland. Marine Geology, v. 22, pp. 175–188.

[3] Normark, W. R., and Piper, D. J. W., 1972. Sediments and growth pattern of Navy deep-sea fan, San Clemente Basin, California Borderland. Journal of Geology, v. 80, pp. 192–223.

[4] Shepard, F. P., and Dill, R. F., 1966. Submarine canyons and other sea valleys. Rand McNally and Co., Chicago, IL.

[5] Normark, W. R., Piper, D. J. W., and Hess, G. R., 1979. Distributary channels, sand lobes, and mesotopography of Navy submarine fan, California Borderland, with applications to ancient fan sediments. Sedimentology, v. 26, pp. 749–774.

[6] Piper, D. J. W., and Normark, W. R., 1983. Turbidite depositional patterns and flow characteristics, Navy submarine fan, California Borderland. Sedimentology, v. 30, pp. 681–694.

[7] Bowen, A. J., Normark, W. R., and Piper, D. J. W., 1984. Modelling of turbidity currents on Navy submarine fan, California Continental Borderland. Sedimentology, v. 31, pp. 169–185.

[8] Spiess, F. N., and others, 1976. Fine scale mapping near the deep-sea floor. Proceedings Oceans '76 Marine Technology Society–In-

stitute of Electrical and Electronic Engineers Annual Meeting, pp. 8A1–8A9.

[9] Normark, W. R., 1971. Minitopography of deep-sea fans: geometric consideration for facies interpretations in turbidites: In: Geologic Guidebook Newport Lagoon to San Clemente, Orange County, California. Pacific Section Society of Economic Paleontologists and Mineralogists, pp. 22–36.

[10] Damuth, J. E., 1975. Echo character of the western equatorial Atlantic floor and its relationship to the dispersal and distribution of terrigenous sediments. Marine Geology, v. 18, pp. 17–45.

III

Modern Submarine Fans

Passive Margin Setting

CHAPTER 15

Amazon Fan, Atlantic Ocean

John E. Damuth and Roger D. Flood

Abstract

The Amazon Deep-Sea Fan began to form in the Early Miocene and is characterized by a highly meandering distributary channel system. On the middle fan, these leveed channels coalesce to form two broad levee complexes. Older, now buried levee complexes are also observed within the fan. These levee complexes grow through channel migration, branching, and avulsion. Probably only one or two channels are active at any given time. Sediments reach the fan only during glacio-eustatic low stands of sea level. Coarse sediments largely by-pass the upper and middle fan via the channels and are deposited on the lower fan.

Introduction

The Amazon Fan (or Amazon Cone) is the third largest modern deep-sea fan and extends from the continental shelf off northeast Brazil to depths in excess of 4700 m (Figures 1 and 2). The morphology, sedimentation processes, age, and structure of the fan were first examined in detail [1,2] using conventional seismic and sediment data. Additional studies have indirectly addressed some aspects of the fan age and development in less detail [3–10]. More recently, a new series of Amazon Fan studies has been initiated which utilizes more sophisticated, state-of-the-art oceanographic instruments and techniques (e.g., side-scan sonar, bathymetric swath mapping) to address problems of sedimentation processes and growth pattern of the fan that could not be resolved using the conventional geologic/geophysical techniques and data of previous studies. The first of these new studies was a survey of the upper and middle fan using the GLORIA long-range side-scan sonar [11]. The GLORIA enabled mapping of entire trends and bifurcation patterns of individual channels for the first time. This study revealed much new information about channel morphology (e.g., highly meandering) [11] and relative age relationships [12]. It also raised a number of new problems that must be addressed during upcoming studies. Below, we briefly summarize the present state of knowledge of the morphology, sedimentation processes, and growth pattern of the Amazon Fan.

Regional Setting

The Foz do Amazonas is the sedimentary basin that underlies the Amazon River mouth, continental shelf, and upper Amazon Fan. Sediments range from Middle Cretaceous to Quaternary in age; thicknesses range from 5 km onshore to possibly 14 km at the shelf edge and under the upper fan [10].

The Amazon Fan extends downslope from the shelf break for as far as 700 km (Figures 1 and 2) and has an average gradient of 6.6 m/1000 m (~1:15 or 0.4°). The fan is about 250 km wide at the shelf break, but it quickly widens downslope to a maximum width of approximately 650 to 700 km. This elongated radial pattern covers approximately 330,000 km^2. The maximum thickness of the fan is 4 to 5 km [10], and the total volume of sediments in the fan is probably in excess of 700,000 km^3.

Fan Morphology

We divide the Amazon Fan into three morphologic divisions (upper, middle, and lower) based on morphologic and acoustic characteristics. The *upper fan* extends from the shelf break

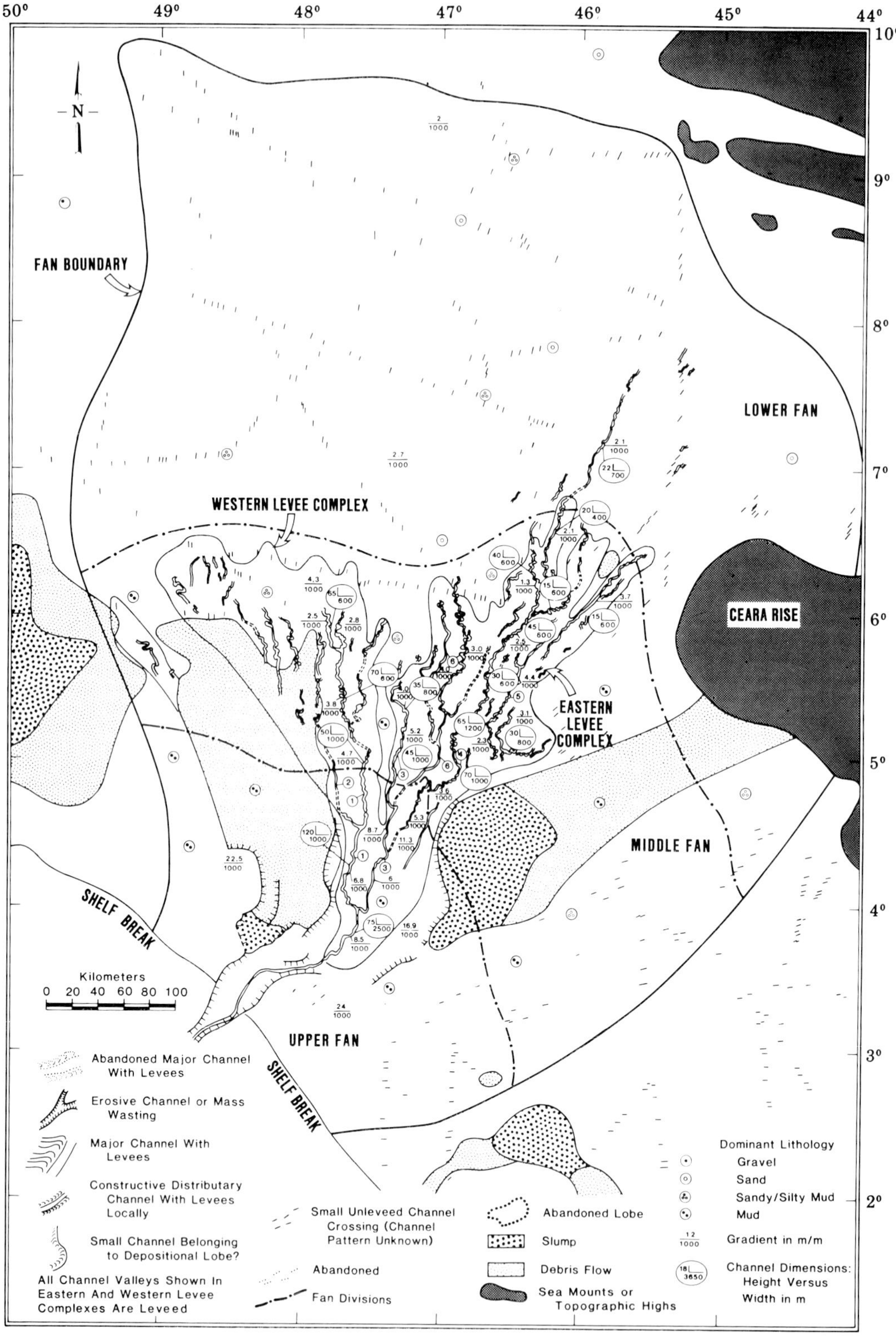

Figure 1. Morphology of Amazon Fan. Distributary channels on upper and middle fan were mapped from GLORIA side-scan sonographs (uncorrected for slant range) [11] and conventional PDR echograms [1,2]. Small channels on lower fan are mapped from PDR only. Numbers 1 through 6 beside major distributary channels indicate relative ages of channels in order of increasing age (see text).

to about 3000 m (Figure 1) where the bathymetric contours show a noticeable break in slope (Figure 2). Gradients range from about 25 m/1000 m (1:40 or 1.4°) near the shelf break to 10 m/1000 m (1:100 or 0.6°) near 3000 m; the average gradient is 14 m/1000 m (~1:70 or 0.8°). The upper fan surface is often rugged, and steep scarps of up to a few hundred meters relief are observed (Figures 1 and 3, Profile WX). The major morphologic features of the upper fan are the Am-

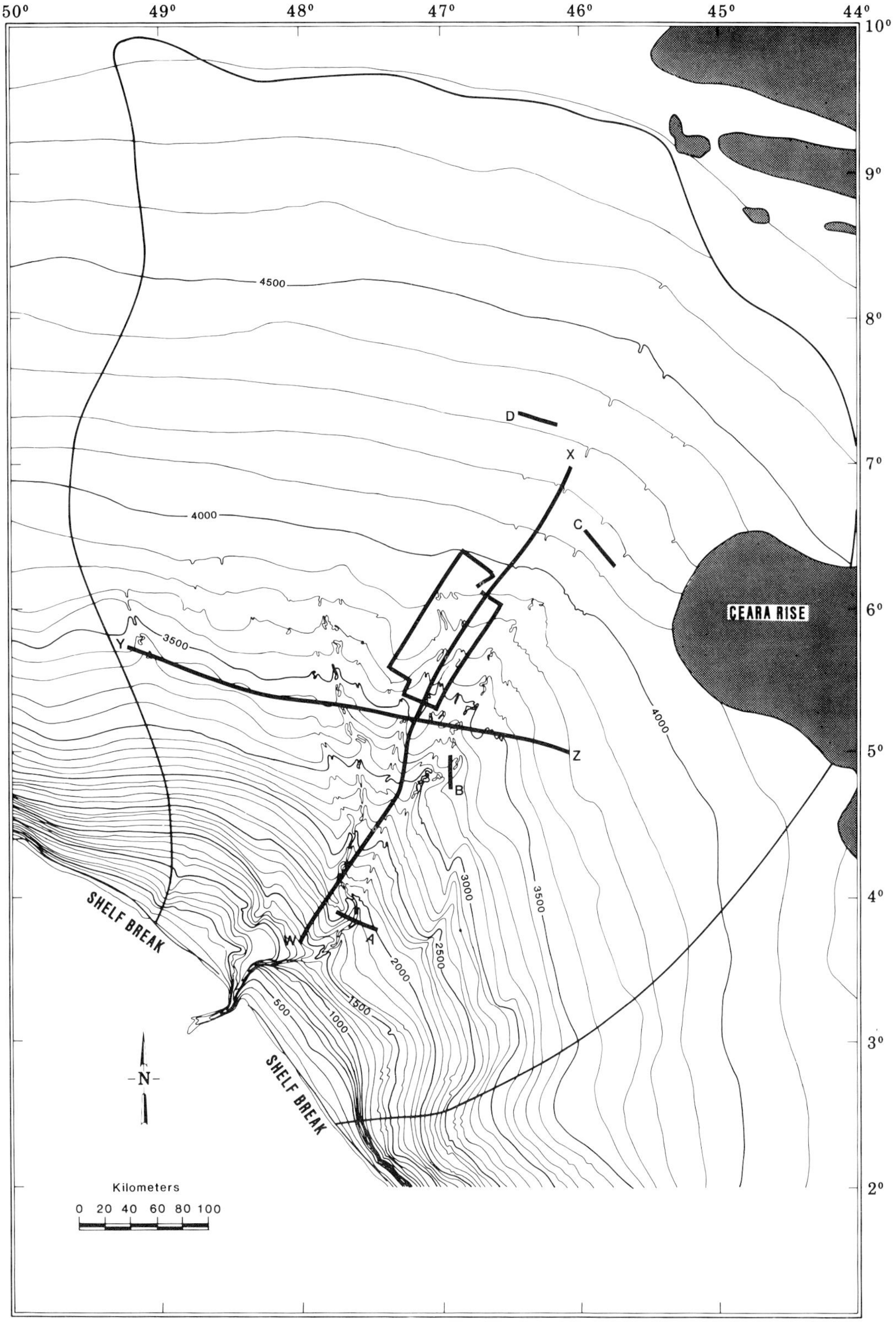

Figure 2. Bathymetry of Amazon Fan. Contours in corrected meters. Heavy lines WX and YZ show locations of seismic profiles in Figure 3; lines A to D show locations of 3.5-kHz echograms in Figure 4. Box outlines location of GLORIA sonograph in Figure 5.

azon Submarine Canyon and a large leveed central distributary channel that divides into four prominent leveed distributaries between 2000 and 3000 m (Figures 1–3, 4A).

The *middle fan* extends from about 3000 m to 4000–4200 m (Figure 1) where a subtle change in gradient occurs (Figure 2). Gradients range from 10 m/1000 m to 4 m/1000 m (1/250 or 0.2°); the average gradient is about 5 m/1000 m (1:200 or 0.3°). The middle fan is characterized by numer-

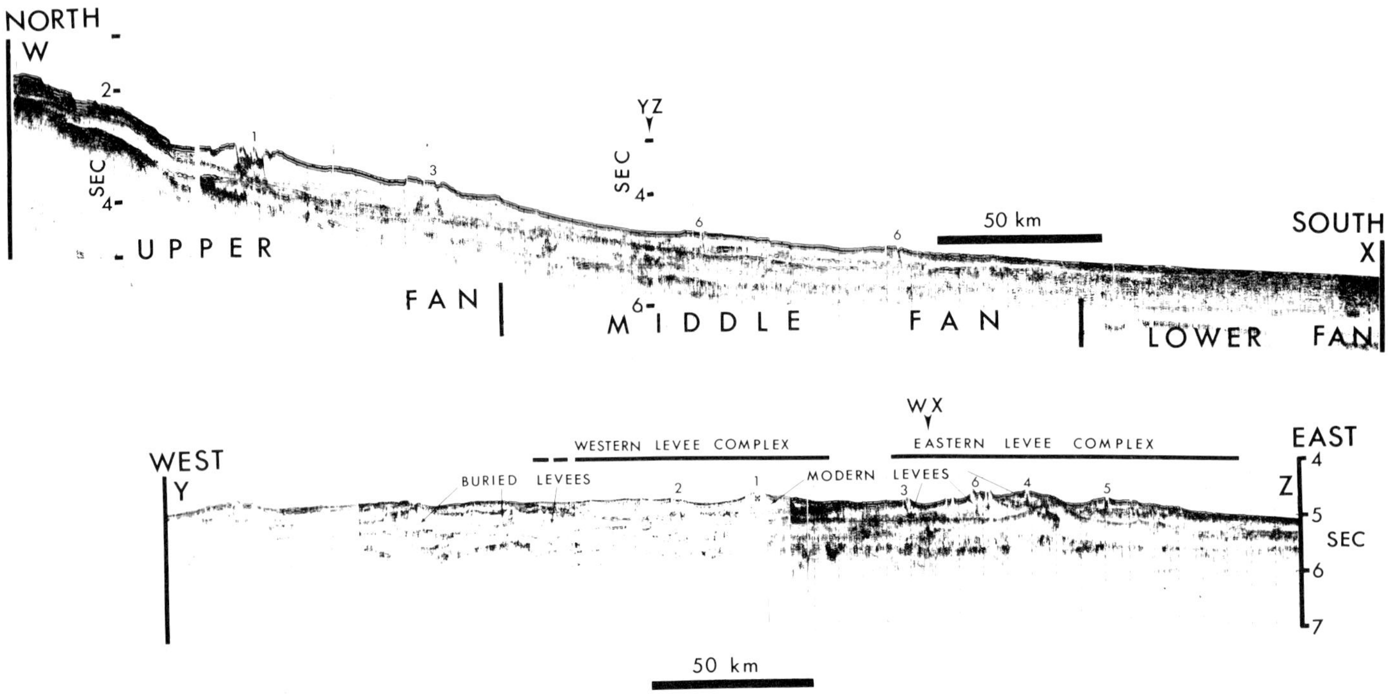

Figure 3. Seismic-reflection profiles (100 Hz) from Amazon Fan. Locations in Figure 2. Numbers 1 through 6 identify major distributary channels in order of increasing relative age (see Figure 1 and text). Black lines above YZ show lateral extent of Western and Eastern Levee Complexes (Figure 1).

ous large, leveed distributary channels (Figures 1–3, 4B, and 4C). The individual levee systems coalesce to form two large, distinct levee complexes (the Western and the Eastern complexes, Figure 1) whose downslope ends define the boundary between the middle and lower fan.

The *lower fan* is smooth and very gently sloping with an average gradient of 2.3 m/1000 m (1:430 or 0.1°). Numerous small, unleveed distributary channels cross the lower fan (Figures 1 and 4D). On low-frequency seismic records, the lower fan is acoustically distinct (Figure 3). The upper and middle fan are characterized by acoustically transparent to semitransparent sediment wedges and prisms that are recognizable as buried channel-levee systems and former fan surfaces (Figure 3). In contrast, the lower fan uniformly returns a highly reflective, conformably stratified pattern (Figure 3, Profile WX). Transparent channel-levee systems are not observed.

Distributary Channel System

The distributary channel system of the Amazon Fan radiates outward from the Amazon Submarine Canyon (Figures 1 and 2). The canyon is up to 600 m deep and extends from the outer shelf (40-m contour) to at least 1400 m. Near 1400 m, the canyon abruptly widens, and a large, leveed central channel or fan valley arises (Figures 1 and 2). The levee system associated with this channel is up to 50 km wide, 1 km thick, and rises up to 300 m above the surrounding fan surface; the channel itself is perched atop this levee system and is up to 250 m deep and 2500 m wide (Figure 4A). This central channel trends downfan (gradient ~8.5 m/1000 m) to about 2200 m, where GLORIA sonographs show that it bifurcates; the deeper (up to 200 m) main branch curves sharply northwestward, whereas the shallower (40 to 60 m) eastern branch curves northeastward (Figure 1).

Using GLORIA sonographs and PDR echograms, we identified and mapped six major distributary channels (1 through 6 in Figure 1) and their associated branches on the upper and middle fan [12]. Channels 1 and 3 arise from the bifurcation of the central channel. Channels 2 and 4 arise between 2500 and 2800 m, apparently from bifurcation of channels 1 and 3, respectively (Figure 1). Below 3000 m, the trends and branching patterns of the observed channel segments (including 5 and 6) are less certain because most appear as discontinuous segments on GLORIA sonographs. Although some of these discontinuities are due to small gaps in GLORIA coverage (e.g., directly beneath ship or the absence of slant-range correction), many others appear to represent channel cutoff and abandonment with subsequent partial burial or destruction of channel segments (Figure 1).

Each distributary channel on the upper and middle fan is perched atop a wide levee system (e.g., Figure 4B) that builds upward and laterally by overbank spilling [1,11,12]. Individual levee systems appear as thick (up to 1 km), semitransparent, wedge-shaped deposits on seismic profiles (Fig-

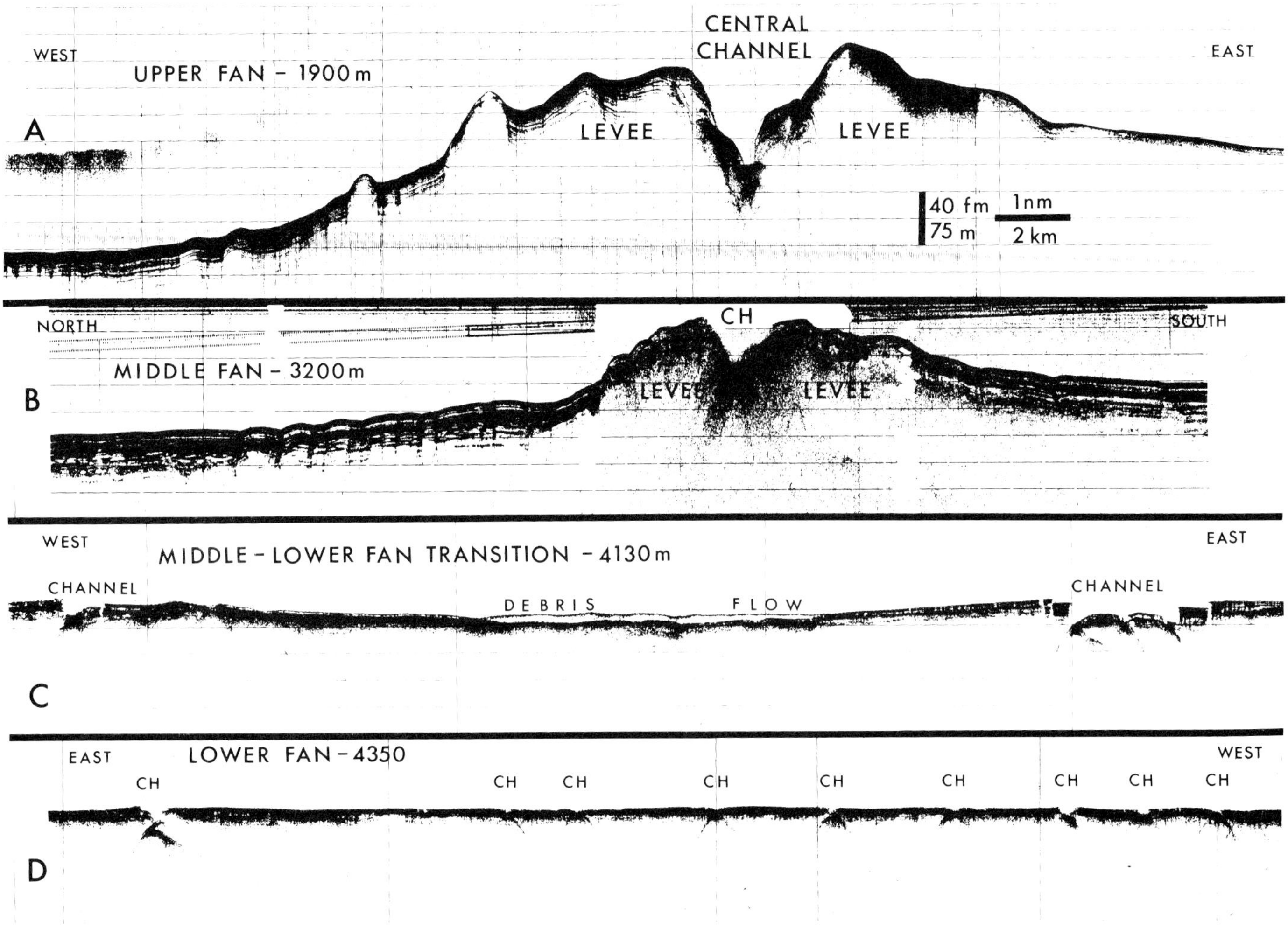

Figure 4. The 3.5-kHz echograms of distributary channels that show examples of channel size and morphology at various locations downfan (locations shown in Figure 2). A. Leveed central channel (1) on upper fan. B. Leveed meandering channel (4) typical of middle fan. Average gradients on backsides of such levees are generally 15 m to 40 m per 1000 m (~1° to 2°); locally gradients range up to 166 m/1000 m (9.5°). C. Small channels with low levees characteristic of the middle-to-lower fan transition. Small debris flow is from a nearby basement knoll. D. Small channels (CH) less than 20 m deep that are typical of lower fan.

ure 3). High-amplitude reflectors are commonly observed beneath the channel axis in each levee deposit and may be reflected from coarse sediments deposited in the channel floor as the levee sequences built upward. Numerous older, now buried channel-levee systems are observed on seismic profiles (Figure 3).

On the lower fan, PDR echograms and seismic profiles show numerous, small (5 to 25 m deep) distributary channels (black dots, Figure 1) that have little or no associated levee development (Figure 3, Profile WX and Figures 4C and 4D). GLORIA sonographs show these channels to be highly meandering. One 20 m deep channel was traced continuously from 4100 m to 4450 m (Figure 1).

On the middle fan (3000 to 4200 m), two distinct, separate levee complexes, each composed of several individual, coalescing channel-levee systems, diverge down the middle fan (Figure 1). These were designated the Western Levee Complex and the Eastern Levee Complex [11,12]. On the upper fan, these two complexes merge, and only one large levee complex is associated with the central channel and its two branches (1 and 3). Beneath the present levee complexes, older buried levee complexes consisting of ancient channel-levee systems are observed on seismic records (e.g., "buried levees" in Profile YZ, Figure 3).

The most striking characteristic of the distributary channels revealed by GLORIA [11] is their extensive and intricate meander patterns (Figures 1 and 5). Nearly all channels deeper than 2500 m exhibit high sinuosity (1.5 to 2.5) with well-developed, recurving meanders; channels shallower than 2500 m have lower sinuosities of 1.0 to 1.5. In addition, cutoffs, abandoned meander loops (oxbows) and scars, and other floodplainlike features are observed on the GLORIA sonographs [11]. The sinuosity, meander patterns, and associated morphologic features of these channels appear to be quite similar to meander patterns and associated floodplain features of mature fluvial systems on land. For example, the dimensions of these features on the middle fan (e.g., meander wave length, amplitude, and frequency; channel and levee dimensions) are equal to or larger than those of similar features of the lower Mississippi River [11].

The GLORIA sonographs show channel bifurcation or branching at several locations, and other bifurcations can be inferred even though they are not directly observed (Figure 1). At most locations, bifurcation appears to result from breaching of channel walls and levees, especially on the outside curves of meander loops [11]. Bifurcation apparently does not occur as frequently as predicted by previous studies [1,2], and many channels extend for long distances down the fan without apparent bifurcation (Figures 1 and 5). However, discontinuous channel segments are also observed that have no apparent upslope connection (Figure 1) and thus may be abandoned segments. At present, we are unable to determine with certainty the nature of most of the observed bifurcations. In some cases, both channel branches may remain active whereas in others the original channel may be abandoned as a result of channel avulsion. Certainly avulsion might be expected where bifurcation is caused by breaching of a meander loop. No evidence of channel braiding has been observed.

Sedimentation Processes

Mass-Transport Deposits

The distributary channel system on the upper and middle fan is bounded on either side by large slump and debris-flow deposits that cover at least 46,000 km^2 or about 14% of the fan's surface [2] (Figure 1). The upslope portions of these deposits consist of hummocky slump or slide deposits that are generally bounded by scarps. Downslope from these hummocky zones, the deposits consist of thin (10 to >50 m), acoustically transparent debris flows. These debris flows have traveled down the fan for distances of up to 300 km on slopes with gradients as low as 2.5 m/1000 m (0.14°). The total amount of fan sediments displaced may exceed 2500 km^3. Piston cores show zones of disturbed, contorted bedding and suggest that the age of these mass movements is probably Late Wisconsin [2].

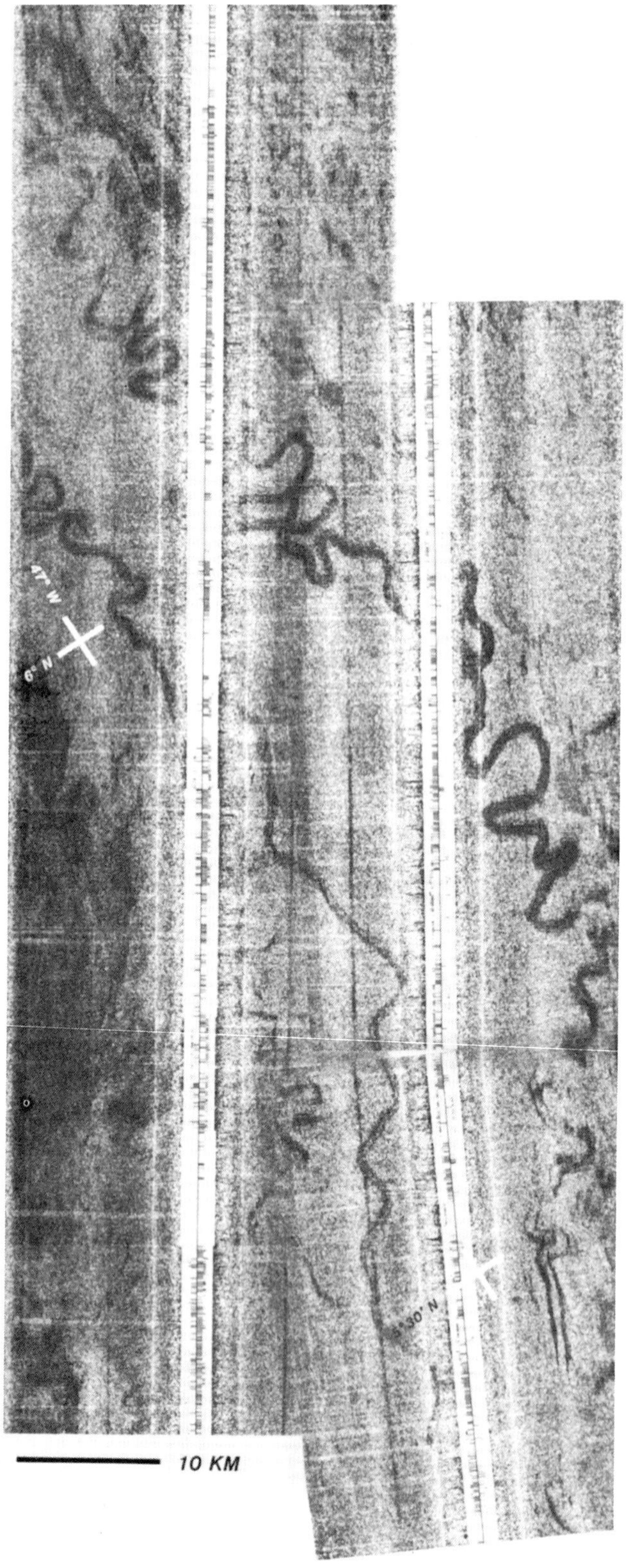

Figure 5. GLORIA long-range side-scan sonograph mosaic showing examples of meandering distributary channels on the middle Amazon Fan (location in Figure 2). Reflective areas are black. The processing routines corrected for slant range, ship speed, and beam pattern; a time-varying gain was applied and the data were contrast stretched. Ship tracks are along the white bands down the center of each sonograph. The 10-km scale bar applies to distance along ship track as well as distance laterally from track.

Sediments

Forty-four sediment cores have been collected from the Amazon Fan by Lamont-Doherty [1,2,4] and French scientists [9]. Unfortunately, only six of these cores were raised from the channel-levee complexes of the upper and middle fan, and only one is from a channel floor. Seven cores have been taken in the Amazon Submarine Canyon. The rest of the cores are from the lower fan, the slump-debris flow complexes, and areas peripheral to the fan.

Nearly all Amazon Fan cores have an upper meter or less of light-brown pelagic foram ooze or marl of Holocene age. The Pleistocene–Holocene boundary is generally marked by a rusty-colored iron-rich crust [1,2,4,13]. The remainder of each core is latest Wisconsin in age and consists of gray hemipelagic clay, often with interbeded silt-sand turbidites [1,4]. The hemipelagic sediment is silty terrigeneous clay with abundant organic detritus (organic carbon content = 0.5 to 0.75% [13]) and was apparently deposited slowly but continuously throughout the latest Wisconsin by gravity-controlled bottom flows [1,4]. Minimum sedimentation rates exceed 15 to 40 cm/10^3 yr and may reach 100 cm/10^3 yr [1]. The interbedded turbidites are less than 1 cm to a few meters thick and consist mainly of silt to medium sand; however, grain sizes from clay to fine gravel are observed. Quartz (>60%) and feldspar (>30%) are the dominant minerals. Organic detritus (wood and leaf fragments) is commonly disseminated throughout the silt/sand and occasionally forms discrete beds.

Sediment Distribution and Cycles

The pelagic Holocene sediments indicate that the Amazon Fan has been temporarily inactive during the last 11,000 years; the high sea-level stand traps Amazon River sediments on the inner continental shelf [1,2,4,5]. Presumably the fan was also inactive during previous interglacial sea-level stands. In contrast, sea-level lowering during the Wisconsin and previous glacials permitted the Amazon River to discharge sediments directly into the Amazon Canyon, hence sediments could easily be transported to the fan by turbidity flows via the canyon and distributary channels.

The sediment cores reflect the dispersal patterns of coarse terrigenous sediment across the fan during latest Wisconsin [1]. Cores from the upper and middle fan contain little or no bedded silt/sand. When present, beds are generally <5 cm thick and are often only laminae. In contrast, cores from the lower fan and the adjacent Demerara Abyssal Plain contain numerous, thick (20 cm to >5 m) beds of silt-to-gravel-sized particles [1]. This distribution pattern of coarse sediments is further evidenced by variations in 3.5-kHz echo character across the fan [1,3]; the upper and middle fan generally return distinct echos with continuous parallel subbottoms (indicative of little or no coarse sediment), whereas the lower fan and adjacent abyssal plains generally return indistinct prolonged echos with no subbottoms (indicative of abundant coarse sediment).

The apparent absence of abundant or thick silt/sand beds from the upper and middle fan implies that most coarse sediment by-passes these regions, presumably via the distributary channels, and is deposited across the lower fan and adjacent abyssal plains by turbidity flows. Zones of high-amplitude reflectors are generally observed directly below each channel axis on low-frequency seismic records (Figure 3) and may represent residual coarse sediment trapped in the channel axis as the channel and associated levees built upward through time.

Age, Thickness, and Average Sedimention Rate

Damuth and Kumar [1] estimated an age of 8 to 15 m.y. (Middle to Late Miocene) for the Amazon Cone by extrapolating the age (2.2 m.y.) of a prominent acoustic reflector on the Ceara Rise that can be traced into the fan; calculating an average sedimentation rate (~100 cm/10^3 yr) for the fan sediments above this reflector; and then extrapolating this rate to the maximum thickness of the fan (9.7 to 13.7 km) as calculated from sonobuoy, seismic refraction, and sediment-compaction data. Kumar [8] subsequently revised the age to approximately 22 m.y. (Early Miocene) when DSDP drilling revealed the prominent reflector to be 6 m.y. old. More recently PETROBRAS scientists ([10] and Kowsmann, personal communication, 1983) have established an age of 16.5 m.y. (Early Miocene) for the fan by correlation of seismic facies beneath the Foz do Amazonas and upper fan with borehole data from the shelf and Amazon estuary. These studies suggest that the maximum thickness of the upper fan may be only about 4.2 km. This yields an average sedimentation rate of 25 cm/10^3 yr.

Growth Pattern

Seismic-reflection profiles (for example Profile YZ in Figure 3) across the upper and middle fan reveal that the individual levee systems associated with each major channel (1 through 6, Figure 1) stratigraphically overlie one another and are thus of relatively different ages [12]. For example, channel 4 in Profile YZ (Figure 3) clearly overlies the flank of channel 6, and is thus stratigraphically younger. By determining the stratigraphic succession of the major channel-levee systems in this manner, a tentative age relationship has been established [12]; the major channels are labeled 1 through 6 in order of increasing age (Figures 1 and 3). Channel 1 on the Western Levee Complex is the youngest channel. Channel 3 is the youngest on the Eastern Levee Complex but is older than channels 1 and 2. Channels 2, 4, 5, and 6 are apparently older, now abandoned channels.

These relationships suggest that for each levee complex (Western and Eastern) only one major channel was active at any given time [12]. This active channel deposits a large levee sequence that eventually partially or completely buries the flanks of adjacent levees (e.g., Figure 3). Eventually this active channel is abandoned, probably by avulsion, and a new channel segment develops at the edge of the levee complex where gradients are steeper. Through time, the formation and abandonment of a succession of channel-levee systems builds a levee complex (such as the Western Levee Complex) that grows laterally and downslope.

Occasionally, the course of a newly established channel may be so far from the previous channel that the entire levee complex is abandoned, and a new, separate levee complex begins to develop. This appears most likely to take place when a channel is abandoned well upslope near its head where gradients are highest and the channel can diverge well away from other channels as it builds downfan [12]. For example, the formation of channels 1 and 2 apparently resulted in abandonment of the Eastern Levee Complex and establishment of the Western Levee Complex (Figure 1).

The formation of a succession of discrete, commonly overlapping levee complexes (such as the Western and Eastern Complexes), each composed of several abandoned channel-levee systems, gradually builds the fan upward and radially outward downslope. Additional evidence for this type of growth pattern is observed deeper within the fan on seismic profiles that show older, buried levee complexes beneath the present Western and Eastern Complexes (e.g., the series of "buried levees" beneath the western flank of the Western Levee Complex in Figure 3, Profile YZ).

The growth pattern for Amazon Fan described here and in [12] is not entirely certain because it is possible that two or more channels on a levee complex, or even two or more levee complexes are active simultaneously. For example, GLORIA sonographs of the bifurcation point of the central channel into channels 1 and 3 (~2000 m, Figure 1) appear to show branching, with possibly both channels remaining open. Thus it is possible that during fan growth more than one major channel, and, possibly, even more than one levee complex, may have been active simultaneously. Further detailed studies will be required to resolve this problem and thus verify the growth pattern proposed here.

Conclusions

Amazon Fan studies to date, especially the recent GLORIA survey, have provided important new information on the distribution and morphologic characteristics of the various features of the upper and middle fan, including individual distributary channels and associated levee systems, broad coalescing levee complexes of several channel-levee systems, mass-transport deposits, and the Amazon Submarine Canyon. We are now beginning to understand the growth pattern and sedimentation processes of the fan as well as the details of distributary-channel anatomy (e.g., meander pattern, bifurcation pattern, etc.). However, these studies have also raised important new problems. For example, the discovery of high-sinuosity meandering channels and associated floodplainlike features (Figure 5) has important ramifications for the type, volume, and continuity of the turbidity flow that forms, maintains, and modifies such channels. Such meander formation may require hydrodynamic conditions within channels similar to those of fluvial channels on land. Possibly flows through the fan channels must be relatively steady, long term, and high volume, rather than short, intermittent, sporadic flows generally associated with traditional turbidity-current events [11]. On the other hand, how could such fluviallike traction flow transport coarse sediments (presumably as bed load) for hundreds of kilometers down a fan channel, then deposit these sediments in the form of classic turbidites as observed on the lower fan? Our future studies with Sea Beam, high-resolution seismics, midrange side-scan sonar, and close-interval coring during the upcoming years will be aimed at solving these and similar critical problems by focusing on more detailed and quantitative studies at critical locations on the fan.

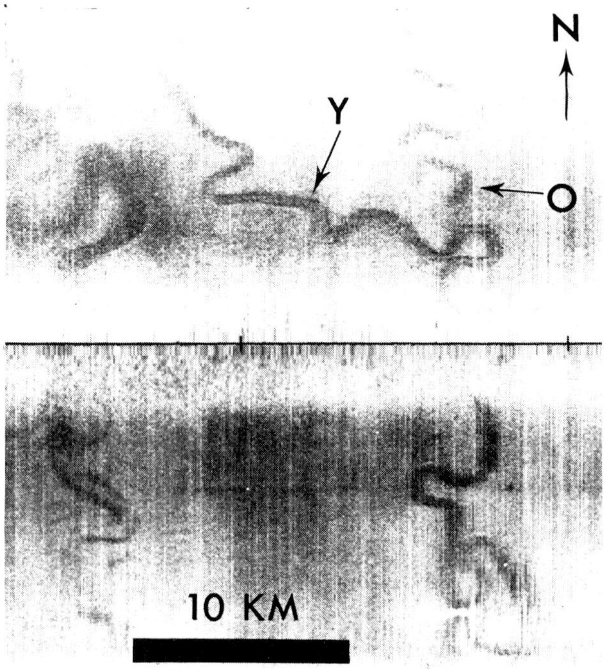

Figure 6. GLORIA long-range sonograph showing channel bifurcation. The sonograph is corrected for slant range and ship speed; ship track is E-W along center black line. Highly reflective areas are black. Scale bar applies in any direction. Location is at the end of channel labelled 1 in Figure 1 near 5° 45′ N; 47° 45′ W (water depth ~ 3600 to 3800 m). Y is younger, active branch, whereas 0 is older, abandoned segment.

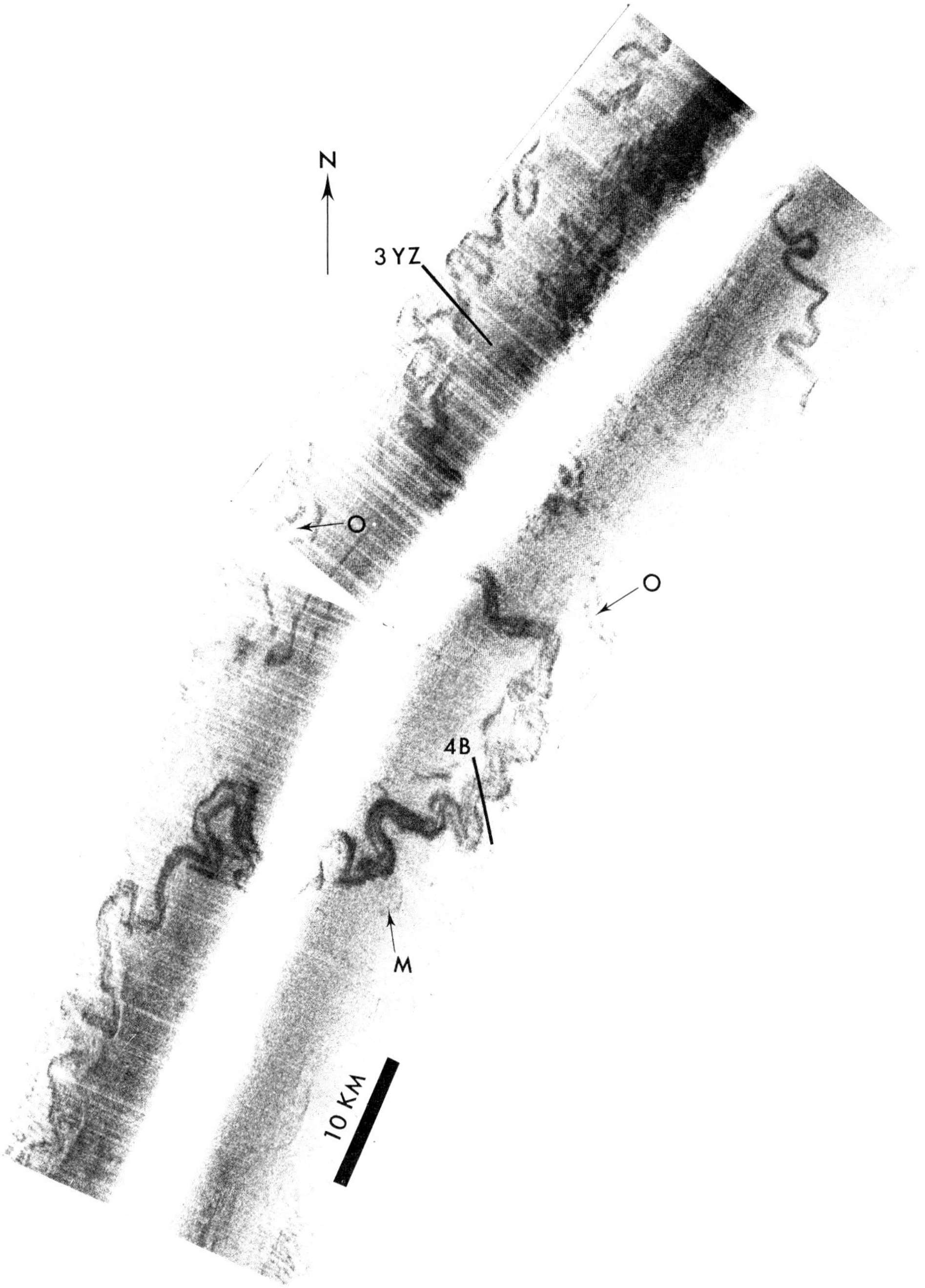

Figure 7. GLORIA long-range side-scan sonograph of the upslope portion of a major distributary channel (labelled 4 in Fig. 1). The sonograph is corrected for slant range and ship speed; ship track is SW-NE along center white area. Highly reflective areas are black. Scale bar applies to any direction. The location of this sonograph is centered near 5°N, 47°W (Fig. 1) in water depths of 3000 to 3500 m. Segments of older, partially buried channels (arrows labelled 0) are seen as less reflective targets east and west of the channel; and another young, highly reflective channel enters the sonograph on the upper right (northeast). Lines 4B and 3YZ show locations of seismic profiles in Figures 3 and 4. M identifies a probable cut-off, buried meander loop.

Addendum

Figures 6 and 7 are GLORIA sonographs showing additional examples of meandering distributary channels on the middle Amazon Fan (see the preceding section on Distributary Channel System). Figure 6 is an example of the type of channel bifurcation most commonly observed; that is, breaching the channel wall and levee of a meander loop to form a new channel, with subsequent cut-off and abandonment (avulsion) of the former channel. Bifurcation occurs on the upper right (east) where the prominent channel (Y) meanders sharply westward. A smaller, less reflective channel (0) apparently branches off from the tight meander loop and trends northward. At present it is not entirely certain which segment is older and abandoned. However, detailed examination of the GLORIA sonographs and 3.5-kHz echograms across these segments suggests that the western segment (Y) is new and that the eastern segment (0) is older and abandoned. Segment 0 appears to be disconnected (perhaps by filling) from the meander loop.

Figure 7 shows the upslope portion of the channel labeled 4 (Fig. 1). This is one of the major distributaries and has been continuously mapped downfan for 200-km between 2800 and 3500 m (Fig. 1). The channel is up to 100-m deep and 1.5-km wide. Profile B (Fig. 4) shows a 3.5-kHz echogram across this channel, and the entire levee system associated with the channel can be seen on seismic profile YZ (labeled 4 in Fig. 3). This channel levee sysem overlaps the eastern levee of channel 6 on the east side of the profile.

Acknowledgments

The GLORIA study was supported by the National Science Foundation (OCE-81-17469), PETROBRAS, and the Institute of Oceanographic Sciences. Other Amazon Fan studies were supported by the National Science Foundation and the Office of Naval Research. A. Shor and K. Kastens critically reviewed the manuscript. Lamont-Doherty Geological Observatory Contribution No. 3619.

References

[1] Damuth, J. E., and Kumar, N., 1975. Amazon Cone: morphology, sediments, age, and growth pattern. Geological Society of America Bulletin, v.86, pp. 863–878.

[2] Damuth, J. E., and Embley, R. W., 1981. Mass-transport processes on Amazon Cone: western Equatorial Atlantic. Bulletin of the American Association of Petroleum Geologists, v.65, pp. 629–643.

[3] Damuth, J. E., 1975. Echo character of the western Equatorial Atlantic floor and its relationship to the dispersal and distribution of terrigenous sediments. Marine Geology, v.18, pp. 17–45.

[4] Damuth, J. E., 1977. Late Quaternary sedimentation in the western Equatorial Atlantic. Geological Society of America Bulletin, v.88, pp. 695–710.

[5] Milliman, J. D., Summerhayes, C. P., and Barretto, H. T., 1975. Quaternary sedimentation on the Amazon continental margin—a model. Geological Society of America Bulletin, v.86, pp. 610–614.

[6] Milliman, J. D., and Barretto, H. T., 1975. Relict magnesian calcite oolite and subsidence of the Amazon shelf. Sedimentology, v.22, pp. 137–145.

[7] Milliman, J. D., 1979. Morphology and structure of Amazon upper continental margin. Bulletin of the American Association of Petroleum Geologists, v.63, pp. 934–950.

[8] Kumar, N., 1978. Sediment distribution in the western Atlantic off northern Brazil—structural controls and evolution. Bulletin of the American Association of Petroleum Geologists, v.62, pp. 273–294.

[9] Coumes, F., and Le Fournier, J., 1979. Le Cône de l'Amazone (Mission Orgon II). Bull. Cent. Rech. Explor.—Prod. Elf-Aquitaine, v.3, pp. 141–211.

[10] Castro, J. C., Miura, K., and Braga, J. A. E., 1978. Stratigraphic and structural framework of the Foz do Amazonas Basin. 10th Annual Offshore Technology Conference, Houston, v.3, pp. 1843–1847.

[11] Damuth, J. E., Kolla, V., Flood, R. D., Kowsmann, R. O., Monteiro, M. C., Gorini, M. A., Palma, J. J. C., and Belderson, R. H., 1983. Distributary channel meandering and bifurcation patterns on Amazon Deep-Sea Fan as revealed by long-range side-scan sonar (GLORIA). Geology, v.11, pp. 94–98.

[12] Damuth, J. E., Kowsmann, R. O., Flood, R. D., Belderson, R. H., and Gorini, M. A., 1983. Age relationships of distributary channels on Amazon Deep-Sea Fan: implications for fan growth pattern. Geology, v.11, pp. 470–473.

[13] McGeary, D. F. R., and Damuth, J. E., 1973. Post-glacial iron-rich crusts in hemipelagic deep-sea sediments. Geological Society of America Bulletin, v.84, pp. 1201–1212.

CHAPTER 16

Bengal Fan, Indian Ocean

F. J. Emmel and J. R. Curray

Abstract

Bengal Submarine Fan, with or without its eastern lobe, the Nicobar Fan, is the largest submarine fan known. Most of its sediment has been supplied by the Ganges and Brahmaputra Rivers, probably since the Early Eocene. The "Swatch-of-No-Ground" submarine canyon connects to only one active fan valley system at a time, without apparent bifurcation over its 2500-km length. The upper fan is comprised of a complex of huge channel–levee wedges of abandoned and buried older systems. A reduction of channel size and morphology occurs at the top of the middle fan, where meandering and sheet flow become more important.

Introduction

Although initial collision of India with a subduction zone south of Asia occurred in the Early Eocene [1,2], the Himalayas did not start uplifting until hard collision between the continental masses of India and Asia during the Mid-Miocene. These events, combined with the separation of the Rajmahal Hills and the Shillong Plateau [3], set the stage for the modern Bengal Delta and its seaward extension, the Bengal and Nicobar Fans, really lobes of a single fan separated by the Ninetyeast Ridge (Figs. 1, 2). Both fans are underlain by oceanic crust. The western boundary of the Bengal Fan is the continental slope of India and Sri Lanka; the eastern boundary north of 10°N is the accretionary prism of the Sunda subduction zone, and south of 10°N the Ninetyeast Ridge forms the eastern boundary.

Our interpretations regarding the modern Bengal Fan are based on our 3.5-kHz bottom-penetrating echo sounder and analog airgun seismic reflection records. In this paper, a general description is presented on the morphology and valleys of the fan, and its sediments and sedimentation.

Morphology

The Bengal Fan proper has an area of approximately 2.8 to 3 × 10^6 km^2. The length is between 2800 and 3000 km, extending from 20°10′N to 7°S latitude. Its greatest width is 1430 km at 15°N, and its narrowest part is at 6°N, 832 km, between Sri Lanka and the Ninetyeast Ridge. Depth for its apex and base are 1400 m and 5000 m, respectively. These dimensions make it the largest fan in the world; projected on a chart of western North America, it would extend from Vancouver to well into Baja California (Fig. 3). The overall surface of the fan is smooth and has gradients varying from about 6 m/km on the uppermost fan to less than 1 m/km on its lower reaches. Several valleys mark the surface of this fan and extend for various distances along its length. Presently only one active valley (A.V., Fig. 2) is connected to the "Swatch-of-No-Ground" canyon, but it has been cut off from its supply of sediment since about the time sea level rose above the head of the submarine canyon, probably about 7000 to 10,000 years ago.

Based on the gradients of the active valley and of the fan surface (Fig. 4), a threefold division of the fan is proposed.

(1) Upper fan, with an average valley gradient of 2.39 m/km and a fan gradient of 5.7 m/km; the lower boundary of the upper fan is placed where the fan surface becomes higher than the valley thalweg.
(2) Midfan, where both the fan and fan valley gradients average 1.68 m/km and the valleys are smaller in cross-sectional area.
(3) Lower fan, where the gradients drop to less than about 1 m/km, except locally, where the gradient may increase because of valley fill.

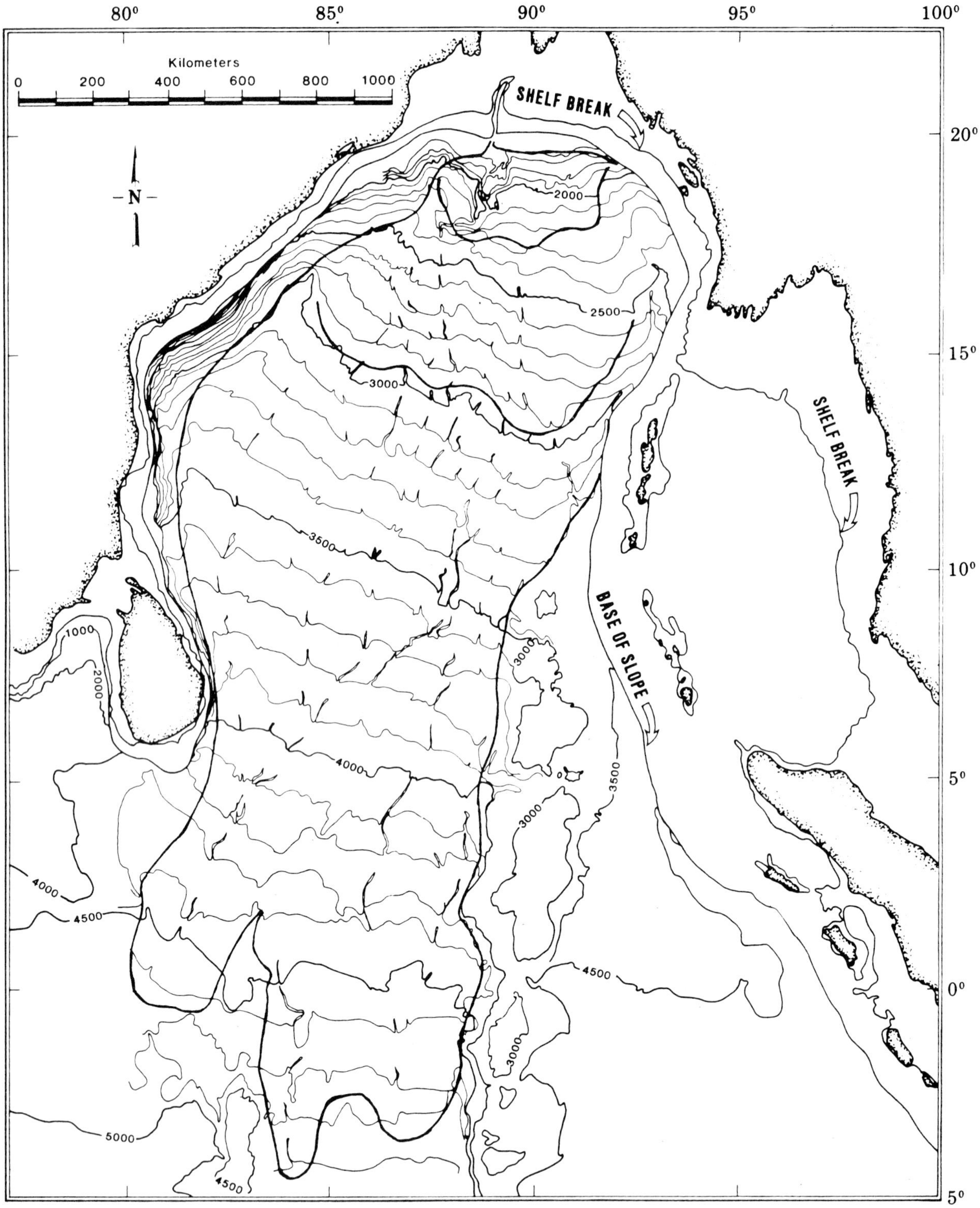

Figure 1. Contour chart of the Bengal Fan in uncorrected meters based on a large number of track lines. Approximate boundaries between upper, middle, and lower fan are indicated by heavy line.

The boundaries between the upper, middle, and lower fan can be approximated by the 2250 and 2900-m contours, respectively. The size, morphology, and structure of the upper-fan valleys contrast with those of the middle- and lower-fan regions. Correlation of valleys below the upper fan has been aided by the recognition of distinctive valley types and the

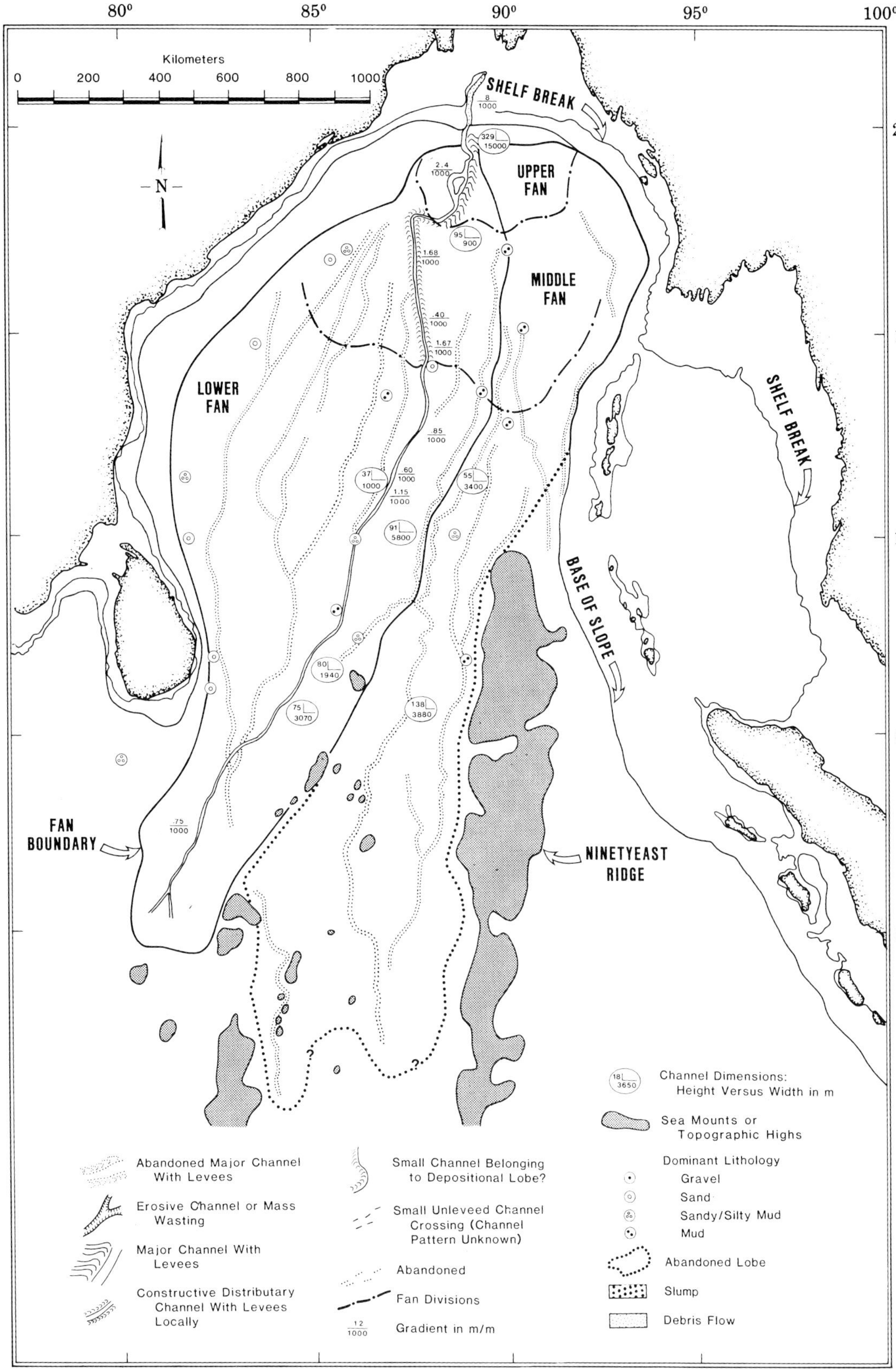
80°
85°
90°
95°
100°
20°
15°
10°
5°
0°
Kilometers
0 200 400 600 800 1000
N
SHELF BREAK
UPPER FAN
MIDDLE FAN
LOWER FAN
FAN BOUNDARY
BASE OF SLOPE
NINETYEAST RIDGE
SHELF BREAK
Abandoned Major Channel With Levees
Erosive Channel or Mass Wasting
Major Channel With Levees
Constructive Distributary Channel With Levees Locally
Small Channel Belonging to Depositional Lobe?
Small Unleveed Channel Crossing (Channel Pattern Unknown)
Abandoned
Fan Divisions
Gradient in m/m
Channel Dimensions: Height Versus Width in m
Sea Mounts or Topographic Highs
Dominant Lithology
Gravel
Sand
Sandy/Silty Mud
Mud
Abandoned Lobe
Slump
Debris Flow

Figure 2. Morphometric map of the Bay of Bengal.

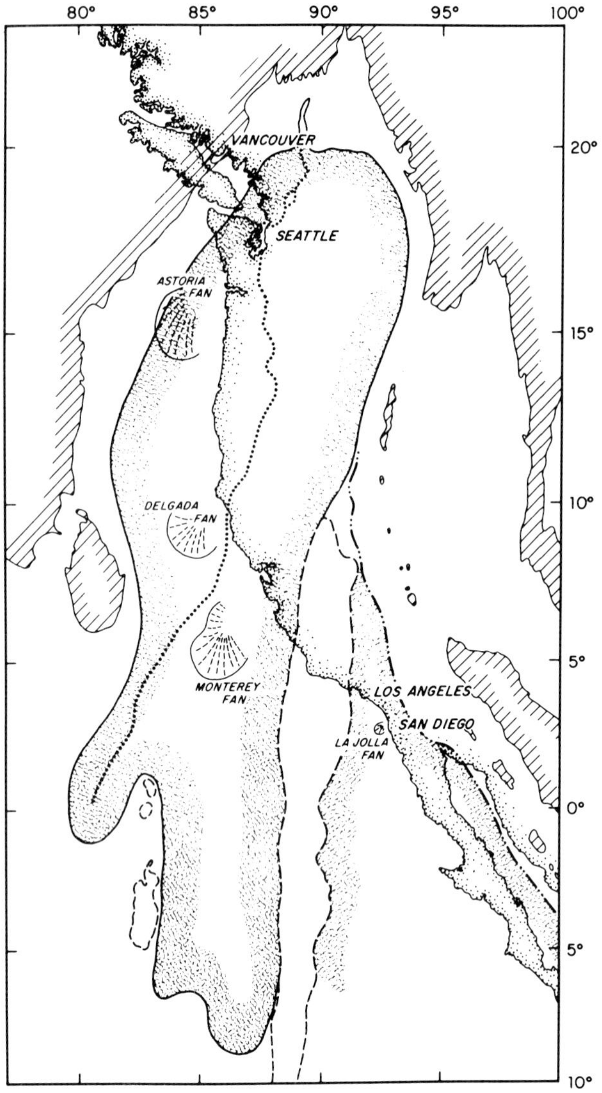

Figure 3. Size comparison between the Bengal Fan and Northwest America. Dots represent the most recent valley. Note that it extends from Vancouver to beyond San Diego.

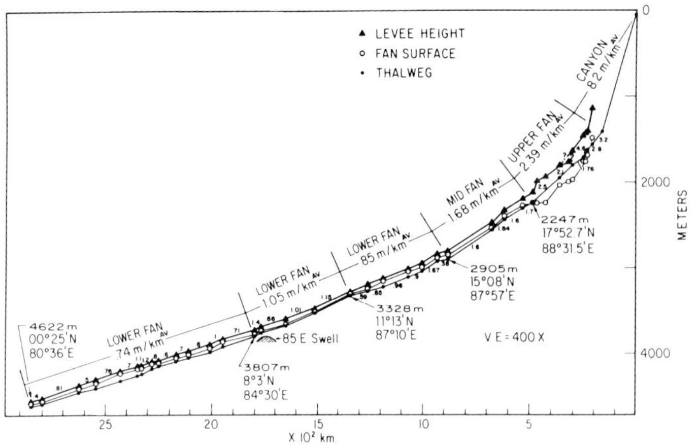

Figure 4. Gradients of Active Valley and the fan surface on which the division of the fan is based.

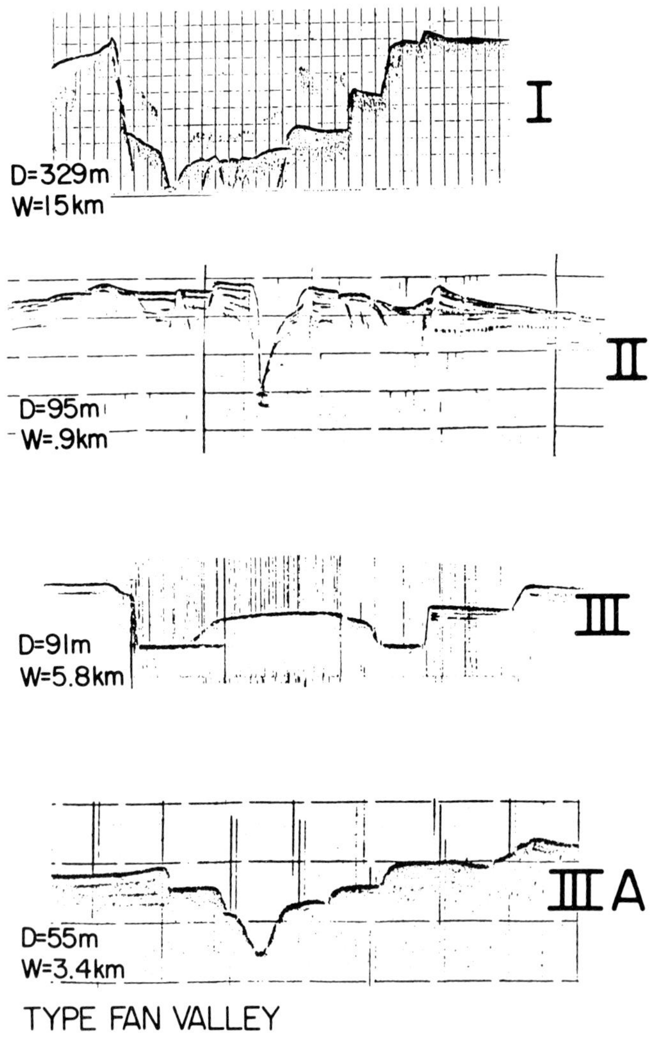

Figure 5. Valley types on the Bengal Fan. Type I is only found on the upper fan. Type II typifies the most recently active valley on the middle and lower fan; Types III and IIIA are older valleys on the middle and lower fan.

realization that throughout much of the fan, valleys retain their type and run more or less parallel to each other. The following valley types are present on the Bengal Fan:

Type I. Elevated valleys on the upper fan, directly connected to the canyon (Fig. 5). They are incised into their own depositional lobe of sediments and are characterized by:

(a) very pronounced levees, built well above surrounding sea floor. The entire channel–levee complex may exceed 50 km in width.
(b) one or more terraces.
(c) depth exceeding 100 m below its levee crests.

(d) inner channel V-shaped and relatively steep-walled.
(e) thalweg higher than adjacent fan surface.
(f) width from levee to levee often greater than 10 km. Channel depth/width ratio ~1:50.

Type II. Valleys lie on the midfan and lower fan and are connected to the active upper-fan valley. They are characterized by:
(a) the V-shape of the incised active inner channel.
(b) occasional narrow terrace.
(c) depth/width ratio >1:50.
Practically the entire length of the active valley can be classified under this type, except in several crossings where it is reoccupying an older valley. This type might be typical for younger valleys not yet adjusted to grade or flow regime.

Type III. These valleys occur on the midfan and lower fan and are no longer connected to a present source. They are characterized by:
(a) wide, flat-bottomed valleys.
(b) smooth and rounded levees, if present.
(c) terraces always present.
(d) depth/width ratio <1:50.
(e) usually relatively steep walls.
This valley type is older and commonly occurs as abandoned channels or oxbows, apparently adjusted to grade and flow regime before they were abandoned.

Type IIIA. Valleys that formerly were Type III and have been rejuvenated or reoccupied. They are characterized by:
(a) wide terraces, sometimes paired (mature valleys).
(b) levees as in Type III.
(c) inner channel shaped in a wide V.
(d) depth/width ratio <1:50, as in Type III.

Filled and Buried Valleys

In addition to the valleys that mark the surface of the Bengal Fan, large numbers of valleys are now filled in at the sea floor or lie buried beneath the fan surface. The filled valleys can be recognized on 3.5-kHz records, while buried valleys can be identified only on deeper penetration seismic reflection records.

Older Pleistocene valleys that are deeply buried can only be recognized on the upper fan, where the wedge shapes of their large natural levees show up well on the reflection records. On the midfan and lower fan, only the remnants of the large channels can be distinguished, and only if burial has not been too deep.

Filling and subsequent burial of fan valleys on the upper fan are closely related to valley migration by channel jumping, which is probably a rapid event.

Sediments of the Bengal Fan

Coring during the three Scripps Institution of Oceanography expeditions into the Bay of Bengal was not very successful. Yount (personal communication, 1981) furnished most of these descriptions. Kolla (personal communication, 1981) provided information on core results from the lower part of the fan. One deep hole, DSDP 218 [4], was drilled at 08°00.04′N, 86°16.97′E, and reached a depth of 772 m. Three sandy turbidite pulses were found in this boring: The deepest is mid-Miocene, the second is of Late Miocene–Pliocene, and the uppermost from 9 to 70 m is of Pleistocene age. Sand in the Pleistocene section ranges from fine to medium. The result of our core samples (maximum 5-m length) indicates that in most cases mud in varying thicknesses (20 to 90 cm) overlies a thin sand layer. In one instance, a second, thicker sand layer was found at about 3 m. Several thin sand layers were encountered in longer cores [5]. A high percentage of sand layers was found in the core taken in the bottom of Active Valley, compared with other samples in abandoned valleys. The thickest pelagic mud, >50 cm, was found in our sample G-42 located at approximately 7°N, 88°54′E, which is on the older pre-Wisconsin part of the fan. A core from the bottom of Active Valley at approximately 10°N, 86°E had only 5 cm of mud on top of poorly sorted sand (W. Ryan, unpublished report, 1980).

Reflection records across valleys of the upper fan and the upper midfan at 17°N usually show a faintly bedded and acoustically transparent top layer, while deeper down the fill is chaotic and strongly reflective in character. Other crossings of Active Valley just north of 15°N show the same reflection character, although both the chaotic reflector and the transparent fill are thinner in a downfan direction. We have interpreted the transparent layer as consisting of mud and silt and the chaotic reflectors as lenses of sand. One reflection record on the lower fan at approximately 5°30′N, 84°–85°15′E shows several buried valleys, the fill of which is chaotic and thus might represent sand or sandy deposits. This assumption does not contradict shallower core results and suggests that this part of the fan was very active in the past and that sand was transported to the lower part of the lower fan.

Previously [6], we discussed that backfilling started north of 11°N at the time when sea level rose. This backfilling jumped to north of 15°N because of a levee break near the boundary of the upper and middle fan. This implies that the mud cover on the bottom of the active valley is much thicker north of 11°N than to the south, an observation substantiated by the core at 10°N, 85°E.

Conclusions

Results of cores from the Deep Sea Drilling Project, piston cores, and a study of our reflection records indicate that the surficial cover of the Bengal Fan consists of mud ranging from 5 cm to several meters in thickness. The highest rates of deposition are on the upper fan, with coarser sediments in the valleys and finer levee and intervalley deposits. Sheet flows of finer sediments are important below the border of the upper and middle fan, where a large reduction of valley size occurs.

As on the upper fan, sand is limited to the valleys, and probably near-valley regions, on the midfan and lower fan, as evidenced by the percentage of coarser layers in the active valley compared to other valleys [5]. Sandy layers can attain thicknesses of tens of meters, even 1300 km away from the canyon mouth [4].

We propose that during sea level lowering, sand was directly introduced via the extended river systems and transported to different sectors on the fan. This was controlled by migration on the upper fan.

Muddy sheet flows covered sandy deposits in abandoned valley systems during sea level lows. During rising sea level, when flow was more restricted to the then-active valley, the previous coarser valley deposits were buried by mud.

Acknowledgments

Contribution of the Scripps Institution of Oceanography, La Jolla, California. Cruises into the Gulf of Bengal were funded by the Office of Naval Research. We gratefully acknowledge the assistance of our colleagues at sea and the captains and crews of the vessels.

References

[1] Curray, J. R., and others, 1982. Structure, tectonics, and geological history of the northeastern Indian Ocean. In: A. E. M. Nairn and F. G. Stehli (eds.), The Ocean Basins and Margins, Vol. 6: The Indian Ocean. Plenum Pub. Corp., New York-London, pp. 399–450.

[2] Curray, J. R., and Moore, D. G., 1974. Sedimentology and tectonic processes in the Bengal Deep-Sea Fan and geosyncline. In: G. A. Burk and C. L. Drake (eds.), Geology of Continental Margins. Springer-Verlag, New York, pp. 617–627.

[3] Evans, P., 1964. The tectonic framework of Assam. Journal Geological Society India, v. 5, pp. 80–96.

[4] Von der Borch, C. C., Sclater, Y. G., and others (eds.). 1974. Initial Reports of the Deep Sea Drilling Project, Leg 22, Site 218. U.S. Government Printing Office, Washington, D.C., pp. 325–332.

[5] Venkatarathnam, K., Moore, D. G., and Curray, J. R., 1976. Recent bottom current activity in the deep western Bay of Bengal. Marine Geology, v. 21, pp. 255–270.

[6] Emmel, F. J., and Curray, J. R., 1981. Dynamic events near the upper and mid-fan boundary of the Bengal Fan. Geo-Marine Letters, v. 1, pp. 201–205.

CHAPTER 17

Cap-Ferret Fan, Atlantic Ocean

Michel Cremer, Patrick Orsolini, and Christian Ravenne

Abstract

Cap-Ferret Fan (Bay of Biscay) is a complex submarine fan fed by two main systems: the Cap-Ferret system and the Capbreton system. During the Pliocene and Quaternary, the fan developed mainly from slope derived sediments that bypassed the partially filled Cap-Ferret depression. The clayey-silty nature of the sediment supply, together with the effect of the Coriolis force, resulted in thick overbank deposits on the right side of channels that migrated to the left and prograded downfan. This resulted in a large sedimentary levee, producing the elongated and asymmetric shape of the present fan.

Introduction

The Comité d'Etudes Pétrolières Marines, in collaboration with the University of Bordeaux I, has studied the Cap-Ferret Fan since 1978. Numerous techniques have been used, including Seabeam bathymetric surveying, direct observations of the bottom using the submersible Cyana, piston coring, 3.5-kHz echosounding, and high resolution seismic profiling using Miniflexichoc [1–3]. This chapter presents some of the main characteristics of the Cap-Ferret Fan and the inferred sedimentary processes and controls.

Regional Setting

The Cap-Ferret Fan is a complex sedimentary body located in the southeastern part of the Bay of Biscay (Fig. 1). Two main canyon-channel systems, the positions of which are structurally controlled, account for most of the sediment supply to this area. The Cap-Ferret system is located in a broad tectonic depression west of the 60-km wide Aquitain Continental Shelf. The northern flank of the depression is bounded by the passive Armorican Margin, and the southern flank by the Landais marginal plateau which is covered by prograding Aquitain Margin deposits. The Capbreton system, the canyon head of which starts close to the shore, forms a narrow and deep depression between the Landais marginal plateau and the northern Spanish Margin. The Spanish Margin is a westward extension of the Pyrenees and was formed during the Eocene when the Spanish plate moved northward. After a phase of extension followed by a phase of compression during the Oligocene, the present morphologic configuration was developed [4].

Fan Morphology

Because this fan morphology (Fig. 2) is different from that of other fans described in this volume, the classical fan division will not be followed [5–6]. The discussion is based on the main characteristics of two different areas: the Cap-Ferret depression and the sedimentary accumulation on the continental rise between 3800 and 4400-m water depth.

The Cap-Ferret depression is an east-west elongated trough closed to the east by a steep slope (≅ 7.8%) that is cut by numerous V-shaped valleys. These valleys, together with the main Cap-Ferret Canyon, converge and form the U-shaped Cap-Ferret Channel. This channel is 60-km long, 4 to 6-km wide, and has an average gradient of 1.3%. To the north, between the channel and the steep slope of the outer Armorican Margin, is a more gentle-sloping area forming a relief of 300 to 400-m above the channel floor. This area, the "northern terrace," is cut by V- to U-shaped channels that join the Cap-Ferret Channel. To the south,

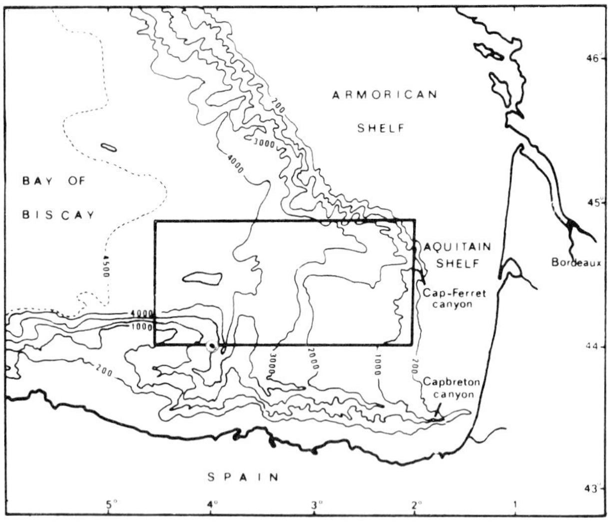

Figure 1. Map of Eastern Bay of Biscay with the general location of the Cap-Ferret Fan. Rectangle presents the position of Figure 2. Bathymetry in meters.

the Cap-Ferret Channel flows along the foot of the Landais marginal plateau with no equivalent to the northern terrace.

The sediment accumulation on the continental rise can be subdivided into three morphologic units:

1) The northern unit is formed by a very large sedimentary levee that extends over more than 100 km. Its upstream width of 15 km increases to 40 km at the western limit of the bathymetric map (Fig. 2), whereas its height decreases from 350 to 50 m and its longitudinal gradient from 1.25 to 0.5%. A concentric arrangement of gentle and steep slopes, related to migrating waves, characterizes the upcurrent part of this levee. At the downcurrent end, it has a more asymmetric shape with a steep slope toward the channel and a long gentle slope on the overbank side.

2) The central area corresponds to a morphologic depression located where the southwestward-flowing Cap-Ferret Channel and the northward-flowing Capbreton Channel merge. Their confluence gives rise to an irregular topography with mounds and closed depressions. There is no direct continuation between these feeding channels and the channel flowing westward (i.e., a headless channel).

3) The southern area is characterized by south to north trending channels. The lower Capbreton Channel displays two distributaries. The eastern distributary is sinuous and U-shaped. Its right flank rises abruptly to a narrow terrace, which bends downcurrent to the northeast and evolves into a levee-like feature. Its left flank presents a more typical but low amplitude levee. The western distributary is V-shaped and displays a straighter path with very low amplitude levees. This configuration shows that the western distributary is a more recent feature than the eastern one, which is separated from the parent channel by a 40-m high step. The Torrelavega Channel flows south to north, then turns and passes south of the Aquitania Seamount. It is a wide valley flanked on its right by a levee-shaped feature that forms the continuation of a structural headland.

In general, the Cap-Ferret Fan shows a complex morphology characterized by several feeding channels, a structural control of the location of the channels across the continental slope, and marked asymmetry of the sedimentary accumulations that are thicker on the right side of the channels.

Sediments

Three sedimentary facies are recognized in piston cores. They are, in order of volumetric importance: 1) muds and silty muds, 2) silts and sandy silts, and 3) pebbly muds and pebbly sands.

Muds and silty muds are the main facies in the surficial sediments. They have a mean grain size of 2 to 10 μm, with less than 10% sand-sized material. The carbonate content is less than 15% in the Pleistocene muds, but may reach as much as 30% in the Holocene muds. They are either homogeneous or finely laminated and normally graded. The texture, the low pelagic content, the fact that the grain size characteristics of these muds change according to the topography, and their relationship with underlying coarser beds showing normally graded sequences suggest that a significant part of this facies was deposited by turbidity currents.

Silts and sandy silts form beds from a few millimeters to a few tens of centimeters in thickness. The thinner beds tend to be silty and the thicker beds more sandy, but only a few beds have a mean grain size in excess of 150 μm. The sedimentary structures (sharp base, normal grading, transitional upper contact with mud) suggest deposition from turbidity currents. Complete BOUMA $T_{a\text{-}e}$ sequences of structures are rarely observed, while numerous beds start with the upper T_c and T_d divisions.

Pebbly muds and pebbly sands are less common and are comprised of: 1) muds constituted only of clay balls, and 2) muds or coarse sands containing gravel, bioclast fragments, and clay rip-up clasts. These facies are found close to steep slopes and in channel axes. They are interpreted as deposits from poorly evolved gravity flows derived from slumping of channel wall and from the Armorican and northern Spanish margins.

Sedimentation Processes

Turbidite Distribution

Based on sedimentary features (Fig. 3) and echosounding [2], three morphosedimentary environments can be distinguished on the Cap-Ferret Fan. These are channels, low-relief areas, and high-relief areas.

Channel deposits (except those of the Cap-Ferret Chan-

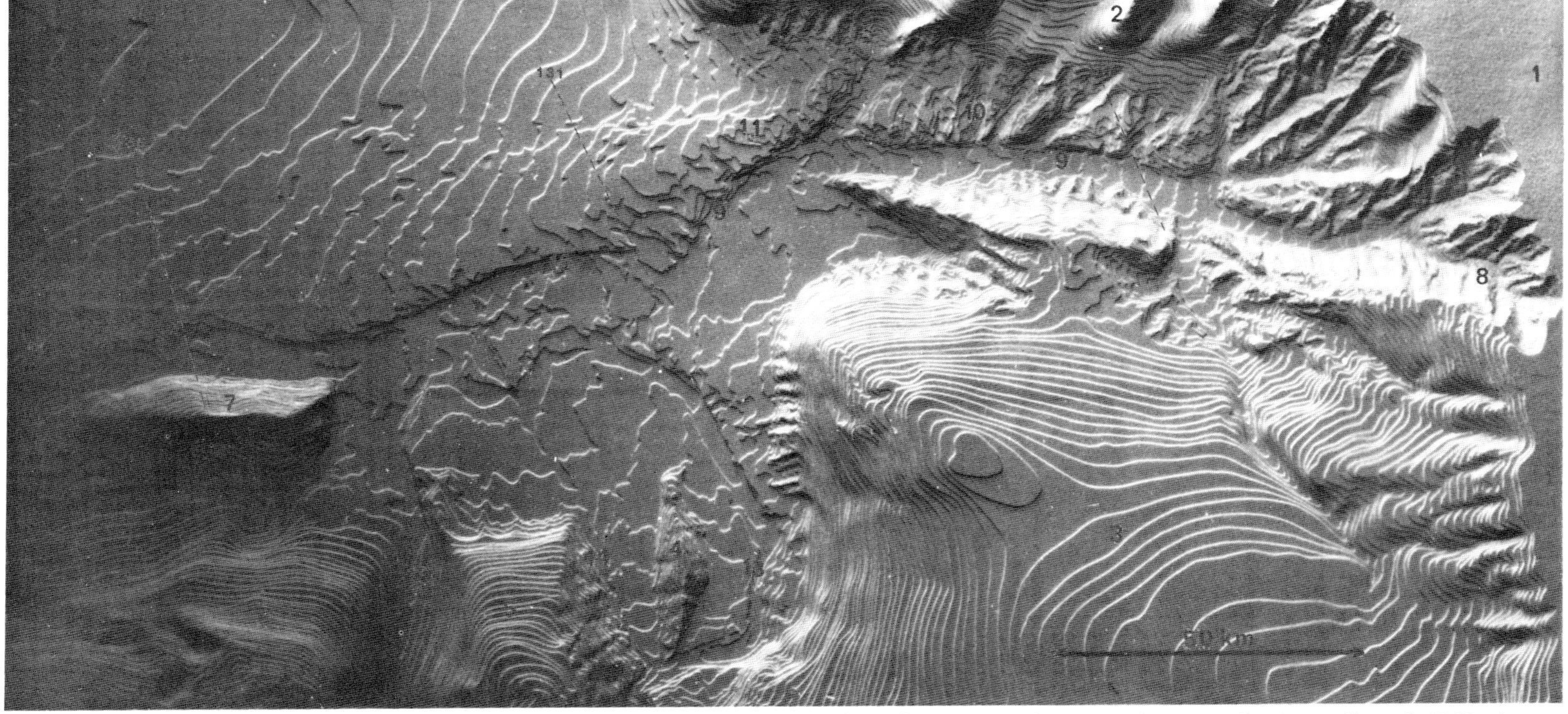

Figure 2. Cap-Ferret Fan model constructed from the Seabeam bathymetric chart (bathymetric curves are 50-m equidistant). 1) Aquitain shelf; 2) Armorican slope; 3) Landais Marginal Plateau; 4) Landes Sea-mount; 5) Santander headland; 6) Le Danois bank; 7) Aquitania Sea-mount; 8) Cap-Ferret Canyon; 9) Cap-Ferret Channel; 10) Northern terrace; 11) Northern levee; 12) Headless channel; 13/14) Eastern and western branches of the Capbreton Channel; and 15) Torrelavega Channel. For location, see Figure 1.

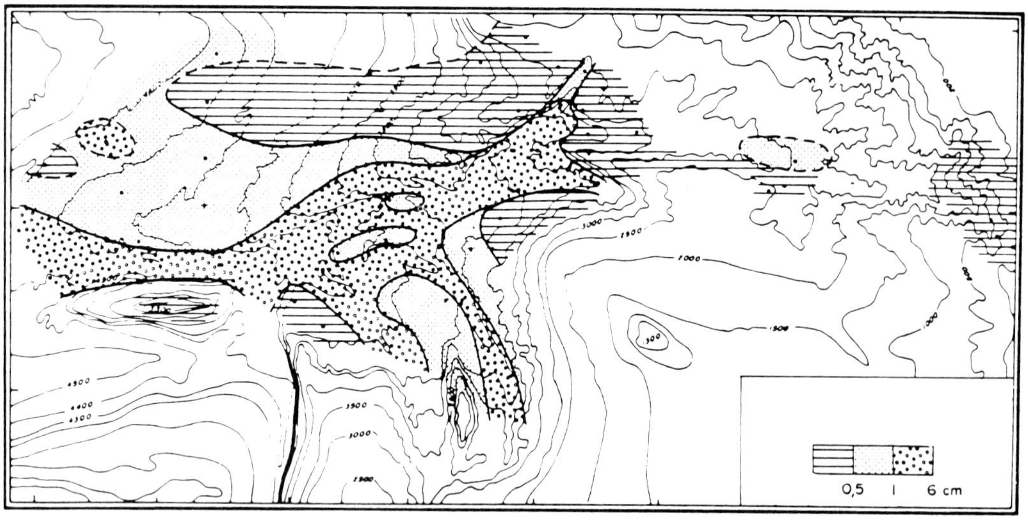

Figure 3. Average thickness variation of silty-sandy turbidites in the superficial deposits. For location, see Figure 1.

nel) contain the thicker-bedded coarser-grained turbidites, with up to 60% silt and sand. Bed thickness is very variable (a few millimeters to 2 m). The channels clearly funnel the lower denser parts of turbidity currents, and the variation in bed thickness most likely results from their variable transport capacity.

Areas of low relief adjacent to channels (levees, terraces) consist of 10 to 20% silty turbidites, normally less than 5-cm thick. These areas probably received only the upper, less dense parts of turbidity currents that overflow the channel walls.

Areas of high relief (greater than 500-m above the channel floor) contain mainly mud with less than 7% silt in beds less than 0.2-cm thick. These areas can only be reached by the thickest turbidity currents that overflow the channels and by hemipelagic or nepheloid suspensions.

The sediments on the northern levee show a marked downfan increase in the thickness of turbidites. This is interpreted as a result of the downcurrent decrease in the levee relief, which enabled turbidity currents coming from the Cap-Ferret and central area channels to overflow more easily [7]. As correlary deposits in the lower part of the Cap-Ferret Channel and in the central area contain more and thicker silty-sandy turbidites than do the deposits in the middle and upper parts of the channel. This likely results from a bypassing and winnowing of the coarse sediment load of turbidity currents toward the channel downdip edges.

Sediment Thickness Distribution

Sediments deposited at the end of the late glacial Wisconsin stage (oxygen isotope stage 2) are presented as an example of the relationship between fan morphology and sediment thickness distribution (Fig. 4). The highest sedimentation rates are observed along the crest of the upper part of the northern levee at the end of the northern terrace and on the levees of the Capbreton and Torrelavega Channels. In contrast, the channel deposits are very reduced in thickness; in the Cap-Ferret Channel and in the central area, sedimentation hiatuses are observed. This sediment thickness distribution is in accordance with a lateral build-up of levees by turbidity currents and with sediment bypassing or lack of sufficient deposition in the channels. The high sedimentation rates on the upfan part of the northern levee imply turbidity currents exceeding 400-m in thickness.

The good relationship between topography and sediment distribution—particularly well expressed on the northern levee—shows that the present fan morphology is the result of sedimentary processes similar to those that occurred during the latest glacial stage and which are characterized by thick overbank deposits and thinner channel fill deposits.

Sea-Level Changes

Important changes in the characteristics, thickness, and distribution of deposits occurred between the late glacial and postglacial stages. This can be clearly seen on the northern levee where the sedimentation rates are low and the transverse and longitudinal trends are weak (Fig. 5). Deposits on the downdip part of the levee are thinner than those on the adjacent channel. In this northern area and in the Cap-Ferret depression, sediments are mainly fine-grained turbidites, whereas the latest Holocene deposits have a more hemipelagic appearance. The highest sedimentation rates are linked to the Capbreton feeding system. Channel deposits in the eastern distributary include turbidites up to 60-cm thick. In this channel and in the central area, a general fining-upward trend is also observed.

These changes are related to the sea level rise between the last glacial and postglacial stages. At the head of the Cap-Ferret depression, the continental shelf became wider, which caused a strong reduction in the sediment supply. However, because the head of the Capbreton Canyon is

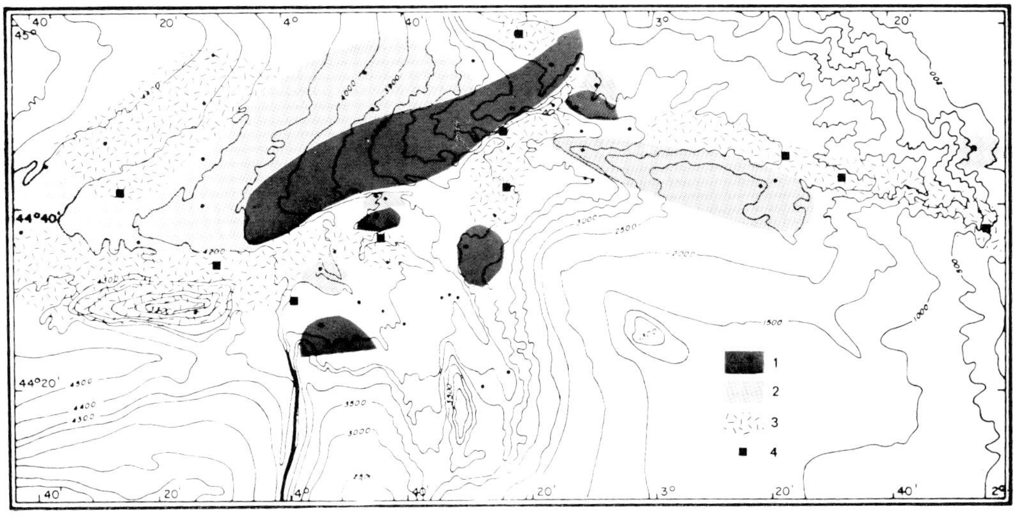

Figure 4. Distribution of the sedimentation rates during the late glacial Wisconsin stage. 1) 20 to 50 cm/1000 years; 2) 5 to 20 cm/1000 years; 3) less than 5 cm/1000 years; 4) sedimentation hiatus.

close to the shore and because the shelf along the Spanish Margin is very narrow, the activity of this feeding system was only partially reduced.

Growth Pattern

The main features of the Cap-Ferret Fan growth pattern were determined from Miniflexichoc high resolution seismic reflection data [2–3].

Cap-Ferret Depression

The late Eocene tectonic phase left a deep elongated through-shaped depression that was isolated from major sediment supplies. From late Oligocene, sediments eroded from the Pryrenees reached the Aquitain Margin, which resulted in progradation. In the upper part of the Cap-Ferret depression, the late Oligocene and early Miocene deposits (equivalent to bottomset deposits) represent the first well-defined channel and overbank deposits. During late Miocene and Plio-Pleistocene time, the progradation of the Aquitain Margin accelerated in response to continuous sediment supply and glacio-eustatic changes. The steep slopes of the head of the Cap-Ferret depression were overrun by prograding slope sediments. This unstable sediment cover induced slope failures which, together with those from the less-supplied Armorican Margin, formed a turbidite infilling in the Cap-Ferret trough. This infilling of over 1000-m thick represents the structurally confined upper part of the Cap-Ferret Fan.

Successive channels giving birth to the present Cap-Ferret Channel are characterized by chaotic reflectors having either high or weak amplitudes, whereas overbank deposits are characterized by more continuous reflectors (Fig. 6a). In the early stages of the Cap-Ferret Fan construction, the

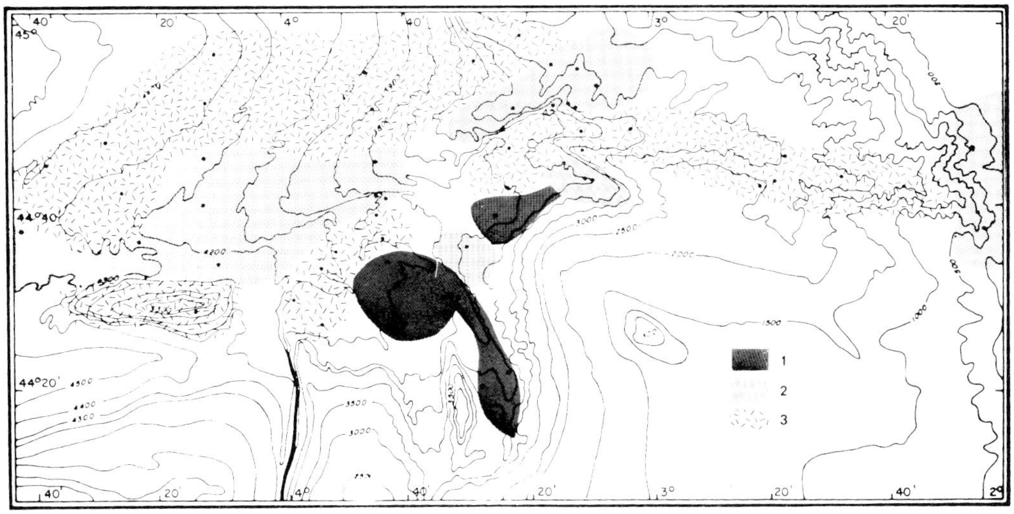

Figure 5. Distribution of the sedimentation rates during the post-glacial stage. 1) More than 20 cm/1000 years; 2) 5 to 20 cm/1000 years; 3) less than 5 cm/1000 years.

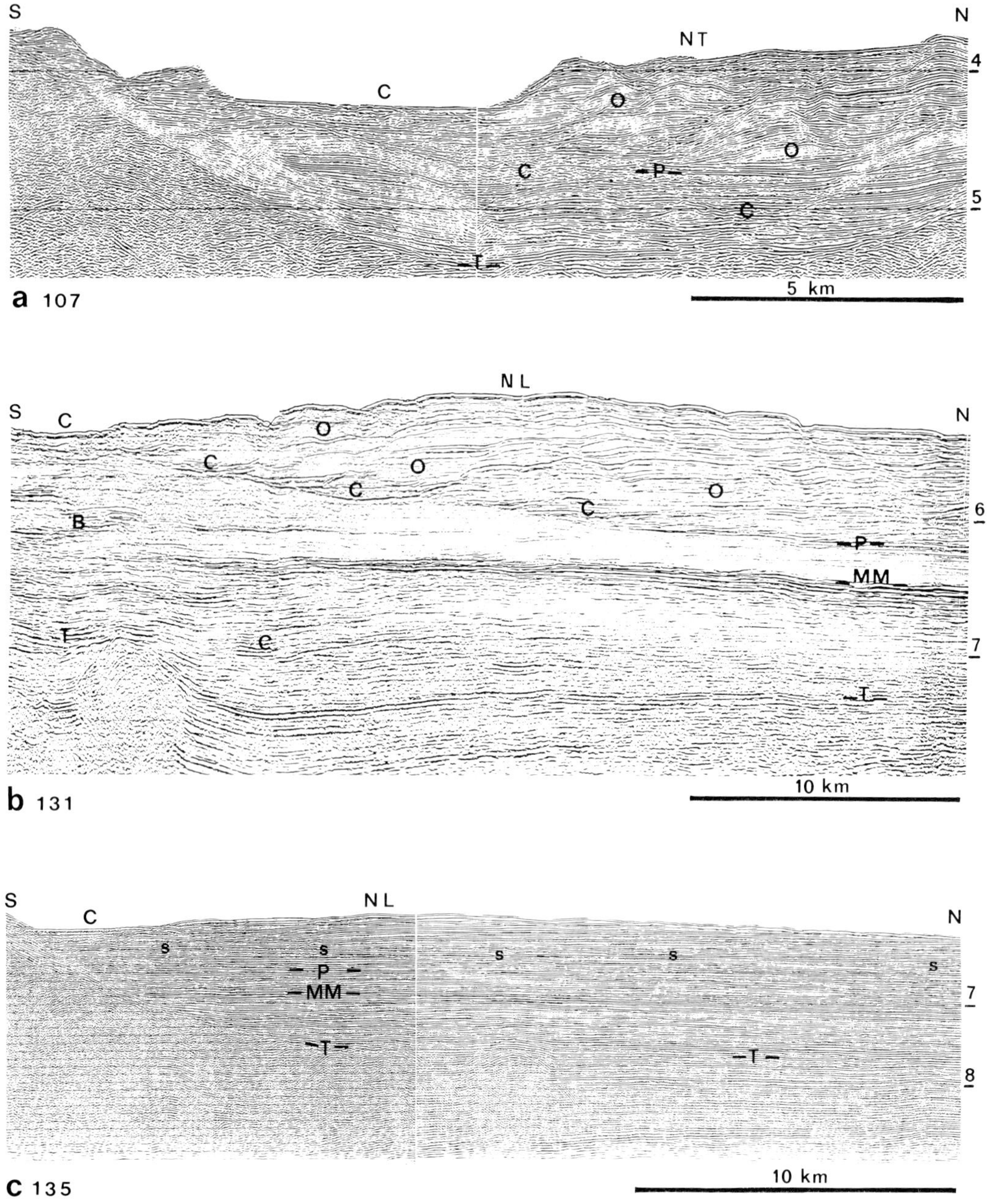

Figure 6. High resolution seismic reflection profiles from the Cap-Ferret Fan. See Figure 2 for location; vertical scale in seconds (two-way travel time). T = late Eocene discordance; MM = Middle Miocene; P = Pliocene base. Line 107 = Turbidite infilling of the Cap-Ferret depression; overbank deposits (O) built up the northern terrace (NT) on the right of the Cap-Ferret Channel (C); they are less developed on the left. Line 131 = Cross-section of the upper part of the northern levee (NL); it built by overbank deposits (O) issuing from the southward-migrating Cap-Ferret Channel (C) and overlaying a levee (transparent facies) developed from the Capbreton Channel (B). An older Cap-Ferret Fan is observed under the high amplitude reflectors of middle Miocene. Line 135 = Cross-section of the lower part of the northern levee overlaying lower fan deposits; its construction above a surface of migrating channel(s) started later than in the upper part of the fan.

paleochannel was shallow, wide, and flanked by lateral deposits showing a transverse lenticular shape and a longitudinal wavy structure. Gradually, the high accretion rate of the overbank deposits on the right side of the channel caused narrowing and deepening of the channel and the build-up of the northern terrace with its internal tabular structure.

The average thickness of the deposits is greater at the head of the Cap-Ferret depression where deep channels are flanked by partially draping deposits that continue into slope deposits, and in its lower part where well-developed overbank deposits form an onlap infilling pattern and where the channel deposits display upcurrent onlaps. This organization supports the concept of turbidity currents derived from slope failure. These gravity flows move through the convergent canyons at the head of the Cap-Ferret depression into the Cap-Ferret Channel, where a progressive decrease in their transport capacity leads to an increase in sediment deposition. However, the fact that the width and depth of the modern channel remain nearly constant down to the mouth of the Cap-Ferret depression is evidence that the entrenched Cap-Ferret Channel has become more of a bypassing channel than a depositional channel.

Continental Rise

The large northern levee was emplaced on the continental rise during Pliocene and Quaternary time by fine-grained sediment-charged turbidity currents coming from the partially filled Cap-Ferret depression. The earlier development of this area can be divided into three phases: 1) during Eocene to middle Oligocene time, the sediments coming from the newly formed Spanish Margin formed an onlap infilling in the abyssal plain; 2) from late Oligocene to early Miocene time, channel-levee systems extended from the Cap-Ferret depression and from the Capbreton Channel, but died out rapidly toward the north and west; and 3) during the middle Miocene, sediments accumulated mainly on the margin slopes. The increase in sediment supply by late Miocene time again resulted in a south-north channel-levee system downdip of the Capbreton Canyon. At the same time or slightly later, the Cap-Ferret channel-levee system started to develop on the backside of the Capbreton right levee (Fig. 6b). During the Pliocene and Quaternary, it became a predominant feature with the build-up of the northern levee.

A seismic cross-section on the upper part of this levee (Fig. 6b) shows its construction with thick overbank deposits that accumulated on the right side of the southward-migrating Cap-Ferret Channel. At the beginning, the channel migration was rapid and the levee was low. With time, as the left hook of the channel increased, channel migration decellerated, levee amplitude increased, and successive channels were more marked. It is possible to rule out any structural control on the channel movement outside the Cap-Ferret depression. The upslope channel migration is obviously linked to sedimentary processes. Moreover, almost all channels on the Cap-Ferret Fan, whatever their direction, display thicker deposits on their right side and indicate leftward migration paths. Thus, the channel migration appears to be influenced by the Coriolis force [8]. The fact that the influence of the Coriolis force is well expressed in the Cap-Ferret Fan may be based on: 1) a low sand/shale ratio of the sediment supply, which results predominantly in a suspension mode of transport within large turbidity currents that have low velocities and, therefore, can be more influenced by the Coriolis force; and 2) a structural control of the position of the main channels all the way down the continental rise, which prevents frequent channel shifts and, thus, increases the possibility of asymmetric deposition of overbank deposits.

Seismic lines west of the confluence of the Cap-Ferret and Capbreton Channels (Fig. 6c) reveal that the construction of the northern levee started later. The well-defined channel-levee system observed upfan changes downfan into deposits characterized by discontinuous slightly undulating reflectors and scattered chaotic and transparent seismic facies. They are interpreted as lower-fan deposits spread by a network of shallow channels flanked by small levees. Upward, the segregation between channel and levee facies increases. Channels, which migrated southward, emerged into a main channel located in the extension of the central confluence area. Overbank deposits from this channel then gave rise to the present levee. This constructional organization shows that, in addition to the lateral spreading of deposits by channel migration, a westward progradation of mid-fan facies on lower-fan facies characterizes the growth pattern of the fan. This is related to the entrenchment of feeding channels by thick overbank deposits that enabled turbidity currents to transport sediments farther downfan.

Conclusion

The more outstanding feature of the Cap-Ferret Fan growth pattern is the build-up of the large northern levee. It gives the present fan an asymmetric and elongated shape. This levee has been constructed on the continental rise during Pliocene and Pleistocene time when the wide intraslope Cap-Ferret depression became the predominant feeding system sourced from the prograding Aquitain margin deposits.

The northern levee build-up, as a well as the construction of the northern terrace that is an equivalent feature in the structurally controlled upper fan, are characterized by: 1) thick overbank deposits that are very well developed on the right side of leftward-migrating channels, 2) the thinness of channel fill deposits, 3) an increase in channel depth with time, and 4) a downfan progradation of the deposits.

It is inferred that the low sand/shale ratio of the sediment supply is the main reason for these characteristics. Such terrigeneous input is likely to cause turbidity currents that are sufficiently thick to easily overflow the channel walls and then feed the overbank areas, and are sufficiently transport-efficient to prevent considerable channel fill. As a result, overbank deposits grow faster than the channel fill. The channel depth increases as well as its funneling efficiency, which in turn leads to an active basinward move of deposition.

The Coriolis force appears to be the second main control on the growth pattern of the Cap-Ferret Fan. The preferential development of overbank deposits on the right side of channels and their correlative leftward migration are related to the overflowing and unchannelized portion of slow-moving turbidity currents which, thus, become strongly influenced by the Coriolis force. In addition, the structural control of the position of channels all the way down the continental rise prevents sudden channel shifts and thereby increases the asymmetry in the depositional patterns.

Acknowledgments

This is a short summary of numerous studies that are part of the Cap-Ferret Project funded by the Comité d'Etudes Pétrolières Marines and completed in collaboration with the University of Bordeaux I. The authors thank the project contributors for sharing data and ideas. D. A. V. Stow and the editors reviewed the manuscript.

References

[1] Coumes, F., and others, 1979. Etude des éventails détritiques profonds du Golfe de Gascogne. Analyse géomorphologique de la carte bathymétrique du Canyon du Cap-Ferret et de ses abords. Bulletin de la Société Géologique de France, Paris v. 21, n. 5, pp. 563–568.

[2] Coumes, F., and others, 1983. Cap-Ferret Deep Sea Fan (Bay of Biscay). In: Watkins J. S., and Drake C. L. (eds.). Studies in Continental Margin Geology. American Association of Petroleum Geologists Memoir 34, pp. 583–590.

[3] Cremer, M., 1983. Approches sédimentologique et géophysique des accumulations turbiditiques. L'éventail profond du Cap-Ferret, la série des Grés d'Annot. Thèse Doctorat ès-Sciences, Université de Bordeaux I. Editions Technip, Paris réf IFP 32036, pp. 344.

[4] Deregnaucourt, D., and Boillot, G., 1982. Structure géologique du Golfe de Gascogne. Bulletin du Bureau de Recherches Géologiques et Minières, Paris, Sect. 1, n. 3, pp. 149–178.

[5] Normark, W. R., 1970. Growth patterns of deep-sea fans. American Association of Petroleum Geologists Bulletin v. 54, pp. 2170–2195.

[6] Normark, W. R., 1978. Fan valleys, channels, and depositional lobes on modern submarine fans: characters for recognition of sandy turbidite environments. American Association of Petroleum Geologists Bulletin v. 62, pp. 912–931.

[7] Piper, D. J. W., and Normark, W. R., 1983. Turbidite depositional patterns and flow characteristics, Navy Submarine Fan, California Borderland. Sedimentology v. 30, pp. 681–694.

[8] Menard, H. W., 1955. Deep-sea channels, topography, and sedimentation. American Association of Petroleum Geologists Bulletin v. 39, pp. 236–255.

CHAPTER 18

Ebro Fan, Mediterranean

C. Hans Nelson, Andres Maldonado, Francis Coumes, Henri Got, and Andre Monaco

Abstract

The Ebro Fan System consists of en echelon channel–levee complexes, 50 × 20 km in area and 200-m thick. A few strong reflectors in a generally transparent seismic facies identify the sand-rich channel floors and levee crests. Numerous continuous acoustic reflectors characterize overbank turbidites and hemipelagites that blanket abandoned channel–levee complexes. The interlobe areas between channel complexes fill with homogeneous mud and sand from mass flow and overbank deposition; these exhibit a transparent seismic character. The steep continental rise and sediment "drainage" of Valencia Trough at the end of the channel–levee complexes prevent the development of distributary channels and midfan lobe deposits.

Introduction

The Ebro Deep-Sea Fan system lies in a restricted basin on the northwestern Mediterranean continental rise between the Ebro River and the Valencia Trough west of the Balearic Islands (Fig. 1). The fan is of particular interest because its morphology and development differ from typical systems. The Ebro system consists of several channel–levee complexes, each associated with a separate slope valley and terminating at Valencia Trough. This is in contrast to typical fan development, in which a main inner fan valley continues into distributaries within midfan or lower-fan depositional lobe areas.

In this study we correlate the near-surface (upper 100 m) seismic stratigraphy with detailed morphology and core stratigraphy. By defining the lobe history and morphological development, it is possible to formulate a growth pattern for the system.

Geologic Setting

The continental margin off the Ebro River is a young, passive margin initiated in Late Paleogene time as a result of the rifting and spreading of the western Mediterranean basins [1]. Listric normal faults striking northeast-southwest created deep-seated graben systems parallel to the margin that control basin configuration. The Pliocene through Quaternary stratigraphy and morphology of the continental margin reflect the seaward progradation of deltaic deposits [2]. On the continental slope these well-stratified deposits are cut by shallow slope canyons, some of which are filled with sediment.

The acoustic units on the continental rise (1200 to 1800-m water depth) are capped by a well-stratified Pleistocene sequence of deep-sea fan and base-of-slope deposits [3]. From the Ebro River and south for about 110 km, the continental rise is covered by channel–levee complexes. Each of these lenticular units is approximately 15-km wide, 40 to 50-km long, and 150 to 200-m thick [4].

The Ebro River, which feeds these deep-sea fan deposits, is one of the four major sediment sources of the Mediterranean Sea. It drains one-sixth of Spain (85,835 km^2); its annual sediment discharge is 3 to 4 million t [5]. Fine-grained sand is trapped in the coastal environments of the delta while the prodelta is built by organic-rich silt and clay. Relict outer-shelf sand and gravel and the lack of turbidites in the Holocene sediment of the fan show that, at present, little Ebro sediment reaches the fan system.

In contrast, greater river discharge during low stands of sea level in the Pleistocene is suggested by coarse-grained,

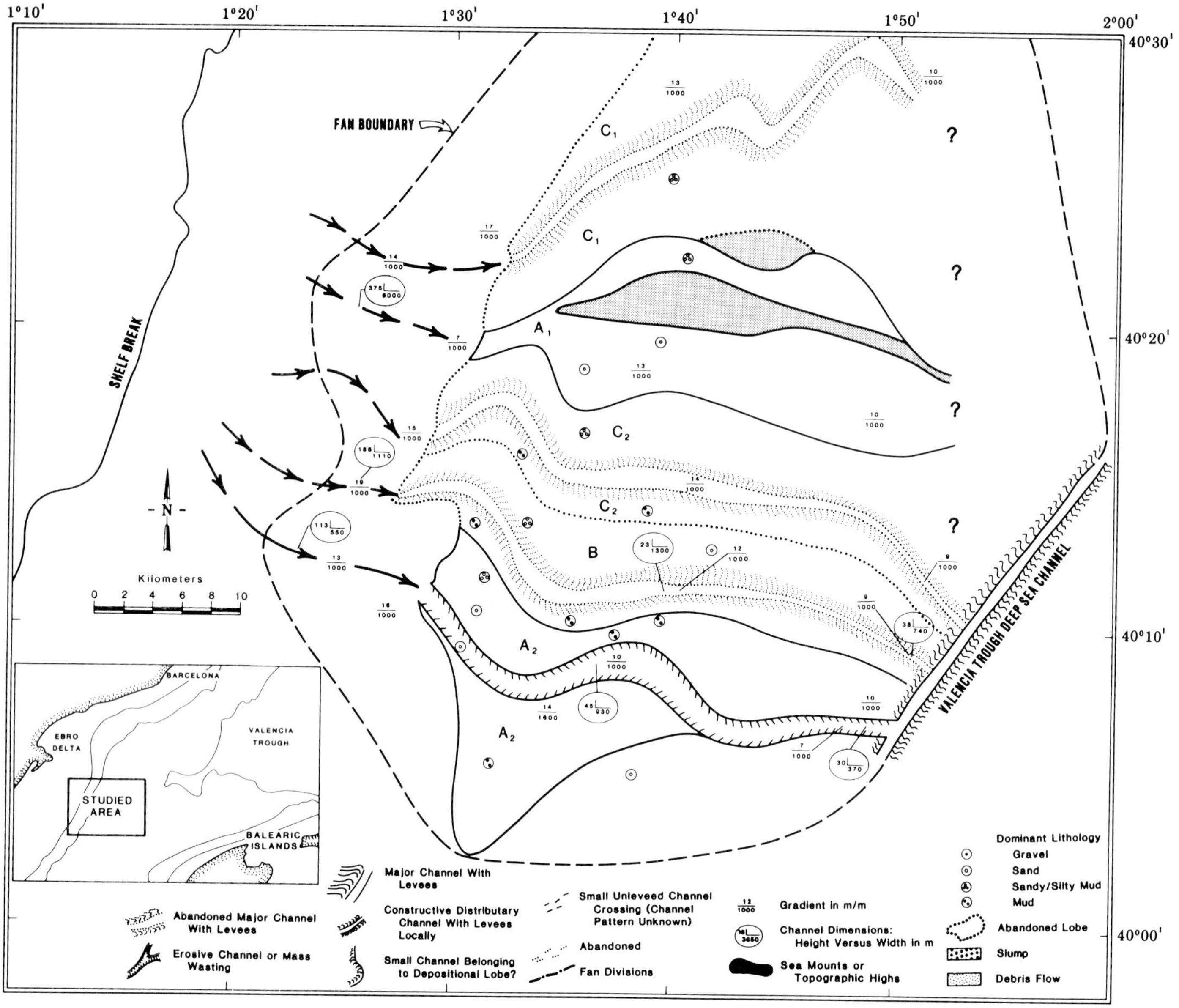

Figure 1. Ebro Fan system morphometry and generalized lithology. Gradients show that there are no generalized breaks in slope. Sand-rich cores have a sand/mud ratio greater than 1:1, sandy–silty cores have sand/mud ratios of 1:2–1:6, and mud-rich cores have sand/mud ratios of less than 1:10. Mud thickness is reduced to one-third for calculations so that these ratios for cores simulate sandstone/shale ratios.

polygenetic gravels both in the Ebro fluvial valley [5] and on the continental shelf [2]. The Ebro deep-sea fan system developed during the Pleistocene as a result of this abundant sediment supply and the concomitant displacement of the river depocenter to the outer shelf [3].

Methods

The *R.V. Cornide de Saavedra* was used in 1979 to take 1000 km of 3-kJ seismic profiles. In 1981, 60 piston cores were collected using this ship and the *R.V. Garcia del Cid.* Satellite and Loran A navigation systems with a precision of 1 to 2 km were used on both vessels. Our tracklines were adjusted to match unpublished seismic data collected with precision Shoran navigation. We surveyed the north half of the entire Ebro Fan system with a line spacing of 5 to 10 km over an area of 2000 km^2.

Morphology

The continental shelf off the Ebro River is wide (up to 70 km) in comparison with other shelf areas in the western Mediterranean (Fig. 1). The lobate Ebro Delta extends offshore, producing an arcuate prodelta bulge in the shelf. The con-

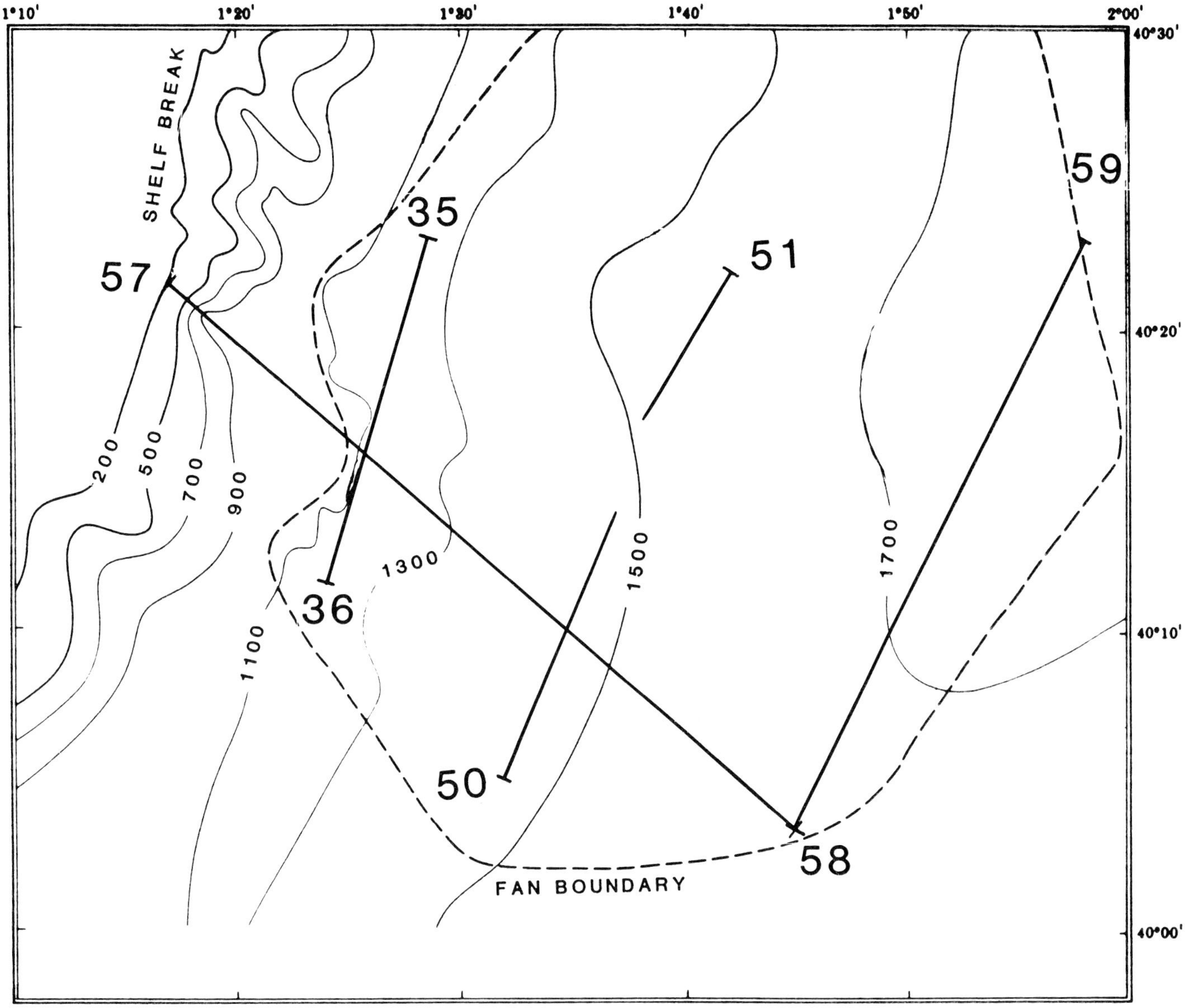

Figure 2. Bathymetric map of the northern half of the Ebro Fan system with contours. Bathymetry based on hydrographic charts of the region with limited updating from our sparker profiles having a 5 to 10-km east-west Spacing. Channel locations have been accurately located with precision navigation and are shown on Fig. 1. Numbered lines indicate position of seismic profiles illustrated in Fig. 3. Dashed line gives outline of fan.

tinental slope is steep (4.5° average), extremely narrow (10 km), and cut by a number of short, steep canyons (Figs. 1, 2). These canyons are less than 1-km wide across their floors and have reliefs of 200 to 600 m (Fig. 3, line 35–36). Canyon-floor gradients range from 3° to 4° and the wall slopes vary from 5° to 7°. The change in gradient from the base of the continental slope to the continental rise is less than usual because the rise is very steep, averaging 1° [3].

Each of the canyons crossing the lower continental slope continues on the continental rise as one of the isolated leveed fan valleys spaced 5- to 10-km apart. The fan valleys traverse the entire rise individually without converging or breaking into distributaries prior to intersecting the Valencia Trough. The Valencia Trough is deeply incised along its length by an erosive deep-sea channel, with levees, whose low gradient is similar to midoceanic canyons or deep-sea channel systems (Figs. 1, 2) [6]. This trough channel extends eastward 160 km, where it divides into smaller, distributary channels that appear to feed fan depositional lobes of the Valencia Fan [8].

The entire Ebro Fan system thus consists of separate channel–levee complexes that extend 50 km from the slope base to the Valencia Trough where sediment is apparently transported for 150 km, perhaps continuing on to depositional lobes. Because of their individual canyon sources plus independent nature in the inner fan and their discontinuity downfan, the

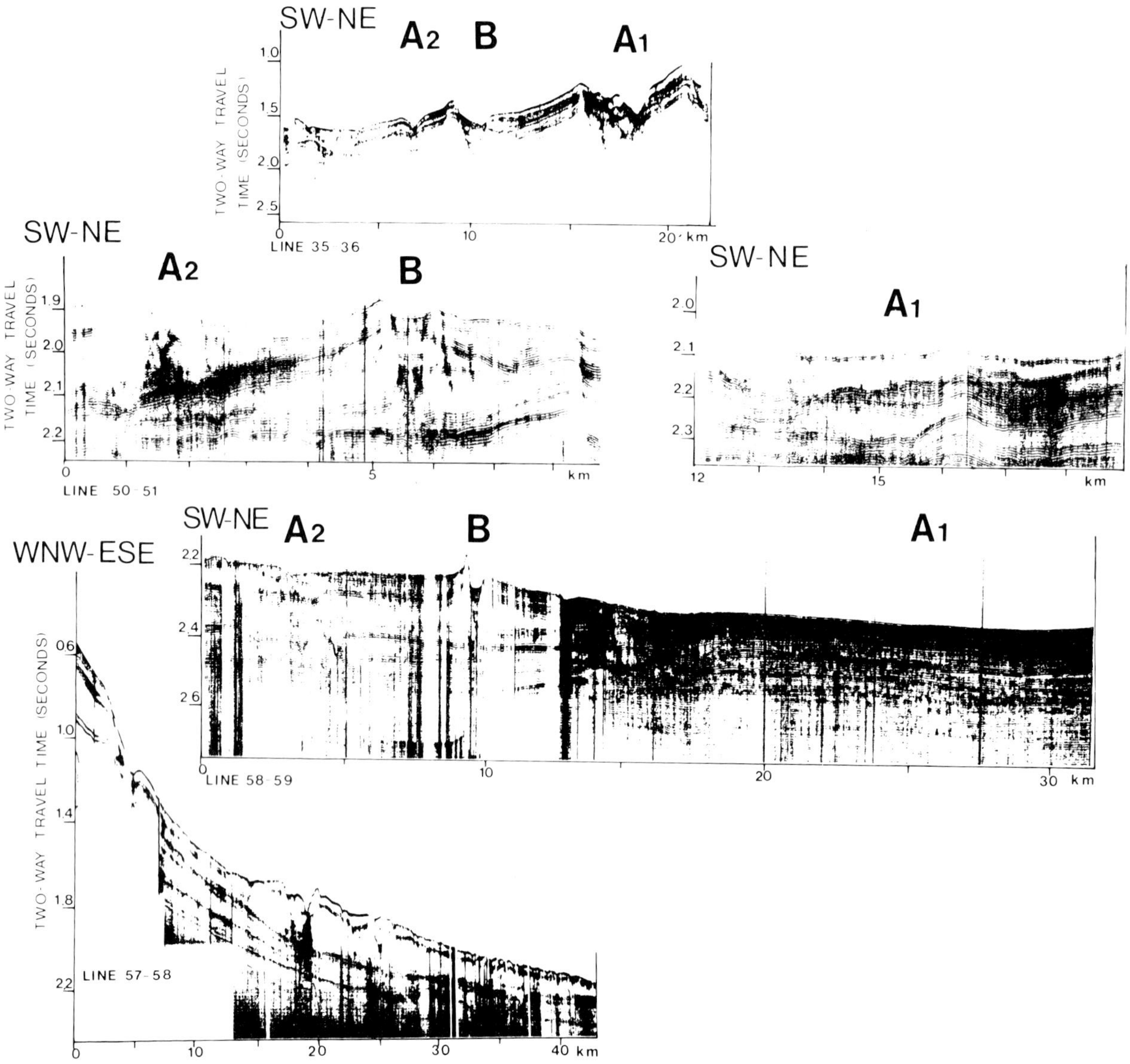

Figure 3. Characteristic 3000-J continuous seismic profiling transects in the Ebro Fan system. Line 35–36: transverse profile in the inner-fan area; line 50–51: transverse profile in the middle-fan area; line 58–59: transverse profile in the outer-fan area; line 57–58: longitudinal profile in the A_2 channelized lobe of the Ebro Fan system. See Fig. 2 for location of seismic sections.

channel–levee complexes are considered separate channelized lobes. These morphologic elements define an Ebro Fan system, albeit one that is atypical, and a system that has been incompletely studied except for the northern channel–levee complexes.

Both the morphology of the channelized lobes and the lobe-surface gradients are consistent from one channelized lobe to the next. Each channel–levee complex is lenticular in cross section (Fig. 3), and longitudinal profiles reveal no abrupt change in downfan gradients (Fig. 1). Gradients of the entire channelized lobe system fall within typical inner-fan gradients of 1:100 or steeper (Fig. 1) [7].

Stratigraphy

On the basis of the reflector densities in seismic profiles, three acoustic facies types can be defined in the upper 100 m of sediment. Type I is a transparent facies with no consistently parallel reflectors (Fig. 3, line 50–51A, A_1 upper unit). Type II is a semitransparent facies with fewer than 10 (typically, 5–7) discontinuous high-amplitude reflectors per 100 m (Fig. 3, lines 50–51 and 58–59, A_2 upper unit). Type III is a facies with more than 25 (typically, 25–27) continuous, parallel, but lower amplitude reflectors per 100 m (Fig. 3, line 58–59, A_1 upper unit).

Using these seismic facies and the relative superposition of the several channel–levee complexes and ponded sediment areas between them, three kinds of lobe systems can be defined (Fig. 3, A_1, A_2, B). The ponded lobe, A_1, has a feeding canyon with extensive slumping (Fig. 3, line 35–36), has no channels in the main flat-surfaced lobe area (Figs. 1, 3), consists of type I facies with no internal reflectors (Fig. 3, line 50–51, A_1), and at its distal end grades to type III facies (Fig. 3, line 58–59, A_1). The youngest channelized lobe, A_2, consists of type II acoustic facies and is characterized by a lenticular profile and an erosive valley (Figs. 2, 3, see A_2). All B and older channelized lobes have filled valleys smoothed by sediment of type III acoustic facies that blankets the underlying channel–levee complex. The buried complex itself is characterized by type II acoustic facies (Fig. 3, B, all lines).

All lobes, channelized or nonchannelized, can be traced into canyon sources (Fig. 1); the seismic stratigraphy of these canyons, however, varies in different lobes (Fig. 3, line 35–36). The seismic profile of the slope valley of the A_1 lobe shows a series of broken reflectors suggesting slumps on its northern side (Fig. 3, line 35–36). The next profile (line 50–51), 15 km to the east, shows that the A_1 valley has broadened and is filled with type I sediment almost to the distal end of the lobe. The A_2 valley shows active erosion along its entire length, whereas all the other valleys exhibit partial filling with type III sediment.

The core stratigraphy varies not only between lobes of different age but also for different environments within the same lobe (Fig. 1). In the ponded lobe (A_1) south of the ridge, sediment in the upper 5 m is predominantly sand, whereas north of the ridge it is predominantly mud. In the youngest channelized-lobe complex (A_2), high sand/mud ratios prevail throughout except for the lower, distal flanks of levees. Except in the lower channel regions of B, the older B and C channelized lobes (Figs. 1, 3) exhibit near-surface sediment that is dominated by mud with interbedded thin-bedded turbidites. The Holocene stratigraphy of all areas is dominated by hemipelagic and turbidite muds with rare thin laminae of thin-bedded sandy turbidites.

Discussion

The youngest lobe (A_1), a ponded lobe, has different morphology, seismic stratigraphy, sediment facies, and resultant sedimentary history than the other channelized lobes. No channel development is apparent in A_1, though the feeding canyon is the largest in this part of the continental slope (Figs. 2, 3). The transparent acoustic facies, lack of channel development, and variable lithology of lobe A_1 all suggest that the otherwise typical channelized turbidite sedimentation fed by a delta distributary source did not take place in this lobe. Perhaps large-scale slumping (Fig. 3, line 35–36, A_1) during Holocene high sea level has resulted in the ponding of unchannelized sediment gravity-flow deposits in this swale between other older channelized lobes (Figs. 1, 3, 4).

The active, erosive fan valley and levees, together with the high sand content throughout the system, suggest that lobe A_2 (Figs. 1, 3) was the most recently active fan valley; apparently it developed in the Late Pleistocene, when a large quantity of sand was fed from river distributaries directly to its canyon head source at the shelf.

Lobe B (Figs. 1, 3) possesses the largest fan valley observed adjacent to the Ebro Delta region. The smoothness of the valley floor, however, and the significant proportion of valley fill, as well as the superposition of lobe A_2 (Figs. 1, 3), show that B is an older fan valley system. Morphology and stratigraphy typical of upper fan valleys extend to the distal end of the B system at Valencia Trough, much as they do in the younger A_2 channelized lobe.

Channelized lobe C_1 lies on the northern margin of the study area and exhibits a well-developed surface fan valley morphology even though there is a significant cover of younger overbank deposits (Fig. 1) and this channel–levee complex is older than A or B lobes. In places, the southern levee is sandier than the northern levee, and the channel floor appears to be newly eroded, perhaps indicating renewal of turbidity–current activity in the channel after its long period of inactivity (Fig. 1).

Channelized lobe C_2 (Fig. 1) is almost completely covered by levee flanks from other, younger channelized lobes, particularly on its southern flank near lobe B. Because of the deep subsurface burial of its formerly active channel and its poor morphologic expression at the sea floor, lobe C_2 is considered to be the oldest channelized lobe system.

It appears that the Ebro Fan complex consists of lobes of different types, with varying ages and depositional history. When canyon slumping or overbank suspension flow feed into a confined area between older channelized lobes, a ponded lobe of heterogeneous sediment is deposited (Fig. 4). A canyon that extends from an active deltaic depositional zone and feeds into an unconfined area on the continental rise develops a sand-rich channel-levee complex or channelized lobe. Older channelized lobes remain exposed on the surface until they are covered by younger overbank and hemipelagic deposits. Older slope and fan valleys may be rejuvenated as activity on the delta shifts and slumping changes slope morphology.

The general growth process for the Ebro Fan system can be summarized as follows: Erosive valleys evolve into filled valleys in channelized lobes; later, ponded lobes develop between the older channelized lobes. The surface morphology eventually will exhibit all aspects of these underlying older lobes, the new active lobes, and the evolutionary stages of erosive, filled, and rejuvenated valleys.

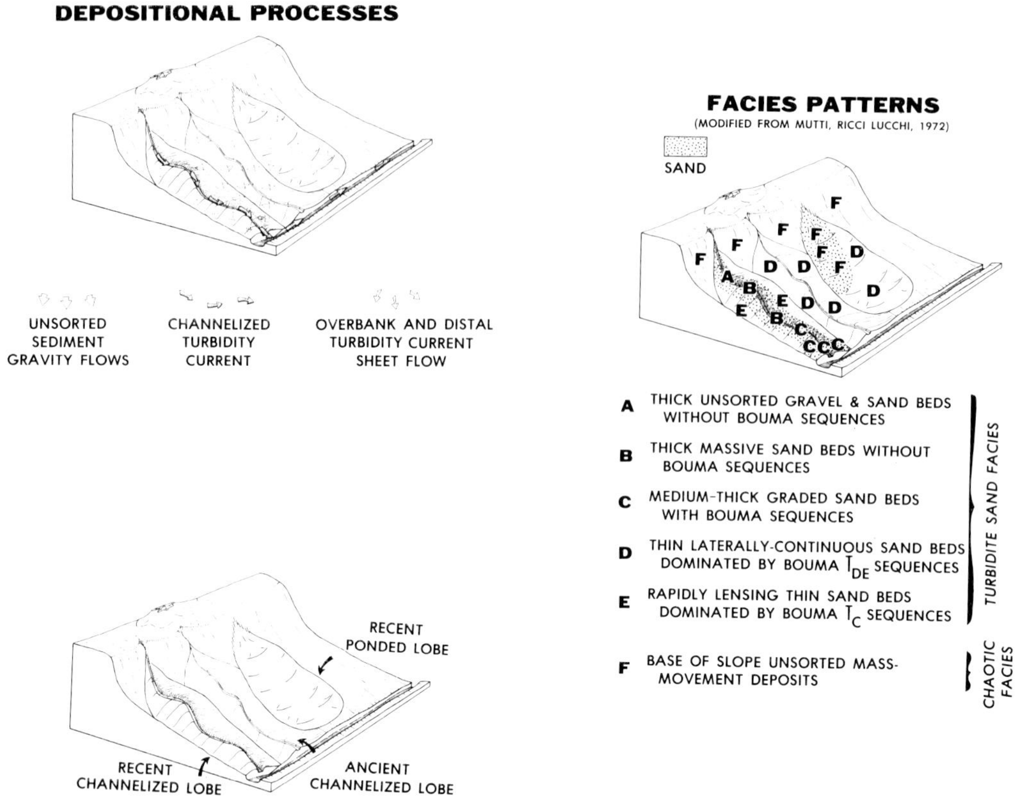

Figure 4. Growth pattern, processes, and facies of the typical Ebro Fan ancient channelized, recent channelized, and recent ponded lobes. Facies classification modified from [11].

Geologic Significance

The morphology, the stratigraphy, and the growth patterns in the Ebro Fan systems are different in many ways from those observed in typical fans whose growth patterns have not been disrupted by tectonic activity. Generally, gradients throughout the channelized lobes of the Ebro Fan system are comparable to upper-fan gradients of similar-sized systems [6]. There are no inner- to middle- to outer-fan morphologic breaks in slope, only convex upward bulges at those places where single channelized sediment bodies are developed. There is no sign of the typical development of a leveed inner fan valley dividing into middle- and lower-fan distributary channels, nor do valley shiftings have a leftward pattern [6]. Sand is not deposited in either middle or lower suprafan depositional bodies (see Normark, Navy Fan, this volume), even though large fan valleys capable of transporting significant quantities of sediment downfan are present. New studies suggest that a large amount of sand has been funneled from Ebro channels through Valencia Trough to Valencia Fan [8]. The proportion that remains in the channel–levee complexes and ponded areas between the lobes cannot be predicted with the available data base.

Fan-growth history in the Ebro system reveals no persistent, single canyon and inner fan valley like that observed in the Rhone Fan system [3]. Several separate canyon to channel–lobe systems continue across the entire continental rise to the floor of the Valencia Trough. This is unlike typical systems such as the Rhone Fan in which a main fan valley evolves into midfan distributary channels and downfan depositional lobes [3,4]. In the Ebro Fan system, large amounts of sand are deposited in a series of inner-fan, channelized sediment bodies that develop sequentially, side by side from north to south.

The grain size type of the sediment source and the resulting efficiency of sand transport have been proposed to be a main control of fan growth patterns [9]. The Ebro and Rhone systems have similar amounts, types, and styles of river-mouth sedimentation; however, differences between their growth patterns suggest that source and transport efficiency have not been the main controlling factor in Ebro Fan depositional patterns.

Near the Ebro delta, the continental margin is dominated by east-west structural control, with subsidence occurring along trends perpendicular to fan growth [1]. Because of this, gradients of the entire rise are similar to those typically found only on the inner fan of most fan systems [6]. Apparently, margin-parallel structural trends near the Ebro Delta [1] and the resulting steep topography enable individual inner-fan valleys to persist as single channelized sediment bodies across the entire continental rise. The crosscutting graben valley of the Valencia Trough then terminates Ebro Fan growth in this part of the system because sediment is transported down the deep-sea channel in the Valencia Trough.

The importance of the structural control of modern fan growth patterns is evident in the Ebro Fan system. Data from some thoroughly drilled systems in rocks [10] also indicate that structural control of a depositional basin results in considerable local variety in fan growth pattern as compared with the pattern predicted by models for typical fan systems.

Acknowledgments

We dedicate this paper to the memory of Maribel Hoyos, whose good humor and assistance persisted throughout the project, but whose life was cut tragically short in an accident a few days after the completion of her laboratory work. We also thank the crews and scientific staffs of the vessels *R.V. Cornide de Saavedra* and *R.V. Garcia del Cid*. This work has been funded by the Comision Asesora de Investigacion Cientifica y Tecnica (Spain) under Grant Number 3678-79. Kirk Johnson and Fidelia Portillo kindly assisted with drafting of model diagrams. N. Bogen, T. Nilsen, and W. Normark provided constructive reviews. Any use of trade names is for descriptive purposes only and does not imply endorsement by the U. S. Geological Survey.

References

[1] Biju-Duval, B., Letouzey, J., and Montadert, L., 1978. Structure and evolution of the Mediterranean basins. In: K. J. Hsu, L. Montadert, and others, Initial Reports of the Deep Sea Drilling Project, v. 42 (part 1). U. S. Government Printing Office, Washington, DC, pp. 951–984.

[2] Maldonado, A., and others, 1981. Mecanismes sedimentaires et edification du plateau progressif sud-catalan (Mediterranee nord-occidentale). Rapports et Proces-verbaux des Reunions Commission Internationale pour l'Exploration Scientifique de la Mer Mediterranee, v. 27, Fasc. 8, Monaco, pp. 25–28.

[3] Bellaiche, G., and others, 1981. The Ebro and the Rhone deep-sea fans: first comparative study. Marine Geology, v. 43, pp. M75–M85.

[4] Aloisi, J. C., and others, 1981. L'eventail sous-marin profond du Rhone et des depots de pente de l'Ebre: essai de comparaison morphologique et structurale. In: F. C. Wezel (ed.), Sedimentary Basins of the Mediterranean Margins. C.N.R. Italian Project of Oceanography, Tecnoprint, Bologna, pp. 227–238.

[5] Maldonado, A., 1972, El delta del Ebro. Estudio Sedimentologio y estratigrafico. Boletin de Estratigrafia, Universidad de Barcelona, v. 1.

[6] Nelson, C., and others, 1970. Development of the Astoria Canyon—Fan physiography and comparison with similar systems. Marine Geology, v. 8, pp. 259–291.

[7] Nelson, C. H., and Kulm, L. D., 1973. Submarine fans and deep-sea channels. In: G. V. Middleton and A. H. Bouma, (eds.), Turbidites and Deep Water Sedimentation. Society of Economic Paleontologists and Mineralogists, Pacific Section, Short Course, Anaheim, pp. 39–78.

[8] Maldonado, A., and others, 1985. The Valencia Fan (Northwestern Mediterranean): a variation on a model for distal deposition. Marine Geology, v. 62, pp. 295–319.

[9] Mutti, E., 1979. Turbidites et cones sous-marins profonds. In: P. Homewood (ed.), Sedimentation Detritique (Fluviatile, Littoral et Marine). Institut Géologique Université de Fribourg, Switzerland, pp. 353–419.

[10] Seimers, C. T., Tillman, R. W., and Williamson, C. R., 1981. Deep-water clastic sediments, a core workshop. SEPM Core Workshop No. 2, Society of Economic Paleontologists and Mineralogists, Tulsa, OK.

[11] Mutti, E., and Ricci Lucchi, R., 1972. Le torbiditi dell'Appennino settrentrionale: introduzione all'analisi di facies. Societa Geologica Italiana Memorie, v. 11, pp. 161–199. (Translation in T. H. Nilsen, 1978. International Geology Review, v. 20, pp. 125–166).

CHAPTER 19

Indus Fan, Indian Ocean

V. Kolla and F. Coumes

Abstract

The upper Indus Fan is characterized by an average 1:500 gradient, channels with 100 m high levees, several continuous subbottom reflectors on 3.5-kHz records, and generally fine-grained sediments. Multichannel seismics show the levee complexes typified by overlapping wedge-shaped reflection sets and channel axis by high-amplitude discontinuous reflections. The middle fan has 1:500–1:1000 gradients and channels with ~20 m high levees. The lower fan has gradients less than 1:1000, channels with 8–20 m high levees, few or no subbottom reflectors on 3.5-kHz records, and high sand content. Besides the dominant unchannelized turbidity currents, channelized and overbank flows also played a significant role in the sedimentation of the lower fan.

Introduction

The Indus Fan, with its 1500 km length, 960 km maximum width, and 1.1×10^6 km^2 area, is the most extensive physiographic province of the Arabian Sea in the northwest Indian Ocean (Fig. 1). It is bounded by the continental margin of India-Pakistan and Chagos-Laccadive Ridge on the east, by the Owen and Murray Ridges on the west and north, and by the Carlsberg Ridge on the south. The shelf-break occurs at about 100-m depth along the India-Pakistan margin. The shelf width, however, is variable and is about 350 km off Bombay and 150 km between the Indus River confluence and the Gulf of Kutch. The shelf width is much less south of Bombay.

The Indus River system, draining the Himalaya mountains, has been the dominant supplier of sediments in the Indus Fan. The Indus River, after leaving the Himalayas, travels a distance of about 1200 miles across the plains of Pakistan before it enters the Arabian Sea. The Hindukush-Sulaiman mountains bound the western side of the Indus River; however, these mountains do not contribute a significant amount of sediment to the river. The Narmada and Tapti Rivers, which drain the peninsular shield of India, although minor in importance, have also contributed sediments to the eastern Arabian Sea.

The Pakistan-India margin and the adjacent Arabian Sea have evolved through a two-stage evolutionary process: rifting and sea-floor spreading [1]. The Chagos-Laccadive Ridge and its partly buried northward extension, the Lakshmi Ridge, is more or less parallel to the India-Pakistan margin [1,2]. The region between the Chagos-Laccadive-Lakshmi Ridge and the India-Pakistan margin evolved as a result of rifting during the Late Cretaceous. The Arabian Sea south and west of the Chagos-Laccadive-Lakshmi Ridge evolved as a result of sea-floor spreading since Paleocene. The Indus Fan sequence was deposited in the Arabian Sea since about Oligocene-Miocene time, coincident with the major uplift of Himalayas. Buried and exposed ridges within the fan and the ridges bounding the fan suggest that the shape of the fan has been controlled structurally as well as depositionally.

The gross acoustic character and sediment-thickness distribution of the Indus Fan have been discussed by several workers [1–5]. Distributions and origins of mineral components in the surficial sediments of the Arabian Sea have also been reported [6,7]. However, these studies did not deal with the fan sequence and sedimentation as such. The objective of this paper is to discuss briefly the morpho-acoustic and sedimentologic characteristics and sedimentation pat-

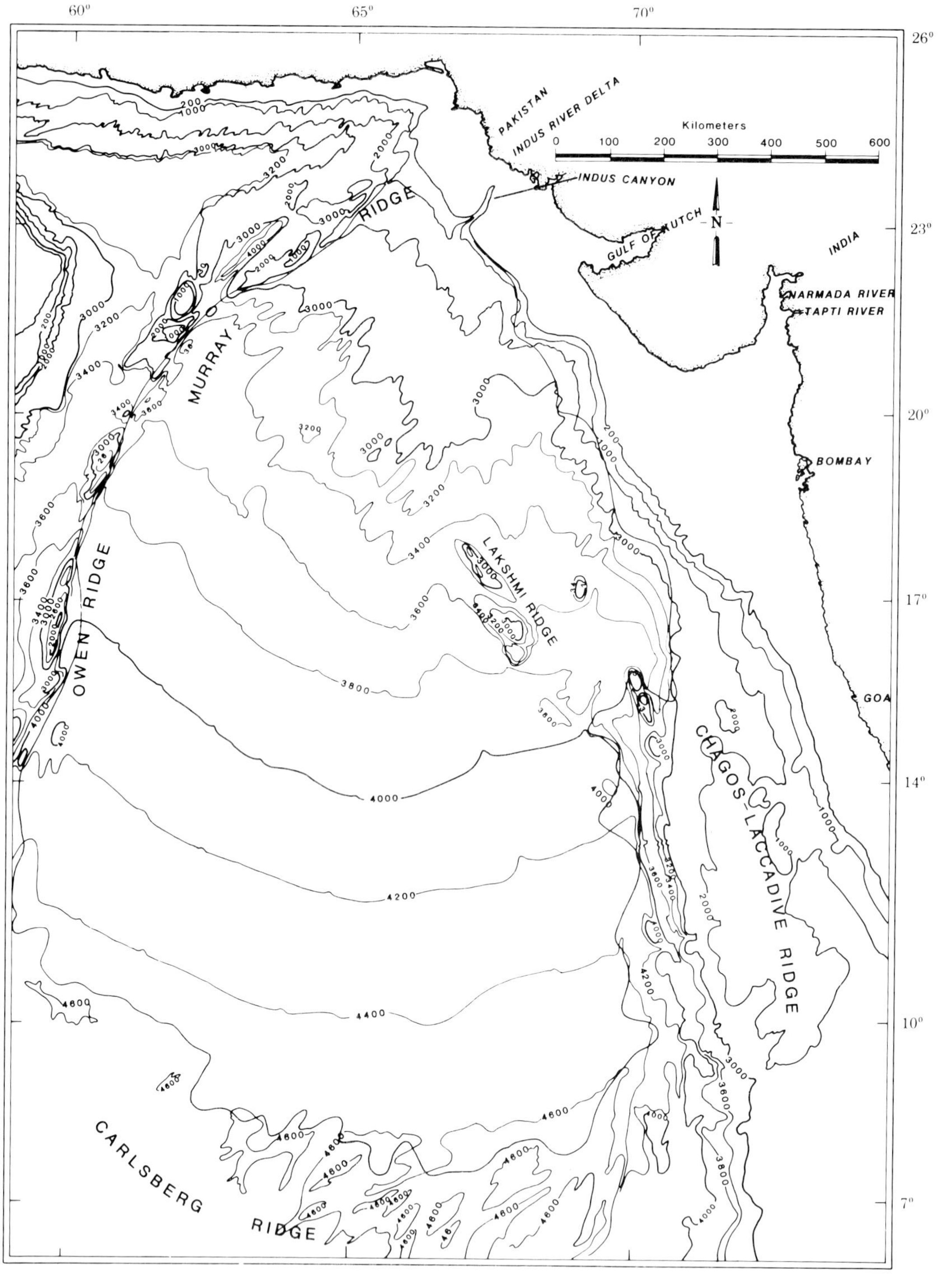

Figure 1. Bathymetry of the Indus Fan and surrounding regions, compiled by the U.S. *Navoceano*.

terns of the Indus Fan. A more detailed contribution on this subject is in preparation.

Data Base

The bathymetric map shown in Fig. 1 is based on U.S. *Navoceano* compilations of all the soundings to date. The morphologic, acoustic, and sedimentologic characteristics (Figs. 2–5) are based on the studies of seismic (3.5 kHz, single and multichannel) data and cores collected by Lamont-Doherty Geological Observatory, U.S. *Navoceano,* and Elf Acquitaine (France) in 1974, 1977, and 1981.

Results and Discussion

The morphologic, acoustic, and sedimentologic characteristics vary gradually down the fan. However, the combined use of all these characteristics allow us to divide the Indus

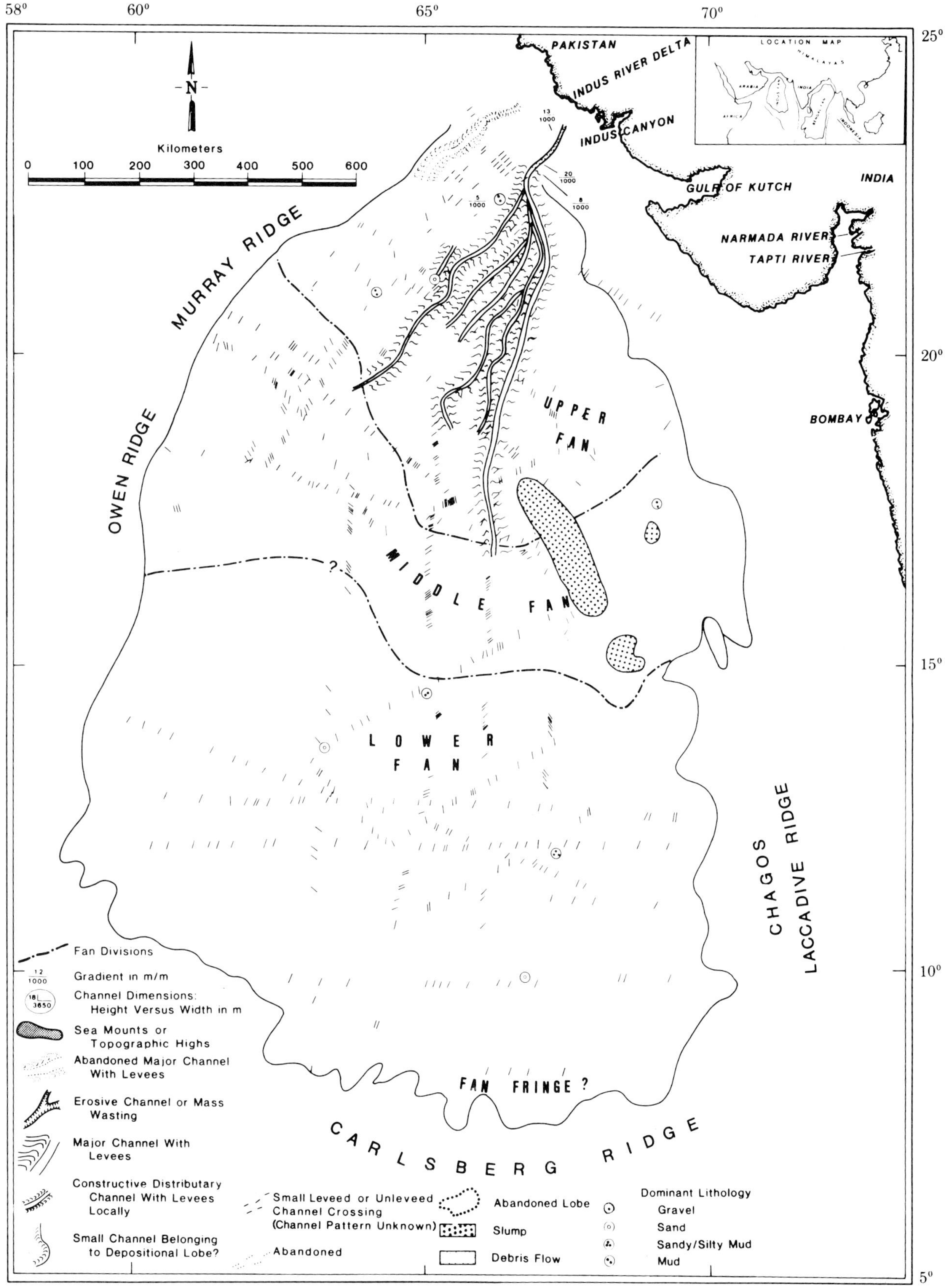

Figure 2. Morphology of the Indus Fan showing channels and general lithology in the upper, middle, and lower fan. Only the pattern of channels in the upper fan corresponding to the most recent Indus Canyon is shown.

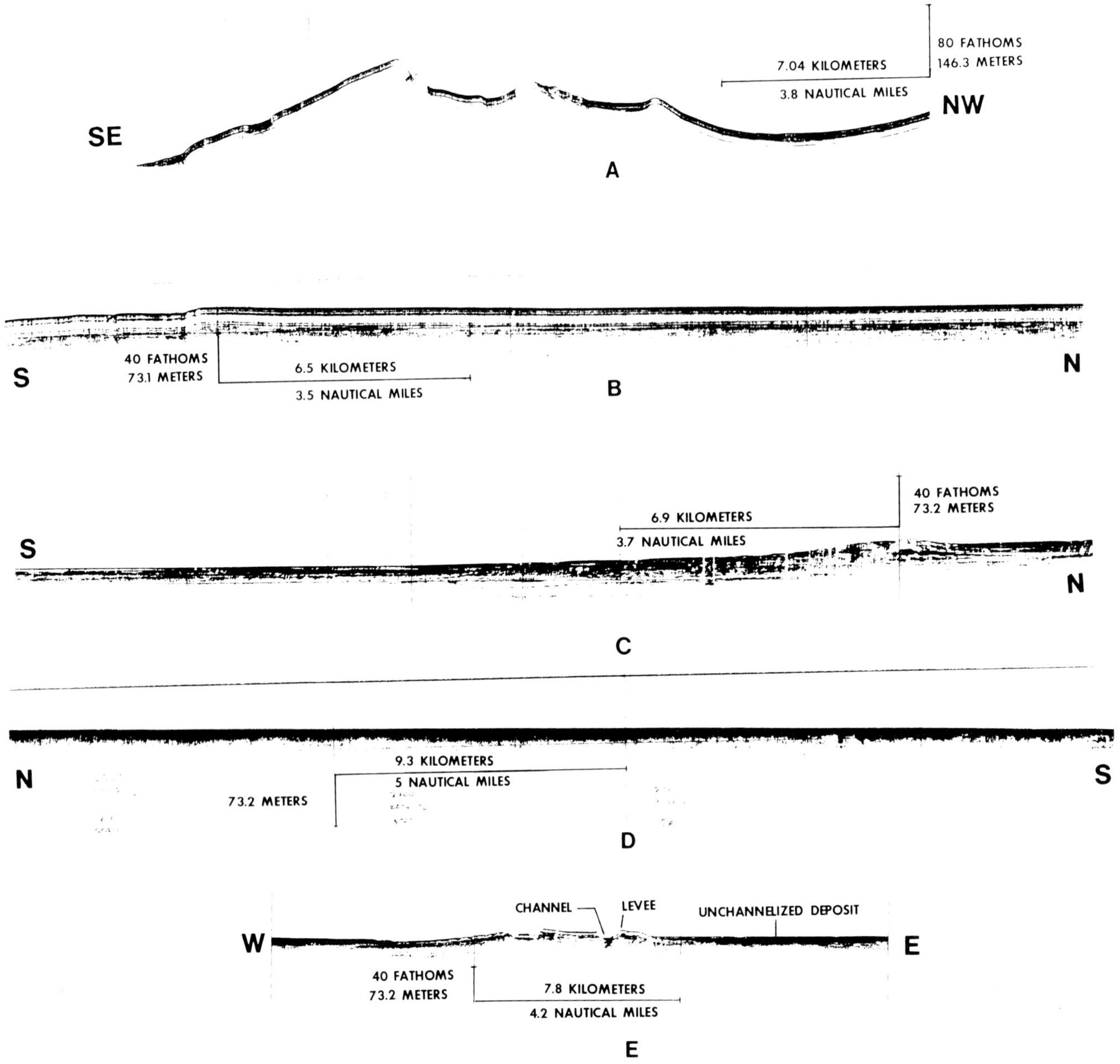

Figure 3. Examples of 3.5-kHz acoustic character from upper, middle, and lower fan. A and B: upper fan; C: middle fan; D and E: lower fan.

Fan into upper, middle, and lower fan regions. The characteristics of these different regions are summarized in Table 1.

The Indus Canyon was erosional on the shelf and upper continental slope (Fig. 2). On the lower slope and upper fan, the canyon and its connecting channels were depositional as well as erosional.

The bathymetry of the upper fan is very irregular with relief in several hundreds of meters (Fig. 1). The seaward bathymetric bulges (Fig. 1) reflect primarily progradational buildups by channel-levee complexes. Channels with widths up to 8 km and levees that are up to 100 m or more in height characterize the upper fan.

The general gradient of the upper fan is about 1:500. The sea floor of this region is characterized by distinct echoes with several continuous subbottom reflections on 3.5-kHz records (Fig. 3A, B). The sediments are generally fine-grained muds (T_{d-e}) except in channels where coarse-grained turbidites (T_{a-e}) may be present (Fig. 4). However, the turbidite sands are usually not coarser than fine-grained. Deposition by channelized turbidity currents and by overbank-spilling of these currents constituted the dominant mechanism in the upper fan. Slumping was not widespread except in restricted zones of channels and levees and along continental slopes at the peripheries of the upper fan.

The levees show up as wedge-shaped reflection packages

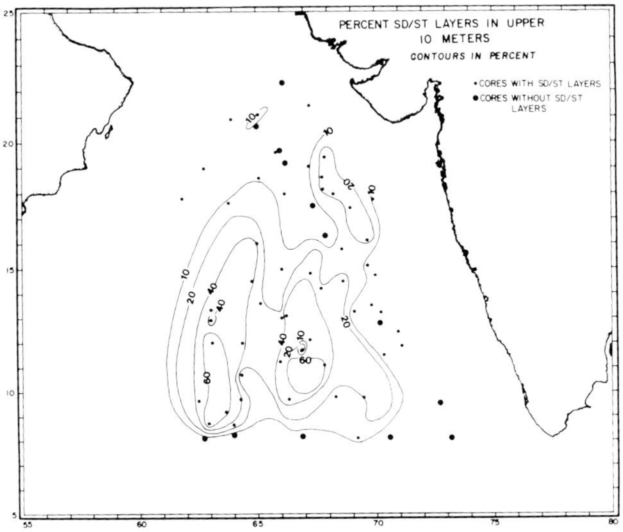

Figure 4. Percent sand-silt layers in the upper 10 m of piston cores.

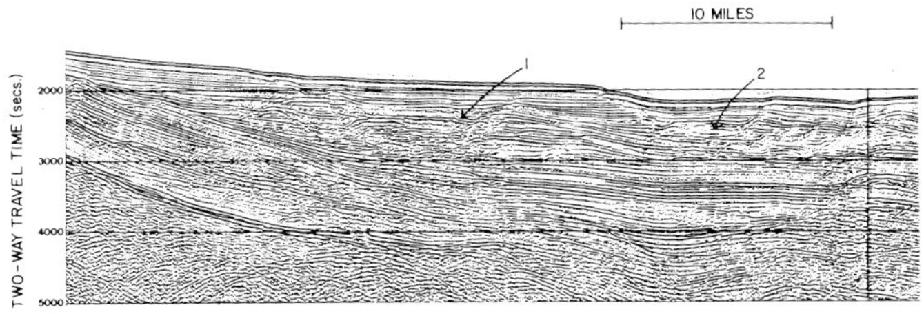

Figure 5. Example of multichannel seismic record showing channel-levee characteristics and types of channel migration in the upper fan.

on multichannel seismic records (Fig. 5). The channel axis itself is characterized by high-amplitude discontinuous reflections that may be indicative of coarse sediments. Two types of channel migration have been observed on the upper fan. One type consisted of gradual and consistent channel migration in one direction as banks on one side of the channels receded because of erosion and the banks on the other side advanced resulting from deposition (Fig. 5). Another

Table 1. Summary of Morpho-acoustic and Sedimentologic Characteristics of the Indus Fan

		Upper Fan	Middle Fan	Lower Fan
Morphology	Relief	>100 M	<50 M	Smooth
Morphology	Gradient	1:500	1:500 to 1:1000	<1:1000
Morphology	Longitudinal Profile	Irregular	Convex upward	Neither convex nor concave
Morphology	Channels and Levees	Width up to 5 naut. miles; depth >100 m; levee height, 100 m	Channels: width <1 naut. mile, depth <50 m, levee height 20 m	Many channels: width <0.5 naut. mile, depth <20 m, levee height 8–20 m
Seismics	3.5 kHz	Distinct echoes with continuous subbottom reflectors, well stratified	Distinct to indistinct echoes with moderate stratification; coalescence of many levee wedges	Prolonged echoes with little or no subbottom reflectors; levees: more stratified; away from channels: sea floor opaque
Seismics	Singlefold	Strong reflections and widespread transparent patches	Seismic relief: small; reflectors more continuous than in upper fan	Smooth, continuous reflectors
Seismics	Multifold	*Channel axis:* strong and discontinuous reflectors and transparent patches *Levees:* wedges with continuous reflectors and transparent patches Coalescence of levee wedges Intense channel migration	Seismic relief: small; reflectors more continuous than in upper fan	
Lithology and Facies		Dominance of fine-grained sediments; channelized sand of miles extent Fining- and thinning (?)-upward vertical sequence	Sand and fine-grained sediments intermediate between upper and lower fan Channel: sand; levee: fine-grained sediment; fining- and thinning-upward sequence	Sand dominance Vertical sequence unknown
Sedimentary Processes		Slumping: within channels and on the levees; along broad fronts in unchannelized regions. Turbidity currents: channelized and overbank spilling	Small scale slumping Channelized and overbanking deposits 70%. Unchannelized deposits 30%. No supra-fan development	Slumping: rare Channelized and overbank deposits 30%; unchannelized sheet-flow deposits 70% (Lobe?)

type of migration consisted of channel abandonment and opening of new channels at different locations. For example in Fig. 5, a channel labeled "2" was opened on the levee of a channel labeled "1" after the latter one was abandoned. We believe that the gradual migration represents channel meandering and the gradual growth of the inward convex bend analogous to a point-bar development in a river meander.

In addition to the present Indus Canyon (Figs. 1 and 2), multichannel seismic data show at least two other older canyons cut into the shelf and slope and many corresponding channels in the upper fan that may have migrated in space and time. The pronounced seaward bathymetric bulge slightly east of the center of the upper fan has been the site of the most recent channel-levee build-up corresponding to the present canyon (Fig. 2). Thus, the sediment source for the Indus Fan was the migrating point-type, and each time the sediments were funneled through a canyon and corresponding channels onto the upper, middle, and lower fan regions.

The middle-fan gradients are 1:500 to 1:1000. The relief decreases considerably, and the levee heights in the middle fan are of the order of 20 m. By analogy with the Amazon Fan, the apparent increase in number of channels (Fig. 2) in the middle Indus Fan may be the result of channel meandering, channel abandonment, and channel avulsion [8]. The sea floor in the middle fan is characterized by distinct to indistinct echoes with moderate subbottom reflectors on 3.5-kHz records (Fig. 3C).

The lower fan has a gradient of about 1:1000. The sea floor is characterized by prolonged echoes with few or no subbottom reflectors except near levees of channels where subbottom reflectors can usually be seen (Fig. 3D, E). Many channels have levees with 9 to 20 m heights although there are channels without levees. Coinciding with low penetration on 3.5-kHz records, sands with T_{a-e} sequences dominate in the lower fan (Fig. 4) compared to other regions of the fan. The turbidite sands are medium- to fine-grained.

In the middle and lower fan, with less availability of fine-grained muds and lower heights of levees, the turbidity currents became progressively less confined to the channels. In addition, several channels that may have been active in the middle fan were probably terminated before reaching far on the lower fan. Because of these factors, deposition by unchannelized turbidity sheet flows was the dominant mode of sedimentation in the lower fan. However, channelized and overbank deposition also played a significant role in the lower fan.

Comparison of Indus Fan to Amazon, Mississippi, and Bengal Fans

The decreasing acoustic penetration seen on 3.5-kHz records and the increasing sand concentrations from the upper to lower fan are similar in all these large fans. This is due to the dominance of mud in the sediments of the river systems that supplied the sediments to these deep-sea fans [9]. The rivers that fed the Indus, Amazon, and Mississippi Fans apparently supplied comparable loads of suspended sediments annually in recent times [10]. However, the dimensions of these fans are markedly different (see this volume). Both the size of the original depositional basins and the tectonic history of the region are important in influencing the dimensions of these fans, in addition to the sediment source. Thus, the size of the original basin of the Indus Fan bounded by ridges, which was available for down-slope and turbidity-current sediment transport, was more extensive, and hence this fan was larger than the Amazon and the Mississippi Fans.

Slumping is extensive on the Amazon and Mississippi Fans; it is largely absent on the Indus Fan, except locally and along continental slopes. Salt tectonism in the upper Mississippi Fan and shale diapirism in the upper Amazon Fan (R. O. Kowsmann, 1981, personal communication) may be partly responsible for their wide-spread slump features. The absence of these processes in the Indus Fan may explain the near-lack of widespread slump features. Also, the content of smectite, which tends to occur in finer grain sizes and has the capacity to carry more water than other clays, is generally higher in the sediments of the Amazon and Mississippi Fans than that in the sediments of the Indus Fan.

In the lower Indus and Mississippi Fan regions, channels with fairly well-developed levees are common whereas in the lower Amazon Fan channels largely lack levees [11,12]. The Amazon Fan gradients are steeper than those of the Indus and Mississippi Fans. The gradients may have resulted from sedimentary and tectonic processes peculiar to each fan. Compared with the Indus River, the Amazon River drainage basin has a rigorous tropical climate, and the river has to travel great distances after leaving the uplifted source region. These factors may have resulted in texturally and compositionally mature sands and clays, with relatively low amounts of silt size factions in the sediments of the Amazon River. In addition to the sand and clay size fractions, the more arid climate prevailing in the drainage basin of the Indus River dictates more silt-size material to be contributed by this river. These possible differences in grain size distribution and the fan gradients could have caused the differences in channel-levee characteristics of the lower Amazon and Indus Fan. However, the climates that prevailed during the Pleistocene times and the resulting grain-size distributions may have been different from what could result under the present climate.

The Bengal Fan is the largest of the deep-sea fans and the sediment loads supplied by the Ganges and Brahmaputra Rivers are the highest of any river [10]. The large size of the Bengal Fan was undoubtedly influenced by the original extent of the basin and the huge sediment loads of the rivers. Although the number of channels in the fan appear to be few, the lower Bengal Fan channels have well-developed levees

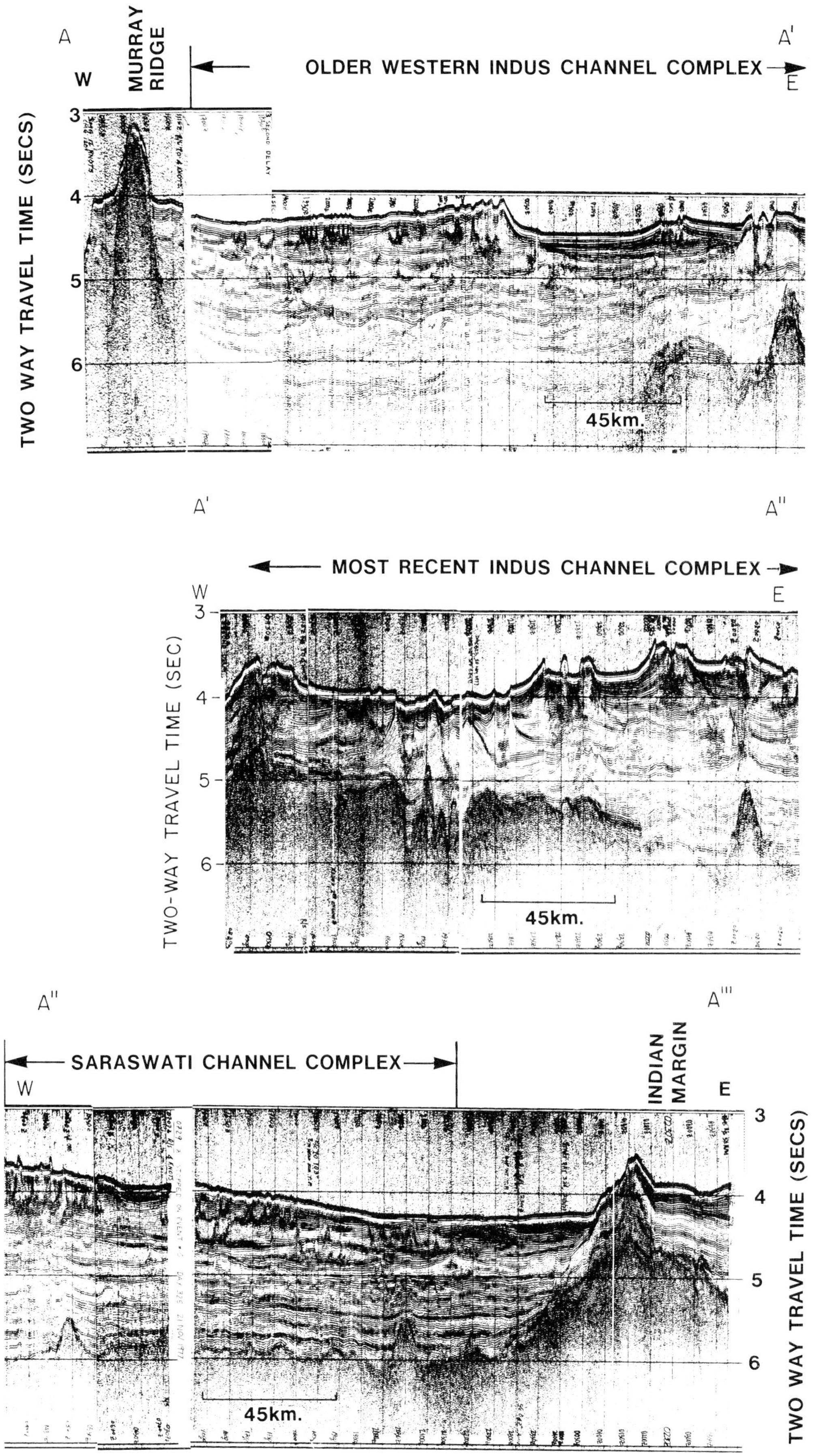

Figure 6. An east-west sparker line from the upper Indus Fan. The most recent Indus Channel complex (Fig. 2), the older channel complex to the west, and the Saraswati Canyon, east of the Indus Canyon and off the Gulf of Kutch, are shown, The U.S. Naval Oceanographic Research Office (NAVOCEANO) provided the sparker line.

(Emmel and Curray, this volume). The flat gradients and the expected higher amounts of silt-size fractions in the sediments of the rivers as a consequence of the nearness of uplifted source regions to the ocean might have resulted in the levee development associated with the lower Bengal Fan channels.

From the above, it is apparent that, although the Indus Fan differs in some respects from other large fans discussed here, it is in general comparable to them.

Addendum

Figure 6 shows an east-west sparker line from the upper Indus Fan, extending from the Murray Ridge on the west to the Indian Margin on the east. This line is split into three segments—AA′, A′A″ and A″A‴ starting from the west and extending progressively to the east.

Acknowledgments

This study was carried out primarily at Lamont-Doherty Geological Observatory under NSF Grant OCE 76-83382 and ONR Contract N00014-80-C-0098. V. K. is grateful to Superior Oil Company for giving an opportunity to finalize the paper. A. H. Bouma, W. Schweller, and D. W. Ford read the manuscript. Lamont-Doherty Geological Observatory Contribution No. 3646. F. C. thanks Elf Aquitaine for permission to publish this paper.

References

[1] Naini, B. R., and Talwani, M., 1982. Structural framework and evolutionary history of the continental margin of Western India. In: J. S. Watkins and C. L. Drake (eds.), Studies in Continental Margin Geology. American Association of Petroleum Geologists, Memoir 34, pp. 168–191.

[2] Naini, B. R., and Kolla, V., 1982. Acoustic character and thickness of sediments of the Indus Fan and the continental margin of Western India. Marine Geology, v. 47, pp. 181–195.

[3] Neprochnov, Y. P., 1961. Sediment thickness of Arabian Sea Basin. Doklady Akademiya Nauk. SSSR, 139, pp. 177–179 (in Russian).

[4] Ewing, M., and others, 1969. Sediment distribution in the Indian Ocean. Deep-Sea Research, v. 16, pp. 231–348.

[5] Bachman, R. T., and Hamilton, E. L., 1980. Sediment sound velocities from sonobuoys: Arabian Fan. Journal of Geophysical Research, v. 85, pp. 849–852.

[6] Kolla, V., and others, 1981. Distributions and origins of clay minerals in surface sediments of the Arabian Sea. Journal of Sedimentary Petrology, v. 51, pp. 563–569.

[7] Kolla, V., and others, 1981. Surficial sediments of the Arabian Sea. Marine Geology, v. 41, pp. 183–204.

[8] Damuth, J. E., and others, 1983. Distributary channel meandering and bifurcation patterns on the Amazon deep-sea fan as revealed by long-range side-scan sonar (GLORIA). Geology, v. 11, pp. 94–98.

[9] Normark, W. R., 1978. Fan valleys, channels and depositional lobes on modern submarine fans: characters for recognition of sand turbidite environments. American Association of Petroleum Geologist Bulletin, v. 62, pp. 912–931.

[10] Listizin, A. P., 1972. Sedimentation in the world ocean. Society of Economic Paleontologists and Mineralogists, Special Publication 17, 218 pp.

[11] Damuth, J. E., and Kumar, N., 1975. Amazon Cone: morphology, sediments, growth pattern. Geological Society of America Bulletin, v. 86, pp. 863–878.

[12] Moore, G. T., and others, 1978. Investigations of Mississippi Fan, Gulf of Mexico. In: J. S. Watkins, L. Montadert, and P. W. Dickenson (eds.), Geological and Geophysical Investigations of Continental Margins. American Association of Petroleum Geologists, Memoir 29, pp. 383–402.

CHAPTER 20

Laurentian Fan, Atlantic Ocean

David J. W. Piper, Dorrik A. V. Stow, and William R. Normark

Abstract

The 0.5- to 2-km thick Quaternary Laurentian Fan is built over Tertiary and Mesozoic sediments that rest on oceanic crust. Two 400-km long fan valleys, with asymmetric levees up to 700-m high, lead to an equally long, sandy, lobate basin plain (northern Sohm Abyssal Plain). The muddy distal Sohm Abyssal Plain is a further 400-km long. The sediment supplied to the fan is glacial in origin, and in part results from seismically triggered slumping on the upper continental slope. Sandy turbidity currents, such as the 1929 Grand Banks earthquake event, probably erode the fan-valley floors; but thick muddy turbidity currents build up the high levees.

Introduction

The Laurentian Fan is located off the eastern continental shelf of Canada, in the bight between the rifted continental margin of Nova Scotia and the (inactive) transform margin of the southwest Grand Banks (Fig. 1). This margin developed during the Early Jurassic opening of the central Atlantic Ocean. The bathymetrically defined modern fan comprises up to 2 km of Quaternary sediment. It rests on a thick Jurassic to Tertiary continental margin sequence [1], principally of older deltaic and fan sediments derived from the ancestral St. Lawrence River [2].

The Laurentian Fan lies seaward of the Laurentian Channel, a 700-km long, 80-km wide, glacial trough that crosses the continental shelf and has been excavated some 300 m below the regional depth of the shelf. The shelf break at the southern end of the Laurentian Channel lies at about 400 m water depth; consequently it would not have been emergent even during maximum glacio-eustatic lowerings of sea level. However, the Laurentian Channel was probably the main discharge route for much of the ice in southern Quebec and the Atlantic Provinces of Canada [3], either as grounded ice, an ice shelf, or icebergs. Rapid melting would have occurred where the Laurentian Channel fed into the open ocean, not far from the Gulf Stream [4], over the continental slope above the Laurentian Fan.

The southern end of the Laurentian Channel and adjacent continental slope is an area of moderate seismic activity and was the site of the magnitude 7.2 Grand Banks earthquake that occurred in 1929 [5]. This seismicity has provided a triggering mechanism for the resedimentation of at least some of the glacial till and glaciomarine sediments [6].

The Laurentian Fan leads southwards to the Sohm Abyssal Plain (SAP). This abyssal plain receives sediment not only from the Laurentian Fan but also from the Northwest Atlantic Mid-Ocean Channel (NAMOC), and from an unnamed channel leading from the slope off Northeast Channel at the entrance to the Gulf of Maine [7] (Fig. 1). Most of the Wisconsinan sediment on the distal SAP has been transported across the Laurentian Fan [8,9].

Sediment accumulation on the SAP is confined between the Bermuda Rise in the west and the flanks of the midocean ridge to the east (Fig. 1). The Bermuda Rise also diverts sediment from the Georges Bank and Nova Scotian continental margins eastwards to the SAP. The SAP is divided into a northern and southern half by the Corner Seamounts and the southeastern part of the New England Seamount Chain (Fig. 2). Abyssal hills and seamounts rise above the southern SAP. The J-anomaly Ridge and the Fogo Seamount Chain divert turbidity currents at the southeastern edge of the Laurentian Fan (Fig. 3).

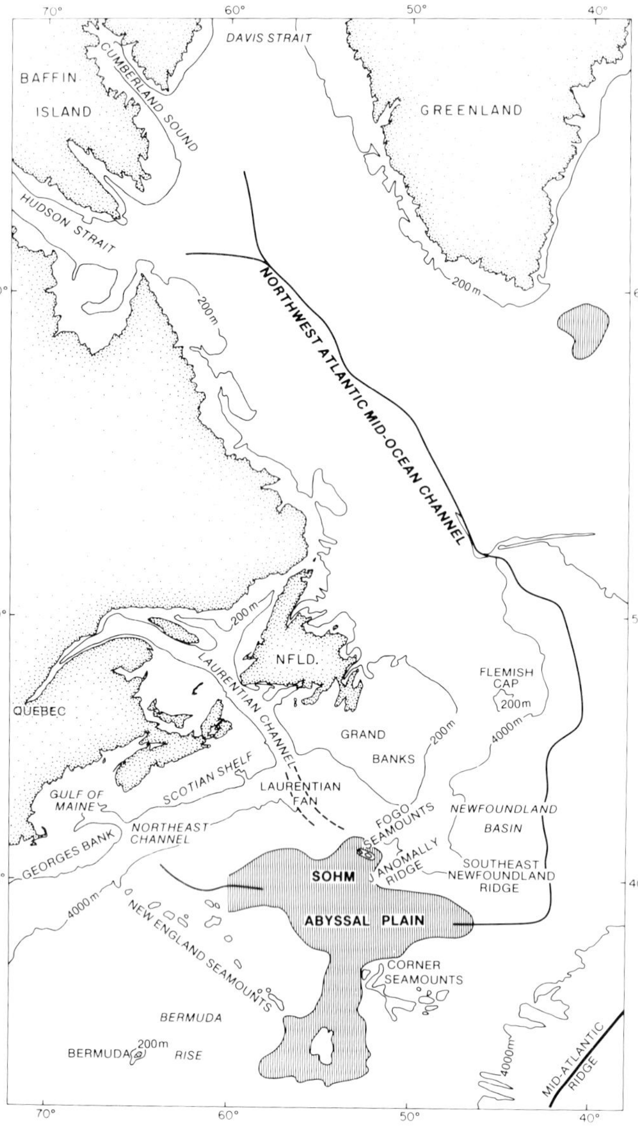

Figure 1. Map of northwestern Atlantic Ocean showing regional setting of Laurentian Fan and Sohm Abyssal Plain.

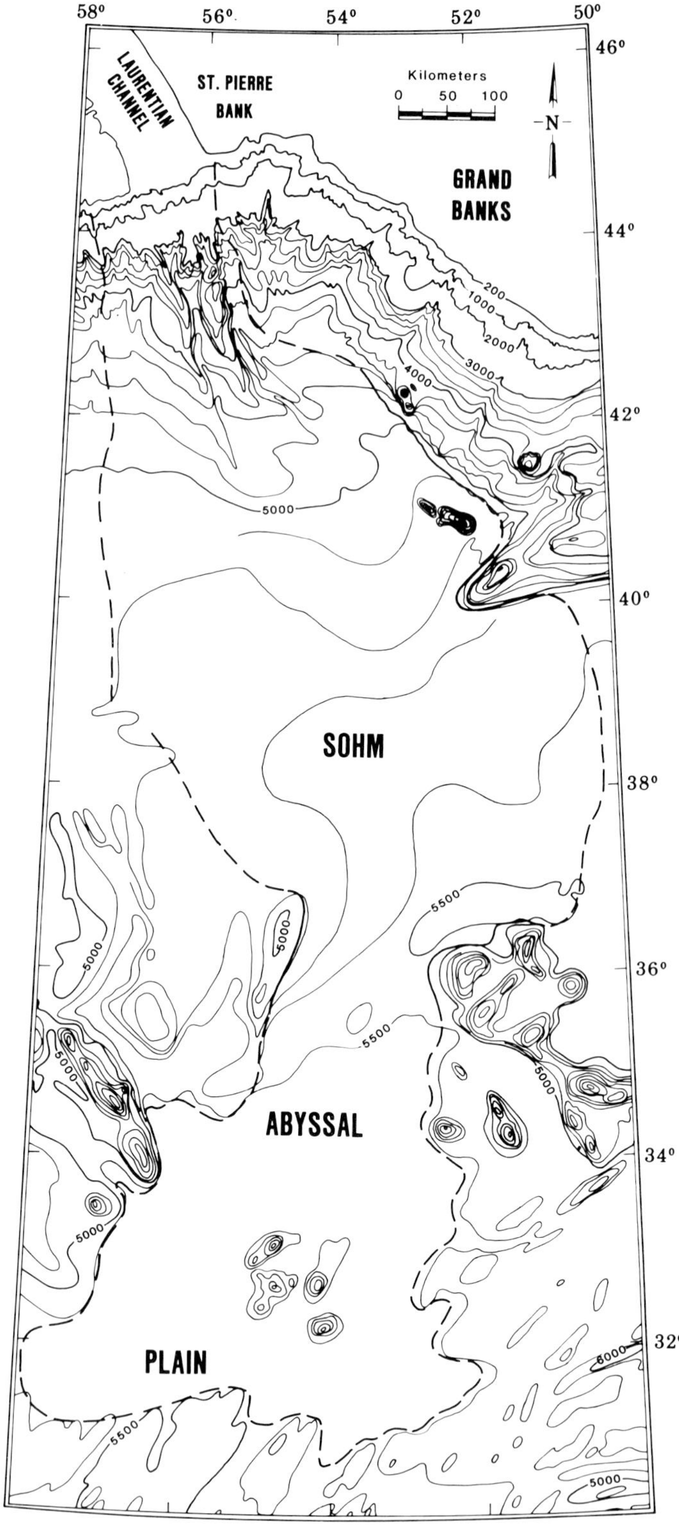

Figure 2. Bathymetric map of Laurentian Fan and Sohm Abyssal Plain. Laurentian Fan and continental margin contoured at 200-m interval, based on Uchupi and Austin [1] with modifications from Bedford Institute of Oceanography (B.I.O.) ship tracks. Southern SAP and adjacent areas contoured at 500-m interval, based on GEBCO chart [21] with modifications from B.I.O. ship tracks. (Track density for compilation is approximately as shown on GEBCO chart [21]).

Morphologic Divisions, Size and Shape

Bathymetric profiles with an average line spacing of 20 km are available for the Laurentian Fan [1,10], but data are more sparse and less well navigated on the SAP. GLORIA coverage has been obtained for the entire fan above the 4500-m isobath [11] but is as yet unpublished.

We recognize the following morphologic subdivisions of the Laurentian Fan from north to south (Fig. 3)

Continental Slope: an area of irregular relief resulting from substantial erosion of Late Quaternary sediments, extending from the shelf break at 400 m to the 2000-m isobath [6].

Slope-valley Transition: the area between the 2000- and 3000-m isobaths. This is an area of complex erosional valleys with some intervalley accumulation of Late Quaternary

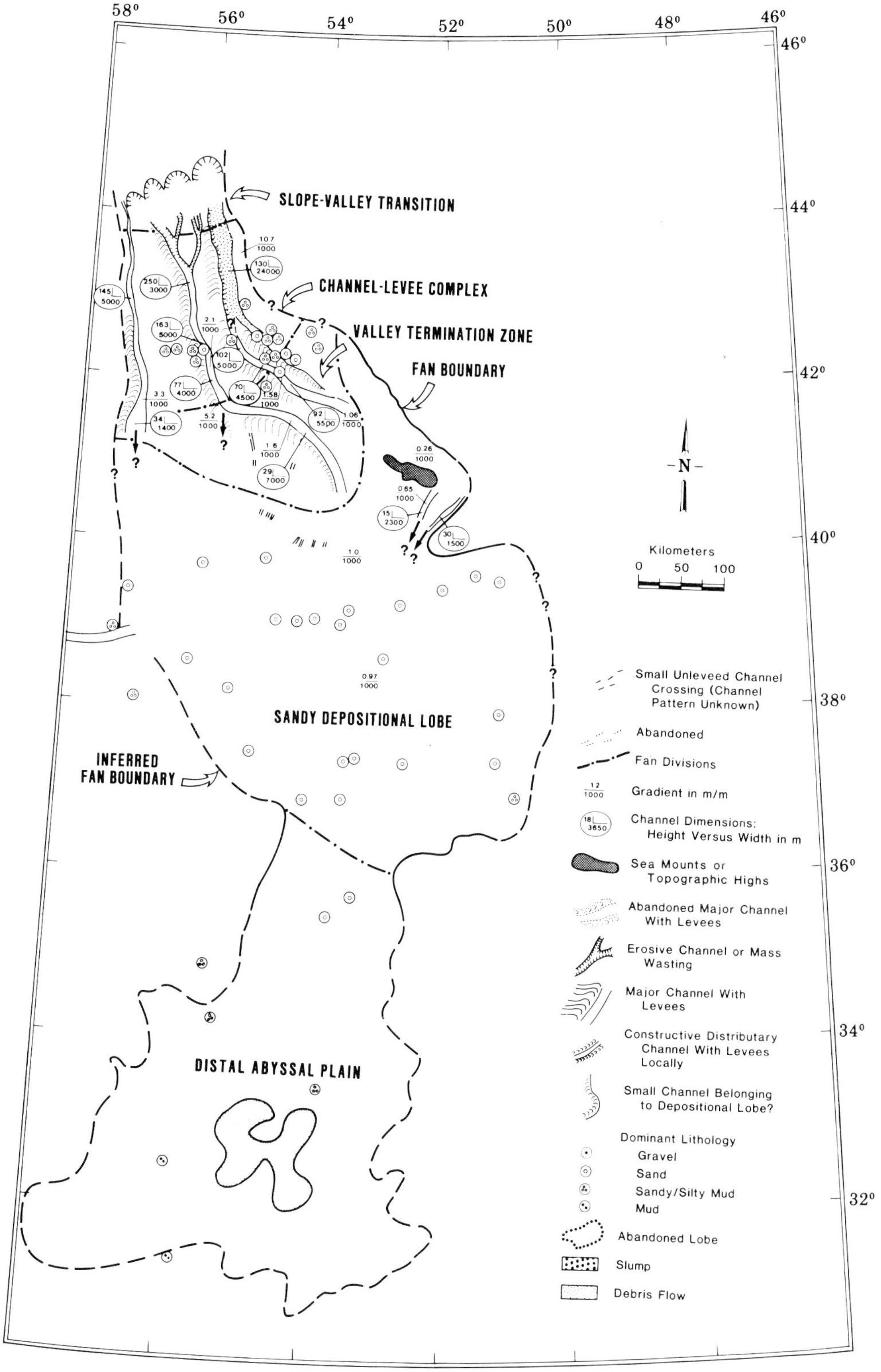

Figure 3. Morphometric map of Laurentian Fan and Sohm Abyssal Plain.

sediment and some large slide blocks [2].

Channel-levee Complex: the area extending from the slope-valley transition zone to the 4600-m isobath. This area is tranversed by two major fan valleys (eastern and western valleys) with erosional western walls and highly asymmetric levee systems (Fig. 2). The higher west levees are up to 1000-m above the valley floor. Both seismic profiles [12] and cores [4] show that the levees accumulated overbank sediment during the Late Quaternary. Their asymmetry is probably the result of the southwestward flow of the Western Boundary Undercurrent in this area [13].

Valley Termination Zone: the area from 4600 to 5000 m. Relief of both the fan valleys and levees becomes progressively more subdued [13], and the eastern valley bifurcates.

Sandy Depositional Lobe: from 5000 to 5500 m. Channels are very shallow or absent in this area. There is a pronounced overall lobate morphology to around 5200 m, but to the south the gradient decreases to less than 1:1000, and morphology appears influenced by the NAMOC and the channel leading from the Gulf of Maine.

Distal Abyssal Plain: this is the virtually flat (?ponded) area southwest of the Corner Seamounts at 36°N, in water depths of more than 5500 m.

Stratigraphy

About 25 piston cores have been collected from the Laurentian Fan [4], 25 from the sandy depositional lobe on the northern SAP [7,14] and 15 from the distal abyssal plain [7,8,15,16]. Biostratigraphic studies allow the distinction of Holocene from Wisconsinan (Late Pleistocene) sediment [4,8]; in most cores, Holocene sediments are about 1 m thick, giving an average sedimentation rate of about 0.1 $m/10^3$ yr. Most cores do not completely penetrate the Wisconsinan sequence. On the Laurentian Fan, Holocene sediments are interpreted as mostly muddy contourites [4]. Holocene turbidites are found on the SAP: turbidity currents apparently flowed through the fan valleys and sediment thus by-passed the fan (Fig. 4).

Wisconsinan mud-accumulation rates on the Laurentian Fan are estimated at 0.1 to 0.3 $m/10^3$ yr, with one turbidity current every 100 to 300 years depositing on the levees [4]. Rates of accumulation on the distal SAP are probably higher [8]. Sand-accumulation rates are not known.

A distinctive acoustic reflector can be mapped beneath much of the Laurentian Fan [1,12,13] and has been tentatively assigned to around the Pliocene–Quaternary boundary, based on a change in depositional style and correlation with stratigraphy in wells on the shelf edge and upper slope. Where there has been no subsequent erosion, this acoustic reflector is found about 1.5 to 2.0 secs subbottom at the 3000-m isobath and about 0.7 secs deep at the 4500-m isobath, giving Quaternary sediment accumulation rates of the order of 0.5 to 1 $m/10^3$ yr (Fig. 5). Acoustic stratigraphy of the channel-levee complex [12] suggests that the Laurentian Fan was a simple progradational system with a single fan valley in the Early Quaternary, but the fan has experienced widespread erosion with the development of two major valleys in the Late Quaternary [12]. There are no suitable data to extend this acoustic stratigraphy to the SAP.

Distribution and Source of Sediment

The acoustic character of the surficial sediments on the fan is mapped from widely spaced 3.5-kHz [1,6,17] and high-frequency, filtered, sparker profiles [12,13]. Acoustically transparent facies occur on levees in both the valley-levee complex and the valley termination zone, and in the distal SAP. Highly reflective sediments occupy the valley floors and the sandy depositional lobe. Valley terraces show intermediate acoustic penetration. Within 100 km of the epicenter of the 1929 earthquake, surface stratified sediments are very discontinuous, indicating widespread surface slumping [6], and valley walls are mantled by material giving many surface hyperbolic reflections [12]. Large slide blocks of dimensions of hundreds of meters are rare [12]. A large debris flow in the eastern valley is distinguished by characteristic surface morphology [6] and extends across the valley termination zone [17]. There is also an acoustically transparent debris flow on the west side of the westernmost levee [17].

Pleistocene cores from the Laurentian Fan and SAP yield a distinctive red-brown mud (5 YR 4/4 to 10R 4/2) interbedded with silt, sand, and gravel. Silts and muds occur on levees and interchannel areas of the fan and on the distal abyssal plain. Sands and gravels are found in the fan valleys. Sand is the dominant sediment on the sandy depositional lobe of the northern SAP; thin sands also reach the distal abyssal plain and the levees. The gravels are composed of clasts largely derived from the Gulf of St. Lawrence, and the mineralogy of sands and clays confirms that this is the dominant source [4]. Holocene sediment contains relatively more montmorillonite transported by the Western Boundary Undercurrent [4].

The principal Late Quaternary sediment source is the slumping of glaciomarine sediment and till [6,18] from the upper slope. Much of this slumping is probably seismically triggered; a zone of recent seismicity extends east-west across the Laurentian Channel and the upper continental slope off St. Pierre Bank (Fig. 4). Turbidites are much more frequent during glacial periods [4,8], suggesting that slumping may in part have been the result of very rapid sediment supply, or even direct flow of cold turbid water from beneath an ice shelf [4]. This slumped material moved across a highly eroded continental slope cut by irregular gullies that coalesce into distinct valleys around the 3000-m isobath.

The 1929 magnitude 7.2 Grand Banks earthquake had an epicenter [19] on the continental slope above the Laurentian Fan (Fig. 4). Surficial sliding occurred within a zone of about

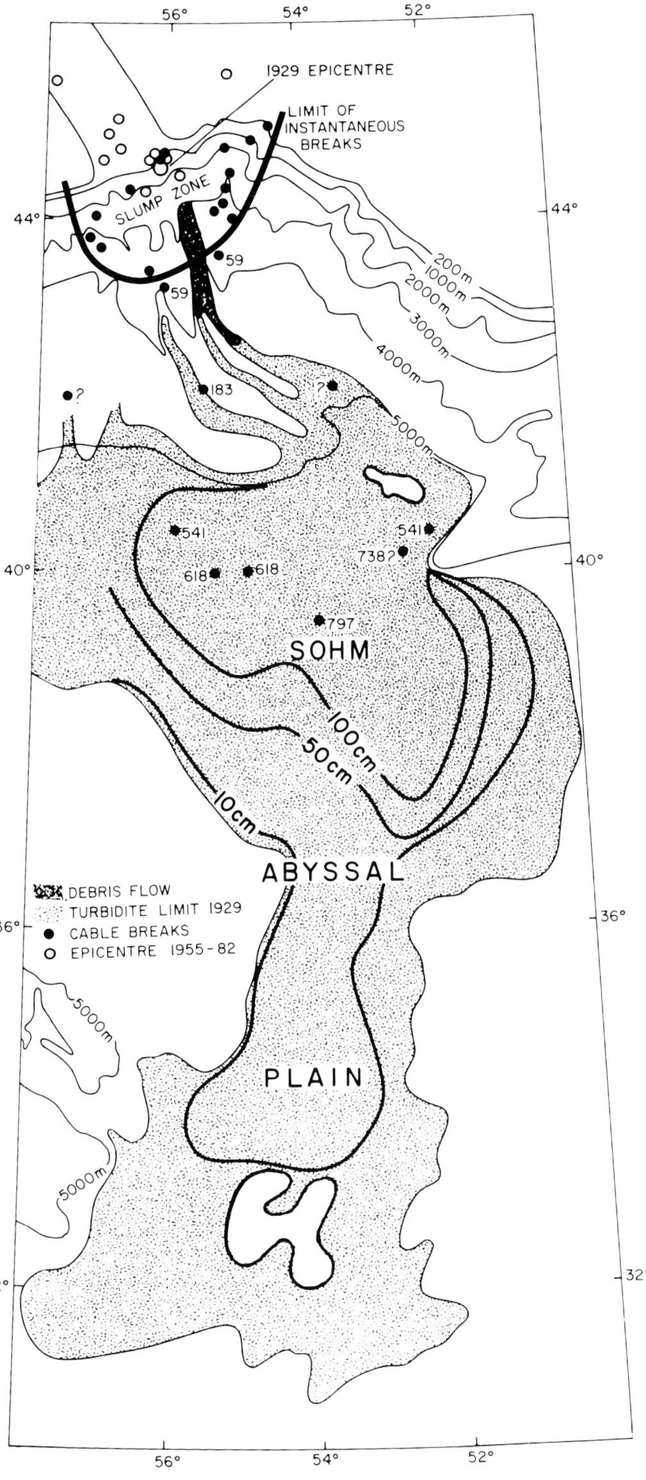

Figure 4. Map showing geologic effects of 1929 Grand Banks earthquake. Epicenter of 1929 earthquake from Basham and Adams [19]. Location and time (minutes after main shock) of cable breaks from Doxsee [5]. Limit of 1929 turbidite on Laurentian Fan interpreted from data of Stow [4]; on SAP from data of Fruth [14]. Isopachs of 1929 turbidite from Fruth [14]. Extent of debris flow based on 12-kHz soundings, and is probably underestimated (Jacobi, personal communication [17]).

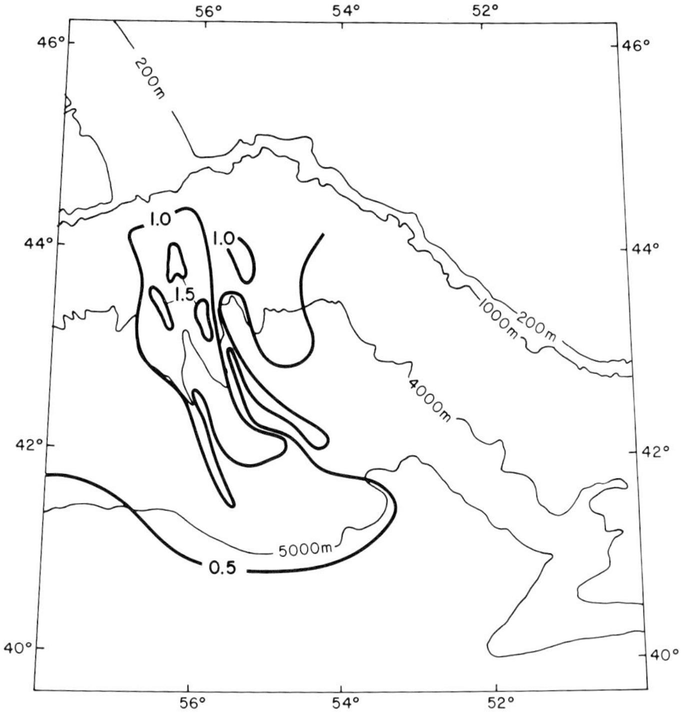

Figure 5. Isopach map of (?) Quaternary sediment on Laurentian Fan (in km) (after Uchupi and Austin [1] with minor modifications from [12]).

a 100-km radius around the epicenter, and a large debris flow moved down the Eastern Valley [6]. Cable breaks [5] indicate that turbidity currents flowed down both the eastern and western valleys at velocities of between 73 and 41 km/hr. The currents appear to have been slowed by ponding behind the southern Fogo Seamounts. Cores from levees from the valley termination zone do not contain turbidities on the surface [4], but thick, sorted, and graded gravels occur in the valleys in this zone [7]. According to Fruth, over a meter of sand was deposited on the sandy depositional lobe [14], whereas only a few centimeters of silty mud may be present in places on the distal abyssal plain [14].

Many Wisconsinan turbidity currents were of quite different character from the 1929 event. They were hundreds of meters thick, overtopped the levees, were probably principally of mud, had low suspended sediment concentrations, and velocities of only 0.10 to 0.15 m/s [20].

Conclusions

The Laurentian Fan resembles other large passive-margin fans by having large and long fan valleys. It has a short line source of slumped glacial material at the end of the Laurentian Channel, the major Pleistocene glacial ice outlet in southeastern Canada. Some of the slumping is seismically triggered. Irregular slope gullies coalesce to form two active fan valleys. Pronounced levee asymmetry is related to the Western Boundary Undercurrent interacting with thick, low-density, muddy turbidity currents. The glacial source supplied large proportions of sand and gravel, which have been resedimented to form a vast sandy lobe ($\approx 10^5$ km^2) with a gradient of about 1:1000, and also abundant mud, which has built up most of the thick channel-levee complex. Sandy turbidity currents are probably responsible for the pronounced Late Quaternary erosion of the Laurentian Fan.

Acknowledgments

We thank the following colleagues for sharing data and ideas and/or reviewing the draft manuscript: John Adams, Dale Buckley, Jim Gardner, Al Grant, Bob Jacobi, Tor Nilsen, and Dave Roberts.

References

[1] Uchupi, E., and Austin, J., 1979. The stratigraphy and structure of the Laurentian cone region. Canadian Journal of Earth Sciences, v. 16, pp. 1726–1752.
[2] Parsons, M. G., 1975. The geology of the Laurentian Fan and the Scotian Rise. Canadian Society of Petroleum Geologists Memoir, no. 4, pp. 115–167.
[3] Prest, V. K., and Grant, D. R., 1969. Retreat of the last ice sheet from the Maritime Provinces of Canada. Geological Survey of Canada Paper 69-33, pp. 1–15.
[4] Stow, D. A. V., 1981. Laurentian Fan: morphology, sediments, processes and growth pattern. American Association of Petroleum Geologists Bulletin, v. 65, pp. 375–393.
[5] Doxsee, W. W., 1948. The Grand Banks Earthquake of November 18, 1929. Publications of the Dominion Observatory, Canada, v. 7, pp. 323–336.
[6] Piper, D. J. W., and Normark, W. R., 1982. Effects of the 1929 Grand Banks earthquake on the continental slope off eastern Canada. Geological Survey of Canada Paper, v. 82-1B, pp. 147–151.
[7] Horn, D. R., and others, 1971. Turbidites of the Hatteras and Sohm Abyssal Plains, western North Atlantic. Marine Geology, v. 11, pp. 287–323.
[8] Vilks, G., Buckley, D. E., and Keigwin, L., in press. Quarternary sedimentation on the southern Sohm Abyssal Plain. Sedimentology.
[9] Hacquebard, P. A., Buckley, D. E., and Vilks, G., 1981. The importance of detrital particles of coal in tracing the provenance of sedimentary rocks. Bulletin des Centres de Recherches Exploration Production Elf-Aquitaine, v. 5, pp. 555–572.
[10] Edgar, D. C., and Piper, D. J. W., 1979. A new bathymetric map of the middle Laurentian Fan. Maritime Sediments, v. 15, pp. 1–3.
[11] Institute of Oceanographic Sciences. 1979. Cruise report Discovery 119.
[12] Piper, D. J. W., and Normark, W. R., 1982. Acoustic interpretation of Quaternary sedimentation and erosion on the channelled upper Laurentian Fan, Atlantic Margin of Canada. Canadian Journal of Earth Sciences, v. 19, pp. 1974–1984.
[13] Normark, W. R., Piper, D. J. W., and Stow, D. A. V., 1983. Quaternary development of channels, levees and lobes of the middle Laurentian Fan. American Association of Petroleum Geologists Bulletin, v. 67, pp. 1400–1409.
[14] Fruth, L. S., 1965. The 1929 Grand Banks turbidite and the sediments of the Sohm Abyssal Plain. Unpublished M.Sc. thesis, Columbia University, New York.
[15] Buckley, D. E., 1981. Geological investigation of a selected area of the Sohm Abyssal Plain, Western Atlantic: CSS Hudson cruise 80-016. Atomic Energy of Canada Ltd., Technical Report 168.
[16] Buckley, D. E., 1982. Canadian report to Site Selection Task Group 1981. In: D. R. Anderson (ed.), Proceedings of Seventh International NEA/Seabed Working Group Meeting, La Jolla, California, March 15–19, 1982. Sandia National Laboratory, SAND 82-0460, pp. 69–73.
[17] Jacobi, R. D., personal communication. Echo character of the Laurentian Fan.
[18] Wang, Y., Piper, D. J. W., and Vilks, G., 1982. Surface texture of turbidite sand grains, Laurentian Fan and Sohm Abyssal Plain. Sedimentology, v. 29, pp. 727–736.
[19] Basham, P. W., and Adams, J., 1982. Earthquake hazards to offshore development on the eastern Canadian continental shelves. Proceedings of the 2nd Canadian Conference on Marine Geotechnical Engineering.
[20] Stow, D. A. V., and Bowen, A. J., 1980. A physical model for the transport and sorting of fine-grained sediment by turbidity currents. Sedimentology, v. 27, pp. 31–46.
[21] GEBCO, 1982. General Bathymetric Chart of the Oceans. 1:10,000-000. Sheet 5.08. 5th edition. Canadian Hydrographic Service.

CHAPTER 21

Mississippi Fan, Gulf of Mexico

Arnold H. Bouma, Charles E. Stelting, and James M. Coleman

Abstract

The Mississippi Fan is a Quaternary accumulation composed of more than seven elongated fanlobes. Isopach and structure maps show frequent shifting of these lobes. The Mississippi Canyon, formed by retrogressive slumping, connects to the youngest fanlobe. The upper fanlobe is characterized by a large, incised, partially infilled, leveed channel. The middle fanlobe is aggradational, convex in cross section, with a channel-levee complex on its apex. The lower fanlobe contains a recently active small channel and several abandoned ones. Depositional patterns can be explained by aggradational processes and shifting of depositional sites.

Introduction

The Mississippi Fan, located in the eastern Gulf of Mexico, is a broad, arcuate accumulation of Pleistocene sediments with a thin Holocene cover. It extends basinward from the Mississippi Canyon and causes a gentle seaward bulging of the bathymetric contours (Fig. 1). The fan is bounded by the West Florida Escarpment (east), by the Sigsbee Escarpment (northwest), and by the Campeche Escarpment (south). To the southeast and to the west, the fan terminates in the Florida and Sigsbee Abyssal Plains, respectively, in water depths of approximately 3200 to 3400 m [1].

The Mississippi Fan covers an area in excess of 300,000 km^2 and has a volume of at least 290,000 km^3. Moore and others [2] divided the fan into three physiographic units—upper, middle, and lower—each defined by distinct morphology and channel characteristics. Recent data, specifically from side-scan sonar, suggest that these subdivisions are too generalized and reflect a lack of detailed data.

Based on data collected over the last 2 years, we present a new seismic facies model of the Pleistocene-Holocene Mississippi Fan and revise concepts pertaining to the surface characteristics and sedimentary processes (Fig. 2).

Internal Mississippi Fan Model

Eight prominent acoustic reflectors of regional extent were identified across most of the Mississippi Fan. Construction of structural contour (paleo bathymetry) and isopach maps of these reflectors defines the developmental patterns of this geologically relatively young deep-sea fan. Correlation of the basal horizon with Deep Sea Drilling Project sites to the west coincides with the base of the Pleistocene. Each structure and isopach map shows the relief and shape of an elongated body, herein called fanlobes (Fig. 3). The juxtaposition of the fanlobes is controlled by the relief of the older ones and the position of sediment input. Consequently, the Mississippi Fan clearly is not built by a vertical stacking of successive sediment bodies, but by lateral shifting of fanlobes that developed in local topographic lows.

Comparison of the structures and isopach maps shows another important aspect of fan construction relative to the location of the upslope sediment source. The successive source areas (canyons incised into the shelf) migrated during the Pleistocene. Seismic studies have indicated several completely filled canyons on the outer shelf and upper slope. Data coverage is insufficient to allow us to connect the individual fanlobes to their respective canyons.

Most of the structure and isopach maps reveal a certain degree of complexity rather than simple cut-and-fill struc-

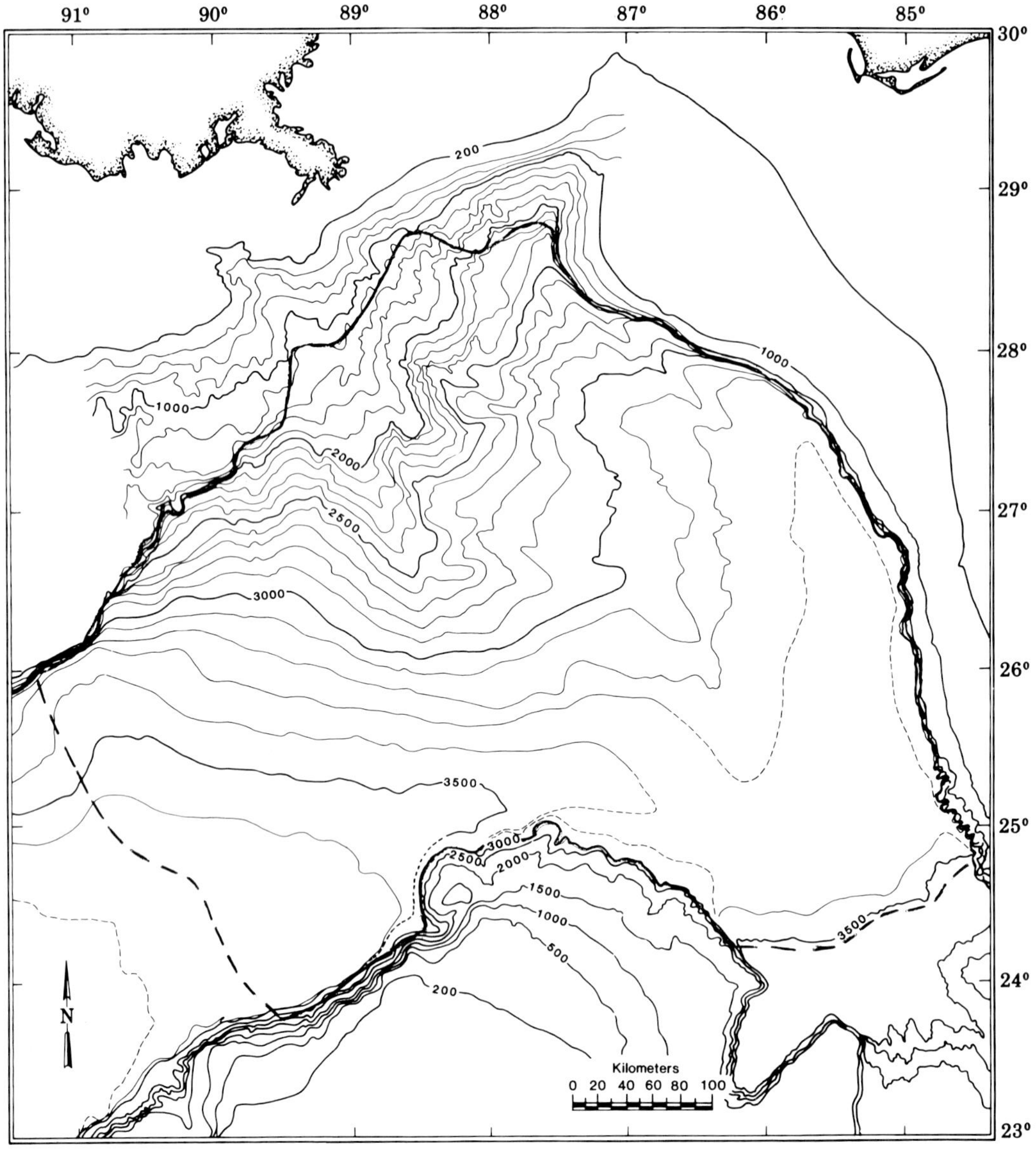

Figure 1. Bathymetric map of the Mississippi Fan (in meters).

tures in the area of the continental slope. Particularly, they show a convex upward shape on the rise that decreases in relief downfan. Data control is insufficient to sort out these complexities; however, it indicates that more than one fanlobe is combined in most of the maps (Figs. 3 A–H).

A comparison of the successive structure and isopach maps in Figure 3 demonstrates the general shape characteristics of the fanlobes and the shifting of sediment source points. Furthermore, comparison of all maps constructed from the eight reflectors shows a systematic progradation of fanlobes during the Pleistocene. The area of greatest thickness, found in the upper part of a fanlobe, tends to shift basinward for each progressively younger horizon. This can be observed to some degree in the constructed longitudinal section given in Figure 4.

Modern Fan Lobe

By combining high- and medium-resolution seismic data with different types of side-scan sonar (GLORIA, SeaMARC [3,4], EDO), we were able to describe in more detail the characteristics of the youngest fanlobe and to use it as a model for the underlying fanlobes. Sediments transported to the Gulf of Mexico have a low sand:clay ratio, and it is assumed that the transport processes may move the majority of the sand-sized material to deeper water (sand-efficient system).

The youngest fanlobe of the Mississippi Fan can be divided into four major regions (Fig. 3): (1) an upslope erosional canyon: Mississippi Canyon, (2) upper fanlobe at base of slope with a nearly filled large channel, (3) middle

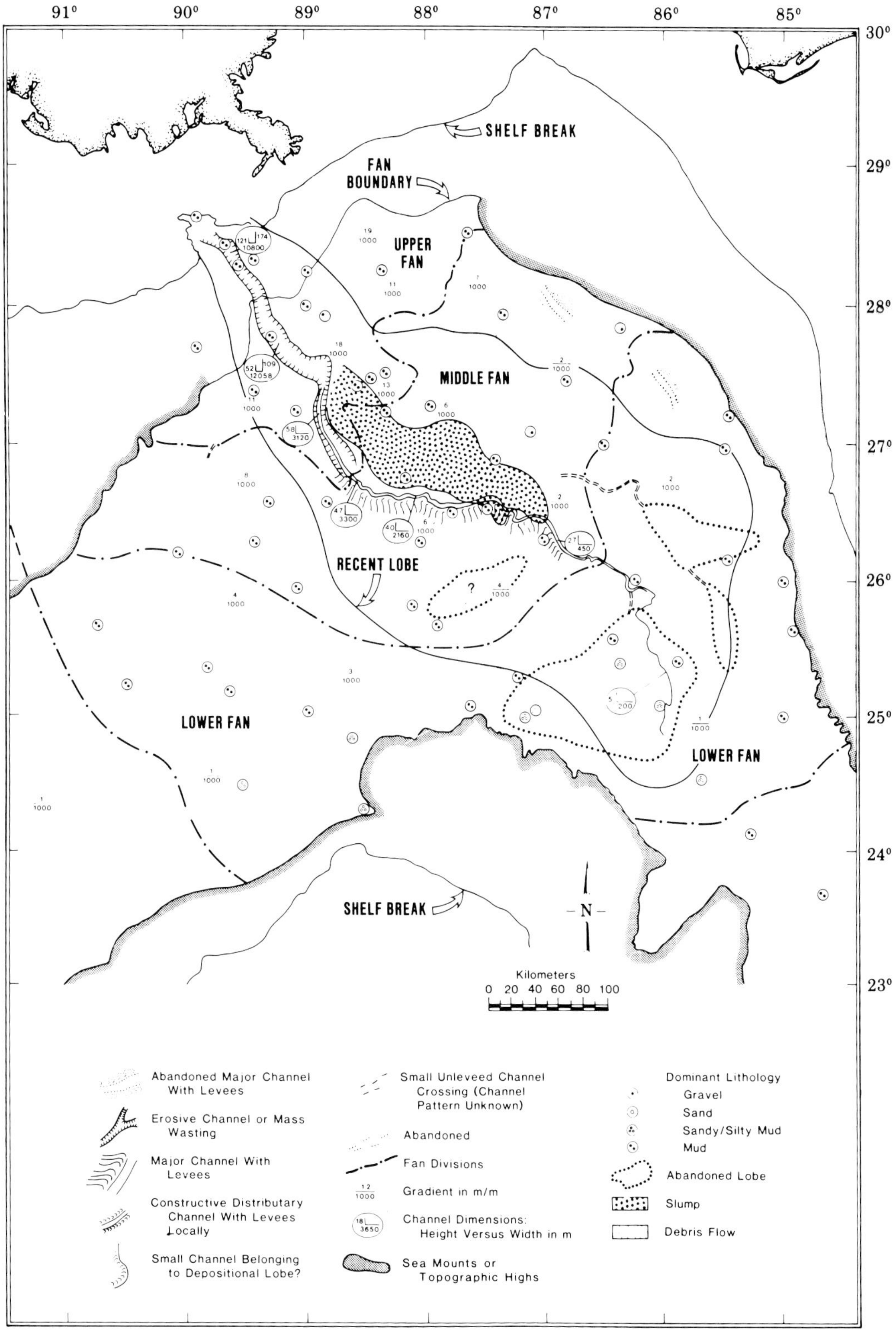

Figure 2. Morphometric map of the Mississippi Fan showing fan divisions, gradients, channel pattern and dimensions, and dominant lithology of the upper meters. Solid line denotes schematic outline of the youngest fanlobe.

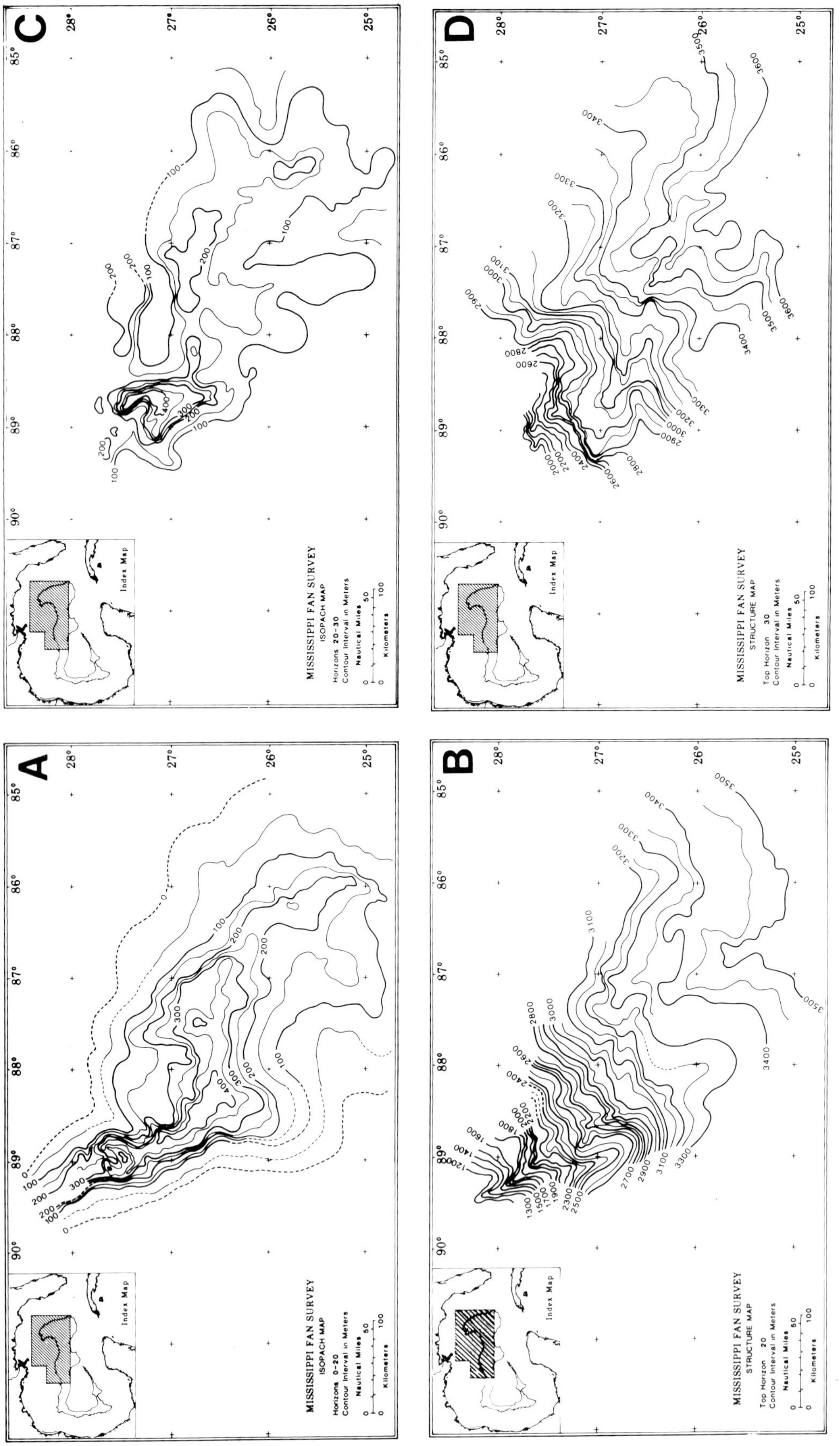
C
MISSISSIPPI FAN SURVEY
ISOPACH MAP
Horizons 20-30
Contour Interval in Meters
Index Map
D
MISSISSIPPI FAN SURVEY
STRUCTURE MAP
Top Horizon 30
Contour Interval in Meters
Index Map
A
MISSISSIPPI FAN SURVEY
ISOPACH MAP
Horizons 0-20
Contour Interval in Meters
Index Map
B
MISSISSIPPI FAN SURVEY
STRUCTURE MAP
Top Horizon 20
Contour Interval in Meters
Index Map

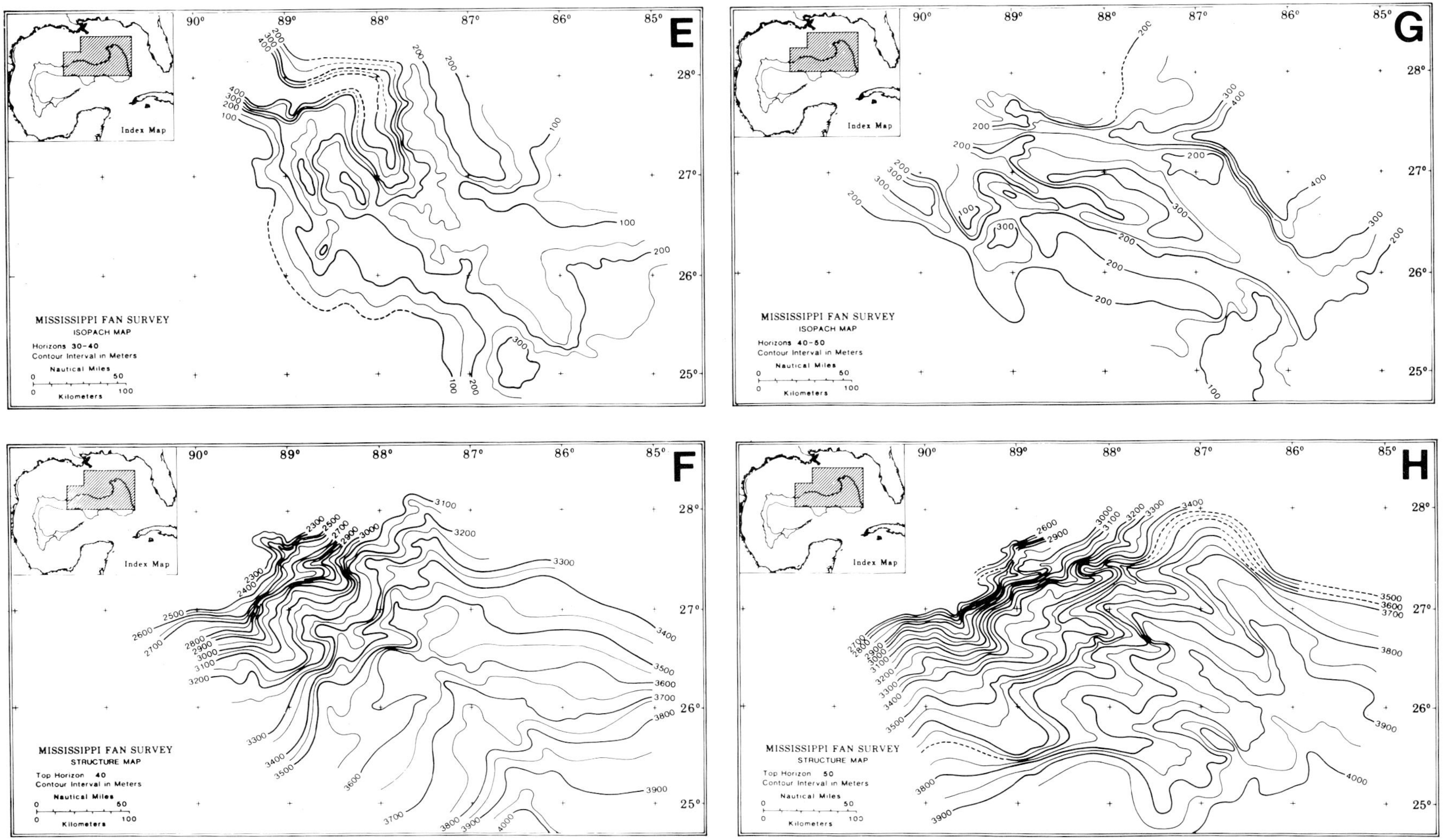

Figure 3. Isopach and structural contour maps of seismic reflector horizons that can be correlated across the Mississippi Fan. (A) Isopach map of the youngest fanlobe. (B) Structure map of "Horizon 20." This horizon represents the slump fill from older material from the sides. (C) Isopach map of Horizons 20-30. (D) Structure map of Horizon 30. (E) Isopach map of Horizons 30-40. (F) Structure map of Horizon 40. (G) Isopach map of Horizons 40-50. (H) Structure map of Horizon 50. This horizon is tentatively identified as Late-Mid Wisconsin (after [5]). See Figure 4 for a longitudinal cross-section of the acoustical horizons.

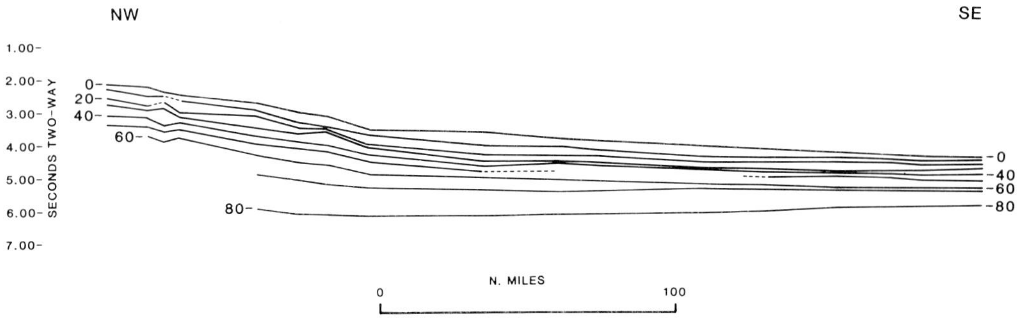

Figure 4. Constructed longitudinal cross-section showing the major acoustic reflection horizons found in the Mississippi Fan. The points used to construct this section come from reflection profiles that cut this section. The numbers of the horizons are selected arbitrarily for communication purposes. Horizon 80 is the base of the Pleistocene.

fanlobe, aggradational in character and convex in cross section with a sinuous axial channel, and (4) aggradational lower fanlobe with one recently active channel and several abandoned ones.

Mississippi Canyon

Extensive seismic coverage, foundation and deeper industrial borings, sediment analyses, and C-14 and stratigraphic dating allows the reconstruction of Upper Pleistocene events during the formation of the canyon [5]. The canyon formed post 25,000 to 27,000 years B.P., and infilling commenced about 20,000 years B.P. The canyon, therefore, formed in a very short time, removing a minimum of 1500 to 2000 km^3 of materials in about 7000 years. Retrogressive slumping from an initial mid to upper slope failure was responsible for the forming of this major submarine channel.

The role of the canyon in fan development cannot be understated. First, material removed during formation of the canyon was transported to deeper water; second, it likely acted as a conduit for large amounts of continentally derived sediments.

We believe that large-scale slumping on an unstable continental shelf-slope area represents the best explanation for the forming of this canyon, especially when considering the shape of the scars along the walls of the canyon; some of the sediment masses on the present canyon floor can be traced back to those scars [5]. Thus, the Mississippi Canyon resulted from shelf edge failures, followed by retrogressive slumping of shelf material around the canyon margins. Rapid shelf progradation during lowering sea level probably produced a large mass of rather weak sediments. During the period 20,000 to 10,000 years B.P., a series of Late Wisconsin delta lobes almost filled in the canyon with prodelta clays [5]. Hemipelagic sedimentation has been dominant during the last 10,000 years.

Upper Fanlobe

The upper part of the youngest fanlobe connects to the Mississippi Canyon at a water depth of approximately 1200 m. The upper fanlobe is comprised of an elongate mound of sediment, with a major cut-and-fill structure along its center, averaging 25 km in width. On the seafloor only a minor depression (valley) is left, having an average depth of 300 m. The width of the valley is variable and its width:depth ratio is extremely high (32:1). Additionally, the channel course in the northern part is influenced by diapirs.

The southern (lower) half of the upper fanlobe shows different characteristics. The width:depth ratio of the valley changes to 50:1. Additionally, a smaller channel developed inside the major cut and fill, and levees become more pronounced. Typically, the lower part of the fill shows an acoustically chaotic pattern overlain by more regular, often broken, discontinuous reflectors that are topped by distinct parallel reflectors similar to those described from the Louisiana intraslope basins [6]. The fill of the smaller channel inside the larger one generally shows the same acoustical pattern as the major fill. We suggest that the chaotic pattern either represents slump deposits or rather thick sandy deposits, possibly representing channel lag deposits.

Middle Fanlobe

The transition between the upper and middle parts of the youngest fanlobe occurs in a water depth between 2000 and 2400 m. Its boundary with the lower fanlobe is placed at a water depth of about 3100 m. The middle part of the fanlobe is convex in shape, about 200 km wide and approximately 400 m thick. An aggradational channel complex, about 10 to 20 km wide, is located along the apex of the midfan lobe. The most striking feature within the channel complex is a leveed, sinuous channel, about 2 to 4 km wide and 25 to 45 m deep. Sequences similar to "levee-overbank" deposits are very prominent on high-resolution seismic profiles. The channel complex decreases in thickness (width:thickness ratio changes from about 375:1 to 650:1) and in sinuosity downfan.

GLORIA, SeaMARC [3,4] and EDO deep-towed sidescan sonar and high-resolution seismic profiling over a portion of the midfan lobe detail the sinuous pattern of the channel and show numerous features interpreted as abandoned channel scars. They also show "ridge and swale" structures of varying orientations within the major channel complex. Side-scan sonar images show transverse "sandwaves" and "longitudinal bars" on the floor of the sinuous central channel. The "bars" are restricted to the channel bends.

The channel complex on the middle fanlobe is very similar to the surficial morphology of fluvial flood plains. Drilling into the channel deposits indicates that extremely coarse sediments, including basal gravels, are present within the base of the channel deposits (see Chapters 36–41, this volume).

Lower Fanlobe

At approximately 3100 m water depth, the central channel has a width of about 500 m and a depth of 5 to 10 m. The decrease in channel dimensions is accompanied by a considerable reduction in sinuosity. The poorly defined channel complex is about 5 km wide. Linear highs, oriented along the channel axis and comparable to fluvial midchannel islands, are quite common. Besides the central channel, several indistinct features that are interpreted as abandoned channels are observed within the channel complex. This implies that only one channel is active at a given time. Minor channel bifurcations are also observed.

Low topographic ridges (overbank levee?) border the channel complex. Outside the ridges are areas that display irregular, minor topographic relief. Side-scan sonar images suggest small "pressure ridges," sometimes in a diamond configuration, that may result from local mass movements. Low conical shapes with a central hole occur frequently and may indicate dewatering volcanoes. Because the topography is very complex on a small scale, it is impossible to observe larger, low-relief features such as sand sheets that were predicted at the end of channels.

Suggested Sedimentary Processes

Present data strongly suggest that slumping of weak shelf edge deposits, followed by retrogressive slumping of shelf sediments, formed the Mississippi Canyon. Carbon-14 dating shows that this process started less than 50,000 years B.P. on the continental slope, had retrogressed shelfward by 27,000 years B.P., and that most of the infilling occurred in less than 7,000 years. We feel that density currents are infrequent events and, because of the short time in which the canyon formed, density currents should not be considered a main erosional mechanism on the continental slope in this area. Subaerial erosion is very unlikely because it would require the river to erode to more than 1000 m below sea level. Local currents and local small-scale slumping are also ruled out as major causes.

The upper part of the youngest fanlobe basically is an infilled sediment conduit. The origin and timing of the channel are still unknown, but it can be assumed that they are contemporaneous with the erosion of the Mississippi Canyon. One possibility is that slumps grooved the canyon and that resulting currents added to the sediment removal. Another concept is that slumps moving through the canyon built up pressure ridges in the underlying material; these ridges became detached and incorporated into the slump.

It can be postulated that the slumps gradually transferred into debris flows and that the fill of the conduit results mainly from slumps and debris flows. As the gradient onto the midfan decreased, each successive transfer occurred further updip producing a backfilling sequence in the conduit. Finally, these deposits are overlain by hemipelagic sediments. High-resolution seismics indicate that this depositional sequence is cyclic and likely occurs during a rise in sea level.

Several different, but not exclusive, mechanisms can be suggested for aggradation of the midfan. Because the surface morphologies are comparable to meandering fluvial systems, the depositional history can be explained in fluvial terms. The channel complex with its sinuous central channel is comparable to a meander belt. Fluvial-type activity is suggested by the "ridge and swale" structures. The flanks of the fanlobe outside the belt are constructed by overbank deposits and local splays. Slumping on a local scale can account for the complex morphology.

D. B. Prior (personal communication, 1983) suggests debris flows as another mechanism. Subaerial equivalents typically have a central, sinuous, channel resulting from freezing of movement on the sides, leaving a mobile central core. Local deep-sea currents or currents generated by the debris flows cause surficial modification resulting in some of the bedforms observed.

A third possibility is a combination of debris flows, turbidity currents, and locally generated currents. Slumps generated by successive shelf edge failures would tend to originate further upslope and to decrease in size as the sediment stability is approached. As the slumps incorporate water, they transform into debris flows and finally into turbidity currents. Generally, the smaller the slump, the earlier the transformations are likely to take place. Larger events may enter the midfan as debris flows, while smaller events may have been transformed into turbidity currents by the time they reach the midfan. Sedimentary deposits resulting from this set of transport phenomena would then be consistent with the typical channel described from ancient rocks [7]. The central channel in each debris flow deposit guides the overall direction of the next transport unit. Stripping of surficial sediments, especially from debris flows, would result in low-density turbidity currents that flow toward the flanks of the fanlobe depositing the thin-bedded turbidites frequently observed in piston cores [8].

The core of the fanlobe thus consists of debris flow deposits and turbidites; the flanks of thin-bedded turbidites.

It seems that the lower fanlobe consists of one active channel that bifurcates near its terminus. Many of the turbidity currents generated higher up the fan likely continue onto the lower fan. The presence of even a minor channel will steer a turbidity current, but once channel restrictions are removed, the load will spread out and be deposited as a sheet.

If the channel system, described earlier, mainly functions as a conduit it would suggest a sand-efficient system in which most of the sand is transported to the deeper part of the fan. If, on the other hand, the channel is migratory in nature, a high percentage of the sand would have accumulated on the midfan.

Sea-level variations likely cause the initiation of slumping. Construction of a fanlobe may take place during the latter part of a sea-level lowering and the first part of a rise [6]. Deposition in deep water during high sea level is mainly hemipelagic and pelagic. During initial sea-level lowering, maximum progradation of the shelf takes place, emplacing large amounts of weak, failure-prone sediments at the outbuilding shelf edge [5]. This reasoning also suggests that subtle variations in sea level can have significant effects. The major reflectors we used for making isopach and structure maps probably represent rather long-duration high sea-level pelagic and hemipelagic deposits. Complexities found in those maps probably result from changes in the depositional pattern as the system responds to short, minor sea-level variations. The discussed framework and sedimentary processes can only be used as a general guideline. Refinement of the "model" must come from detailed grids of seismic reflection profiles, deep-towed side-scan sonar with acoustic profiling, and drilling.

Conclusions

The Mississippi Fan consists of a number of elongated fanlobes. The position of the source points as well as of the fanlobes shifted during the Pleistocene.

The Mississippi Canyon connects to the youngest fanlobe. It formed about 25,000 to 27,000 years B.P., resulting from large-scale failures on the upper slope. The large channel in the upper part of the fanlobe acted as a conduit for sediment that constructed the middle and lower parts of the youngest fanlobe.

Several transport processes may account for the depositional units seen on seismic reflection profiles. We believe that fluctuations in sea level are responsible for the slope failures and dictate the timing of deposition of the fanlobes.

References

[1] Stuart, C. J., and Caughey, C. A., 1976. Form and composition of the Mississippi Fan. Gulf Coast Association of Geological Societies Transactions, v. 26, pp. 333–343.

[2] Moore, G. T., and others, 1978. Mississippi Fan, Gulf of Mexico—physiography, stratigraphy, and sedimentation patterns. In: A. H. Bouma, G. T. Moore, and J. M. Coleman (eds.), Framework, Facies and Oil-Trapping Characteristics of the Upper Continental Margin. American Association of Petroleum Geologists Studies in Geology No. 7, pp. 155–191.

[3] Garrison, L. E., Kenyon, N. H., and Bouma, A. H., 1982. Channel systems and lobe construction in the Mississippi Fan. Geo-Marine Letters, v. 2, pp. 31–39.

[4] Kastens, K. A., and Shor, A. N. 1985. Depositional processes of a meandering channel on the Mississippi Fan. American Association of Petroleum Geologists Bulletin, v. 69, pp. 190–202.

[5] Coleman, J. M., Prior, D. B., and Lindsay, J. F., 1983. Deltaic influences on shelf edge instability processes. in: D. J. Stanley and G. T. Moore (eds.), The Shelf Break, Critical Interface on Continental Margins. Society of Economic Paleontologists and Mineralogists Special Publication 33, pp. 121–137.

[6] Bouma, A. H., 1981. Depositional sequences in clastic continental deposits, Gulf of Mexico. Geo-Marine Letters, v. 1, pp. 115–121.

[7] Mutti, E., and Ricci Lucchi, F., 1972. Le torbiditi dell'Appennino settentrionale: introduzione all'analisi di facies. Societa Geologica Italiana Memorie, v. 11, pp. 161–199.

[8] Bouma, A. H., 1973. Leveed-channel deposits, turbidites, and contourites in deeper part of Gulf of Mexico. Gulf Coast Association of Geological Societies Transactions, v. 23, pp. 368–376.

CHAPTER 22

Rhone Fan, Mediterranean

William R. Normark, Neal E. Barnes, and Francis Coumes

Abstract

The Rhone Fan is a large Plio-Pleistocene turbidite deposit in the western Mediterranean Sea. The fan is fed from the broad Rhone River delta, but only one canyon, the Petit-Rhone, has fed most of the major turbidite depositional sequences that have been mapped. Slumping of sediment from intercanyon areas on the delta slope also has provided much sediment for the fan. The lack of Recent turbidite deposition on the fan suggests that turbidite sedimentation dominates during glacial low stands of sea level, building major leveed-valley sequences, while surficial slumping of the valley levee deposits and pelagic sedimentation seem to mark high stands of sea level during interglacial periods.

Introduction

The Rhone Fan, with a radius of at least 300 km, is the largest fan in the western Mediterranean Sea [1]. Turbidity currents generated on the submarine slopes of the Rhone River delta travel more than 400 km to the ponded Balearic Abyssal Plain (Fig. 1), and the combined area of fan and abyssal plain exceeds 10^5 km^2 [1]. The Rhone Fan sequence began forming in earliest Pliocene time as terrigenous sediments started to cover salt and related evaporitic sequences of the Messinian interval following the refilling of the Mediterranean basin [2]. Fan sedimentation continued through the Pleistocene, but as is true of many modern fans, it received primarily hemipelagic and pelagic sediment during interglacial periods with high sea level [2].

The Rhone Fan is elongated in a north-south direction. Its shape is, in large part, controlled both by the configuration of the basin, which is bounded by Corsica and Sardinia to the east and the Balearic Islands to the west, and by sedimentation. Turbidite sediments from the Ebro Fan system in Valencia Trough mark the western limit of the Rhone Fan. The morphology and structure of the fan are affected by growth faults extending from the evaporite sequence underneath the fan; these faults have been active throughout fan growth [2,3]. The fan growth pattern, especially the channel systems, has been affected by numerous salt diapirs that pierce the middle- and lower-fan sediment [4]. In addition, the channel pattern is also affected by local surficial slides and slumps inducing changes in the direction of turbidity current flow [3].

Much of the northern half of the Rhone Fan has been thoroughly surveyed using numerous seismic-reflection systems as well as SEABEAM bathymetric mapping [2–6]. Three major surveys providing more than 3800 km of seismic reflection profiles of the upper and middle part of the fan have used both "vaporchoc" sound sources and high-resolution systems. South of 41°30′N latitude, however, relatively little data have been published in the open literature. This lack of data, together with the disruption of the fan sedimentation by diapiric structures south of 41°40′N, make it difficult to construct a comprehensive morphometric map for the entire fan (Fig. 2). Most of the discussion below is based on surveys of the northern part of the fan, which is the main depocenter of the fan, because only a few echo-sounding profiles are available from the southern part [1].

Morphology

Although at least four canyons lead to the area of the upper Rhone Fan, two channels corresponding respectively to the Grand-Rhone and Petit-Rhone, and another channel located farther west seem to have the major role in the buildup of

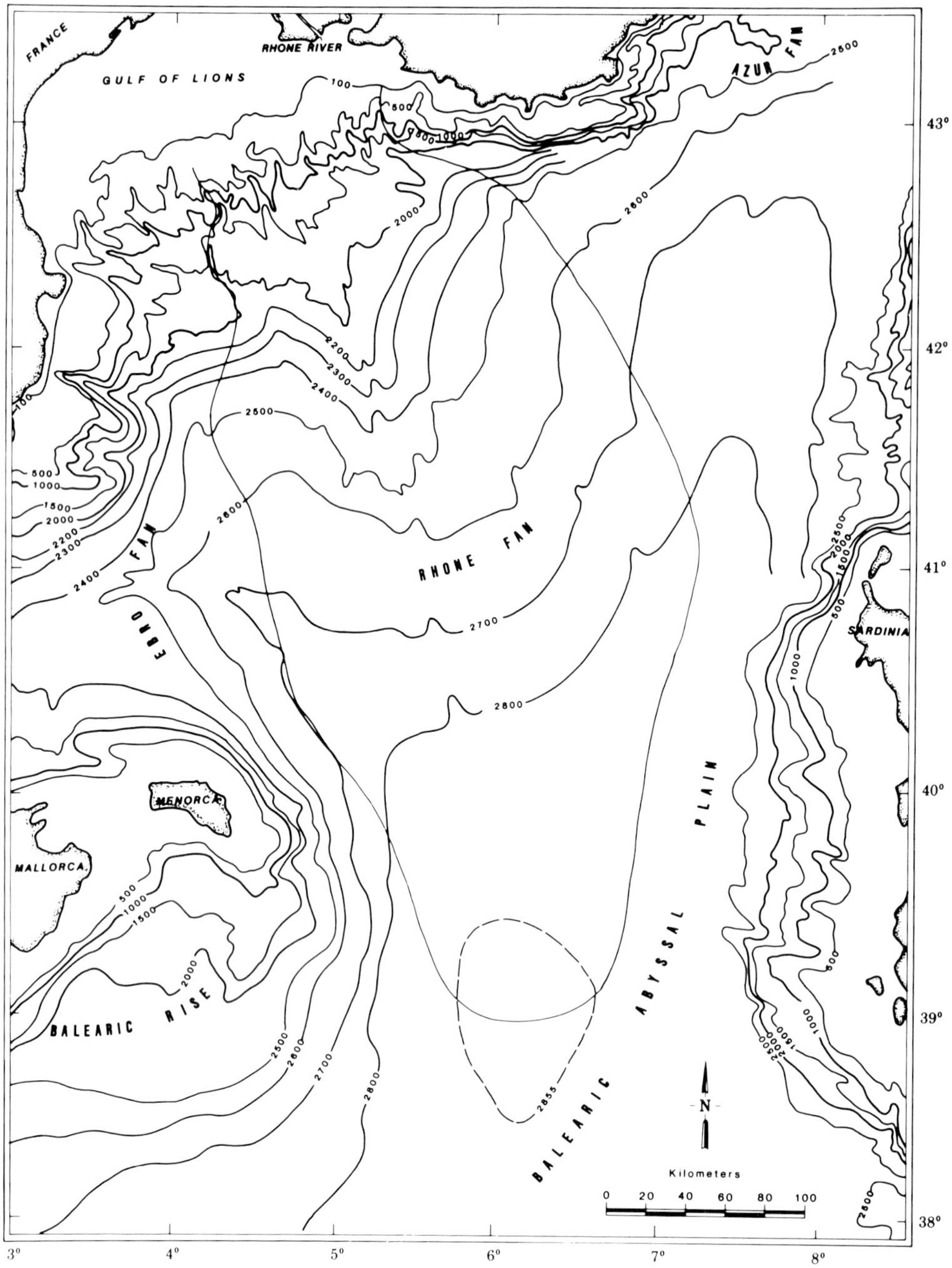

Figure 1. Bathymetric map of Rhone deep-sea fan modified from [1, 6]. Thin solid line shows inferred limits of fan. Contours in meters.

the fan [2,3]. The SEABEAM bathymetry (Fig. 3) shows at present that the Petit-Rhone has a more complex, sinuous shape than do the other canyons on the slope south of the Rhone delta [3,4]. The slopes between the canyons are marked by numerous concave-downslope head scarps indicative of major mass movement events.

The only major fan-valley system on the Rhone Fan extends from the Petit-Rhone canyon. This valley is characterized by a flat floor between broad levees that are 400 m high on the uppermost fan; levee relief decreases rapidly downfan. The total width of the channel-levee complex is as much as 70 km. A sinuous erosional channel lies within the broad depositional valley (Fig. 4). This erosional channel is clearly defined in the Petit-Rhone canyon, where it is 2 km wide and as much as 100 m deep, and it narrows downfan within the fan valley to only 0.25 km. Also, the meander loops of this inner channel have smaller radii of curvature downfan (Fig. 3). The major fan valley itself continues southeast into the diapir area where several possible distributary channels that are seen farther south may form [1].

Near latitude 42°12′N, a breach in the western levee of the main valley reflects a major change of direction in the

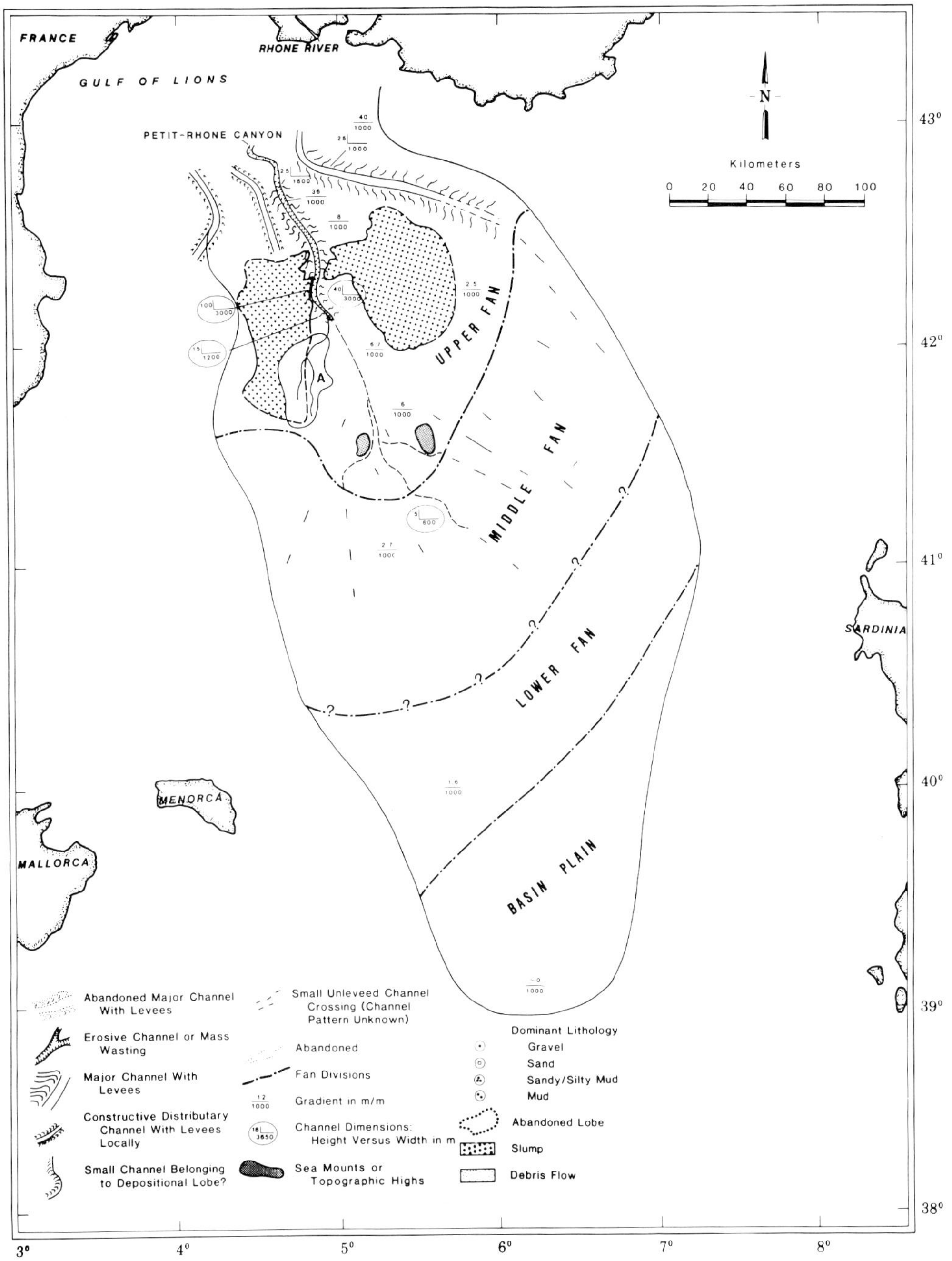

Figure 2. Morphometric map of Rhone deep-sea fan. Salt diapirs are shown by same pattern as for seamounts. Letter A denotes youngest depositional lobe.

meandering inner channel. This breach feeds a small depositional lobe about 40 km long. The depositional lobe and channel that feeds it are the youngest turbidite sequence on the fan [2].

The channels that have been recognized on the outer part of the Rhone Fan are presumably continuations of those entering the diapir area [1]. Small channels from the fan extend at least to the 2800-m depth contour, nearly 300 km from the Rhone delta slope [1,2]. These channels are a few tens of meters deep and about one kilometer wide; however, their shape in plan view and their extent are unknown.

Structure and Depositional Sequences

The extensive geophysical surveys from the northern part of the Rhone Fan provide a relatively clear picture of the growth pattern for the fan. A thick sedimentary megasequence onlaps the Messinian salt. This megasequence may be divided into two sequences separated by a high-amplitude seismic reflector. The lower sequence is characterized by small indistinct channels. The upper sequence is built by major channel-levee complexes [2].

The upper part of the fan sequence is dominated by the

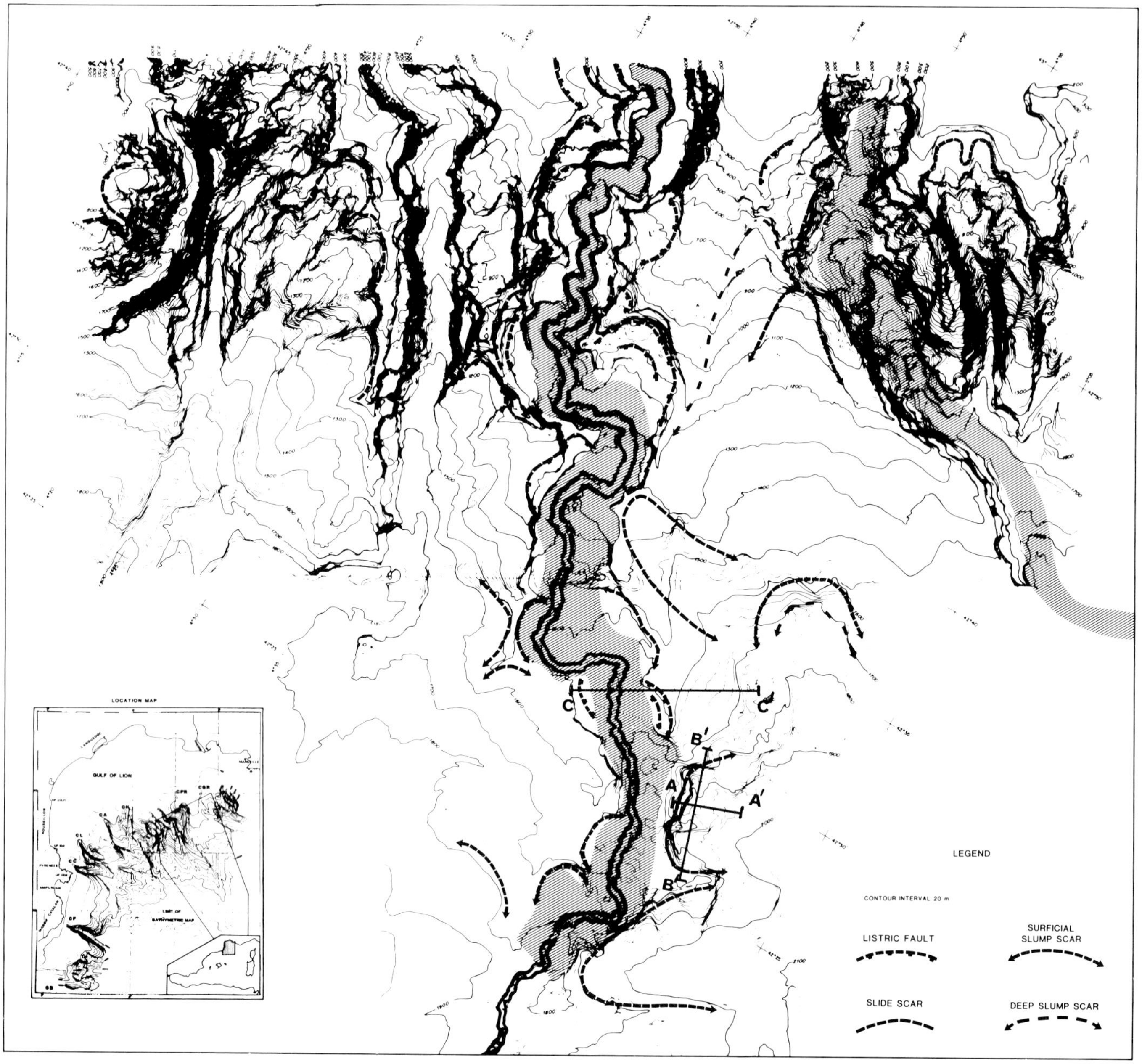

Figure 3. Detailed bathymetry of northern fan (from [3, 6]). Hachure pattern delineates floor of main canyon/valley features. Profile locations for Fig. 5 are shown.

broad depositional leveed valley extending from the Petit-Rhone canyon (Fig. 4). Individual depositional units within the valley complex are up to 70 km in width and may reach 125 m in thickness under the levee crest/valley floor area. These lenticular, mustache-shaped units are very characteristic of depositional valley systems. Isopach maps of individual depositional units show that the area of maximum thickness tends to shift downfan in successive units toward the recent breach in the western levee. The *composite* thickness of the stacked units, however, thins systematically downfan. Eight separate depositional units, all fed principally by the Petit-Rhone canyon, have been mapped [2].

In addition to the lenticularly bedded leveed-valley sequences, the Rhone Fan is composed of several intervals of acoustically chaotic facies interpreted as slide/slump deposits [3]. Two major periods of turbidite deposition on the fan are separated by an interval of chaotic facies. Chaotic units cover extensive areas of the fan, especially at the surface. Some of these youngest slide deposits can be related to the large arcuate slump scars recognized in the inter-canyon areas on the slope (Figs. 2 and 3).

The seismic-reflection records also show that the levees of

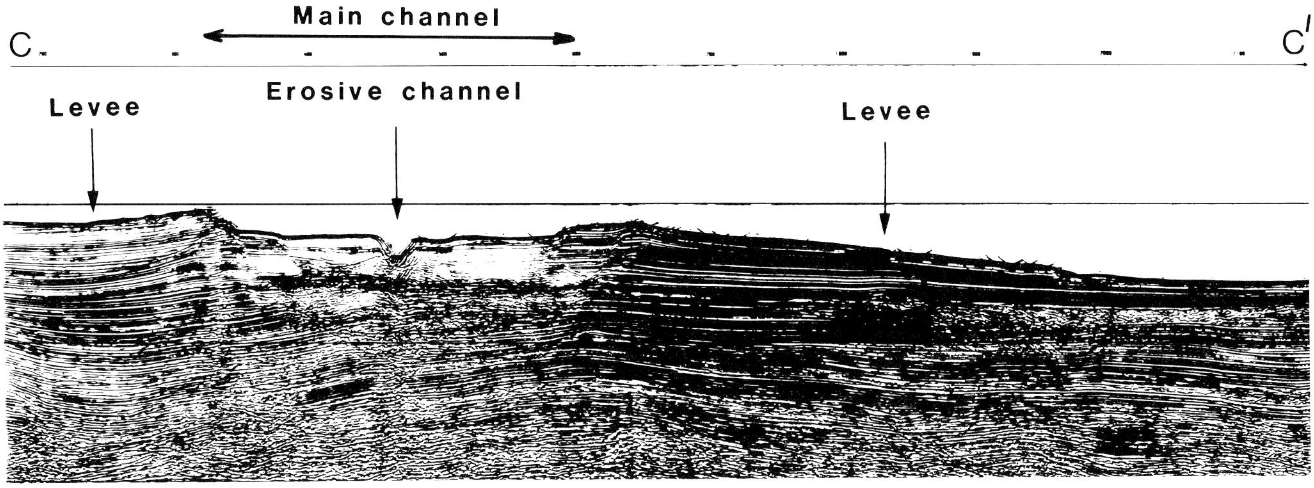

Figure 4. Seismic reflection profile across depositional valley on upper fan (from [3]). Location is shown on Fig. 3.

the main valley are extensively modified by slumping of the uppermost sediment (Figs. 3 and 5). Small superficial slumps toward the valley floor are more common, but outward-directed slumps of larger size occur as well. The slumps directed onto the valley floor modify the topography of the main-channel valley floor and force the erosional channel to meander. The sliding and slumping events are thought to be characteristic of periods of low sedimentation rates on Rhone Fan during high sea level stands. The leveed-valley sequences of turbidite deposition developed during the major low stands [2,3].

Growth Pattern

From the geophysical analysis of Droz and Bellaiche [2] and Nezondet [3], the major events during the growth of the inner part of the Rhone Fan are relatively well understood. Beginning in earliest Pliocene time, the fan was fed through a number of slope gullies resulting in a sediment sequence characterized by unorganized and chaotic sediments. These deposits infilled the previous Messinian paleotopography and were followed by turbidite deposits fed by small channels. This was followed by the development of the Petit-Rhone

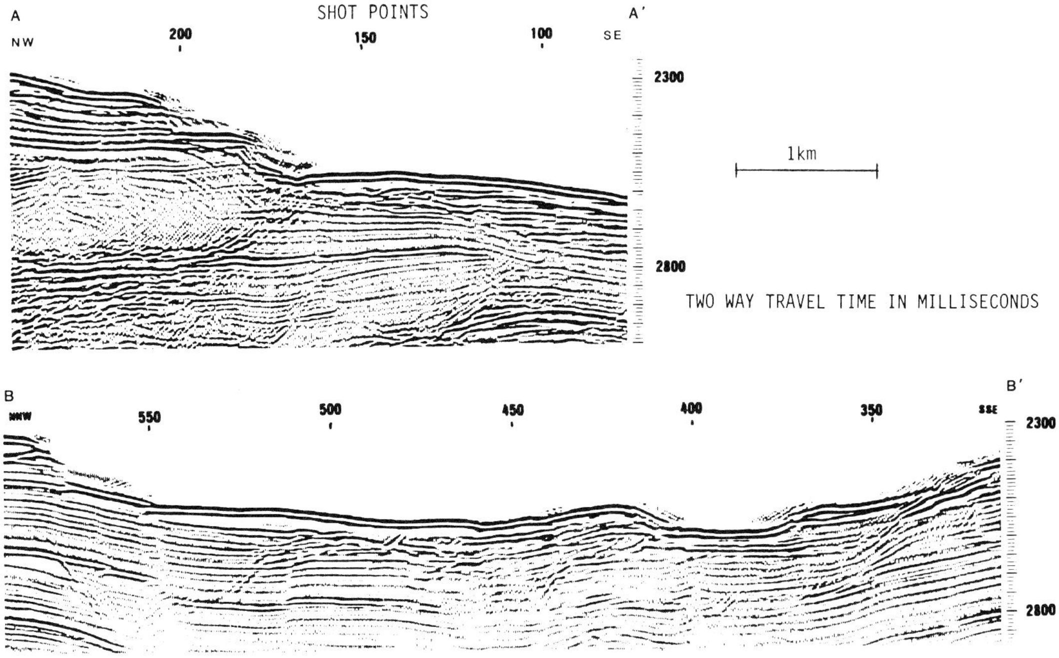

Figure 5. Seismic reflection profile of depositional valley showing surficial slumps along inner levee slope (from [3]). Line locations are shown on Fig. 3.

canyon, which was the main conduit for a series of turbidite sequences forming major depositional valley features. Above a major unconformity, a series of eight successive depositional units was formed. These units, also fed by the Petit-Rhone canyon, are divided by an interval of chaotic reflectors. The units above this chaotic zone lie west of the earlier series [2]. A breach in the main valley formed at the end of this turbidite depositional interval. Pelagic or hemipelagic sedimentation, together with surficial slumping of the uppermost valley units, are the primary sedimentologic processes active at present.

Addendum

Attention is drawn to the detailed bathymetric map recently completed by Orsolini and others [6]. Much of the detail in Figure 1 and the map of Figure 3 was based on preliminary versions [3,5] of this map, which had not been released in final form until recently.

References

[1] Menard, H. W., Smith, S. M., and Pratt, R. M., 1965. The Rhone deep sea fan. In: W. L. Whittard and R. Bradshaw (eds.), Submarine Geology and Geophysics, Butterworths Scientific Publications, London, pp. 271–285.

[2] Droz, L., and Bellaiche, G., in press. The Rhone deep-sea fan: morphostructure and main growth pattern.

[3] Nezondet, M., 1982. Geologie et tectonique de l'eventail detritique sous marin du Rhone. Institut de Physique du Globe, Universite de Strasbourg (unpublished).

[4] Monaco, Andre, and others, 1982. Essai de reconstitution des mecanismes d'alimentation des eventails sedimentaires profonds de l'Ebre et du Rhone (Mediterranee Occidentale). Bulletin de l'Institut de la Geologie Bassin d'Aquitaine, Bordeaux, no. 31, pp. 99–109.

[5] Bellaiche, G. and others, 1981. The Ebro and the Rhone deep sea fans: first comparative study. Marine Geology, v. 43, no. 3/4, pp. M75–M85.

[6] Orsolini, P., and others, 1983. Carte bathymétrique de la marge continentale au large du Delta due Rhône, Golfe du Lion. Sociéte Nationale Elf Aquitaine, map. Edition and orders BEICIP, 92500 Rucil Malmaison, France.

CHAPTER 23

Wilmington Fan, Atlantic Ocean

William J. Cleary, Orrin H. Pilkey, and Jeffrey C. Nelson

Abstract

The Wilmington Fan consists of three subdivisions. These are the upper fan, a truncated mid-fan, and the lower rise hills province. Wilmington Fan differs from most fans in two important aspects: 1) at the present, most sand crossing this portion of the passive continental margin escapes to the adjacent Abyssal Plain, and 2) bottom currents are important in fan development. Over the entire fan, most sediment has been deposited by bottom current-related processes. The lower rise hills formed by bottom currents restricts fan lobe formation on the lower continental rise by channeling the turbidity currents through numerous troughs between the rise hills.

Introduction

The North American Continental Rise off the Northern American Mid-Atlantic States starts at the base of the continental slope where a sharp change in slope occurs at a water depth of about 2500 m. The average seaward slope of the rise is less than 1°. The seaward boundary of the Continental Rise is in the vicinity of the 5300 m isobath; however, it is difficult to delineate because of the lower fan systems that merge with the Hatteras Abyssal Plain (Fig. 1). The Wilmington Fan consists of the portion of continental rise between the Hudson and Hatteras canyon systems (Fig. 1). The Wilmington, Baltimore, Washington, and Norfolk Canyons feed this area. The Norfolk and Washington Canyons merge on the upper rise at about 3200-m water depth, while the Baltimore and Wilmington Canyons coalesce farther north at approximately 2600 m. The combined Norfolk-Washington and Baltimore-Wilmington valley systems converge at approximately 4500 m, where they lose their individual identities in a maze of channels through the lower rise hills. The Wilmington Fan occupies a strip (190 × 600 km) of continental rise of approximately the same dimensions as the Laurentian Fan [1,2], which is the only other large fan system on the North American Continental Rise studied in detail. Due to a lack of sufficiently detailed seismic data, it is difficult to delineate the exact boundaries of the fan system.

This report emphasizes the Late Pleistocene and Holocene development of the Mid-Atlantic Continental Rise (primarily the processes within piston-coring range). Earlier geologic development of this portion of the continental rise has been discussed by Tucholke and Laine [3].

From a bathymetric standpoint, the Wilmington Fan can be divided into: 1) a relatively steep and smooth segment from 2500 to 4250 m, 2) a smooth terrace with a very gentle slope between 4250 and 4600 m, and 3) a segment between 4600 and 5300 m with a slope similar in steepness to that of the first segment and covered by sediment waves known as the lower rise hills, which merge downslope into the Hatteras Abyssal Plain (Fig. 1).

The broad, gentle-sloping continental rise was originally assumed to have been deposited largely from turbidity currents [4]. However, it is now widely recognized that a major portion of the late Quaternary sedimentary column off eastern North America is not of turbidity current origin, but instead consists of layered silts and homogeneous silty clays (contourites) deposited by south-moving contour-parallel bottom currents [5]. Embley [6] suggested that large debris flows also have been important in shaping the rise.

A major problem in studying recent sedimentation on the Mid-Atlantic continental rise has been the distinction of contourites and turbidites in cores. Distinction between layers of sediment derived from contour and turbidity currents is not always unequivocal, but Hollister and Heezen

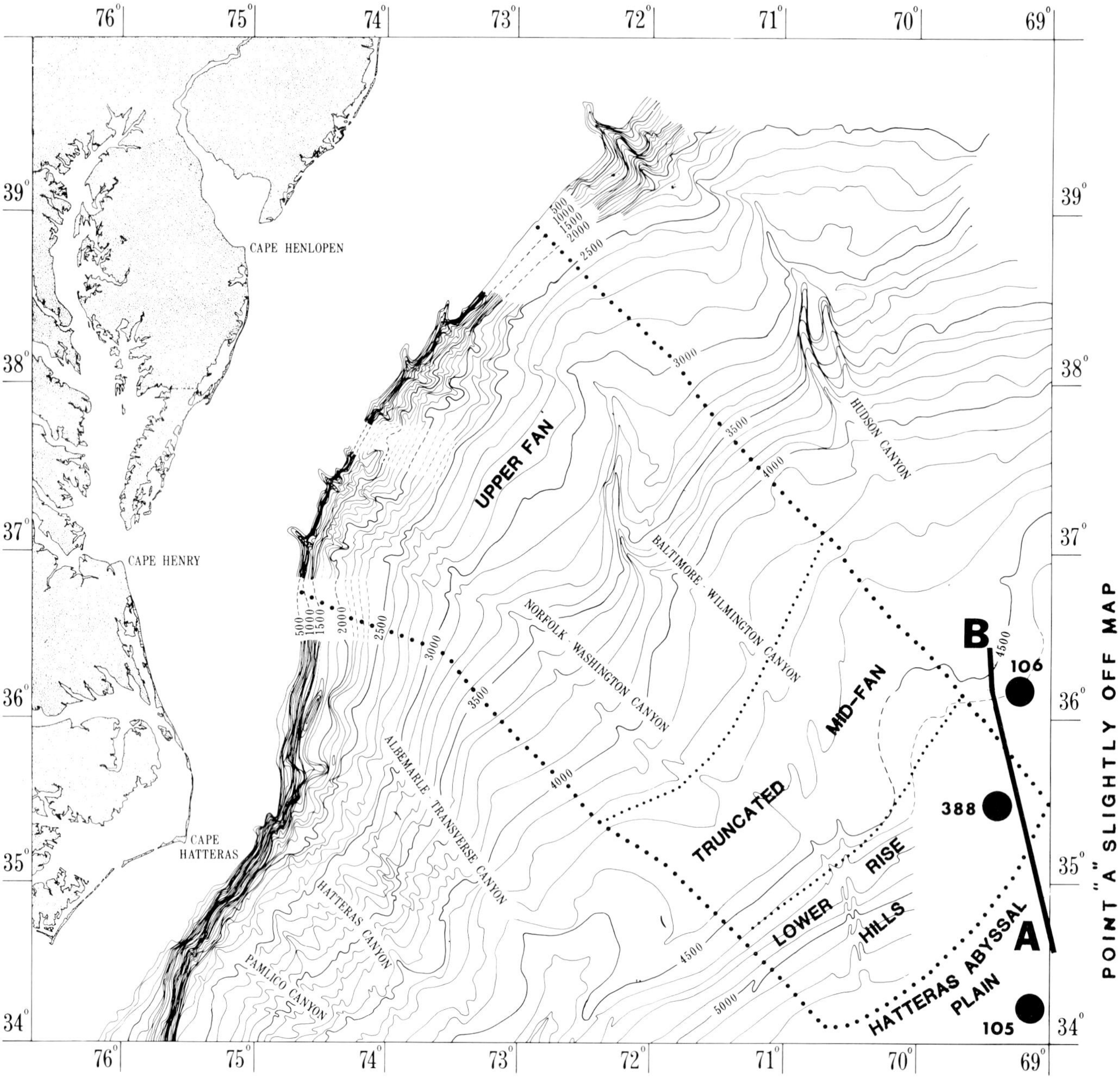

Figure 1. Bathymetric chart of Mid-Atlantic Rise showing major canyons, fan divisions, location of seismic line (Fig. 2), and DSDP bore holes. (After J. Newton, unpublished data.)

[7] and Fritz and Pilkey [8], among others, developed a number of useful parameters for this purpose. These criteria are based largely on a megascopic scale and rely on the abundances and thicknesses of silt/sand beds. Hollister and Heezen [7] indicated that: 1) bedded silt/sand comprises less than 20% of contourite dominant cores, but turbidite dominant cores have typically more than 20% bedded silt/sand; 2) contourite dominant cores have numerous silt/sand beds (50 to 500 beds per 10 m of core) and the beds are always very thin (the thickest bed may be 20 cm, and the average bed thickness is 1 cm); and 3) silt/sand beds in turbidite dominant cores are thicker (usually 20 cm) and not as abundant (< 50 beds per 10 m of core). In addition, turbidite sands are often graded and exhibit diagnostic depositional structures and shallow water faunal components.

Rise Canyon/Channel Systems

The Baltimore-Wilmington and Norfolk-Washington Canyons represent the second and third largest canyon/valley extensions, respectively, into the deep-sea off the eastern United States. These canyons and their valleys extend for

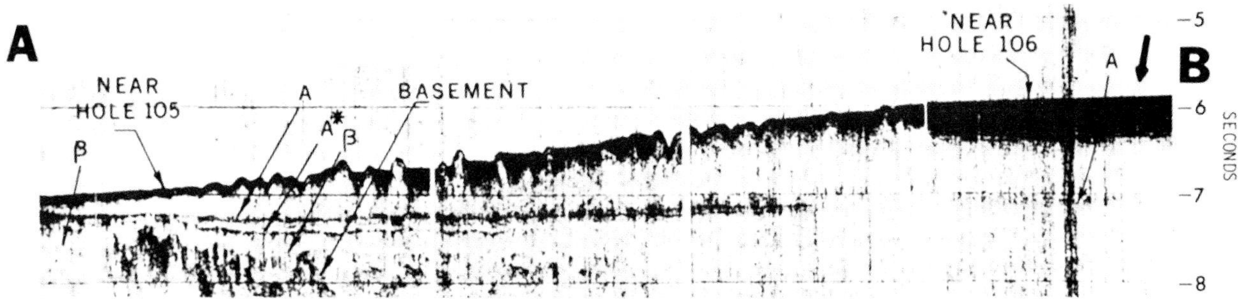

Figure 2. Seismic reflection profile across lower continental rise hills [4]. Arrow indicates the portion of truncated mid fan developed on terrace. Turbidite fill exceeds 450 m. Location indicated on Figure 1. Point A is slightly off Figure 1.

over 360 km from the 2000-m contour to the region of the lower continental rise hills (Fig. 1).

The major tributary of the Wilmington Canyon, the Baltimore Canyon, converges on the right levee (looking downslope) at depths of 2600 m. Farther downslope at 3200 m, the Norfolk Canyon coalesces with the Washington Canyon. The extension channels of these two canyon systems are separated by approximately 80 km at the base of the upper fan at depths of 4200 m. The larger Wilmington Canyon has relief of up to 300 m in a deeply incised 110-km long gorge section.

The two sets of valleys converge on the truncated mid-fan region below 4200 m and are separated by less than 20 km at depths of 4500 m. Channels in this region vary dramatically in size and character. Some of the larger channels have several thalwegs (0.5-1.5-km wide) and a number of the channels are flat-floored with widths up to 7 km. Relief on the larger channels varies from 20 to 50 m. Within the larger valleys, braided channel patterns and step-like terraces occur [9]. On the lower portion (4600 m) of the truncated mid-fan, delineation of the channels becomes uncertain because of the complicated nature of the sediment waves associated with the Hatteras Outer Ridge. These waves have broad outlines with relief of 50 to 75 m and amplitude of 5.5 to 9.0 km.

The shape of the "trough" areas between the sediment waves (lower rise hills) is generally concave upward. Some troughs are extremely flat, suggesting sediment accumulation by ponding. Seismic profiles indicate that the troughs contain horizontally bedded, highly reflective fill indicative of turbidite deposition [3,10,11].

Fan Development

Development of a deep-sea fan in this region commenced in the early Miocene at the terminus of the Norfolk-Washington Canyon. Seaward of this ancestral fan, the Western Boundary Undercurrent (WBUC) ultimately constructed the 500-m high Hatteras Outer Ridge parallel to regional depth contours. Concomitant growth of the outer ridge and the progradation of the ancestral fan resulted in the formation of a 450-m deep basin at the juncture of the two deep-sea features. Seismic data and DSDP drilling results (Figs. 1 and 2) have been used to infer that this basin landward of the outer ridge began to fill with turbidites between 2.8 to 3.0 m.y. ago [3].

Turbidity currents completely filled the basin by 500,000 years B.P. and breached the crestal area of the outer ridge [3]. The ponded turbidites on the western flank of the outer ridge represent the present lower continental rise terrace. The axial portion of the ancient sedimentary ridge lies buried approximately at the 4500-m isoline. Below 4750 m, on the eastern flank of the ridge, topography is complicated because of the large sediment waves (lower continental rise hills) associated with the WBUC. Turbidity currents that enter this region are laterally restricted and are channelized in the troughs between the hills. In a number of areas, they have formed small finger-like ponds and, in many cases, buried smaller hills. The evolution of the fan since the Mid to Late Pleistocene basically involved a progradation of the fan across the lower continental rise hills and the adjacent Hatteras Abyssal Plain [12].

Modern Fan Development

We recognize three distinct subdivisions of the Wilmington Fan system on the basis of sand layer geometry and rise morphology (Fig. 1). Rise physiography, in turn, is largely a function of the long-range rise evolution discussed earlier. The subdivisions of the Wilmington Fan system are different from fan models proposed for other continental margins, largely because of the very strong influence of bottom current sedimentation.

The fan subdivisions we can delineate are: 1) an upper fan, 2) truncated mid-fan, and 3) Lower Rise Hills province (Fig. 3). The upper fan corresponds to the same division recognized on Pacific margin fans [13] and on the Laurentian Fan [1]. The truncated mid-fan is a region formed by ponding behind the ancestral Hatteras Outer Ridge, which accounts for its gentle gradient. Due perhaps to the abrupt reduction in slope at a depth of approximately 4200 m, fan lobe development apparently commences here and extends

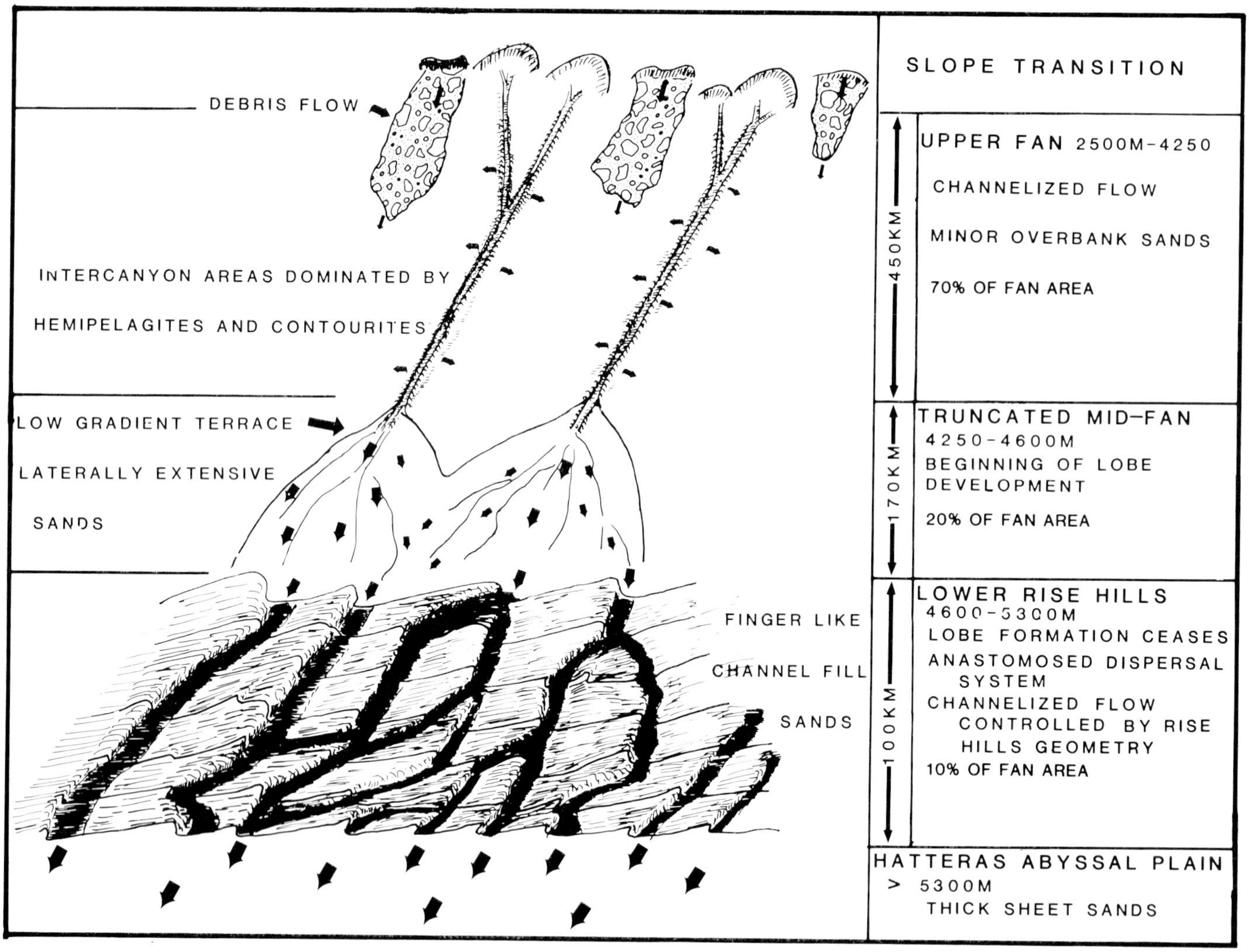

Figure 3. Conceptual model illustrating the three subdivisions of the Wilmington Fan. Figure not drawn to actual scale.

across the gentle slope. However, as flowing sediment impinges on the lower rise hills, fan lobe development is terminated as, once again, turbidity currents are channelized. In effect, the development of the classic mid-fan lobe system is truncated at the landward edge of the lower rise hills province; hence the use of the term "truncated mid-fan" (Fig. 3).

Flows continue seaward through the myriad of channels among the lower rise hills until they eventually disgorge onto the Hatteras Abyssal Plain. Channel systems in the Lower Rise Hills are poorly understood at this point. Probably at least one major through-going channel exists through this entire subdivision in which the very largest flows are transported to the Hatteras Abyssal Plain.

In Late Pleistocene-Holocene times, the Wilmington Fan system has acted as a filter for sediments disgorged onto the Hatteras Abyssal Plain. Evidence for this is the fact that few sand layers less than 10-cm thick and virtually none less than 5-cm thick are found in the upper portions of the abyssal plain sediment column [14,15]. We assume this means that only large or high velocity flows are capable of escaping from or flowing through the Lower Rise Hills. Smaller or low velocity turbidity currents leave the sand behind, either on the truncated mid-fan or within troughs between rise hills.

Sediments

Details of sediment distribution and types were determined by the analysis of more than 100 piston cores (Fig. 4), bottom photographs, and 2000 km of 3.5-kHz P.E.S.R. records. Core stratigraphy of the three fan divisions appears to be related to proximity of canyon axes and depth. For purposes of this discussion, the core sediment columns are considered to consist of lutite layers, silt layers, and sand layers.

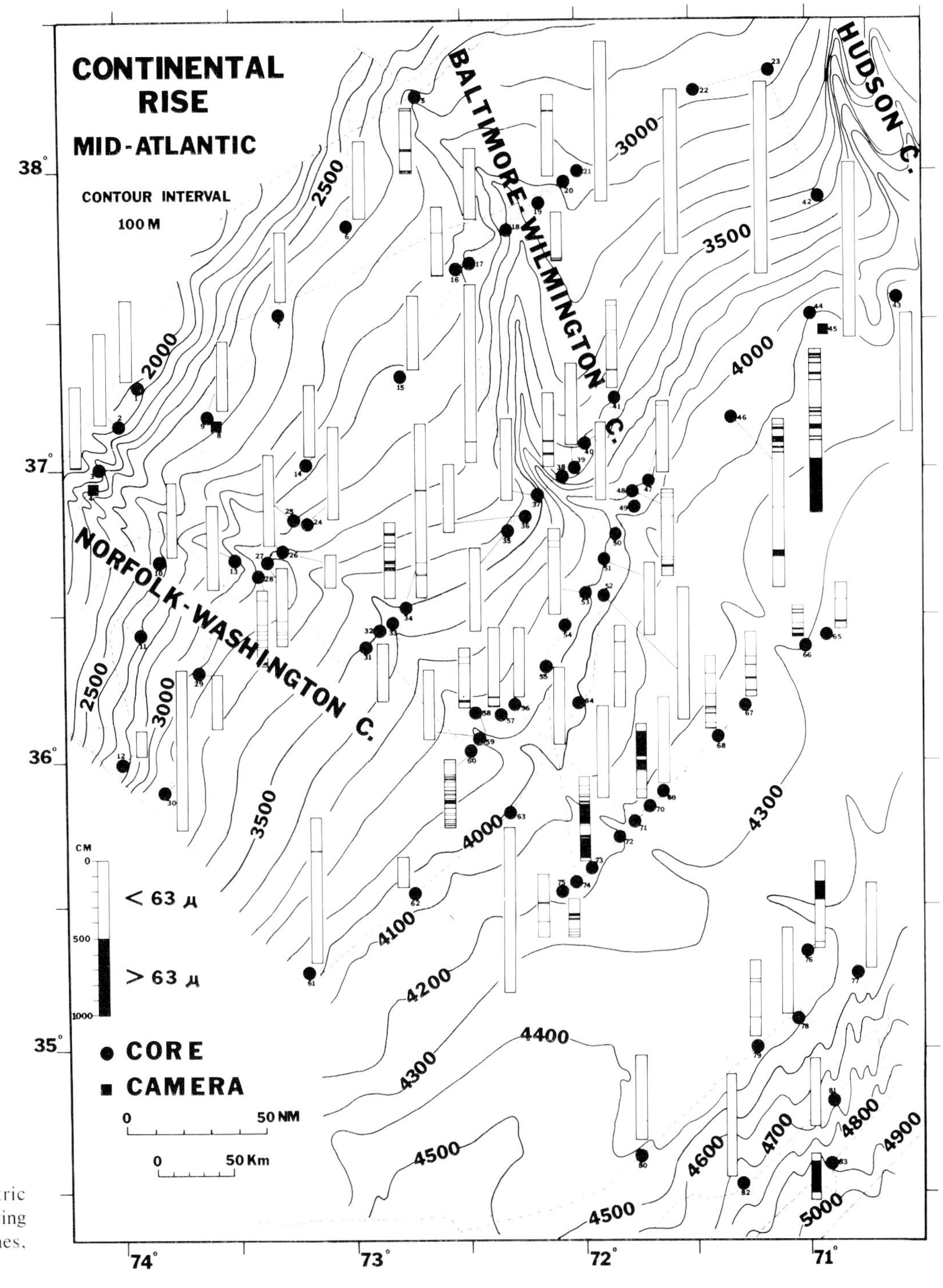

Figure 4. Detailed bathymetric chart of Wilmington Fan showing piston core locations, track lines, and general core stratigraphy.

Lutite Layers

Lutites comprise over 80% of the total core length (Fig. 4). Typically, a 0 to 200-cm thick (average 80 cm), pale yellow-brown carbonate-rich unit (average 35%) caps a lower, olive-gray to gray, carbonate-poor unit (average 15%). Foraminifera form the principal calcareous silt and sand-sized carbonate component.

It is assumed that most of the lutite sections of the piston cores used for this study had been deposited by contour currents. As a rule, only small thicknesses of lutite are obviously color-related to a coarse-grained underlying turbidite layer. Most of the lutite section of a given core exhibits the same color characteristics throughout, except for the above-mentioned ubiquitous light-colored upper core units. Furthermore, on the assumption that bottom current activity is both widespread and long-lived, then even the hemipelagic contribution is probably slightly affected by bottom currents. Hence, distinction between true hemipelagic lutite sections and contour-current deposited sections may be impossible and meaningless. The one lutite variety that is almost certainly of bottom current origin is the so-called rose-gray lutite.

The occurrence of rose-gray lutite has been used as an indicator of transport and deposition by contour currents along the entire eastern U.S. continental margin. The source area for this sediment type is the red bed province of eastern Canada [16]. Rose-gray lutite layers are widespread in all regions across the upper fan surface and are frequently found to be intercalated with thin silt layers and various-colored lutite laminae.

Rose-gray lutite layers are not found in deeper waters than those of the upper fan subdivision. Generally speaking, rose-gray lutite units are more commonly observed below 3500 m (within the upper fan division), although there are no clear trends with respect to frequency and thickness. Rose-gray lutite units average three per 10 m of core length across the entire fan, while the average rose-gray lutite layer thickness for all rise cores is 4 cm. This widespread occurrence is a measure of the degree to which the entire upper fan is affected by contour current activity, although rose-gray lutite layers represent only a small portion of the entire contourite sediment column.

Silt Layers

Good data on silt layer distribution are only available from the upper fan region. Thin silt layers are common and occur as distinct units that are well sorted and generally less than 1-cm thick; we have observed 1.75 m-thick beds. The thickest beds are found on the truncated mid-fan adjacent to the Baltimore-Washington Canyon. Many silt laminae, a few grains thick, occur which are generally too thin to measure accurately for thickness and frequency compilations. No correlation of silt layers was possible even with closely-spaced cores (< 5 km).

For all cores from the upper fan, silt layers average 30 per 10 m. The maximal number recorded (120) occurs within 8 km of canyon axes, close to the confines of levee systems of the Norfolk-Washington Canyon on the lowermost upper fan. On the upper fan, surface silt layers average approximately 22 per 10 m in water depths less than 3000 m and reach a maximum of 45 per 10 m below 4500 m in the truncated mid-fan. The higher numbers in the mid-upper fan region and within the proximity of canyons probably reflects the availability of silt-sized sediment from turbidity currents.

Sand Layers

Data from cores taken on the upper fan combined with previously published data on sand layers of the lower rise [12] reveal a distinct pattern of sand layer distribution related to the effects of bathymetry on turbidity currents (Fig. 5). It is immediately apparent that sand layers are confined to areas near canyon axes on the upper fan. Across the truncated mid-fan region that forms the lower continental rise terrace, sand layers are much more laterally extensive. Within the lower rise hills, sand layers are confined mostly to channels.

Above 4300 m on the upper fan, the sand beds are generally less than 20-cm thick (Figs. 1 and 5). For the most part, those units greater than 10-cm thick, in addition to many thinner units, have the internal characteristics of turbidites. The highest frequency (32 per 10 m) of these sandy turbidite units was found within the proximity of the small low-relief levee of the Norfolk-Washington Canyon on the lowermost upper fan (3850 m).

Generally, sand thickness and frequency of sand layers per meter increase with water depth down the upper fan surface (Fig. 5). On the upper fan at 2500 m, sand layers are completely absent. On the mid-upper fan region (3500 m), sand units are generally less than 5-cm thick within 10 km of the valleys. Downslope at 4200 m on the truncated mid-fan, sand layer thickness increases to an average of 16 cm, with the thickest units located within 10 km of canyons (Fig. 5). Within this region of the fan, a number of 70-cm or more thick sand layers were penetrated. Similarly, a downslope increase in the frequency of sand layers per meter also occurs, from an average low frequency of five per 10 m at 3100 m on the lowermost upper fan to an average of 13 per 10 m at 4200 m (Fig. 5) [12].

Below 4250 m where the canyons emerge onto the 175-km wide lower continental rise terrace, turbiditic sands are much more widespread but are still quite complicated because of the complex distributary system developed on the flanks of the old Hatteras Outer Ridge. Below this depth across the terrace, low-relief channels of the Wilmington Canyon system extend and cross over the former crest of the Hatteras outer ridge near 4500 m. At depths between 4500 and 4750 m, relief is relatively low and the channels are poorly confined to paths between sediment waves that form the lower continental rise hills. Between 4750 and 5300 m on the lower rise hills fan division, the channels are strongly confined within these "troughs" (Fig. 3). Thick-graded and massive sandy (2 m) turbidites locally were deposited among the sediment waves on the eastern flank of the ridge to produce flat-floored sediment ponds. Those currents that had sufficient magnitude and followed the channel extensions through "trough" areas eventually disgorged their sand on the Hatteras Abyssal Plain. Introduction of sand to the Hatteras Abyssal Plain occurs through many troughs across a broad zone at the base of the rise.

The distribution of sand across the mid-Atlantic Continental Rise appears to occur in a long funnel-shaped pattern, widening seaward. The funnel widens most rapidly

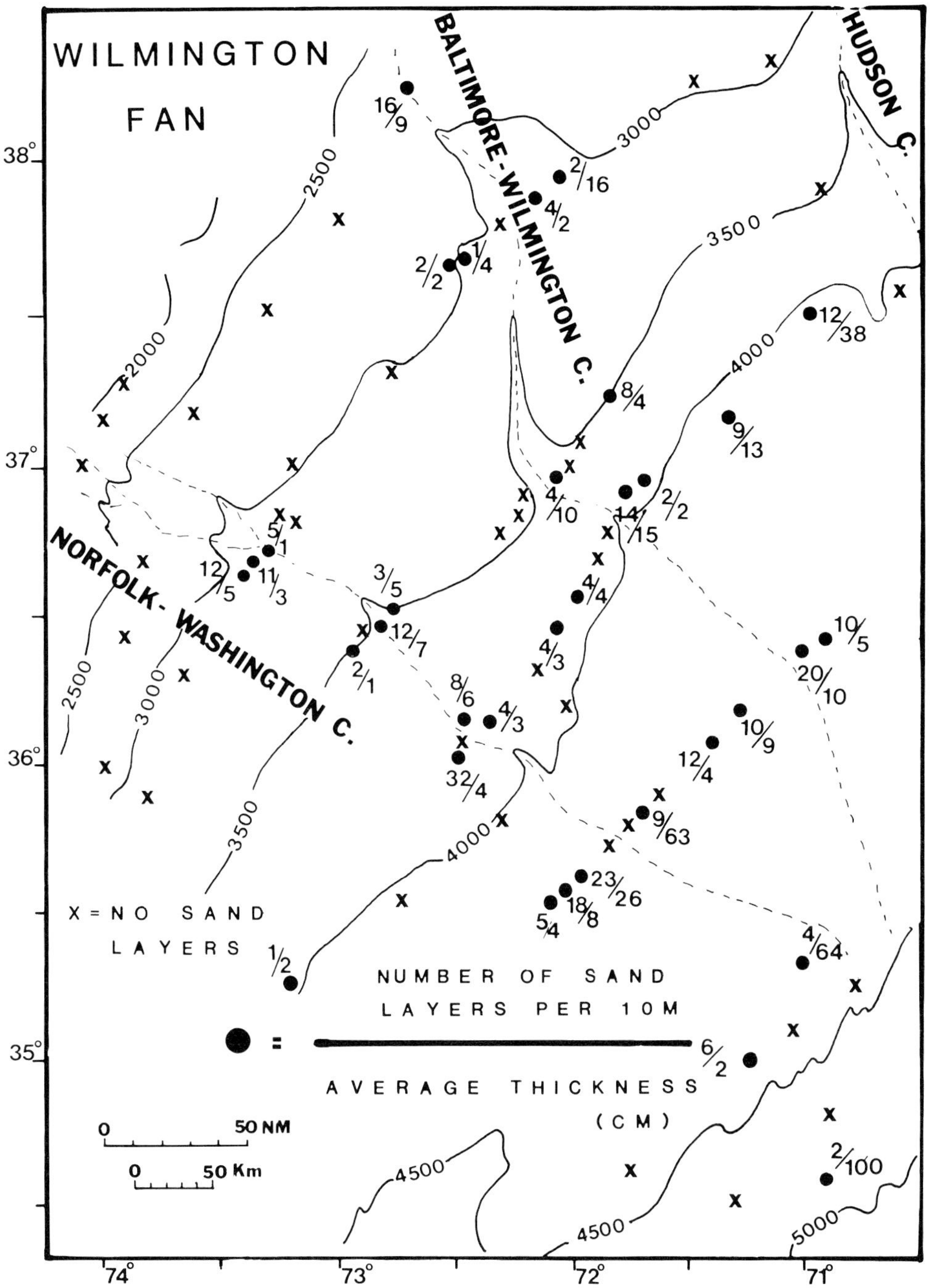

Figure 5. Frequency and average thickness (cm) of sand layers per 10 m. X indicates those locations where no sand units were recorded.

at approximately 4200 m, where dispersal occurs across a broad zone only to be channelized again by lower continental rise hills.

Conclusions

At the present time, on this portion of the North Atlantic Continental Margin, most sand escapes the continental rise and is deposited in thick extensive layers on the Hatteras Abyssal Plain. The Wilmington Fan system acts as a filter, trapping small sand-bearing turbidity currents, but allowing large portions of the large flows to "pass through" to the abyssal plain.

The Wilmington Fan system is strongly affected by bottom current sedimentation in two ways. Much of the lutite contribution to the rise is at least slightly affected by the north-to-south flowing contour-current system. In addition, current-formed features, the Lower Rise Hills, are responsible for channelizing turbidity currents and preventing fan lobe formation.

Three fan subdivisions are recognized. The largest is the

upper fan, which makes up about 70% of the fan area. Here turbidity currents are largely restricted to canyon channels with only minor overbank deposition. Intercanyon areas, which make up the largest area of the continental rise, are dominated by contour-current deposited lutites. On the truncated mid-fan subdivision (between 4250 and 4600-m water depth), fan lobe development commences. Lobe development is rather abruptly terminated when turbidity currents are once again channelized by the lower rise hills beginning at approximately 4600 m.

Acknowledgments

This study was based on work supported by the National Science Foundation under Grant DES 75-14416. We also wish to thank the Duke University Marine Laboratory for the use of R.V. EASTWARD. Additional support was provided by the Program in Marine Sciences, University of North Carolina at Wilmington.

References

[1] Stow, D. A. V., 1981. Laurentian Fan: morphology, sediments, processes and growth pattern. American Association of Petroleum Geologists Bulletin, v. 65, pp. 375–393.

[2] Normark, W. R., Piper, D. J., and Stow, D. A. V., 1983. Quaternary development of channels, levees, and lobes on middle Laurentian Fan. American Association of Petroleum Geologists Bulletin, v. 67, pp. 1400–1409.

[3] Tucholke, B. E., and Laine, E. P., 1982. Neogene and Quaternary development of the lower continental rise off the central U.S. East Coast. In: J. S. Watkins and C. L. Drake (eds.) Continental Margin Geology. American Association of Petroleum Geologists Memoir No. 34, Tulsa, Oklahoma, pp. 295–305.

[4] Emery, K. O., and others, 1970. Continental rise of southern North America. American Association of Petroleum Geologists Bulletin, v. 54, pp. 44–108.

[5] Heezen, B. C., Hollister, C. D., and Ruddiman, W. F., 1966. Shaping of the continental rise by deep geostrophic contour currents. Science, v. 152, pp. 502–508.

[6] Embley, R. W., 1980. The role of mass transport in the distribution and character of deep-ocean sediments with special reference to the North Atlantic. Marine Geology, v. 38, pp. 23–50.

[7] Hollister, C. D., and Heezen, B. C., 1972. Geologic effects of ocean bottom currents: Western North Atlantic. In: A. L. Gordon (ed.) Studies in Physical Oceanography, Volume 2. Gordon and Breach Science Publications, New York, pp. 37–66.

[8] Fritz, S. J., and Pilkey, O. H., 1975. Distinguishing bottom and turbidity current coarse layers on the continental rise. Journal of Sedimentary Petrology, v. 45, pp. 57–62.

[9] Cleary, W. J., Pilkey, O. H., and Ayers, M. W., 1977. Morphology and sediments of three ocean basin entry points, Hatteras Abyssal Plains. Journal of Sedimentary Petrology, v. 47, pp. 1157–1170.

[10] Hollister, C. D., Ewing, J. I., and others, 1972. Site 106, Lower Continental Rise. Initial Reports Deep Sea Drilling Project, v. 11. U.S. Government Printing Office, Washington, D.C., pp. 313–349.

[11] Benson, W. E., and Sheridan, R. E., 1978. Site 388, Lower Continental Rise Hills. Initial Reports of the Deep Sea Drilling Project, v. 44. U.S. Government Printing Office, Washington, D.C., pp. 23–46.

[12] Ayers, M. W., and Cleary, W. J., 1980. Wilmington Fan: mid-Atlantic development. Journal of Sedimentary Petrology, v. 50, pp. 235–245.

[13] Normark, W. R., 1974. Submarine canyons and fan valleys: factors affecting growth patterns of deep-sea fans. In: Modern and Ancient Geosynclinal Sedimentation. Society of Economic Paleontologists and Mineralogists Special Publication No. 19, pp. 56–68.

[14] Cleary, W. J., and others, 1978. Patterns of turbidite sedimentation on a trailing plate margin: Hatteras Abyssal Plain, Western North Atlantic Ocean. In: Proceedings Tenth International Congress on Sedimentology, v. 1, pp. 126–127.

[15] Pilkey, O. H., Locker, S. D., and Cleary, W. J., 1980. Comparison of sand-layer geometry on flat-floors of ten modern depositional basins. American Association of Petroleum Geologists Bulletin, v. 64, pp. 841–856.

[16] Needham, H. D., Habib, D., and Heezen, B. C., 1969. Upper Carboniferous palynomorphs as a tracer of red sediment dispersal patterns in the northwest Atlantic. Journal of Geology, v. 77, pp. 111–120.

IV

Ancient Turbidite Systems

Active Margin Setting

CHAPTER 24

Blanca Turbidite System, California

Hugh McLean and D. G. Howell

Abstract

Blanca fan is a submarine fan composed of Miocene volcaniclastic strata. Parts of the fan system are exposed on Santa Cruz and Santa Rosa Islands, and possibly correlative strata crop out on San Miguel and Santa Catalina Islands.

The Blanca fan and underlying breccia reflect regional transcurrent faulting in the California Continental Borderland and development of a system of rapidly subsiding basins and uplifted linear ridges during early and middle Miocene time. Erosion of uplifted crystalline basement rocks followed by the onset of silicic volcanism created linear sediment sources for the alluvial and submarine fans, respectively.

Introduction

A sequence of early and middle Miocene conglomerate, breccia, sandstone, and siltstone at least 1200-m thick deposited as part of an alluvial and submarine fan system, crops out on Santa Cruz and Santa Rosa Islands, California (Fig. 1). The fan complex, named herein the Blanca fan, consists of strata belonging to, in ascending order, the Vaqueros and Rincon Formations, San Onofre Breccia, and Blanca and Monterey Formations, [1–5]. The Blanca and Monterey Formations constitute the volcanogenic submarine fan while the lower three formations are derived from unroofing of basement rocks, which include the Willows Diorite of Weaver [6] and structurally lower blueschist not now exposed.

Lithofacies relations include a complex assemblage of alluvial fan and paralic and submarine fan facies [7]. Deposition was influenced by (1) the initiation of wrench tectonics in response to Pacific–North America transform faulting, (2) a eustatic rise in sea level, and (3) volcanism, seemingly centered along a post-late Eocene, pre-early Miocene right-slip fault system.

Stratigraphy and Regional Setting

Correlation of sections exposed on Santa Rosa and Santa Cruz Islands is primarily based on three sedimentary petrofacies (Fig. 2). The Vaqueros Formation is stratigraphically lowest and contains diorite and greenschist debris, but no blueschist, silicic volcanic clasts, or tuffaceous beds. The Rincon Formation and San Onofre Breccia [4] contain abundant glaucophane schist debris, but no silicic volcanic clasts. The Blanca Formation is highest stratigraphically and contains abundant andesitic and dacitic clasts mixed with blueschist debris, all in a light-colored tuffaceous matrix. The regional distribution of the three petrofacies suggests that the Blanca fan is a regressive system associated with progressive uplift of basement and onset of explosive silicic volcanism [3].

The Blanca fan complex is well exposed on the southwest part of Santa Cruz Island. Here strata butt against crystalline basement rocks and represent a series of coalescing debris aprons that grade laterally into generally finer grained strata, which crop out to the west along the eastern part of Santa Rosa Island. Dacite-rich strata of similar age and depositional environment crop out on San Miguel Island, and middle and late Miocene strata on Santa Catalina Island may be part of subsidiary fans (Fig. 3). Numerous dart cores taken along the Santa Cruz–Catalina Ridge consist of middle Miocene strata with abundant schist and volcanic detritus [8].

Transcurrent faults in the northern part of the California Continental Borderland became active in early Miocene time, and right-lateral displacement added geometric complexity to existing submarine fan complexes [9]. Dacite-rich marine conglomerate and sandstone on San Miguel Island are stratigraphically coeval with the Blanca fan and may represent a subsidiary submarine fan derived from a nearby, but geo-

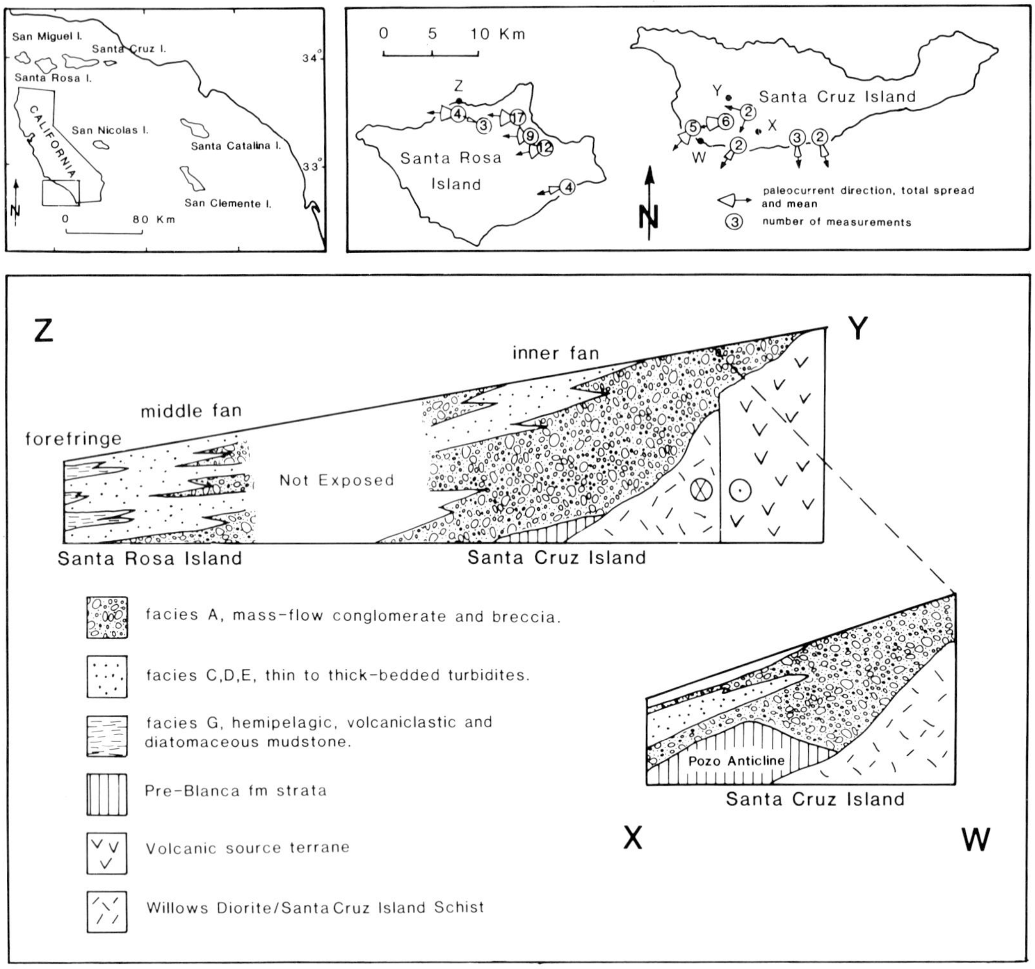

Figure 1. Index map of northern Channel Islands; paleocurrent localities, number of measurements and flow directions and panel cross-section showing distribution of submarine fan lithofacies of Mutti and Ricci Lucchi [13].

graphically separate, source terrane. Areas where dacite was recovered in dart cores are shown on Figure 4 [8]. Volcaniclastic rocks of dacitic composition in the Fisherman's Cove area of Santa Catalina Island [10] resemble the strata of the Blanca fan, but are mostly middle and late Miocene [10,11].

Fan Divisions

At the base of the Miocene section, conglomerate in the Vaqueros Formation consists mainly of an alluvial-fan facies, composed of cobbles and boulders of hornblende diorite, that grades upward into a fossiliferous marine facies [2]. The overlying San Onofre Breccia was deposited in a paralic environment and contains clasts of blueschist, greenschist, and diorite interfingers with deep-water (bathyal) mudstones within a distance of 5 km.

The borderland physiography of deep basins and linear ridges that began forming in early Miocene time was well established by the time the San Onofre Breccia and Rincon Formation were deposited. Basins formed in local regions as a result of tectonic transtension, whereas uplifted blocks resulted from transpression. Submarine fan deposition is best displayed by the Blanca Formation, a sequence of epiclastic and reworked pyroclastic rocks that indicate rapid deposition from a nearby source [12].

On Santa Cruz Island, channelized breccia and conglomerate of both alluvial-fan and paralic marine facies rapidly thin and become finer grained to the southwest across a paleo-high, the Pozo anticline (Fig. 1). Volcanogenic strata of the Blanca Formation also become finer grained across the bathymetric paleo-high in a southwest direction, yet more westerly directed paleo-flows of conglomerate extend as far west as Santa Rosa Island (Fig. 5).

Most of the strata of the Blanca fan reflect mass-flow depositional processes that constructed a debris apron along a linear fault scarp. Although conglomerate clast populations tend to become more evenly bedded and finer grained toward the west, lateral and vertical relations are complex and generally nonsystematic, and deposition was influenced by irregularities of paleobathymetry and catastrophic pulses of volcaniclastic sediment input. A simple fan-model cannot be applied to these strata.

The characteristic pattern is coarse-grained, thick-bedded organized and disorganized in the east, lithofacies *A* predominate (Fig. 4a). However, thinning and fining of strata occur

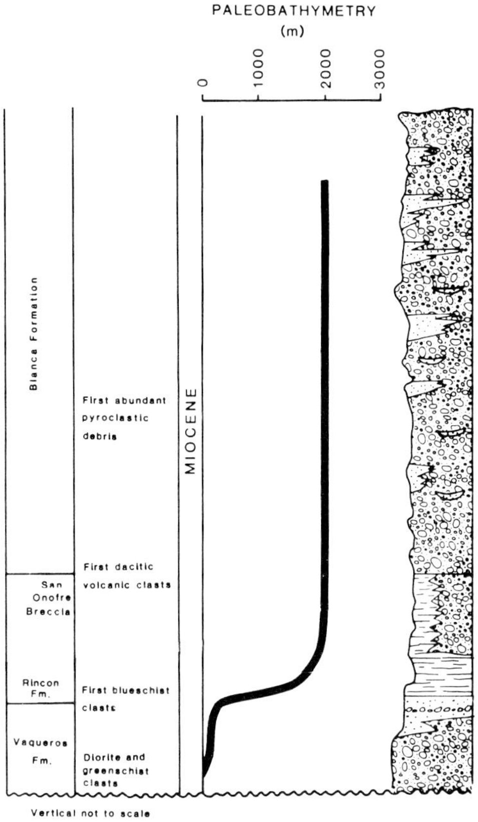

Figure 2. Composite columnar section with stratigraphic distribution of clast assemblages, and estimated paleobathymetry based on biofacies analysis of benthic foraminiferal assemblages. Figure and stratigraphic nomenclature modified from Ingle [15, p. 181].

where the sediment flows cross a paleo-high, and organized, stratified conglomerate and thin-bedded turbidites (lithofacies *E*) [13] predominate (Fig. 4c). Where the proximal debris flows moved into a more distal setting unaffected by bathymetric irregularities, a wide spectrum of depositional lithofacies occurs, principally *B*, *C*, *E*, and *F*. These lithofacies are well developed on Santa Rosa Island where channels of boulder conglomerate interfinger with sandy turbidites that are interstratified with diatomaceous, hemipelagic mudstone, lithofacies *G* of a distal fan-fringe setting (Figs. 5a–c).

So-called inner-fan facies are largely nonchannelized strata of thick-bedded conglomerate reflecting debris aprons that cascaded from fault-controlled line sources bordered by steep gradients. Shelf deposition or flow through submarine canyons is not evident. In more distal middle-fan facies, channeling and lens-shaped beds are common. Bouma sequence turbidites are well developed, and coarse-grained beds are interbedded with thin-bedded (T_{c-e}) turbidites. There are no well-developed sequences where beds thin or thicken upward in systematic patterns. The most distal parts of the Blanca fan contain massive beds of hemipelagic tuffaceous mudstone.

Mass-flow conglomerates of the coalesced debris aprons crop out on shore for at least 40 km from east to west (present coordinates). The average width (direction normal to the line source) is about 10 km. Paleocurrent directions mea-

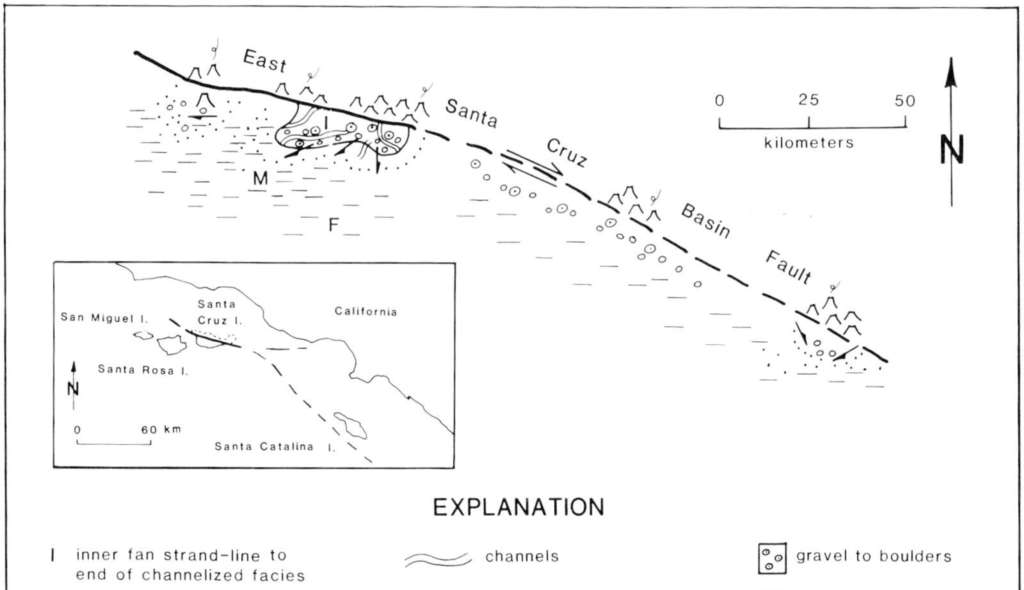

Figure 3. Regional distribution of the Blanca fan and subsidiary fan complexes, and inferred paleogeography during deposition of the Blanca Formation.

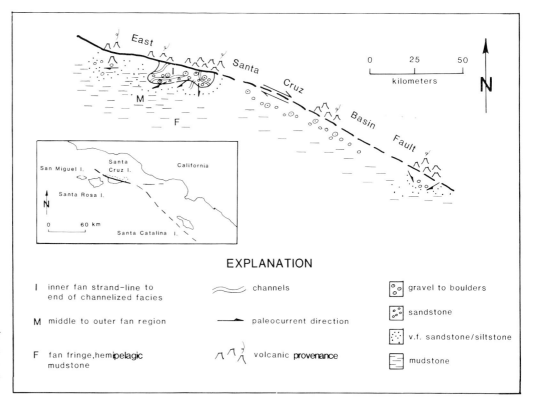

A
B
C

A
B
C

sured from channel axes and imbricated clasts indicate south to southwest flow on Santa Cruz Island [7], which corresponds well with the east-west-trending basement-sediment buttress, and fault contacts. To the west on Santa Rosa Island, paleo-flow is primarily westward directed (Fig. 1), however, paleomagnetic data suggest substantial (about 120°) clockwise rotation for a tectonic block that includes southern Santa Cruz Island and Santa Rosa Island [14], which complicates paleogeographic reconstructions.

Possible middle-fan deposits within the Blanca fan are represented by well-sorted channelized sandstone and pebble conglomerate (T_{a-c} and T_{b-d}) interbedded with thin-bedded sandstone and mudstone (T_{d-e}) that are interpreted as overbank and interchannel deposits. Commonly, large volcanic clasts of relatively low density form discontinuous strings or isolated fragments in the upper portions of the T_b unit. The overbank and interchannel deposits are often involved in slump-folding and soft-sediment deformation.

Channelized middle-fan sandstone is inferred to extend over an area of about 50 km by 15 km, parallel to west-trending paleocurrent directions measured on Santa Rosa Island. Middle-fan sand thickness is estimated to be 200 to 500 m.

Thin-bedded and laminated mudstone, shale, and sandstone, deposited in an outer fan-fringe environment, interfinger with the channelized middle-fan (?) sandstone facies of the Blanca fan. The fan-fringe (?) strata are present on Santa Rosa Island as thin-bedded tuffaceous sandstone and diatomaceous shale of the Monterey Formation.

In summary, the volcaniclastic Blanca fan system on Santa Cruz and Santa Rosa Islands is dominated by organized and disorganized inner-fan conglomerate, and by a channelized middle fan composed of sandstone and well-stratified organized conglomerate. The channelized facies are interbedded with thin-bedded overbank and interchannel sandstone and shale and grade rapidly outward into thin-bedded and laminated mudstone of the fan-fringe (?) facies.

Present paleocurrent indicators suggest southward, southwestward, and westward current flow, but the entire Blanca fan may have been tectonically rotated by as much as 120° in a clockwise direction [14]; therefore, paleogeographic reconstructions must be made with caution.

References

[1] Avila, F. A., and Weaver, D. W., 1969. Mid-Tertiary stratigraphy, Santa Rosa Island. In: D. W. Weaver and others, Geology of the Northern Channel Islands. American Association of Petroleum Geologists and Society of Economic Paleontologists and Mineralogists, Pacific Section Special Publication, pp. 48–67.

[2] Bereskin, S. R., and Edwards, L. N., 1969. Mid-Tertiary stratigraphy, southwestern Santa Cruz Island. In: D. W. Weaver and others, Geology of the Northern Channel Islands. American Association of Petroleum Geologists and Society of Economic Paleontologists and Mineralogists, Pacific Section Special Publication, pp. 68–79.

[3] McLean, H., Howell, D. G., and Vedder, J. G., 1976. Miocene strata on Santa Cruz and Santa Rosa Islands—A reflection of tectonic events in the southern California Borderland. In: D. G. Howell (ed.), Aspects of the geologic history of the California Continental Borderland. American Association of Petroleum Geologists, Pacific Section Miscellaneous Publication 24, pp. 241–253.

[4] Stuart, C. J., 1979. Middle Miocene paleogeography of coastal southern California and the California Borderland—Evidence from schist-bearing sedimentary rocks. In: J. M. Armentrout, M. R. Cole, and H. Ter Best, Jr. (eds.), Cenozoic paleogeography of the Western United States. Society of Economic Paleontologists and Mineralogists, Pacific Section, Pacific Coast Paleogeography Symposium 3, pp. 29–44.

[5] Weaver, D. W., Griggs, G., McClure, D. V., and McKey, J., 1969. Volcaniclastic sequence, south central Santa Cruz Island. In: D. W. Weaver and others (eds.), Geology of the Northern Channel Islands. American Association of Petroleum Geologists and Society of Economic Paleontologists and Mineralogists, Pacific Section Special Publication, pp. 84–90.

[6] Weaver, D. W., 1969. The pre-Tertiary rocks. In: D. W. Weaver and others (eds.), Geology of the Northern Channel Islands. American Association of Petroleum Geologists and Society of Economic Paleontologists and Mineralogists, Pacific Section Special Publication, pp. 84–90.

[7] Howell, D. G., and McLean, H., 1976. Middle Miocene paleogeography, Santa Cruz and Santa Rosa Islands. In: D. G. Howell (ed.), Aspects of the geologic history of the California Continental Borderland. American Association of Petroleum Geologists, Pacific Section Miscellaneous Publication 24, pp. 266–293.

[8] Vedder, H. G., Crouch, J. K., and Lee-Wong, F., 1981. Comparative study of the rocks from Deep Sea Drilling Project holes 467, 468, and 469 and the Southern California Borderland. Initial Reports of the Deep Sea Drilling Project, v. 63, pp. 907–918.

[9] Howell, D. G., and Vedder, J. G., 1980. Structural implication of stratigraphic discontinuities across the California Continental Borderland. In: W. G. Ernst (ed.)., The geotectonic development of Cali-

Figure 4. (left) (A) Santa Cruz Island, disorganized conglomerate, facies *A*, a volcaniclastic debris flow typical of the proximal part of the Blanca fan. (B) Santa Cruz Island, large-scale, cross-bedded conglomerate interbedded with massive debris-flow units. Inner part of the Blanca fan. (C) Santa Cruz Island, thin-bedded turbidites on the west flank of the Pozo anticline. These strata are correlative with coarse-grained, disorganized units of a more proximal facies on the east flank of the Pozo anticline.

Figure 5. (right) (A) Santa Rosa Island, thin- to thick-bedded turbidites, typical of middle-fan associations. Note thin, discontinuous stringers of pumiceous debris in the T_b intervals (top of photo). (B) Santa Rosa Island, thick-bedded, lenticular turbidite and other thin-bedded turbidites (facies *E*) overlying with erosional discordance a sequence of facies *E* turbidites, channel and interchannel deposits of the middle fan. (C) Santa Rosa Island, well-stratified, normally graded conglomerate (facies *A*) interbedded with thin-bedded turbidites (facies *E*), channel deposits of the middle fan.

fornia. University of California, Los Angeles, Rubey, v. 1., pp. 535–558.
[10] Vedder, J. G., and Howell, D. G., 1976. Neogene strata of the southern group of Channel Islands, California. In: D. G. Howell (ed.), Aspects of the geologic history of the California Continental Borderland. American Association of Petroleum Geologists, Pacific Section Miscellaneous Publication 24, pp. 80–106.
[11] Vedder, J. G., Howell, D. G., and Forman, J. A., 1979. Miocene strata and their relation to other rocks on Santa Catalina Island. In: J. M. Armentrout, M. R. Cole, and H. Ter Best, Jr. (eds.), Cenozoic paleogeography of the Western United States. Society of Economic Paleontologists and Mineralogists, Pacific Section, Pacific Coast Paleogeography Sympsoium 3., pp. 239–256.
[12] Fisher, R. V., and Charlton, D. W., 1976. Mid-Miocene Blanca Formation, Santa Cruz Island, California. In: D. G. Howell (ed.), Aspects of the geologic history of the California Continental Borderland. American Association of Petroleum Geologists, Pacific Section, Miscellaneous Publication 24, pp. 228–240.
[13] Mutti, E., and Ricci Lucchi, F., 1972. Le torbiditi dell'Apennino settentrionale: Introduzione all'analisi di facies. Memorie della Societa Geologica Italiana, v. 11, p. 161–199. (English translation in International Geology Review 1978, v. 20, no. 2, pp. 125–166.)
[14] Kammerling, M. J., Luyendyk, B. P., Powell, T. S., and Terres, R., 1980. Tectonic rotation of the northern Channel Islands of the southern California borderland. EOS, Abstracts with Programs, v. 61, p. 1125.
[15] Ingle, Jr., J. C., 1980. Cenozoic paleobathymetry and depositional history of selected sequences within the southern California Continental Borderland. Cushman Foundation Special Publication No. 19 (Memorial to Orville L. Bandy), pp. 163–195.

CHAPTER 25

Butano Turbidite System, California

Tor H. Nilsen

Abstract

The Eocene Butano Sandstone was deposited as a submarine fan in a relatively small, partly restricted basin in a borderland setting. It is possibly as thick as 3000 m and was derived from erosion of nearby Mesozoic granitic and older metamorphic rocks located to the south. Deposition was at lower bathyal to abyssal water depths. The original fan may have been 120- to 160-km long and 80-km wide. Outcrops of submarine-canyon, inner-fan, middle-fan, and outer-fan facies associations indicate that the depositional model of Mutti and Ricci Lucchi can be used to describe the Butano Sandstone.

Introduction

The Butano Sandstone crops out discontinuously over a northwest-southeast distance of about 60 km in the Santa Cruz Mountains of northern California (Fig. 1). It was deposited in the La Honda basin and forms the reservoir for the La Honda and Costa oil fields in the northern part of the basin. The La Honda basin is bounded on the northeast by the right-lateral Pilarcitos and San Andreas faults, which truncate outcrops of the Butano Sandstone, and on the southwest by the right-lateral San Gregorio fault. The northwestern and southern margins of the La Honda basin are defined by Mesozoic granitic and older metamorphic rocks of the Montara Mountain and Ben Lomond Mountain areas, respectively. The La Honda basin forms one of several major Late Cretaceous and Tertiary basins that developed in the Salinian block, a crustal sliver bounded on the northeast and southwest by strike-slip faults and underlain by granitic rocks that accreted to the California continental margin in the Lower Tertiary.

The Butano was named for outcrops of sandstone and conglomerate on Butano Ridge (Fig. 1). Its areal distribution and stratigraphic relations have been studied by workers at Stanford University and the U.S. Geological Survey [1–3]. It ranges in age from Penutian (Early Eocene) to Narizian (Middle Eocene) and accumulated at lower bathyal to abyssal depths in a basin that had unrestricted access to the open ocean [4,5]. The Butano Sandstone unconformably overlies deep-marine shale of the Paleocene Locatelli Formation or rests nonconformably on granitic or metamorphic rocks of the Ben Lomond Mountain area. The Butano is conformably overlain by deep-marine shale of the Twobar Shale Member of the San Lorenzo Formation. Because the Butano Sandstone crops out in separate folded and fault-bounded blocks, no complete section is exposed. Although its total thickness is not known, its minimum thickness is 1500 m and its maximum thickness is 3000 m. Because outcrops are generally present within redwood forests and other areas of dense vegetation and abundant landsliding, beds and groups of beds cannot be traced laterally, and most stratigraphic information comes from measurement of sections in creek bottoms.

Fan Definition

The Butano Sandstone forms the southwestern part of a larger Eocene deep-sea fan that was truncated and dismembered by Late Cenozoic right slip of several hundred kilometers along the San Andreas fault [6]. The northeastern and most distal part of the deep-sea fan crops out east of the San Andreas fault in the Temblor Range (Fig. 1), where it forms the Point of Rocks Sandstone, from which oil is also produced [7]. Evidence that the Butano Sandstone was deposited as a deep-sea fan includes (1) microfauna indicative of deposition in deep-marine environments, (2) abundant turbidite and other sediment-gravity-flow deposits, (3) an outward-radiating paleocurrent pattern, (4) presence of almost all of the facies

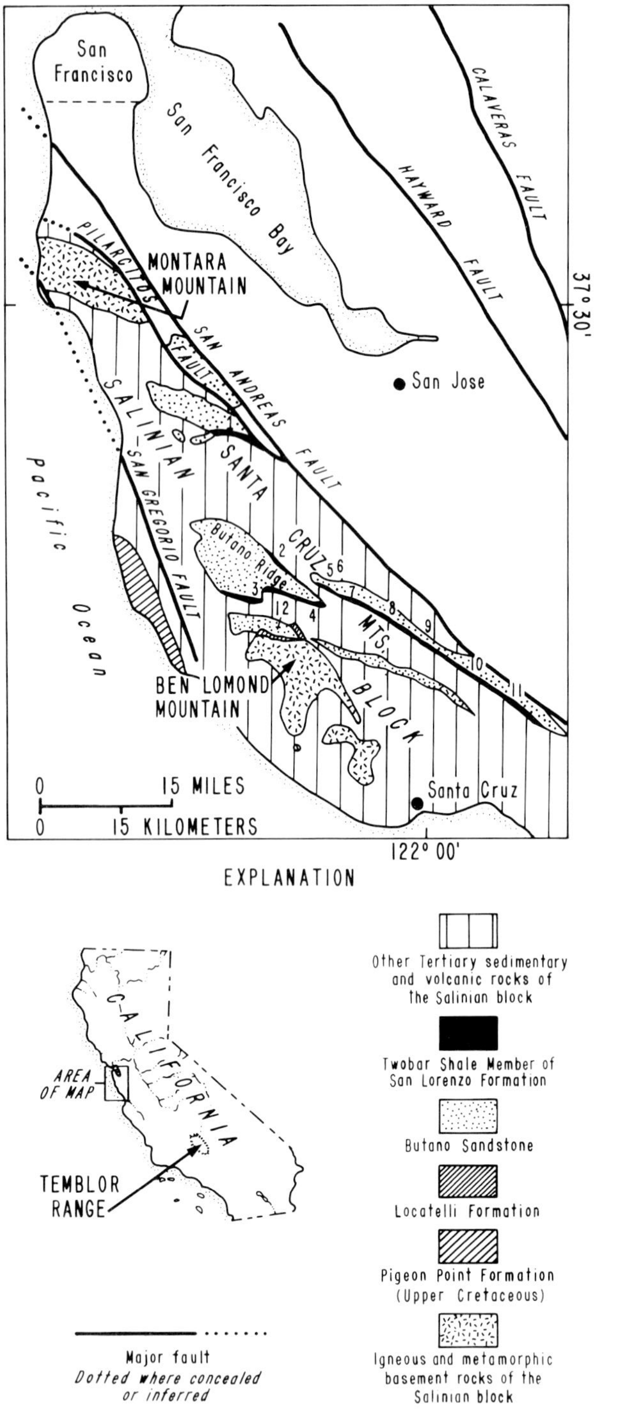

Figure 1. Generalized geologic map of the Santa Cruz Mountains showing the distribution of Butano Sandstone. Numbers indicate locations of measured sections. Geology modified from Brabb [8].

and facies associations defined by Mutti and Ricci Lucchi [8,9] for ancient turbidite systems, (5) strongly defined proximal-to-distal stratigraphic and sedimentologic relations, and (6) prominently developed channelized and nonchannelized deposits [10].

The Butano deep-sea fan, as well as the combined Butano–Point of Rocks deep-sea fan, generally fits the published models of Mutti and Ricci Lucchi [8] for fans of mixed-sediment origin. Although it is quite sandstone-rich and also contains coarse boulder conglomerate, it also has well-developed large inner-fan channels, smaller middle-fan distributary channels, and outer-fan lobes.

Fan Divisions

The three major outcrop belts of the Butano Sandstone, southern, central, and northern, contain the inner-fan, middle-fan, and outer-fan facies associations, respectively (Fig. 2). Lateral relations between facies associations are unclear, and the lateral extent of channels and lobes are uncertain because of the nature of the outcrops. However, the characteristic features of each facies association were determined from 12 measured sections, 1 in the northern, 10 in the central, and 1 in the southern outcrop belt (Fig. 3). Sections 1 to 11 were measured up to the base of the overlying Twobar Shale Member of the San Lorenzo Formation, whereas section 12 has no recognizable stratigraphic boundary at either its top or base. These sections contain various megasequences characteristic of deep-sea fan facies associations.

The fan-fringe and basin-plain facies associations are not present in outcrop, except as thinner bedded turbidites between distinctive thickening- and coarsening-upward outer-

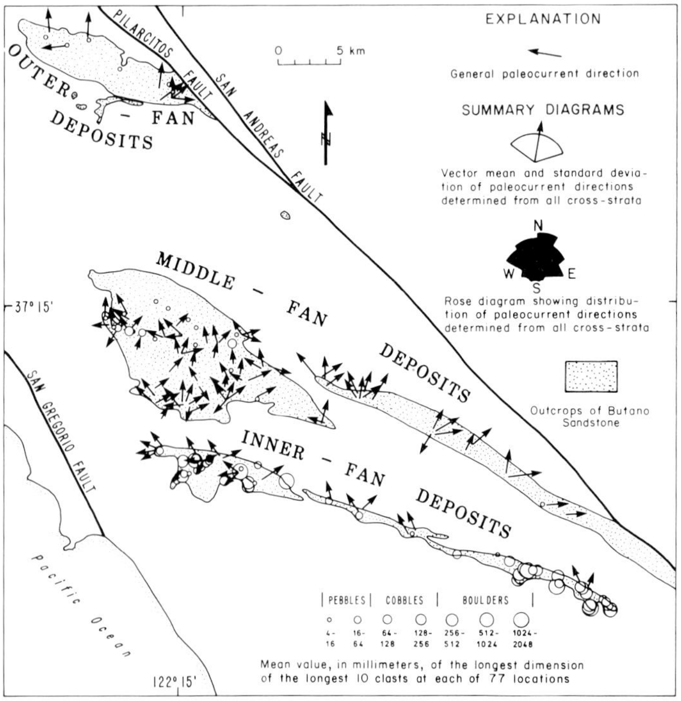

Figure 2. Map showing distribution of deep-sea fan facies associations, paleocurrents, and longest clasts in conglomerate beds of the Butano Sandstone. Modified from Nilsen [11] and Nilsen and Simoni [12].

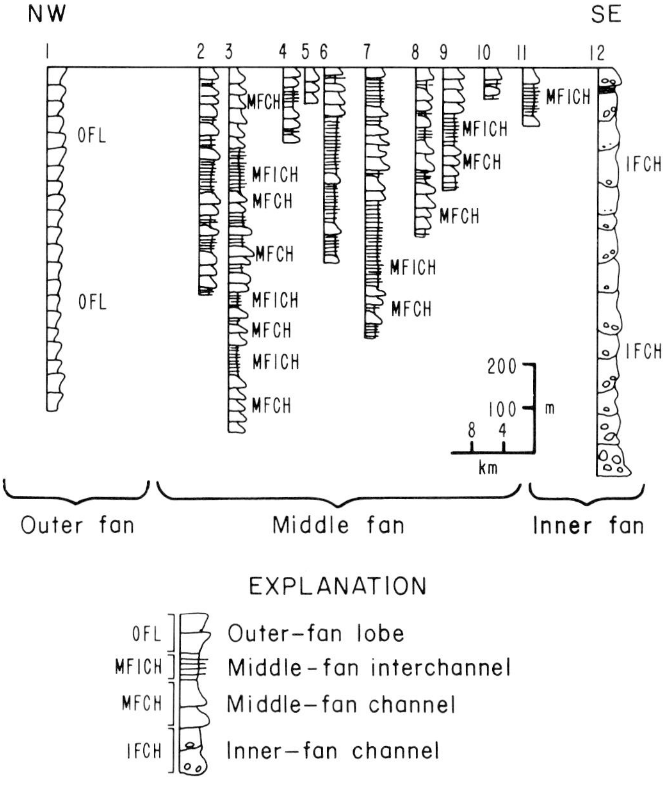

Figure 3. Stratigraphic cross section of the Butano Sandstone showing distribution of channel, interchannel, and outer-fan lobe deposits. The location of the 12 measured sections is shown on Figure 1. Modified from Nilsen [11].

fan lobe megasequences within section 1. However, these facies associations are present in great abundance within subsurface units of Point of Rocks Sandstone east of the San Andreas fault.

Fan Facies

The inner-fan facies association consists chiefly of laterally discontinuous facies A and B beds (Fig. 3, section 12). Thick and massive amalgamated beds of coarse- to very coarse-grained sandstone, conglomeratic sandstone, and conglomerate are the dominant rock types. Sandstone and conglomerate-to-shale and mudstone ratios are about 100:1 in sections of the inner-fan facies association. Interchannel and levee deposits are very sparse and thin in outcrop. However, in one area between the two prominent inner-fan channel deposits north of Ben Lomond Mountain, thin sections of overbank deposits of facies E are characterized by widespread discontinuities produced by synsedimentary slumping. The coarsest conglomerates are generally present in the lower parts of the inner-fan channel deposits; the largest clasts, mostly of granitic composition, are as large as 2 m in longest dimension. The conglomerate beds are generally clast-supported with sandy matrices, crudely stratified, reverse-to-normally graded, and have a preferred orientation of clasts (long axes generally imbricated and oriented parallel to the northward paleotransport direction). Rip-up clasts of shale and siltstone as long as 50 m are abundant within the beds of conglomerate and sandstone. Aside from the overall fining-upward character of the inner-fan channel section, beds of the inner-fan channel deposits are not arranged into either well-developed thickening- or thinning-upward megasequences.

The middle-fan facies association is organized into well-developed thinning- and fining-upward megasequences with channelized bases (Fig. 3, sections 2–11). The megasequences are 10- to 100-m thick and contain very thick, massive, amalgamated beds of medium- to coarse-grained facies A and B sandstone in their upper parts. The type area of the Butano Sandstone along Butano Ridge consists of these repetitive megasequences. Sandstone-to-shale ratios range from about 10:1 to 3:1. Toward the San Andreas fault to the northeast, the middle-fan deposits are finer grained, individual beds are thinner, sandstone:shale ratios are lower, and channelized megasequences are thinner. Conglomerate, limited to pebble-size clasts, is not abundant and where present is generally at the bases of the channelized megasequences. Thick local sequences of shale with few thin-bedded sandstone turbidites represent interchannel and levee deposits. Syndepositional slumping is common within channels and along channel margins in these facies.

The outer-fan facies association is organized into a thick series of thickening- and coarsening-upward megasequences that do not have channelized bases (Fig. 3, section 1). The megasequences are 10- to 35-m thick and contain thin to medium beds of fine- to medium-grained facies C and D sandstone in their upper parts. Sandstone-to-shale ratios range from about 5:1 to 1:3. Interbedded shale-rich sections contain thin-bedded facies C and D turbidites that are locally organized into thin (1- to 5-m thick) megasequences that resemble the fan-fringe facies association.

Paleocurrents and Paleochannels

Paleocurrents from the Butano Sandstone indicate a complex pattern of northward sediment dispersal (Fig. 2). These data include 565 measurements of flute casts, groove casts, ripple markings, convolute laminations, small-scale cross-laminations, flame structures, and clast imbrications that yield a northward paleo-flow direction with considerable radial diversity [12]. An additional 105 measurements of contorted strata yielded a mean direction and standard deviation for the fan paleoslope of 8 ± 63° [12].

The paleocurrent variability of the middle-fan facies association reflects two chief factors. First, meandering, braiding, and irregular trends of channels, which can even be oriented perpendicular to the regional slope of the fan, result in diverse paleocurrent orientations. Second, overflow deposits in the interchannel areas will have paleocurrents directed not only away from adjacent channels but also downslope from them. This additional component of flow results from the adjustment of the overflow current to local orientation of the interchannel slope. The sediment of the outer-fan lobes of the northernmost outcrops was also transported generally northward and partly northeastward, toward the present location of the San Andreas fault.

The largest clasts in the Butano Sandstone are concentrated into two distinct size groups at the southwestern and southeastern edges of the outcrop area (Fig. 2). These boulder accumulations represent deposits in major channels of the inner-fan region. The inner-fan channels were probably connected to submarine canyons to the south that fed sediment northward to the Butano fan. Although the basal contact is obscured by faulting, conglomerates in the eastern channel may rest directly on granitic basement and thus form a lower slope submarine canyon facies association. The western inner-fan channel conglomerates may also rest unconformably on the Locatelli Formation, but the contact is not clearly exposed. Although the two major inner-fan channels may have been active at the same time, limited fossil data indicate that deposits in the eastern channel are older. During growth of the fan, the older eastern channel may have been replaced by the younger western channel.

The maximum clast size decreases markedly both northward and laterally away from these areas; only scattered very small pebbles are found in the northern and central outcrop belts of the middle- and outer-fan facies associations. The generally northward transport direction and arkosic nature of the Butano Sandstone suggest a granitic source terrane located generally to the south.

Paleogeography

The Butano Sandstone was deposited in the La Honda basin, a small but deep basin that developed in the Salinian block, an elongate borderland terrane underlain by granitic crust in western California. Movement of the Salinian block apparently produced a number of small, deep, at least partly restricted basins within the borderland that were filled with arkosic turbidites derived from adjacent uplifted granitic blocks [13]. Adjacent to and east of the northern part of the borderland was the San Joaquin basin, into which some fans spilled from the restricted borderland basins. The fan-fringe and adjacent basin-plain facies of the Butano are present beneath the San Joaquin Valley, at the edge of and southeast, east, and north of the distal limits of the Point of Rocks Sandstone (Fig. 4). The tectonic and physiographic setting of the Butano–Point of Rocks fan was thus similar to that of modern well-studied fans of the present Southern California Borderland.

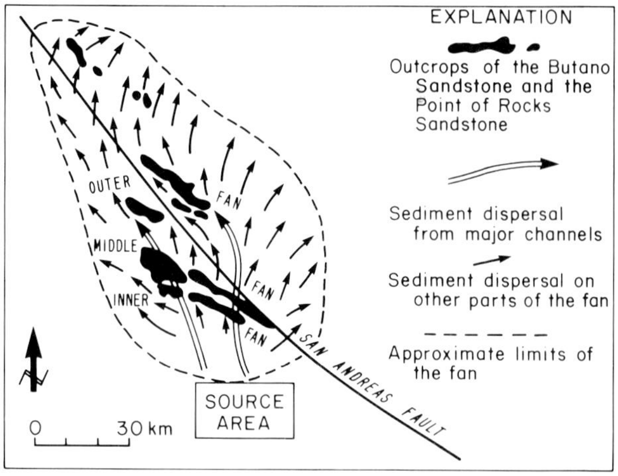

Figure 4. Sediment dispersal pattern and paleogeography of Butano–Point of Rocks deep-sea fan. Modified from Nelson and Nilsen [10].

Sedimentation appears to have been continuous and controlled mainly by tectonic activity. Because of the large volume and coarseness of sediment supplied to the fan, and the general framework of a tectonically active borderland basin, the Butano fan probably prograded rapidly into the basin in the Early Eocene. However, because the measured sections are correlated by the upper contact with the Twobar Shale Member of the San Lorenzo Formation rather than with the basal contact, it is difficult to determine stratigraphically whether the Butano fan was progradational into the La Honda basin. Deposition appears to have been terminated abruptly rather than by gradual retreat.

There is no clearly defined relation of sedimentation to global changes in sea level, although there are numerous minor fluctuations during the Eocene [14]. No unconformities exist, as far as is known, within the Butano Sandstone, a probable reflection of the deep-marine paleoenvironment. The abrupt upward cessation of deposition of coarse clastics of the Butano Sandstone in the Middle to Late Eocene is probably related to tectonic factors rather than eustatic sea level or climatic changes. In the Late Oligocene, a second deep-sea fan, the Vaqueros Sandstone, actively prograded northward over the deep-marine pelites of the San Lorenzo Formation [1].

Conclusions

The Butano Sandstone was deposited as an Eocene continental-borderland deep-sea fan in the La Honda basin of the Santa Cruz Mountains, California. The source area to the south consisted mainly of granitic basement rocks of the Salinian block. Inner-fan, middle-fan, and outer-fan megasequences have been delineated in southern, central, and northern outcrop belts, respectively. Paleocurrents form a northward-radiating pattern from the two inner-fan channels that define the fan apex, indicating that growth of the fan and

Figure 5. Boulder conglomerate of inner-fan channel facies association, Jamison Creek Road; hammer circled for scale.

Figure 6. Conglomerate, sandstone, and shale of middle-fan-channel facies association organized into fining- and thinning-upward cycles, Bear Creek Road near Boulder Creek; notebook circled for scale; arrows indicate vertical extent of cycles.

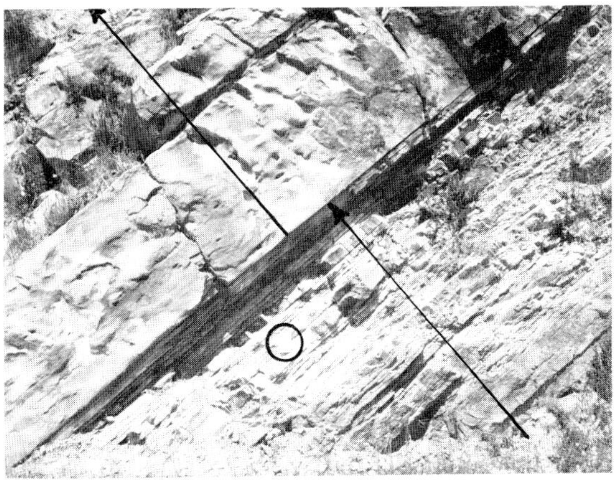

Figure 7. Sandstone and shale of middle-fan-channel facies association organized into fining- and thinning-upward cycles, Bear Creek Road; notebook circled for scale; arrows indicate cycles.

Figure 8. Facies B sandstone of middle-fan-channel facies association organized into fining- and thinning-upward cycle, Bear Creek Road; pencil circled for scale.

Figure 9. Thin-bedded laterally discontinuous facies E sandstone of middle-fan-levee facies association, Bear Creek Road.

Figure 10. Conglomerate-filled scour of middle-fan-channel facies association overlain by sandstone with dish structure, South Fork of Butano Creek.

Figure 11. Conglomeratic sandstone of the middle-fan-channel facies association that grades laterally from left to right from flat-stratified to wavy-stratified, possible antidunes, East Waddell Creek; stratification is truncated erosionally by overlying fine conglomerate.

Figure 12. Cross-stratified lens of sandstone interbedded with middle-fan-channel facies association, East Waddell Creek.

dispersal of sediments resemble that of modern deep-sea fans. Paleoslope measurements from contorted strata within the Butano indicate a regional northward paleoslope for the fan. Neogene right-lateral slip along the San Andreas fault displaced the northeastern part of the fan about 305 km southeastward.

Addendum

Characteristics of the Eocene Butano Sandstone at the outcrop level are presented in Figures 5–12.

Acknowledgments

I gratefully acknowledge helpful review comments by C. H. Nelson and D. G. Howell of the U.S. Geological Survey.

References

[1] Cummings, J. C., Touring, R. M., and Brabb, E. E., 1962. Geology of the northern Santa Cruz Mountains, California. California Division of Mines and Geology Bulletin 181, pp. 179–220.

[2] Brabb, E. E., (compiler), 1970. Preliminary geologic map of the central Santa Cruz Mountains, California. U.S. Geological Survey Open-File map, scale 1:62,500.

[3] Clark, J. C., 1981. Stratigraphy, paleontology, and geology of the central Santa Cruz Mountains, California Coast Ranges. U.S. Geological Survey Professional Paper 1168.

[4] Fairchild, W. W., Wesendunk, P. R., and Weaver, D. W., 1969. Eocene and Oligocene Foraminifera from the Santa Cruz Mountains, California. California University Publications in Geological Sciences, v. 81.

[5] Poore, R. Z., and Brabb, E. E., 1977. Eocene and Oligocene planktonic Foraminifera from the upper Butano Sandstone and type San Lorenzo Formation, Santa Cruz Mountains, California. Journal of Foraminiferal Research, v. 7, pp. 249–272.

[6] Clarke, Jr., S. H., and Nilsen, T. H., 1973. Displacement of Eocene strata and implications for the history of offset along the San Andreas fault, central and northern California. In: R. L. Kovach and A. Nur (eds.), Conference on Tectonic Problems of the San Andreas Fault System. Stanford University Publications in Geological Sciences, v. 13, pp. 358–368.

[7] Clarke, Jr., S. H., 1973. The Eocene Point of Rocks Sandstone—provenance, mode of deposition, and implications for the history of offset along the San Andreas fault in central California. Ph.D. Thesis, University of California, Berkeley, CA.

[8] Mutti, E., and Ricci Lucchi, F., 1972. Le torbiditi dell'Appennino setentrionale—introduzione all'analisi di facies. Societa Geologica Italiana Memorie, v. 11, pp. 161–199.

[9] Mutti, E., and Ricci Lucchi, 1975. Turbidite facies and facies associations. In: Examples of Turbidite Facies and Facies Associations from Selected Formations of the Northern Apennines. Ninth International Congress of Sedimentology, Nice, France, Field Trip Guidebook All, pp. 21–36.

[10] Nelson, C. H., and Nilsen, T. H., 1974. Depositional trends of modern and ancient deep-sea fans. In: R. H. Dott, Jr., and R. H. Shaver (eds.), Modern and Ancient Geosynclinal Sedimentation. Society of Economic Paleontologists and Mineralogists Special Publication 19, pp. 69–91.

[11] Nilsen, T. H., 1979. Sedimentology of the Butano Sandstone, Santa Cruz Mountains, California. In: T. H. Nilsen and E. E. Brabb (eds.), Geology of the Santa Cruz Mountains, California. Cordilleran Section, Geological Society of America, Field Trip Guidebook, pp. 30–39.

[12] Nilsen, T. H., and Simoni, Jr., T. R., 1973. Deep-sea fan paleocurrent patterns of the Eocene Butano Sandstone, Santa Cruz Mountains, California. U.S. Geological Survey Journal of Research, v. 1, pp. 439–452.

[13] Nilsen, T. H., and Clarke, Jr., S. H., 1975. Sedimentation and tectonics in the early Tertiary continental borderland of central California. U.S. Geological Survey Professional Paper 925.

[14] Vail, P. R., Mitchum, R. M., and Thompson, III, S., 1977. Seismic stratigraphy and global changes of sea level. Part 4. Global cycles of relative changes of sea level. In: C. E. Payton (ed.), Seismic Stratigraphy—Applications to Hydrocarbon Exploration. American Association of Petroleum Geologists Memoir 26, pp. 83–97.

CHAPTER 26

Cengio Turbidite System, Italy

Carlo Cazzola, Emiliano Mutti, and Bartolomeo Vigna

Abstract

The Cengio sandstone member of the Tertiary Piedmont Basin in northwestern Italy has a conservatively estimated volume of 2.5 to 3 km^3 (length: 6.4 km; width: 4.8 km; thickness: 170 m). It is interpreted as a sandstone-rich submarine fan deposit. The Cengio member consists of eight tabular depositional sandstone lobes that are 5- to 25-m thick. These lobes filled a submarine structural depression and onlap and/or pinch-out against bounding slope mudstones. The stacking of the lobe units was related to synsedimentary tectonism.

Introduction

The Upper Oligocene–Lower Miocene Cengio member is a small turbidite system deposited within a fault-controlled submarine depression. It forms part of the Piedmont Basin in northwestern Italy [1].

The Cengio member can be interpreted as a deep-sea fan deposit because of its vertical facies arrangement and facies associations. In particular, the Cengio member is mainly represented in outcrop by depositional sandstone lobes that are separated vertically by mudstone units and still show evidence of broad and shallow channeling in their proximal sectors. The feeder channel is not preserved. However, paleocurrent directions indicate the same fixed point source for all the lobes.

The objective of this paper is to describe the geometry and the main facies types of the Cengio turbidite system in order to provide some detailed information about a small, ancient fan system that can be essentially described as a complex of stacked, relatively distal suprafan sandstone lobes in the sense of Normark [2].

Geologic and Depositional Setting

The Tertiary Piedmont Basin, northwestern Italy, consists of a thick sequence of predominantly terrigenous sediments that were deposited unconformably on the top of highly deformed Alpine and Apenninic terranes between Oligocene and Miocene times (Fig. 1).

The present structure of the Tertiary Piedmont Basin is a broad monocline gently dipping to the northwest. This structure is more complex toward the eastern margin of the basin, where these sedimentary rocks are involved in Late Tertiary thrusting in the northwestern Apennines [3].

The lower part of the Tertiary Piedmont Basin, in the study area, includes, from base to top, the Molare, the Rocchetta, and the Monesiglio formations (Fig. 2). The Molare is made up of alluvial fan and fan-delta conglomerates and sandstones [4] that were deposited unconformably on top of pre-Cenozoic rocks. Mudstone of the Rocchetta conformably overlies the Molare and records a phase of basin expansion and deepening. These mudstones enclose a number of lenticular turbidite sandstone and conglomeratic sandstone bodies of which the Cengio unit is one (Fig. 2). The Monesiglio, which overlies conformably the Rocchetta sequence, consists of basinal mudstone and turbidites.

Synsedimentary faulting strongly affected the early phases of deposition of the Tertiary Piedmont Basin and controlled the distribution and the geometry of the turbidite sandstone bodies within the Rocchetta sequence [1,5]. This structural control is clear in the stratigraphically lower part of the Rocchetta sequence, where conglomerate and pebbly sandstone form highly lenticular fills of fault-bounded structural depressions (Fig. 2).

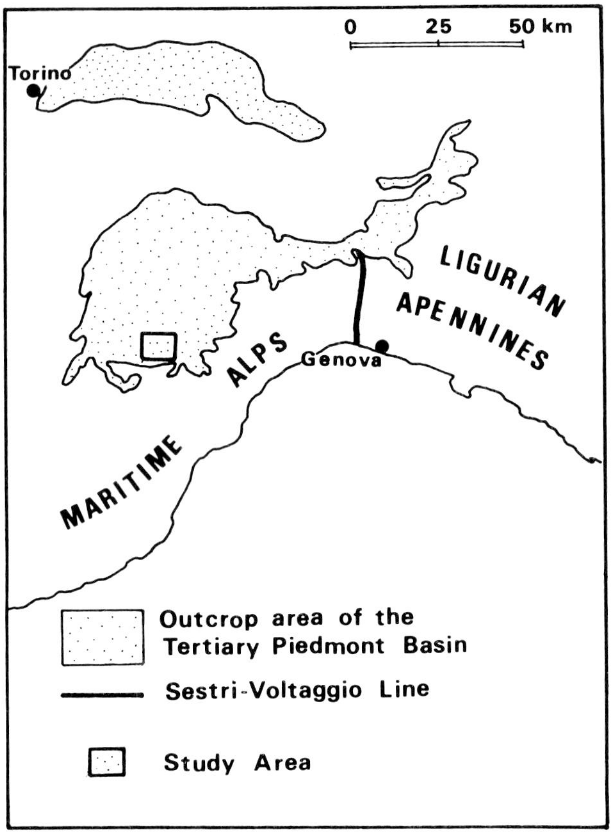

Figure 1. Outcrop area of the Tertiary Piedmont Basin and location of th study area.

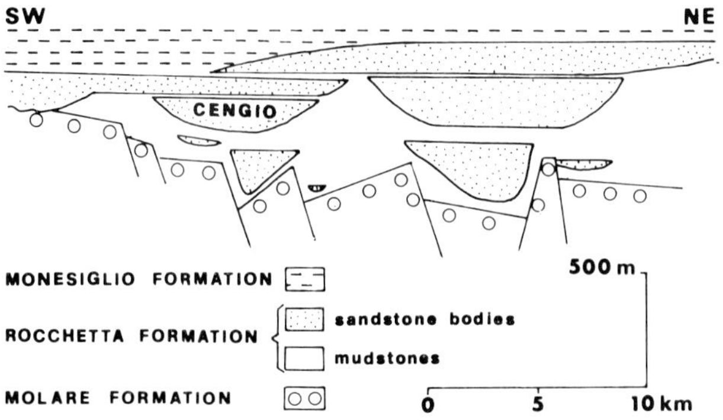

Figure 2. Diagrammatic stratigraphic cross-section of the stratigraphically lower part of the Tertiary Piedmont sequence in the study area (simplified after Cazzola and others [1]). For section location see [1].

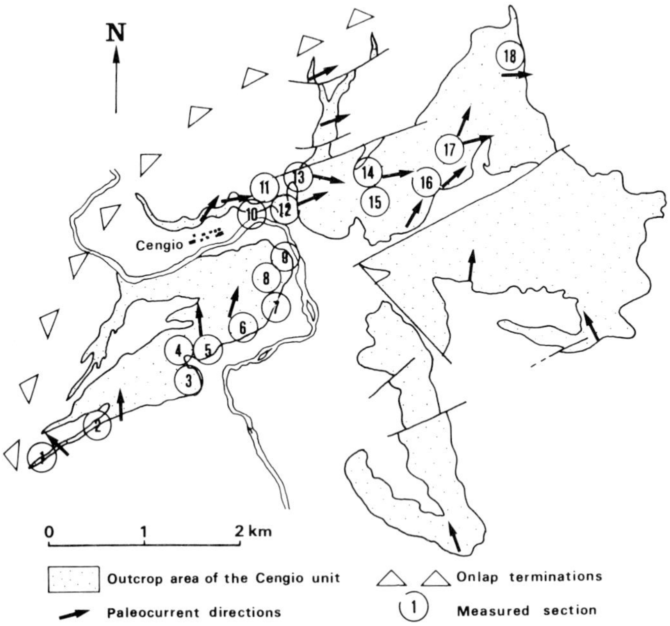

Figure 3. Outcrop area of the Cengio system, showing paleocurrent directions and location of measured sections (modified after Cazzola and others [1]).

Facies and Geometry

The Cengio submarine turbidite system crops out over an area of approximately 13 km^2 and has a maximum thickness of about 170 m. The total sediment volume is about 2.5 to 3 km^3.

The sequence consists of two distinct sandstone units that are separated vertically by a 10-m thick mudstone interval. The lower sandstone unit is very poorly exposed and confined within the narrow axial zone of a structural depression; the upper sandstone unit is more widespread, has good exposures, and is as thick as 110 m. Only the upper unit will be considered in the following sections.

The upper sandstone was investigated in detail by field mapping at a scale of 1:10,000, measurement of 60 paleocurrent directions, bed-by-bed measurement of 30 sections with an aggregate thickness of about 1000 m, and correlation of thicker sandstone units in the field (Fig. 3).

The main facies types observed within the Cengio, in decreasing order of importance, are listed below.

Facies 1 consists of thick to massive beds of crudely graded medium to granule conglomeratic sandstone. These beds are either amalgamated or separated by thinner mudstone partings. They are locally capped by one set of current ripple laminae or, more commonly, by burrowed sandy mudstone. Bedding surfaces are generally even and parallel, although cut-and-fill features may be abundant in amalgamated packets. In the more upcurrent sectors of the system, this facies contains more granule conglomerate and has an abundance of broad scours and shallow channels filled with a limited number of beds. Liquefaction features, synsedimentary small-scale drag folds, and rip-up intraformational clasts are locally abundant.

Facies 2 consists of thin to medium beds of subtly graded medium to fine sandstone with crude to well-developed horizontal laminae and occasional sets of current ripple laminae. These beds, whose tops are generally bioturbated, are separated by subordinate amounts of silty to sandy mudstone.

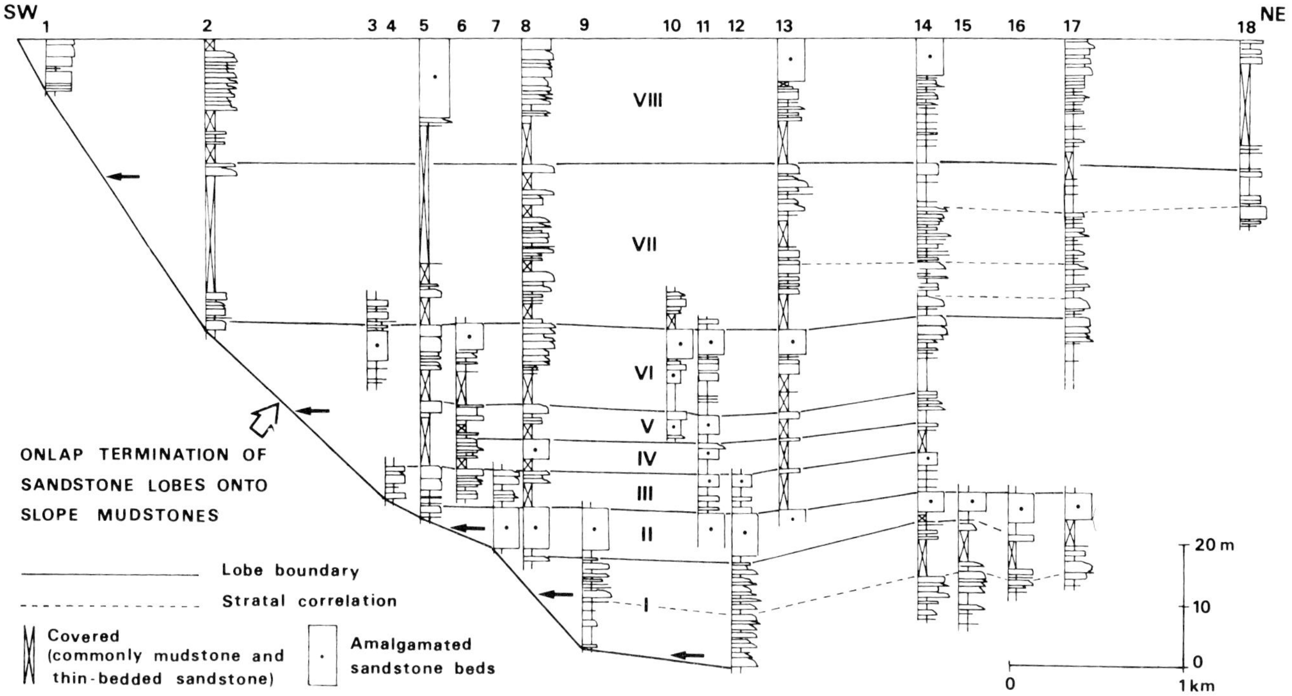

Figure 4. Detailed stratigraphic cross-section of the Cengio sandstone lobes. Location of measured sections is shown in Fig. 3.

The bedding geometry of this facies is even and parallel in the peripheral sectors of the system; in an upcurrent direction, the geometry of the bedding surfaces becomes progressively more irregular, particularly where these sediments are associated, vertically and laterally, with locally broadly channelized deposits of Facies 1.

Facies 3 consists chiefly of mudstone with scattered thin beds of fine-grained sandstone with rippled tops and poorly visible internal current ripple laminae.

Facies 4 consists of silty mudstone virtually devoid of sandstone interbeds, thought to represent "normal" hemipelagic deposits.

Vertical and lateral stratigraphic relations show that facies 1 through 3 are genetically interrelated and were deposited by waning turbidity currents.

Facies 1 and 2 form eight distinct sandstone bodies within the Cengio sequence (Fig. 4). As indicated by paleocurrent directions, stratigraphic relations, and overall facies changes, these sandstone bodies were deposited by south-derived turbidity currents that were deflected toward the northeast within a fault-bounded, southwest-northeast-trending submarine depression (Fig. 3). The southeast-dipping slope that bounded the depression to the northwest is clearly delineated by a series of abrupt onlap terminations of the Cengio sandstone bodies onto slope mudstones (Fig. 5). As inferred from present geometric relationships, the angle of the original slope was between 2°30′ and 5° [1]. The turbidity currents were gradually decelerated by another submarine high that bounded the Cengio trough toward the northeast. Mudstones and minor amounts of thin-bedded turbidites were deposited on and beyond the inferred sill area and may represent fan-related basin plain deposits. These sediments are not described in this paper. However, the volume of these deposits is not significant in relation to the lobe sandstones.

The depositional setting of the Cengio system is illustrated by the cross-section of Figure 4. The section is roughly perpendicular to paleocurrent flow between sections 1 and 6 and becomes approximately parallel to the flow between sections 7 and 22. Therefore, from southwest to northeast, the section depicts (1) the onlap terminations of the different sandstone bodies onto the slope limiting the Cengio system to the northwest, (2) the sandier and amalgamated portions of the sandstone bodies preferentially restricted to the axial zone of the Cengio system, and (3) the progressive thinning and eventual pinch-out of the sandstone facies in a downcurrent direction, toward the high bounding the Cengio system to the northeast.

The Depositional Sandstone Lobes

Eight main sandstone bodies can be traced across the Cengio basin. These bodies, with individual thicknesses between 5 and 25 m, are characteristically devoid of basal channeling, have an overall tabular geometry, and are separated from each

Figure 5. Onlap termination of sandstone lobe (lobe V) onto slope mudstone. Note the abrupt thinning of individual sandstone beds over very short distance. Mudstone partings can be traced across the onlap surface and drape both sandstone beds and slope facies. This onlap termination is not shown on the cross-section of Fig. 4. The overlying sandstone lobe (lobe VI) undergoes the same type of onlap termination farther northward.

Figure 6. Thickening-upward sequence of lobe VIII.

other by finer grained and thinner bedded deposits, most commonly by facies 4 mudstone.

These bodies are here interpreted as depositional sandstone lobes because of their nonchannelized geometry and are thought to represent an original depositional environment located in smooth and distal fan regions where waning turbidity currents deposited the bulk of their suspended sand load. The lobe boundaries are expressed by abrupt basin-wide changes either at the top of thick, amalgamated sandstone facies or at the base of mudstone units (Fig. 4). Several types of lobes were recognized. The lobes will therefore be described by types rather than in stratigraphic sequence.

Lobe II is almost entirely composed of amalgamated sandstone beds; its irregular geometry may have been produced by predepositional differential sinking of basin floor. Lobes VI and VIII comprise distinct thickening-upward sequences whose upper and sandier portions undergo a marked pinch-out in a downcurrent direction. This pinch-out is expressed by a decrease in both overall sandstone content and the thickness of individual sandstone beds. The same general pattern of vertical sequential arrangement and lateral facies variation is also observed within the thinner units comprising lobes III, IV, and V.

All the above lobes show a clear progradational character expressed by the downcurrent migration of thick-bedded and often amalgamated sandstone on top of thinner bedded, stratigraphically equivalent deposits.

Some lobes apparently prograded rapidly enough to form thick-bedded sandstones directly overlying mudstone facies (lobe II); other lobes prograded more gradually, thus permitting the formation of well-developed and basinwide thickening-upward sequences (lobes VI and VII) (Fig. 6).

Lobes I and VII lack distinct overall thickening-upward sequences and consist of alternating packets of thick-bedded and thinner bedded sandstone with variable amounts of interbedded mudstone. Although largely covered, the southwestern portion of lobe VII is probably affected by shallow channeling as indicated by stratal correlation and an abundance of coarse-grained sandstone facies. Both lobes display repetitive thickening-upward sequences that are composed of a limited number of beds and are not separated by laterally persistent mudstone facies. As indicated by stratal correlation (Fig. 4), several of these sequences are of limited areal extent.

The repetitive and laterally unpersistent thickening-upward units that comprise the two lobes may be compensation features [6], produced in virtually flat areas by subtle depositional relief and resulting lateral shifting of the axis of subsequent flows.

Conclusions

The Cengio submarine system was produced by eight main episodes of basinward progradation of turbidite sands. Each episode of progradation is recorded by a depositional sandstone lobe, and the intervening phases of inactivity of the system are recorded by mudstone facies. The lobes were probably fed by the same fan channel, which also underwent alternating phases of activity and inactivity with time. The lobes pinch-out downcurrent over short distances without development of substantial amounts of peripheral thin-bedded and fine-grained turbidites and can be physically traced into upcurrent equivalents containing shallow channels. Both characteristics are typical of "poorly efficient" systems where sand-laden turbidity currents of relatively small volume deposit the bulk of their load near the mouths of distributary channels and do not form significant amounts of thin-bedded lobe-fringe, fan-fringe, and basin-plain turbidites [7,8]. A modern possible analog of these systems is the suprafan depositional lobe as developed by Normark [2,9], although in the case of the Cengio, the combination of an abundant sand supply with a small confined basin resulted in virtually tabular and basinwide lobes without apparent convex-upward expression. In addition, only the smooth and basically nonchannelized outer portions of the original suprafans are preserved in the Cengio depositional lobes.

The Cengio sandstone lobes are basinwide units that cannot be explained simply as a product of channel shifting coupled with basin subsidence. These lobes formed primarily as a response to cyclically repeated periods of abundant sediment availability followed by periods of basin starvation. As indicated by the regional depositional setting [1], the turbidite bodies that occur within the Rocchetta sequence (Fig. 2) were essentially derived from coarse-grained sediment (sand and gravel) originally deposited in fan-delta systems along narrow, fault-controlled coastal zones. The remobilization through gravity flows of these marginal marine and alluvial deposits, and the final deposition of them in deeper water as turbidites, requires lowering of sea level that can be produced either by eustatic changes, by tectonic uplift, or by both.

Based on the general geologic setting, the stacking of the Cengio sandstone lobes is here thought to represent mainly the product of intermittent, synsedimentary tectonic uplift related to a pattern of oblique-slip faulting.

Acknowledgments

We thank W. R. Normark and F. Ricci Lucchi for constructive remarks in the field. This project was supported by the C.N.R., Rome. We are also grateful to C. H. Nelson and T. H. Nilsen for their critical reviews of this manuscript.

References

[1] Cazzola, C., and others, 1981. Geometry and facies of small, fault controlled deep sea fan systems in a transgressive depositional setting (Tertiary Piedmont Basin, Northwestern Italy). In: F. Ricci Lucchi (ed.), International Association of Sedimentologists, 2nd European Regional Meeting, Bologna, 1981, Excursion Guidebook, pp. 7–53.

[2] Normark, W. R., 1970. Growth patterns of deep-sea fans. American Association of Petroleum Geologists Bulletin, v. 54, p. 2170–2195.

[3] Gelati, R., and others, 1974. Evoluzione stratigrafico-strutturale dell'Appennino Vogherese a nord-est della Val Staffora. Rivista Italiana di Paleontologia e Stratigrafia, v. 80, pp. 479–514.

[4] Gnaccolini, M., 1981. Oligocene fan-delta deposits in the northern Italy: a summary. Rivista Italiana di Paleontologia e Stratigrafia, v. 87, pp. 627–636.

[5] Gelati, R., and Gnaccolini, M., 1980. Significato dei corpi arenacei di conoide sottomarina (Oligocene–Miocene inferiore) nell'evoluzione tettonico-sedimentaria del bacino Terziario ligure-piemontese. Rivista Italiana di Paleontologia e Stratigrafia, v. 87, pp. 167–186.

[6] Mutti, E., and Sonnino, M., 1981. Compensation cycles: a diagnostic feature of turbidite sandstone lobes. In: International Association of Sedimentologists, 2nd European Regional Meeting, Bologna, 1981, Abstracts, pp. 120–123.

[7] Mutti, E., 1979. Turbidites et cônes sous-marins profonds. In: P. Homewood (ed.), Sédimentation Détritique (Fluviatile, Littorale et Marine). Institut Géologique Université de Fribourg, Switzerland, pp. 353–419.

[8] Nilsen, T. H., 1980. Modern and ancient submarine fans: discussion of papers by R. G. Walker and W. R. Normark. American Association of Petroleum Geologists Bulletin, v. 64, pp. 1094–1101.

[9] Normark, W. R., 1978. Fan valleys, channels, and depositional lobes on modern submarine fans: character for recognition of sandy turbidites environments. American Association of Petroleum Geologists Bulletin, v. 62, pp. 912–931.

CHAPTER 27

Chugach Turbidite System, Alaska

Tor H. Nilsen

Abstract

Turbidites of the Upper Cretaceous Chugach terrane of southern Alaska were deposited in a trench during northward-directed subduction. The fault-bounded outcrop belt of the Chugach terrane is about 2000-km long and 100-km wide and was accreted to Alaska during the Cenozoic. Turbidites are at least 5000 m thick, are extensively deformed, have been regionally metamorphosed, and have been intruded by anatectic granites. Facies associations indicate an east-to-west progression from inner-fan to middle-fan, outer-fan, fan-fringe, and basin-plain deposits. To the north is a marginal trench-slope facies association and a basin.

Introduction

The Chugach terrane was named for outcrops of graywacke, argillite, slate, conglomerate, volcanic rocks, chaotic melanges, and granitic plutons in the Chugach Mountains of southern Alaska [1]. It consists of two subparallel belts: (1) a landward melange belt of polydeformed and regionally metamorphosed phyllite, metagraywacke, quartzite, metachert, greenstone, amphibolite, and ultramafic rocks, and (2) a seaward flysch belt of graywacke and slate (Fig. 1). The Chugach terrane crops out continuously for about 2000 km from Baranof Island in the southeast to Sanak Island in the southwest, except where covered by large glaciers and marine waters of the Gulf of Alaska [2,3,4]. The outcrop belt is as wide as 100 km but is highly variable where affected by Cenozoic strike-slip faulting, particularly in southeastern Alaska.

The seaward flysch belt of the Chugach terrane, here referred to as the Chugach flysch terrane, appears to represent a Late Mesozoic trench-fill that is possibly the longest, most continuous, and best-preserved ancient trench-fill in the world. The age of the Chugach flysch terrane has been established as Late Cretaceous, primarily Maestrichtian, on the basis of scattered megafossils [5]. Some Early and Late Cretaceous fossils have also been reported from parts of the Chugach flysch terrane in southeastern Alaska [6]. The landward melange belt, herein referred to as the Chugach melange terrane, has been interpreted to be a product of subduction [3]. It contains blocks of chert with radiolarians of Triassic, Jurassic, and Cretaceous (as young as Cenomanian) age [7].

The Chugach flysch terrane is bounded both landward and seaward by major faults throughout most of its extent (Fig. 1). The Eagle River fault separates the Chugach flysch terrane from the Chugach melange terrane. Where the Eagle River fault is not present, the Border Ranges fault forms the northern margin of the Chugach terrane; this fault extends for most if not all the outcrop length of the Chugach terrane, dips landward, and locally juxtaposes the Chugach terrane with Paleozoic and Mesozoic metamorphic rocks [2]. The Chugach flysch terrane is separated from Paleogene turbidites and mafic volcanic rocks to the south by the Contact fault, in most places a high-angle landward-dipping reverse fault. In southeastern Alaska, strike-slip faults such as the Queen Charlotte fault and Fairweather fault form the southwestern margins of the Chugach flysch terrane.

Within the Chugach flysch terrane and on its south flank are extensive but discontinuous outcrops of volcanic rocks that are locally highly deformed and metamorphosed. Pillow lava and bedded chert are locally in depositional contact with the flysch on the Sanak Islands [8]. In the Prince William Sound area, diabasic sheeted dikes, gabbros, and basalts crop out within the Chugach flysch terrane [9]. Late Mesozoic oceanic basalts form a mappable terrane along the southwestern margin of the Chugach flysch terrane in southeastern Alaska [10], and volcanic rocks are interbedded with the

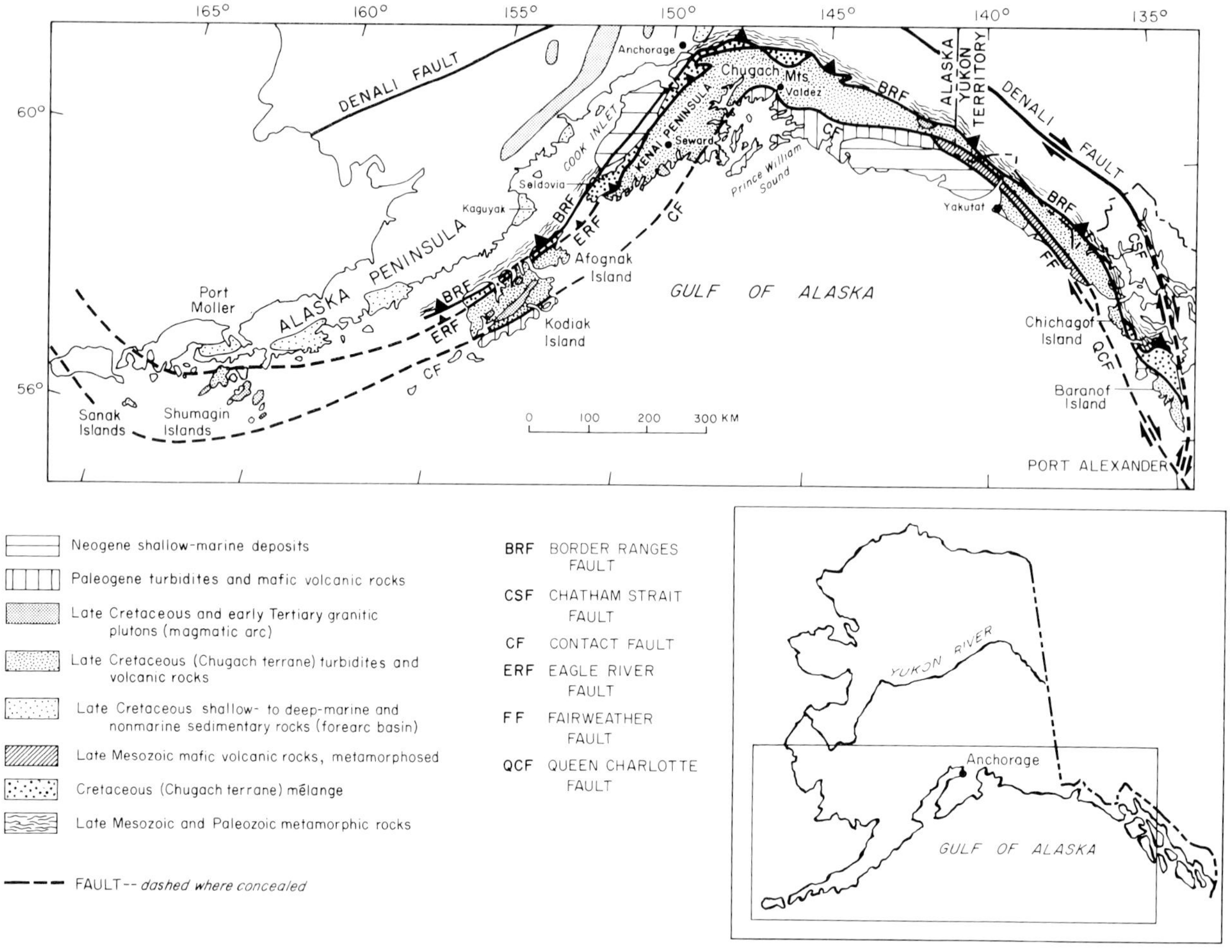

Figure 1. Distribution of Chugach terrane and major faults along margins of the Gulf of Alaska.

Chugach flysch terrane on Chichagof Island [11]. The volcanic rocks appear to be at least in part of oceanic origin [9] and indicate involvement of ocean floor with deposition and deformation of the Chugach flysch terrane.

Granitic batholiths of the Sanak–Baranof belt intrude the Chugach flysch terrane in many areas. They young eastward from about 60 m.y. in the Sanak Islands to about 45 m.y. on Baranof Island [12]. These batholiths have been interpreted to be products of anatectic melting of the subduction complex rather than arc magmatism [13]. In the Yakutat area, regional metamorphism of Paleogene age and penetrative deformation typify the Chugach flysch terrane [14].

Stratigraphy and Structure

The Chugach flysch terrane consists of deformed, thick, repetitively interbedded graywacke and shale or slate with local conglomerate, limestone, and volcanic rocks. The only estimates of its thickness are from the Sanak and Shumagin Islands, where it is at least 3- to 4-km thick [8], and on Kodiak Island, where it is estimated to be at least 5-km thick [4]. Throughout its extent, the Chugach flysch terrane is highly folded and faulted (Fig. 2). Strata generally dip landward, except in southeastern Alaska, and strike parallel to the regional trend of the belt. Faults and axial planes of folds also dip landward and strike parallel to the regional trend.

On Kodiak Island, the flysch consists chiefly of landward-dipping homoclinal sequences of turbidites with landward-facing stratigraphic tops that are separated by landward-dipping high-angle reverse faults. Interpreted directions of underthrusting of 332°, 334°, and 340° on Kodiak Island are perpendicular to the general strike [3]. On the Sanak and Shumagin Islands, interpretations of the structural history of the Chugach flysch terrane indicate initial deformation in a partially lithified state, development of an axial-plane slaty cleavage, and subsequent uplift and landward tilting along high-angle faults [15].

Fan Definition

The Chugach flysch terrane forms a narrow and linear outcrop belt that clearly does not form a fan-shaped depositional

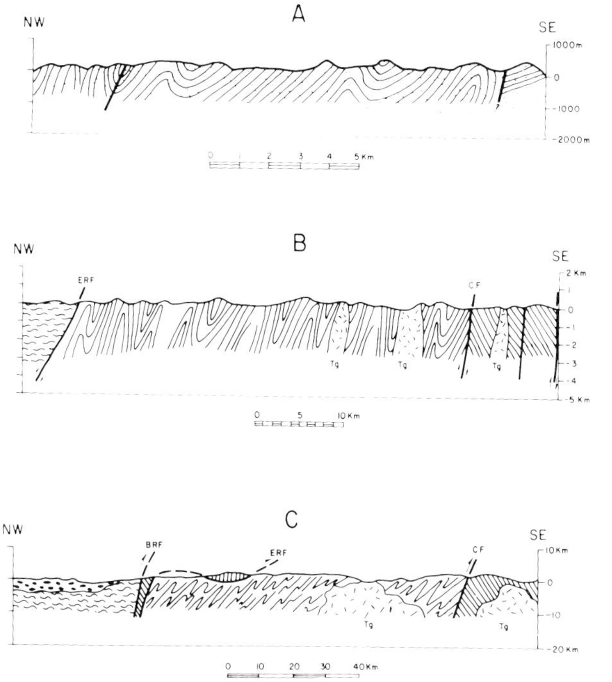

Figure 2. Cross sections showing general structure of Chugach terrane and relations with bounding terranes. Map symbols and abbreviations as in Figure 1 except for Tg, Tertiary granitic plutons. (A) Northwest-southeast cross section of part of Shumagin Islands (from [16]). (B) Northwest-southeast cross section of Kodiak Island (from [15]). (C) Northwest-southeast cross section from Anchorage area to Prince William Sound (from [2]).

body. Its linear trend may result from either original deposition as a very long and narrow body or from tectonic compression of an originally broader and possibly fan-shaped body. The regional distribution of paleocurrents, facies and facies associations, petrographic variations, and the regional paleogeographic framework suggest original deposition as a narrow elongate body in a trench [17]. However, some syndepositional and postdepositional shortening and stacking of tectonic slivers of the flysch terrane also undoubtedly took place during underthrusting at the trench margin.

The flysch terrane is divisible into two distinct groups of facies associations that permit the delineation of possibly two types of deep-sea fans. The northern or landward part of the flysch terrane contains smaller fault-bounded bodies of sandstone with some conglomerate that are tectonically and possibly originally were depositionally surrounded by mudstone. These bodies are as thick as several hundred meters, are sandstone-rich, were deposited by south-flowing turbidity currents, may have been fed by submarine canyons filled with various types of conglomerate and sandstone, are dominated by channelized facies, and are laterally discontinuous. I interpret these small bodies to be remnants of small deep-sea fans deposited in trench-slope basins that were fed by small, intermittently active submarine canyons cut into the landward trench slope (Fig. 3).

The southern or seaward part of the flysch terrane consists of interbedded sandstone and shale that are much more continuous laterally, especially in the western parts of the outcrop belt. The seaward part contains sequences as thick as several thousand meters, has generally west-flowing paleocurrents, and appears to have been fed primarily by a point source originally located to the southeast. It has facies associations that grade progressively westward from inner fan to middle fan, outer fan, and basin plain. The coarsest conglomerate clasts are located on Baranof Island to the southeast. I interpret the seaward part of the Chugach flysch terrane to be a large, elongate depositional body that exhibits most of the characteristics of ancient deep-sea fan deposits, although its original shape and geometry will remain partly uncertain because of the extent of subsequent deformation.

Fan Divisions

The marginal slope facies association, with its small trench-slope fans and irregular submarine canyons, characterizes the landward part of the Chugach flysch terrane (Figs. 3 and 4). Within the seaward part of the flysch terrane, inner-fan, middle-fan, outer-fan, and basin-plain deposits have been identified [4,8,11,17]. The facies associations have been determined mainly from reconnaissance field studies. Few sections have been measured and almost none have been published. The trench-fill body appears to have almost all the facies and facies associations of ancient mixed-sediment deep-sea fan deposits of Mutti and Ricci Lucchi [18,19].

Turbidite Facies

The slope facies association, where examined on Kodiak Island, the Kenai Peninsula, and Chichagof Island, consists

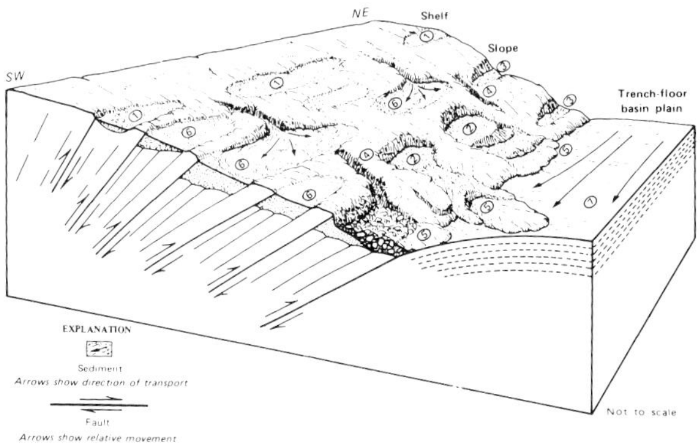

Figure 3. Block diagram showing inferred paleogeography and depositional setting of the Chugach terrane on Kodiak Island, from Nilsen and Moore [4]. (1) Shelf-edge basin; (2) slide-scarp basin; (3) intraslide basin; (4) abandoned canyon basin; (5) basin-of-slope debris flow; (6) structurally produced basin; and (7) basin-plain turbidites. Arrows on the basin plain indicate direction of flow of turbidity currents.

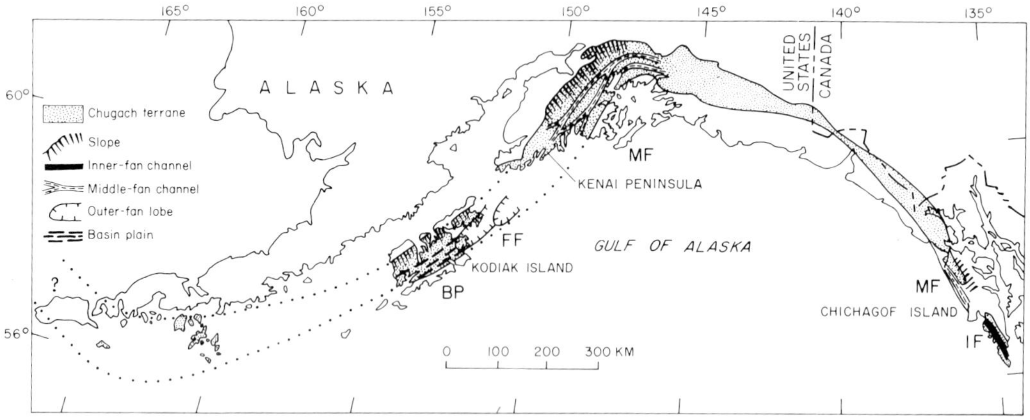

Figure 4. Generalized distribution of facies associations from the Chugach terrane. Abbreviations (from east to west); IF, inner fan; MF, middle fan; FF, fan fringe; BP, basin plain.

primarily of thin-bedded facies D turbidites and massive facies G mudstone [4,11]. These deposits are in many places clearly deformed into facies F by synsedimentary slumping but are in most places also deformed by tectonic shearing. The submarine-canyon deposits consist of facies A conglomerates with some associated thin-bedded facies D and E turbidites. The small fans within the slope facies association contain mostly beds of facies A, B, and C and have sandstone-to-shale ratios that range from 50:1 to 5:1.

The inner-fan facies association on Baranof and Chichagof Islands consists mostly of facies A and B conglomerate and sandstone deposited in channels [11]. The conglomerate contains clasts as long as 25 cm and is typically stratified and locally graded. Channel-margin thin-bedded facies D and E turbidites and mudstone crop out in only a few places. Sandstone-to-shale ratios are greater than 100:1.

The middle-fan facies association on Chichagof Island consists of about 15% thick-bedded to massive facies A and B sandstone and 85% thinly bedded facies D and E sandstone and associated mudstone [11]. Erosive scouring and channeling are common, and both thin and thick beds are generally laterally discontinuous. Sandstone-to-shale ratios generally range from 10:1 to 1:1.

The outer-fan, fan-fringe, and basin-plain facies associations on Kodiak Island consist primarily of facies C and D turbidites [4]. The outer-fan deposits consist of laterally continuous beds of facies C and D with some facies B beds. Channeling is generally absent except for minor scours. Sandstone-to-shale ratios range from 10:1 to 1:1. Fan-fringe deposits are similar but are thinner bedded, consist almost wholly of facies D turbidites, and have lower sandstone-to-shale ratios.

The basin-plain deposits on Kodiak Island consist primarily of fine- to medium-grained facies D turbidites that are laterally very continuous. Some facies C turbidites are included, and locally facies G olistostromes are present in the basin-plain deposits along the northern edge of the outcrops. Channeling is absent and sandstone-to-shale ratios range from 1:1 to 1:10.

Facies Associations

The down-axis changes in facies associations within the seaward part of the Chugach flysch terrane suggest fill by a fan system with proximal channeled deposits and distal nonchanneled deposits rather than by a single large channel depositing a wedge (Fig. 4). The relations suggest penecontemporaneous slope deposition adjacent to a large, elongate, outbuilding fan system in a narrow trough. The longitudinal system of trench filling was undoubtedly supplemented by lateral infilling from canyons that cut across the landward trench slope. However, the facies associations have been determined chiefly from the Sitka area, Chichagof Island, eastern part of the Kenai Peninsula, and Kodiak Island areas; areas in between have only been examined in reconnaissance.

On western Chichagof Island, an inner-fan facies association is present to the southeast, a middle-fan facies association to the northwest, and a slope facies association between the two and along the northeastern margin of the outcrop belt [11]. The inner-fan facies association extends southeastward to Baranof Island. The inner-fan facies association consists chiefly of massive sandstone and conglomerate. The middle-fan facies association consists of channelized bodies of massive sandstone surrounded by thicker sequences of thin-bedded turbidites and shaly channel-margin and interchannel deposits. The slope facies association consists of thick sequences of hemipelagic mudstone with thin interbeds of siltstone and mudstone turbidites, local olistostromes, and slide blocks of shallow-marine sandstone.

Northwest of Chichagof Island, the Chugach flysch terrane consists largely of argillite and graywacke with smaller amounts of conglomerate and mafic volcanic rocks [6,10]. These deposits have been interpreted as an inner- to middle-fan channel and interchannel facies association (J. Decker, oral communication, 1980). The incorporation of conglomerate within thick sequences of argillite and thin-bedded turbidites suggests the presence of channelized deep-sea fan deposits and possibly slope deposits.

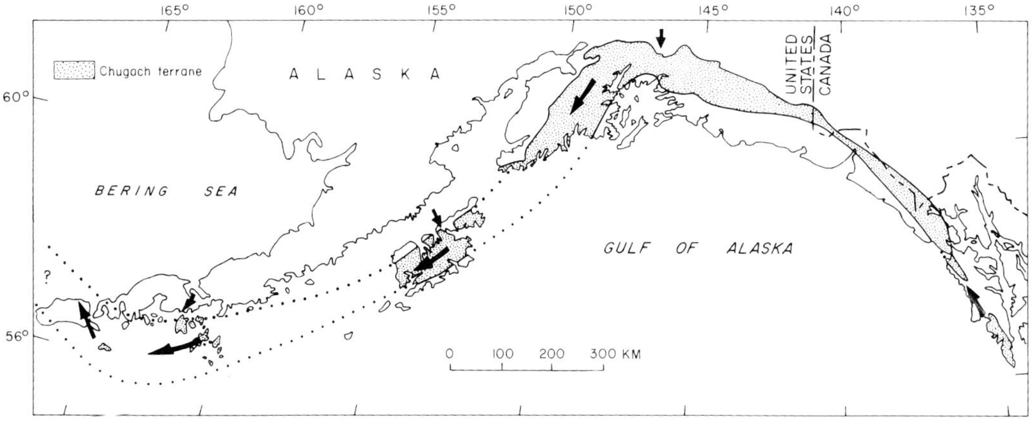

Figure 5. Generalized paleocurrent directions from the Chugach terrane. Longer arrows indicate longitudinal flow, shorter arrows lateral flow.

On the Kenai Peninsula between Anchorage and Seward, a slope facies association is present to the northwest and a middle-fan facies association to the southeast. Part of the slope facies association includes large olistoliths and olistostromes.

On northeastern Kodiak Island, the Chugach flysch terrane consists of two facies associations, a landward slope association and a seaward basin-plain association [4]. The slope association contains abrupt changes in thickness and grain size of deposits, conglomerate deposited in channels and canyons, small trench-slope basins, abundant slumps and synsedimentary folds, and thick pelitic sections. The trench-slope basins contain little-deformed deep-sea fan deposits probably fed by small canyons and fan channels within more highly deformed fine-grained slope deposits. These little-deformed slope-basin deposits are typically more coarse grained than the adjacent trench-fill deposits to the south. The seaward basin-plain association contains fine- to medium-grained sandstone turbidites that extend for the entire length of outcrops without perceptible changes in thickness or grain size and little organization of beds into thickening- or thinning-upward megasequences, although locally outer-fan lobe and fan-fringe deposits can be distinguished.

On the Shumagin and Sanak Islands, descriptions of the turbidites by Moore [8] suggest the presence of slope and deep-sea fan or basin-plain facies associations. Because no major channeling was noted, most of the deposits may represent outer-fan or basin-plain deposits. Conglomerates that contain clasts as large as 20 cm and are described as pebbly mudstone olistostromes probably form marginal slope deposits. Slumps as thick as 135 m are reported, with the directions of slumping subparallel to the paleocurrent flow directions.

Paleocurrents

Paleocurrents from the Chugach flysch terrane can be divided into two groups that are associated with the landward slope facies association and the seaward fan to basin-plain facies association (Fig. 5). Although data have been obtained only from the Chugach Mountains, Kodiak Island, and the Shumagin Islands, the landward slope facies association has paleocurrent directions that are oriented perpendicular to the general strike of the linear outcrop belt of the Chugach terrane and generally indicate flow toward the south. Some of these measurements are from channeled conglomerate thought to have been deposited in submarine canyons.

The seaward inner-fan to basin-plain facies associations have paleocurrent directions that are oriented parallel to the general strike of the outcrop belt and generally indicate flow toward the west. Abundant measurements have been obtained from the Chugach Mountains, Kodiak Island, the Shumagin Islands, and the Sanak Islands, but few data are available from southeastern Alaska.

Paleogeography

North of the Chugach terrane are extensive outcrops of nonmarine and shallow- to deep-marine Upper Cretaceous sedimentary rocks (Figs. 1 and 6). These units, of Campanian and Maestrichtian age, are common in parts of the Alaska Peninsula, Cook Inlet, and northeast of Anchorage [20]. These clastic sedimentary rocks, generally transported southward and derived from volcanic and plutonic sources to the north, are forearc basin or arc-trench-gap deposits coeval with and landward of the Chugach terrane.

North of the forearc basin deposits is a belt of Upper Cretaceous to Lower Tertiary quartz diorite to granite plutons that form the Iliamna–McKinley phase of the Alaska Range–Talkeetna Mountains batholithic complex [12,13]. These rocks have yielded K-Ar ages of 83 to 55 m.y., and plots of K_2O trends suggest that these plutons were generated in response to northwest-directed subduction [3]. Cretaceous and Early Tertiary granitic plutons are also abundant in southeastern Alaska.

Southern Alaska has been reinterpreted in the last few years as a collage of various terranes that have been transported northward and accreted to Alaska. Paleomagnetic data, largely from rocks associated with the Chugach terrane, indicate that

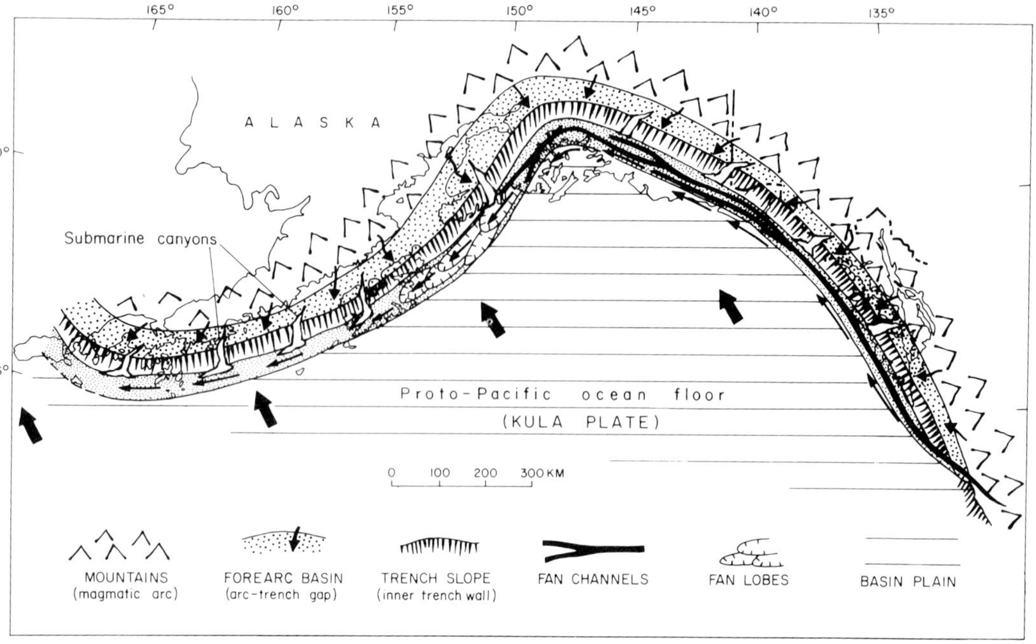

Figure 6. Paleogeographic map of Chugach terrane showing magmatic-arc source area, forearc basin, trench slope, trench, and proto-Pacific ocean floor. Large arrows show approximate direction of seafloor movement. Small arrows show principal directions of sediment transport. Incised areas on trench slope and forearc basin indicate submarine canyons, some of which transport sediment to the trench floor. The Late Cretaceous trench is shown in its present position without palinspastic restoration to a more southerly position, as indicated by paleomagnetic data, or straightening, as indicated by structural studies of the southern Alaska orocline.

it was originally deposited far to the south of its present position. During the Cenozoic the Chugach terrane moved northward, eventually reaching its present latitude and accreting to Alaska. However, the process, timing, and tectonic effects of the accretion process on the Chugach flysch terrane and related rocks are not well understood.

The Chugach flysch terrane has previously been considered to be a trench deposit on the basis of tectonostratigraphic considerations [3,4,8,11,15,17]. Its structural framework suggests early deformation in a trench setting followed by regional uplift and tilting along landward-dipping reverse faults [15], associated with regional metamorphism and intrusion of anatectically derived granitic rocks.

Conclusions

The deformation history, paleocurrents, facies associations, and regional stratigraphic and structural relations suggest that the turbidites of the Upper Cretaceous Chugach flysch terrane represent a trench-fill deposit. Tectonic inclusions of pillow basalt, gabbro, sheeted dikes, and chert suggest that the turbidites were deposited on ocean floor and were subsequently offscraped on the upper plate of the subduction zone. Regional evidence for the trench depositional setting consists of the presence of the following coeval units progressively to the north: (1) slope deposits with south-flowing paleocurrents and trench-slope basins; (2) remnants of a tectonic melange zone [3] separated from the Chugach flysch terrane by a landward-dipping major fault zone; (3) south-transported forearc basin deposits that form a belt parallel to the outcrop trend of the Chugach flysch terrane and contain abundant shallow-marine and nonmarine sequences; and (4) granitic intrusions that form a major subparallel magmatic arc. The complete arc-trench system is at present bounded by major faults, mostly right-lateral strike-slip faults of Cenozoic age.

The trench was filled both longitudinally and laterally. The major paleocurrent trends and facies associations indicate an axial east-to-west change from proximal to distal turbidite sedimentation, suggesting the development of a very elongate and highly channelized deep-sea fan system within the narrow trench (Fig. 6). The principal source areas for these axial deposits were at the southeastern end of the arc-trench system. Lateral infilling from the north is suggested by the presence of south-directed paleocurrents and a slope facies association along the northern edge of the outcrop belt. Rock-fragment petrography of sandstone samples from the Chugach terrane indicates derivation from an older subduction complex and from a magmatic arc that was increasingly dissected eastward [17]. The partly coeval Chugach melange terrane, although severely deformed and metamorphosed by subduction, could also include deformed olistostromes emplaced by large-scale submarine landsliding on the trench slope.

The trench formed in response to north-directed subduction relative to the present distribution of the Upper Cretaceous trench-fill and forearc basin deposits and magmatic arc. It is clear that the Chugach terrane and its related melange terrane, forearc basin, and magmatic arc constitute a coherent larger terrane that moved northward as a single unit [17]. Accretion of the terrane to Alaska, based on paleomagnetic evidence, must have occurred long after the Late Cretaceous subduction zone had ceased to be active. The present Aleutian trench and arc system was superimposed on the Chugach terrane in Cenozoic time. This arc system contains volcanic

Figure 7. Depositional facies and facies associations, Kodiak Formation, Kodiak Island, Alaska. (A) Thinly bedded facies C and D basin-plain turbidites, Mill Bay. (B) Thinly bedded facies C and D basin-plain turbidites, Uyak Bay. Stratigraphic top to right; hammer circled for scale. (C) Channelized facies A conglomerate and lenses of facies B sandstone, submarine-canyon facies association, Uyak Bay. Stratigraphic top to left; hammer circled for scale. (D) Synsedimentary recumbent fold in thinly bedded facies F siltstone and sandstone, slope facies association, Uyak Bay; beds overturned. (E) Channelized facies A shale rip-up-clast conglomerate, submarine-canyon facies association, Uyak Bay. (F) Thinly bedded and cleaved facies D and E slope overbank deposits, Uyak Bay. Sequence overturned with stratigraphic top to lower right.

rocks as old as Eocene, yielding an overlap in timing of the arrival of the Chugach terrane and initiation of the new arc system.

Addendum

Characteristics of the depositional facies and facies associations of the Chugach terrane are presented in a number of outcrop photographs (Fig. 7).

Acknowledgments

I thank George W. Moore, Gary R. Winkler, Arnold H. Bouma, J. Casey Moore, John Decker, George Plafker, and Gian G. Zuffa for helpful discussions about the sedimentology of the Chugach flysch terrane. T. E. Moore and J. C. Yount provided helpful reviews of this paper.

References

[1] Berg, H. C., Jones, D. L., and Richter, D. H., 1972. Gravina–Nutzotin belt—tectonic significance of an upper Mesozoic sedimentary and volcanic sequence in southern and southeastern Alaska. U.S. Geological Survey Professional Paper 800-D, pp. D1–D24.

[2] Plafker, G., Jones, D. L., and Pessagno, Jr., E. A., 1977. A Cretaceous accretionary flysch and melange terrane along the Gulf of Alaska margin. U.S. Geological Survey Circular 751-B, pp. B41–B43.

[3] Moore, J. C., and Connelly, W., 1979. Tectonic history of the continental margin of southwestern Alaska, Late Triassic to earliest Tertiary. In: A. Sisson (ed.), The Relationship of Plate Tectonics to Alaskan Geology and Resources. Alaskan Geological Society Symposium, pp. H1–H29.

[4] Nilsen, T. H., and Moore, G. W., 1979. Reconnaissance study of Upper Cretaceous to Miocene stratigraphic units and sedimentary facies, Kodiak and adjacent islands, Alaska. U.S. Geological Survey Professional Paper 1093.

[5] Jones, D. L., and Clark, S. H. B., 1973. Upper Cretaceous (Maestrichtian) fossils from the Kenai–Chugach Mountains, Kodiak and Shumagin Islands, southern Alaska. U.S. Geological Survey Journal of Research, v. 1, pp. 125–136.

[6] Brew, D. A., and Morrell, R. P., 1979. Correlation of the Sitka Graywacke, unnamed rocks in the Fairweather Range, and Valdez Group, southeastern Alaska. U.S. Geological Survey Circular 804-B, pp. B123–B125.

[7] Karl, S., Decker, J., and Jones, D. L., 1979. Early Cretaceous radiolarians from the McHugh Complex, south-central Alaska. U.S. Geological Survev Circular 804-B, pp. B88–B90.

[8] Moore, J. C., 1973. Cretaceous continental margin sedimentation, southwestern Alaska. Geological Society of America Bulletin, v. 84, pp. 595–614.

[9] Tysdal, R. G., and others, 1977. Sheeted dikes, gabbro, and pillow basalt in flysch of coastal southern Alaska. Geology, v. 5, pp. 377–383.

[10] Plafker, G., and Campbell, R. B., 1979. The Border Ranges fault in the Saint Elias Mountains. U.S. Geological Survey Bulletin 804-B, pp. B102–B104.

[11] Decker, J., Nilsen, T. H., and Karl, S., 1979. Turbidite facies of the Sitka Graywacke, southeastern Alaska. U.S. Geological Survey Circular 804-B, pp. B125–B129.

[12] Hudson, T., 1979. Mesozoic plutonic belts of southern Alaska. Geology, v. 7, pp. 230–234.

[13] Hudson, T., and Plafker, G., 1979. Paleogene anatexis along the Gulf of Alaska margin. Geology, v. 7, pp. 573–577.

[14] Hudson, T., and Plafker, G., 1982. Paleogene metamorphism of an accretionary flysch terrane, eastern Gulf of Alaska. Geological Society of America Bulletin, v. 93, pp. 1280–1290.

[15] Moore, J. C., 1973. Complex deformation of Cretaceous trench deposits, southwestern Alaska. Geological Society of America Bulletin, v. 84, pp. 2005–2020.

[16] von Huene, R., Moore, J. C., and Moore, G. W., 1978. Cross section of Alaska Peninsula–Kodiak Island–Aleutian Trench. Geological Society of America Map Chart Series MC-28A.

[17] Nilsen, T. H., and Zuffa, G. G., 1982. The Chugach terrane, a Cretaceous trench-fill deposit, southern Alaska. In: J. K. Leggett, (ed.), Trench-Forearc Geology: Sedimentation and Tectonics on Modern and Ancient Active Plate Margins. Geological Society of London Special Publication no. 10, pp. 213–227.

[18] Mutti, E., and Ricci Lucchi, F., 1972. Le torbiditi dell'Appennino settentrionale—introduzione all'analisi di facies. Societa Geologica Italiana Memorie, v. 11, pp. 161–199.

[19] Mutti, E., and Ricci Lucchi, F., 1975. Turbidite facies and facies associations. In: Examples of Turbidite Facies and Facies Associations from Selected Formations of the Northern Apennines. 9th International Congress of Sedimentology, Nice, France, Field Trip Guidebook, A11, pp. 21–36.

[20] Mancini, E. A., Deeter, T. M., and Wingate, F. H., 1978. Upper Cretaceous arc trench gap sedimentation on the Alaska Peninsula. Geology, v. 6, pp. 437–439.

CHAPTER 28

Ferrelo Turbidite System, California

D. G. Howell and J. G. Vedder

Abstract

Remnants of an Eocene fan system are preserved onshore at San Diego and in the central part of the southern California borderland. Even though faults and erosion have truncated its margins, geophysical data and exploratory wells indicate that remaining parts of the fan extend beneath an offshore area nearly 400-km long and 40- to 100-km wide. Environments representing fluvial, fan-delta, shelf-channel, overlapping inner- to outer-fan, and basin-plain facies are recognized or inferred. Three progradational cycles onshore and two distinct pulses of sand accumulation offshore are attributable to eustatic low sea-level stands rather than to tectonic uplift or shifts in depositional patterns.

Introduction

The Ferrelo Fan (new name) of Eocene age is defined by outcrops on the mainland and islands off southern California together with exploratory well data and seismic-reflection profiles from the California Continental Borderland. Although approximately 90% of the system is concealed by overlying rocks and water, the fan is recognized by fluvial, fan-delta, and shelf-to-slope channel facies in the vicinity of San Diego as well as by overlapping inner- and middle-fan facies on San Miguel, Santa Rosa, Santa Cruz, and San Nicolas Islands (Figs. 1 and 2). Nine exploratory wells, situated along Santa Rosa–Cortes Ridge and one near Santa Barbara Island, drilled through correlative Eocene strata that presumably represent overlapping middle- and outer-fan facies. Seismic reflection profiles suggest that middle- and outer-fan facies extend southward from beneath Santa Cruz Basin to San Nicolas Basin, and possibly to East and West Cortes Basins. By inference, outer-fan and basin-plain facies constitute the Eocene section beneath Velero Basin (Figs. 1 and 2). The original dimensions of Ferrelo Fan are unknown. Much of the western edge of the fan has been truncated by post-Eocene faulting and erosion, but the remaining part of the system between San Miguel Island and southern Velero Basin is at least 400-km long and 40- to 100-km wide.

The fan is named for Bartolome Ferrelo, pilot and log keeper for the Juan Rodriguez Cabrillo voyage of exploration along the west coast of Mexico and California in 1542. Ferrelo took command after Cabrillo's death and successfully completed the expedition in 1543.

Ferrelo Fan is best characterized by a composite of the models of Ricci Lucchi [1] and Walker [2] as depicted by Howell and Normark [3]. We believe that it is a prime example of a fan that reflects cyclical eustatic changes in sea level during growth.

Eocene strata that constitute the Ferrelo Fan are underlain by Paleocene and Upper Cretaceous submarine fan deposits throughout most of the borderland. One exception may be in the southern part, where it is possible that the Eocene outer-fan and basin-plain facies strata lap onto pre-Upper Cretaceous rocks both west and east of Velero Basin. Another exception is along the faulted east margin west of San Clemente Ridge and the western edge of Blake Knolls where Eocene fan strata may lie directly on pre-Upper Cretaceous rocks.

Oligocene nonmarine strata and shelf-slope deposits overlie the Eocene fan strata. The nonmarine beds are restricted to Santa Rosa Island, northernmost Santa Rosa–Cortes Ridge, and possibly to the northwestern part of Santa Cruz–Catalina Ridge between Santa Barbara and Santa Cruz Islands. At Santa Cruz Island, however, the Eocene strata are unconformably overlain by beds that are related to the Blanca Fan system (McLean and Howell, this volume).

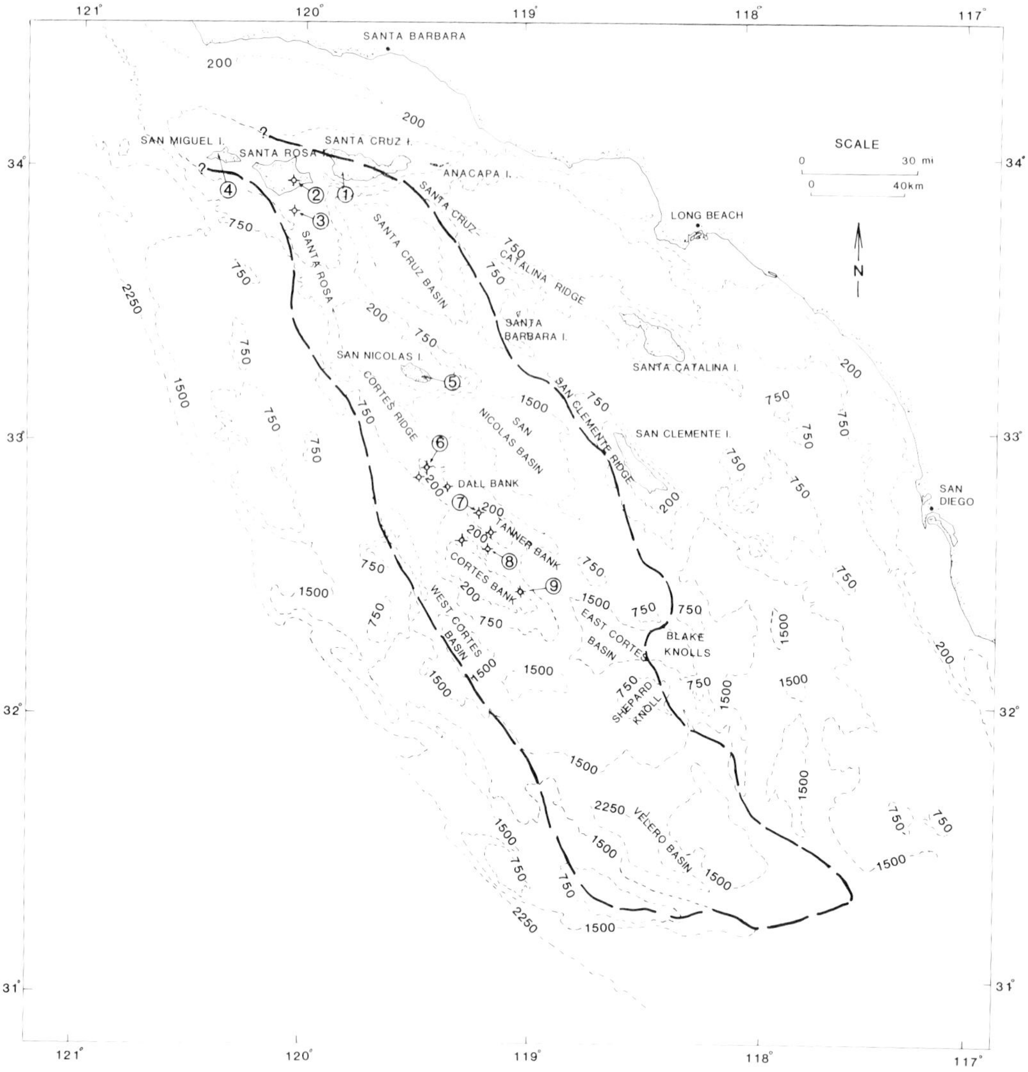

Figure 1. Index map outlining present extent of the Eocene Ferrelo Fan in the California Continental Borderland. Numbered symbols are locations of wells and outcrops shown in Fig. 3. Contours in meters.

Fan Divisions

Channelized Sandstone Bodies

Exposures in the San Diego area include largely conglomeratic fluvial-channel and alluvial-fan facies that grade westward into coastal-plain and fan-delta facies, which contain large channelized sandstone bodies (Friars and Mission Valley Formations of Kennedy and Moore, [4]). Paralic facies exposed near the present-day shoreline also contain channelized sandstone bodies (Torrey Sandstone, and Friars, Mission Valley, and Delmar Formations of Kennedy and Moore [4]). Shelf and submarine-channel facies at San Diego are 50% sandstone, 40% mudstone, and only 10% conglomerate in contrast to the dominantly conglomeratic nonmarine strata. The channel-shaped bodies of the Eocene shelf region are generally 10- to 300-m wide and up to 10-m deep; nested channel deposits may be as much as 50-m thick [5].

The Jolla Vieja Formation of Doerner [6] on Santa Cruz Island is composed of lenticular bodies of sandstone and conglomerate that are enclosed by fine-grained strata of the Canada and Cozy Dell Formations of Doerner [6]. The Jolla Vieja probably represents deposition in a large inner-fan channel. This unit is at least 200-m thick but thins, normal to paleocurrent direction, to less than 5 m within a distance of 2 km.

On San Miguel, Santa Rosa, and San Nicolas Islands, thick-bedded sandstone units are interbedded with mudstone and siltstone units; conglomerate is a minor constituent. These strata are interpreted to be representative of overlapping inner- and middle-fan facies. Sandstone bodies range in thickness from a few meters to 150 m, but limited seacliff exposures preclude estimation of their lateral dimensional variations. Logs from wells on the south Santa Rosa Island shelf and in the Dall–Tanner–Cortes Bank area show sequences of sandstone bodies perhaps as thick as 200 m but of unknown shape. Conglomerate occurs in minor amounts

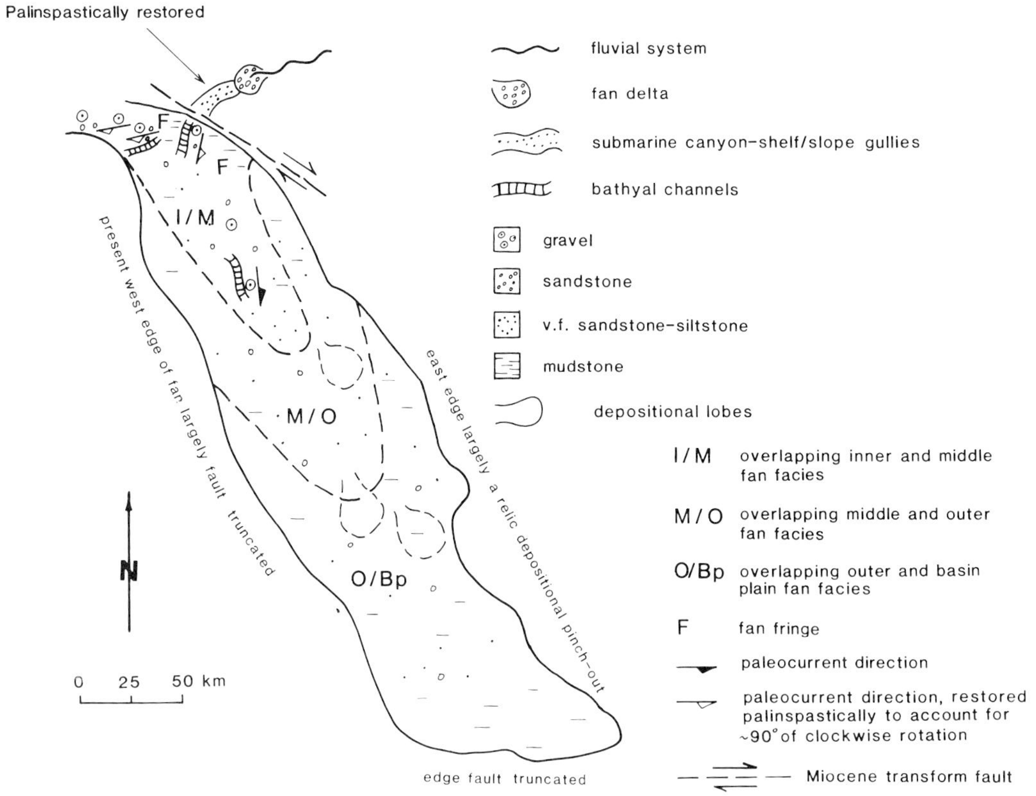

Figure 2. Schematic map showing selected morphometric characteristics of the Eocene Ferrelo Fan. The nonmarine to shelf facies of San Diego are palinspastically restored to their original position (top).

in one well at Tanner Bank (well no. 7, Fig. 3), and small quantities are likely to be present at places between this bank and San Nicolas Island.

Nonchannelized Sandstone Bodies

Although it is not practicable to differentiate channelized and nonchannelized sandstone bodies on the basis of available subsurface and seismic data, nonchannelized sandstone units are assumed to occur within the outer-fan and basin-plain facies beneath the basins and ridges southeast of Cortes Bank (Fig. 2). Until appropriate subsurface and seismic data are acquired, proportions of channelized sandstone versus nonchannelized sandstone cannot be estimated realistically. However, it is noteworthy that conglomerate generally decreases in abundance southward and is known to be present in only one of the wells south of San Nicolas Island (well no. 7, Fig. 3). In this well, the conglomeratic beds are confined to the upper part of the section.

Fan Fringe–Basin Plain

The systematic changes in fan facies that are observable on the mainland and islands, if projected southward by means of well data and seismic interpretations, lead to the conclusion that outer-fan and basin-plain deposits could occur in the vicinity of Velero Basin and possibly in East and West Cortes Basins. The fine-grained strata that enclose the coarse-grained Jolla Vieja channel on Santa Cruz Island represent fan-fringe strata contiguous to the slope or inner-fan setting.

Lithofacies A through F

Each of the lithofacies designations of Mutti and Ricci Lucchi [7] can be recognized in outcrops on the islands and are indicated in the logs of the wells on Santa Rosa–Cortes Ridge. On Santa Cruz Island, facies A and B strata are represented by the lens-shaped Jolla Vieja Formation of Doerner, in which individual, stacked channel-filling, matrix-supported conglomerate units are as much as 5-m thick and show normal as well as inverse grading. Coarse-grained channelized sandstone beds are 1- to 20-m thick and display groove and flute casts, dish structures, disturbed bedding, climbing ripples, grading, and abundant rip-up clasts. Some disrupted conglomeratic sandstone zones represent facies F. The Canada and Cozy Dell Formations of Doerner include facies D, E, and G strata that are interpreted to be inner-fan, overbank, and slope deposits that enclose the Jolla Vieja channels.

Multiple progradational and retrogradational events are indicated by strata on San Miguel, Santa Rosa, and San Nicolas Islands. Facies A and B strata are interbedded with facies C, D, and F strata. Common thinning- and fining-upward sequences at Santa Rosa and San Nicolas Islands range

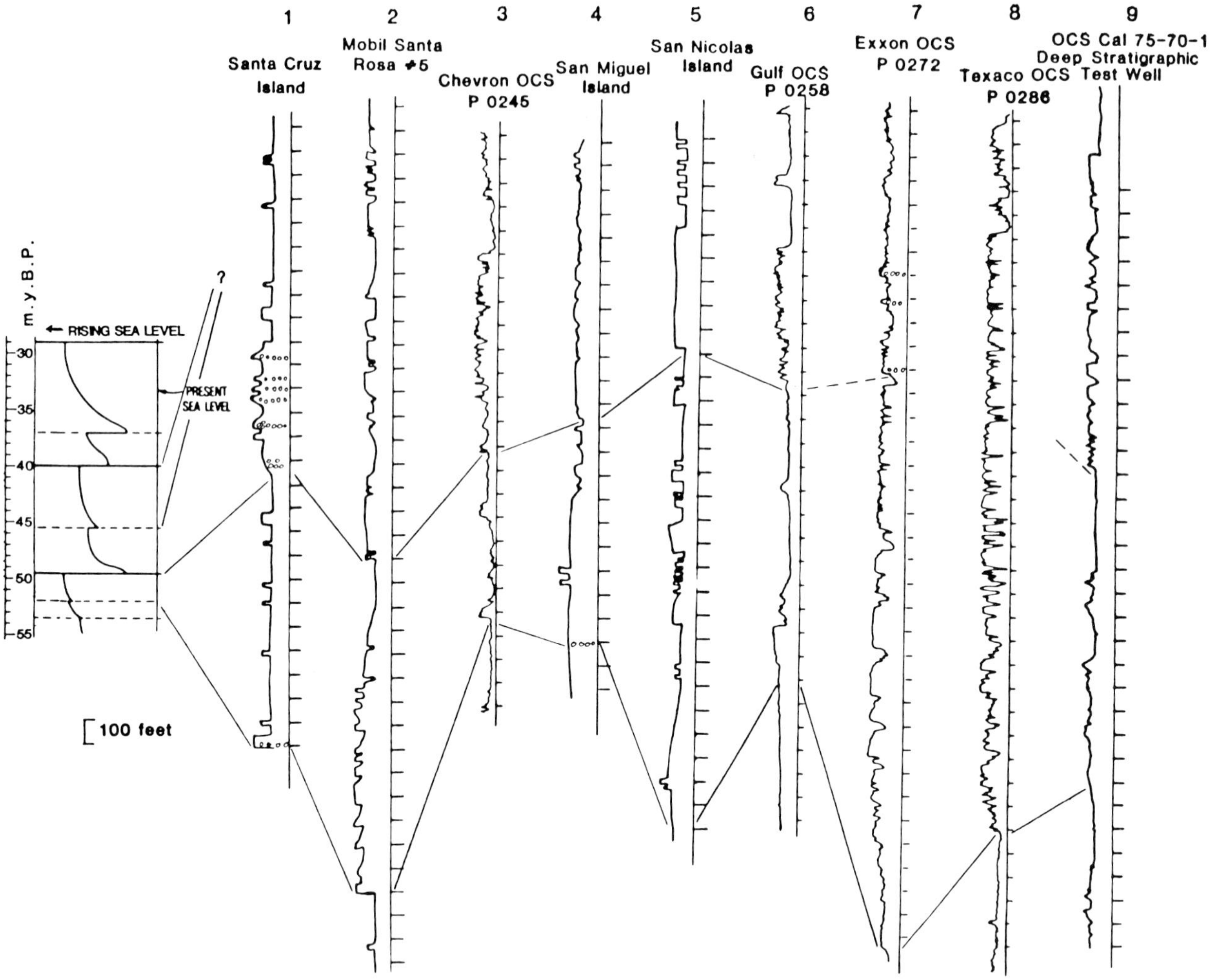

Figure 3. Selected measured sections and electric S-P well logs of Eocene strata, from the California Continental Borderland (Fig. 1 for location). Conglomerate zones and pronounced sandy intervals are correlated with low stands of sea level.

from 25 to 150 m in thickness and may represent deposition in migrating channels and laterally associated levee or overbank deposits. Facies A strata ordinarily are organized and are massive or normally to inversely graded. Facies B strata interbedded with the A strata display grading, dish structures, rip-up clasts, and disrupted bedding, and several flow units are commonly amalgamated. Facies C strata are uncommon and randomly interspersed with facies B, D, and E. Repetitive sequences of D and E strata commonly contain sets of small climbing ripples and starved ripple-drift structures. Facies F and G strata are uncommon. In combination, facies A, and B plus some C strata and their stratigraphic relations to facies E plus some D strata suggest an environment in which anastomosing or braided channels, levees, and interchannel areas overlapped.

From electric logs of wells at the southeast end of Santa Rosa–Cortes Ridge (well nos. 6 and 9, Fig. 3), thickening-upward cycles of facies B, C, and D, E are inferred. These cycles may reflect deposition of progradational lobes in an overlapping outer-fan and basin-plain environment.

Sandstone-to-mudstone ratios seem to change systematically from north to south. In the Santa Rosa Island well (no. 2) the ratio is approximately 60/40, at San Nicolas Island (no. 5) about 50/50, and in the Cortes Bank Stratigraphic Test Well (no. 9) nearly 40/60.

Cyclic Progradational Sequences

Within the depositional sequence represented by the Eocene Ferrelo Fan, three major progradational cycles are detectable. In the nearshore environments these cycles are heralded by pronounced lobes of conglomeratic lithofacies; from oldest to youngest these are: (1) Mount Soledad Formation, (2) conglomerate of the Scripps Formation and the Stadium Conglomerate, and (3) Pomerado Conglomerate of Peterson and

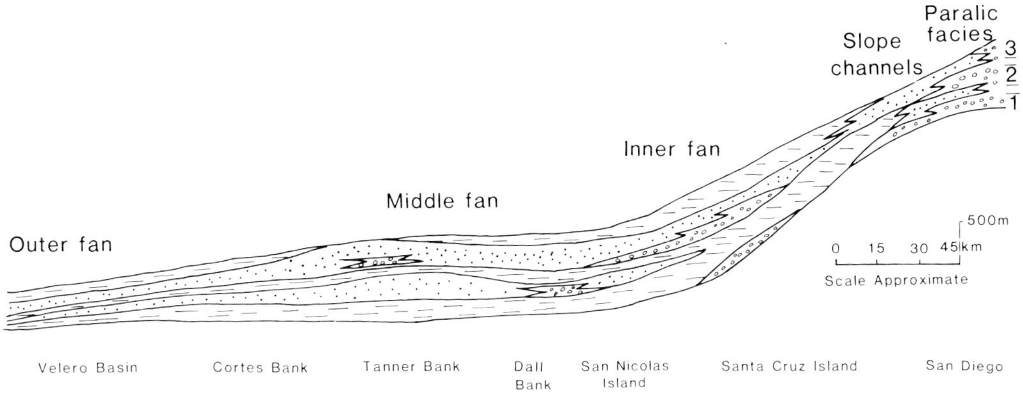

Figure 4. Schematic longitudinal cross section of the Eocene Ferrelo Fan. Note suprafan bulges and conglomerate pulses that correspond to times of low sea level.

Kennedy [4,8]. In the submarine fan environments, the lower progradational cycle is recognized by conglomerate- or sand-filled channels in the proximal inner- to middle-fan settings and by coarsening-upward cycles in more distal settings (Fig. 3). In general, the lower progradational cycle is reflected by the first sequence of coarse-grained strata that blankets broad areas of the entire fan area (Figs. 3 and 4).

Initiation of the second progradational cycle is marked locally by conglomeratic strata (see columns 1 and 7 of Fig. 3). The lateral extent and abundance of sand appears to be most pronounced during this second pulse of deposition (Fig. 3).

The third progradation cycle is not evident in the submarine fan facies. Several thick sand sequences from the middle-fan area may be an expression of the progradation event that corresponds to the seaward-advancing Pomerado Conglomerate of the nearshore setting.

The first two progradational events are traceable from the nonmarine environments through the entire submarine fan complex. The conglomerate is composed principally of siliceous volcanic clasts. A potential source terrane is not known in southern California. These clasts record fluvial transport from an eastern source across an Eocene peneplain that is currently represented by the Peninsular Ranges. The conglomeratic and coarse-grained deposits throughout the submarine fan must therefore reflect eustatic low sea-level stands rather than local tectonics or shifts in the locus of deposition of coarse-grained facies.

Relation to Eustatic Low Sea-level Stands

The cyclic progradational events discussed above indicate fan-wide changes in depositional patterns. These pulses of increased sand influx appear to be correlated with periods of eustatic low sea level. This correlation is constrained by biostratigraphic age determinations for strata that crop out on the islands. Although stratigraphic ages from the well sections are not as precisely known, available data permit the inferred correlation shown on Figure 3.

During episodes of high sea-level stands, sediment is trapped in estuarine and other inner-shelf settings. When sea-level drops to near or below the shelf break, the previously ponded sediment is transported into the deeper marine environments. The duration of the low sea-level stands seems to be relatively short compared to the periods of high standing sea levels [9]. The accumulation of sediment in the submarine fan, however, is greatest during the low stands. Two distinct pulses of sand accumulation are evident in all the wells or measured sections (Fig. 3), except for wells 7 and 8. These two wells, from the middle to outer fan areas, compose the thickest sections of sand strata. The approximately 1000 m of sand does not contain significant muddy intervals that might reflect intervening high sea-level stands. Two possible explanations are offered: (1) The entire sandy section of both wells represents a single low-stand event, probably the more pronounced 49.5 m.y. B.P. low-stand, or (2) these two wells happen to be situated in the midfan area where sand tends to accumulate during periods of both low and high sea level. The occurrence of thick sand packages in a relatively distal part of the fan affects petroleum exploration strategies, and these occurrences confirm the concept of sediment by-pass.

Discussion

In summary, we believe that the combined evidence provided by turbidite lithofacies, submarine-fan models, depositional patterns, and cyclic progradational events indicates that stratigraphic sequences within Ferrelo Fan developed in response to eustatic sea-level changes during Eocene time.

References

[1] Ricci Lucchi, F., 1975. Depositional cycles in two turbidite formations of northern Apenni (Italy). Journal of Sedimentary Petrology, v. 45, pp. 3–43.

[2] Walker, R. G., 1978. Deep water sandstone facies and ancient sub-

marine fans: models for exploration for stratigraphic traps. American Association of Petroleum Geologists Bulletin, v. 62, pp. 932–966.

[3] Howell, D. G., and Normark, W. R., 1982. Sedimentology of submarine fans. In: P. A. Scholle and D. R. Spearing (eds.), Sandstone Depositional Environments. American Association of Petroleum Geologists Memoir 31, pp. 365–404.

[4] Kennedy, M. P., and Moore, G. W., 1971. Stratigraphic relations of Upper Cretaceous and Eocene formations, San Diego coastal area, California. American Association of Petroleum Geologists Bulletin, v. 55, pp. 709–722.

[5] Howell, D. G., and Link, M. H., 1979. Eocene conglomerate sedimentology and basin analysis, San Diego and the Southern California Borderland. Journal of Sedimentary Petrology, v. 49, pp. 517–540.

[6] Doerner, D. P., 1969. Lower Tertiary biostratigraphy of southwestern Santa Cruz Island. In: D. W. Weaver (ed.), Geology of the Northern Channel Islands. American Association of Petroleum Geologists Pacific Section Special Publication, pp. 17–29.

[7] Mutti, E., and Ricci Lucchi, F., 1972. Le torbiditi dell'Apennino settentrionale: introduzione all'analisi di facies. Societa Geologica Italiana Memorie, v. 11, pp. 161–199.

[8] Peterson, G. L., and Kennedy, M. P., 1974. Lithostratigraphic variations in the Poway Group near San Diego, California. Transactions of the San Diego Society of Natural History, v. 17, pp. 251–257.

[9] Vail, P. R., and Mitchum, Jr., R. M., 1979. Global cycles of relative changes in sea level from seismic stratigraphy. In: J. S. Watkins, L. Montadert, and P. W. Dickerson (eds.), Geological and Geophysical Investigations of Continental Margins. American Association of Petroleum Geologists Memoir 29, pp. 469–472.

CHAPTER 29

Gottero Turbidite System, Italy

Tor H. Nilsen and Ernesto Abbate

Abstract

The Cretaceous and Paleocene Gottero Sandstone was deposited as a small deep-sea fan on ophiolitic crust in a trench-slope basin. It was thrust northeastward as an allochthonous sheet in Early and Middle Cenozoic time. The Gottero, as thick as 1500 m, was probably derived from erosion of Hercynian granites and associated metamorphic rocks in northern Corsica. Outcrops of inner-fan channel, middle-fan channel and interchannel, outer-fan lobe, fan-fringe, and basin-plain facies associations indicate that the depositional model of Mutti and Ricci Lucchi for mixed-sediment deep-sea fans can be used. The original fan had a radius of 30 to 50 km.

Introduction

The Gottero Sandstone crops out discontinuously over an east-west distance of about 75 km and a north-south distance of 50 km between Genova and Carrara in the Ligurian Apennines of northern Italy [1]. Outcrops are bounded to the south by the Tyrrhenian Sea, with the best exposures in coastal cliffs that are accessible by boat (Fig. 1). The Gottero forms part of the Vara Supergroup, one of several stacked allochthonous sequences that were thrust northeastward during the Cenozoic to form the core of the northern Apennines [2]. The Vara Supergroup contains an ophiolitic base that is overlain in ascending order by radiolarian chert, deep-marine micritic and siliceous limestone, shale, the Gottero Sandstone, and shale. The basal ophiolite, which contains serpentine, gabbro, and basalt, has yielded Jurassic radiometric dates [3].

The Upper Mesozoic and Lower Tertiary turbidites within the stacked allochthons of the northern Apennines are of two types: (1) calcareous turbidites, such as most of the Helminthoid Flysch sequences, thought to have an Alpine source to the north [4,5], and (2) arenaceous turbidites, such as the Gottero Sandstone, thought to have a Corsican–Sardinian source to the southwest [6]. The arenaceous turbidites generally form relatively small, coarse-grained deep-sea fans that prograded northeastward and locally may interfinger with the southward-transported calcareous basin-plain turbidites.

The Gottero Sandstone was named for outcrops of sandstone in the Mt. Gottero area (Fig. 1), where it is as thick as 1500 m [7]. It contains a microfauna of Albian to Paleocene age that is indicative of deposition at bathyal depths [7–9]. The Gottero is a feldspathic graywacke that contains fragments of metamorphic, volcanic, and sedimentary rocks [10], but only rare fragments of ophiolite [6]. Its composition suggests a continental-block provenance in which large igneous crystalline masses were exposed [11].

The Gottero Sandstone rests conformably on the Albian to Cenomanian Lavagna Shale and to the north pinches out into it [6]. The Lavagna Shale is more than 300 m thick and contains thin graded beds of calcareous sandstone, marlstone, and limestone [6]. However, near its top the Lavagna Shale is commonly tectonically disrupted and also contains intercalated olistostromes. The Gottero is conformably overlain by the Paleocene Giariette Shale, which is as thick as 600 m, contains thin beds of sandstone, siliceous limestone, siltstone, and marlstone, and forms the youngest unit of the Vara Supergroup [6,7].

Fan Definition

The Gottero Sandstone forms a fan-shaped depositional body that was built out to the north and east into a basin underlain by oceanic crust [6]. Evidence that the Gottero was deposited as a fan in a deep-sea setting include: (1) microfauna from

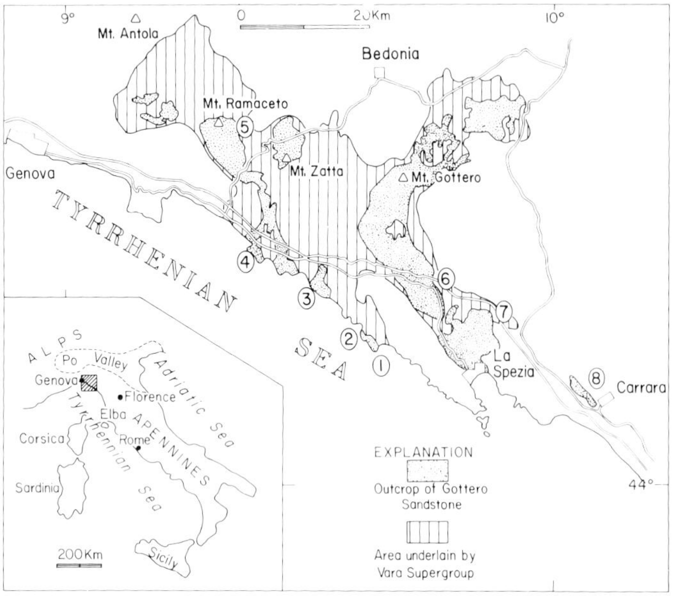

Figure 1. Generalized geologic map of part of the Ligurian Apennines showing the distribution of the Gottero Sandstone and Vara Supergroup. Numbers indicate locations of measured sections. Geology modified from Abbate and others [1].

the Gottero and conformably underlying and overlying shale deposits indicate deposition in deep-marine environments, (2) abundant turbidite and other sediment-gravity-flow deposits, (3) an outward-radiating paleocurrent pattern, (4) presence of almost all of the facies and facies associations defined by Mutti and Ricci Lucchi [12,13] for ancient turbidite systems, (5) distinct proximal-to-distal stratigraphic and sedimentologic changes, and (6) prominently developed channelized and nonchannelized deposits [14].

The Gottero deep-sea fan generally fits the published models of Mutti and Ricci Lucchi [12] for fans of mixed-sediment origin. Although it is sandstone-rich, it contains a prominent inner-fan channel, smaller middle-fan distributary channels, outer-fan lobes, and fan-fringe deposits.

Fan Divisions

Inner-fan, middle-fan, outer-fan, and fan-fringe facies associations and fan divisions have been recognized in the Gottero Sandstone (Fig. 2). Because of the distribution of outcrops and lack of continuous exposures, lateral relations between facies associations, as well as the lateral extent of channels and lobes, are generally uncertain. The characteristic features of each facies association were determined from eight measured sections (Fig. 1). Five of the sections start at the basal contact of the Gottero Sandstone on the Lavagna Shale and three are wholly within the Gottero Sandstone, without stratigraphic markers at the base or top. The thickest measured section is 295 m thick, whereas the maximum reported thickness of the Gottero Sandstone is 1500 m near Mt. Gottero.

Fan Facies

The inner-fan-channel facies association of section 1 rests with channeled contact on the uppermost part of the Lavagna Shale (Figs. 1 and 3). Here the Lavagna is highly disrupted by synsedimentary slumping. The facies association can be divided into 12 megasequences that average 17 m in thickness. These megasequences fine and thicken upward, in contrast to most published descriptions of inner-fan channel deposits.

The conglomerate and sandstone-to-shale ratio is 99:1. The megasequences typically commence with a channelized base overlain by conglomerate that contains mostly rip-up clasts of shale and siltstone but also some lithic pebbles as large as 7 cm. These erosional bases cut down gradually, over several hundred meters, into underlying strata, but locally erosional relief is as much as several meters. The beds of rip-up conglomerate contain large-scale, low-angle, inclined cross-strata that dip in variable directions.

Above the basal rip-up-clast conglomerate are cross-stratified facies E or B_2 beds of medium- to coarse-grained sandstone that are 5- to 60-cm thick. These beds have sharp, generally flat, erosive bases and sharp, wavy tops. The cross-strata suggest tractional reworking of the rip-up-clast conglomerate within the inner-fan channel, probably within thalweg channels. The cross-strata indicate northward transport of sediment.

The uppermost parts of the inner-fan megasequences consist of thick, massive beds of medium- to very coarse-grained facies A and B sandstone that contain abundant dish and pillar structures. These beds, which are typically amalgamated, are as thick as 15 m and display repeated alternation of mas-

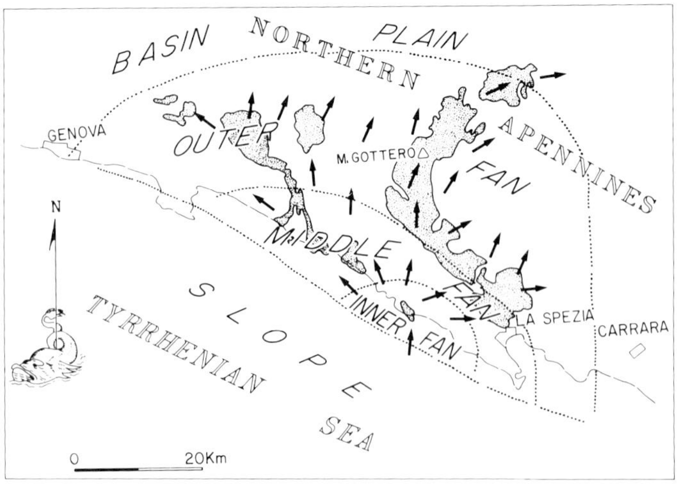

Figure 2. Distribution of turbidite facies associations and generalized paleocurrent directions for the Gottero Sandstone.

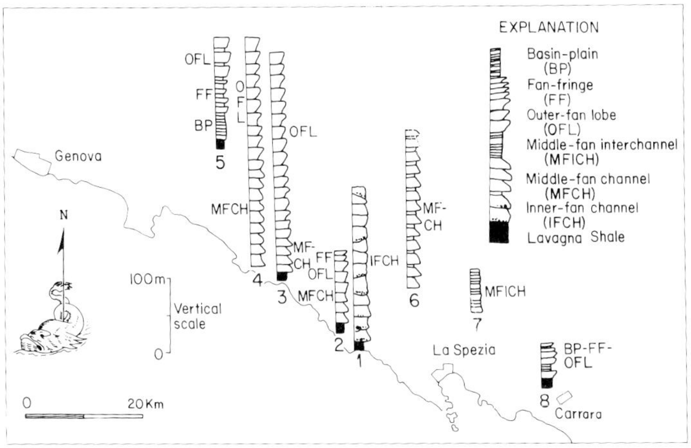

Figure 3. Measured sections and interpretive facies associations of the Gottero Sandstone.

sive layers and layers with diffuse parallel lamination and dish structure.

The middle-fan facies association consists of thinning- and fining-upward megasequences with locally thick fine-grained and thin-bedded levee and interchannel deposits. Middle-fan channel deposits in sections 2, 3, 4, and 6 have sandstone-to-shale ratios that average about 10:1 and megasequence thicknesses that average 15.5 m. The lower parts of the megasequences contain facies *A* and *B* beds and the upper parts facies *E* and B_2 beds. Downcutting of as much as several meters marks the lower contact of the megasequences. Scattered rip-up clasts and lithic pebbles as large as 3 cm are present. Slumped intervals as thick as several meters are present in the upper parts of some megasequences.

The 60-m-thick middle-fan levee and interchannel deposits of section 7 consist of thin beds of very fine- to medium-grained facies *D* and *E* sandstone. The beds are mostly organized into thickening-upward minisequences that are 50- to 650-cm thick and have a sandstone-to-shale ratio of 2.5:1, compared to an overall ratio of 1.3:1 for the entire section. The direction of sediment transport in these overbank deposits is toward the northwest, almost 180° opposite to that determined from nearby middle-fan channel and outer-fan lobe deposits.

The outer-fan lobe facies association is present in sections 2, 3, 4, 5, and 8. These deposits consist of thickening- and coarsening-upward megasequences that average about 15 m in thickness. The sandstone-to-shale ratios average about 11:1 in the outer-fan lobe deposits. Scattered lithic clasts as large as 3 cm are present. The lower parts of the outer-fan lobe megasequences consist of fine- to medium-grained beds of facies *C* and *D* sandstone and the upper parts of medium- to very coarse-grained beds of facies *C* and locally facies *B* sandstone. Slurried beds with rip-up clasts concentrated in their middle parts [15] are very common in the outer-fan lobe deposits, as are minor thickening-upward compensation cycles [16].

The fan-fringe or lobe-fringe facies association is present in sections 2, 5, and 8. It consists of thin bundles that average 2.5 m in thickness of thickening- and coarsening-upward facies *C* and *D* beds of sandstone. The sandstone-to-shale ratios in these deposits average 2.6:1.

The surrounding basin-plain facies association was measured in sections 5 and 8. It consists of beds of facies *C* and *D* sandstone that are not organized into megasequences. The average sandstone-to-shale ratio for these sections is 0.25:1.

Paleocurrents

Parea [17] determined that the Gottero Sandstone was deposited by northerly flowing turbidity currents that had a mean azimuthal orientation of 350°, based on 241 measurements of flute and groove casts. An additional 15 measurements by Abbate [18] near section 1 (Fig. 1) yielded northward transport. We obtained 232 paleocurrent measurements from the measured sections, which also show a dominantly northward direction of sediment transport (Fig. 4). These measurements were taken from flute casts, groove casts, other sole markings, cross-strata, current ripple markings, convolute laminations, primary current lineations, clast long-axis orientations, and channel-wall orientations. There is northwesterly flow in western outcrops, northerly flow in central outcrops, northeasterly flow in eastern outcrops, and southeasterly flow in southeastern outcrops. The data indicate an outward-radiating pattern of flow from the inner-fan channel deposits at section 1, which supports the general conclusions about distribution of facies associations and that the Gottero Sandstone is a fan-shaped depositional body.

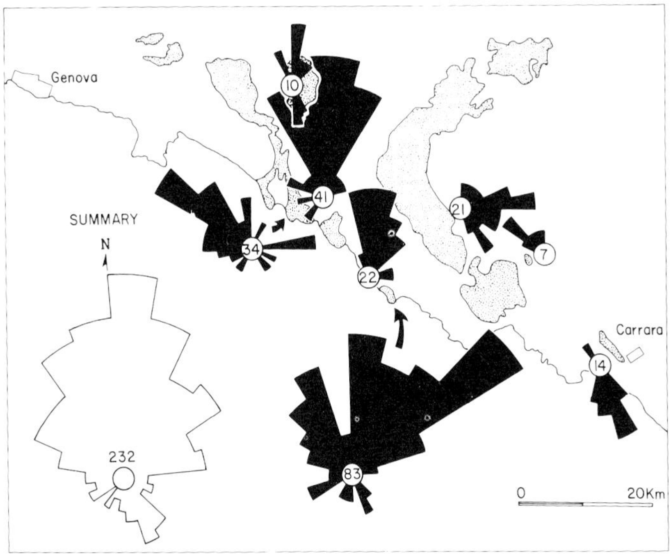

Figure 4. Paleocurrent pattern determined from measured sections of the Gottero Sandstone. Numbers in centers of circles indicate total number of measurements from each measured section.

A.

B.

C.

D.

E.

Figure 5. Photographs of facies associations of the Gottero Sandstone, Ligurian Apennines, Italy. (A) Basal part of inner-fan-channel facies association, Monterosso; section 1 of Figures 1 and 3. (B) Middle-fan-channel facies association with slumped interval at top of thinning- and fining-upward cycle, Levanto; sequence overturned, stratigraphic top to lower right; section 2 of Figures 1 and 3. (C) Middle-fan-interchannel facies association, Ceparana; stratigraphic top to right, hammer circled for scale. (D) Outer-fan-lobe facies association, near Moneglia; stratigraphic top to left; section 3 of Figures 1 and 3. (E) Basin-plain facies association, near Montemoggio.

Paleogeography

The Gottero Sandstone was deposited on the margin of a small Tethyan oceanic basin initiated by rifting and spreading in the Jurassic. The Cretaceous and Paleocene Gottero Fan was probably derived from northern Corsica. Sagri and Marri [19] proposed that the Gottero Sandstone was deposited as a fan in a trench-slope basin located on the western margin of the small ocean basin concurrently with subduction of the basin westward beneath the Corsican massif. By this reconstruction, the Alpine-derived carbonate turbidites to the east were deposited below the carbonate-compensation depth in a bordering trench. Closure of the basin and eastward thrusting of the underlying ophiolite and the Gottero Sandstone onto continental crust of Italy probably began in the Eocene or Oligocene [6]. Thrusting of the large slab that comprised the Vara Supergroup apparently resulted in spatial retention of the chief elements that formed the Gottero Fan, without major foreshortening or disruption by major faulting.

Sedimentation of the Gottero Fan appears to have been continuous, and was probably controlled by tectonic uplift of the active convergent margin in northern Corsica. Influx of sediment apparently continued through the major Late Cretaceous eustatic rise in sea level, without any apparent effects that can be discerned from the available measured sections. The abrupt initiation of fan sedimentation in the Albian may correspond to a global drop in sea level at that time, but this fluctuation is apparently minor in comparison to the long-term rise in sea level through the Cretaceous. Neither the major global lowering of sea level at the end of the Cretaceous nor several fluctuations of sea level in the Paleocene are apparently recorded in the sedimentary record of the Gottero Sandstone.

Progradation of the Gottero Fan into the basin is recorded in some measured sections and retrogradation in others. The general trend of thalweg migration in the inner-fan channel deposits of section 1 suggests initial growth of the Gottero fan to the northwest, followed by later outbuilding to the north and east. Sections 2, 3, and 4 to the northwest record retrogradation, with more distal facies associations overlying more proximal facies associations (Fig. 3). Section 5, however, the most distal, clearly records progradation of fan-fringe and outer-fan lobe deposits over basin-plain deposits. Section 8, to the southwest, contains fan-fringe and outer-fan lobe deposits over the Lavagna Shale, a framework indicative of fan progradation. It thus appears that although the fan clearly grew by progradation into the basin, at any particular time interval some parts were actively prograding and others were retreating.

Conclusions

The Gottero Sandstone was deposited as a small deep-sea fan on ophiolitic ocean crust in a small Mediterranean-type basin. A major inner-fan channel funneled sediment northwestward, northward, and eastward onto the fan, whose radius was 30 to 50 km. Inner-fan, middle-fan, outer-fan, and fan-fringe facies associations have been recognized in measured sections. Hercynian granitic and metamorphic rocks of northern Corsica probably formed the chief source area. The fan prograded out onto the basin floor, and different parts of the fan were active at different times. The fan deposits pinch out distally into shale. Deposition was continuous through several fluctuations in Cretaceous and Paleocene sea level. In later Cenozoic time, the entire fan was thrust eastward onto the Italian peninsula.

Addendum

Characteristics of the Gottero Sandstone in general overview and at the outcrop level are presented in Figure 5.

Acknowledgments

Harry Cook and Bill Normark of the U.S. Geological Survey provided very helpful reviews.

References

[1] Abbate, E., and others, 1970. Geological map of the Northern Apennines and adjoining areas. In: G. Sestini (ed.), Development of the Northern Apennines Geosyncline. Sedimentary Geology, v. 4, scale 1:500,000.

[2] Abbate, E., and others, 1970. Introduction to the geology of the Northern Apennines. In: G. Sestini (ed.), Development of the Northern Apennines Geosyncline. Sedimentary Geology, v. 4, pp. 207–249.

[3] Bortolotti, V., and Passerini, P., 1970. Magmatic activity. In: G. Sestini (ed.), Development of the Northern Apennines Geosyncline. Sedimentary Geology, v. 4, pp. 559–624.

[4] Scholle, P. A., 1971. Sedimentology of fine-grained deep-water turbidites, Monte Antola flysch (Upper Cretaceous), Northern Apennines, Italy. Geological Society of America Bulletin, v. 82, pp. 629–658.

[5] Sagri, M., 1974. Rhythmic sedimentation in deep-sea carbonate turbidites (Monte Antola Formation, Northern Apennines). Societa Geologica Italiana Bolletin, v. 93, pp. 1013–1027.

[6] Abbate, E., and Sagri, M., 1970. The eugeosynclinal sequences. In: G. Sestini (ed.), Development of the Northern Apennines Geosyncline. Sedimentary Geology, v. 4, pp. 251–340.

[7] Ghelardoni, R., Pieri, M., and Pirini, C., 1965. Osservazioni stratigraphiche nell'area dei fogli 84 (Pontremoli) e 85 (Castelnuovo ne' Monti). Societa Geologica Italiana Bolletin, v. 84, pp. 297–416.

[8] Fierro, G., and Terranova, R., 1963. Microfacies fossilifere e sequenze litologiche nelle "Arenarie superiori" dei monti Ramaceto e Zatta. Atti Istituto Geologica, Universita di Genova, v. 1, pp. 473–510.

[9] Passerini, P., and Pirini, C., 1964. Microfaune paleoceniche nella formazione dell'Arenarie del M. Ramaceto e degli Argilloscisti di Cichero. Societa Geologica Italiana Bolletin, v. 83, pp. 211–218.

[10] Malesani, P. G., 1966. Ricerche sulle arenarie, XV; L'Arenaria Superiore. Rendiconti della Societa Italiana di Mineralogia e Petrologia, v. 22, pp. 113–175.

[11] Valloni, R., and Zuffa, G. G., 1981. Detrital modes of arenaceous formations of the Northern Apennines. International Association of Sedimentologists, 2nd European Regional Meeting, Bologna, 1981, Abstracts, pp. 198–201.

[12] Mutti, E., and Ricci Lucchi, F., 1972. Le torbiditi dell'Appennino settentrionale—introduzione all'analisi di facies. Societa Geologica Italiana Memorie, v. 11, pp. 161–199.

[13] Mutti, E., and Ricci Lucchi, F., 1975. Turbidite facies and facies associations. In: Examples of Turbidite Facies and Facies Associations from Selected Formations of the Northern Apennines. Ninth International Congress of Sedimentology, Nice, France, Field Trip Guidebook A11, pp. 21–36.

[14] Nilsen, T. H., and Abbate, E., 1976. The Gottero Sandstone, a Late Cretaceous and Paleocene deep-sea fan complex in the Ligurian Apennines, northern Italy. Geological Society of America Abstracts with Programs, v. 8, pp. 1028–1029.

[15] Mutti, E., and Nilsen, T. H., 1981. Significance of intraformational rip-up clasts in deep-sea fan deposits. International Association of Sedimentologists, 2nd European Regional Meeting, Bologna, 1981, Abstracts, pp. 117–119.

[16] Mutti, E., and Sonnino, M., 1981. Compensation cycles: a diagnostic feature of turbidite sandstone lobes. International Association of Sedimentologists, 2nd European Regional Meeting, Bologna, 1981, Abstracts, pp. 120–123.

[17] Parea, G. C., 1964. La provenienza dei clastici dell'Arenaria del M. Gottero. Atti Memorie Nazionale Science Lettere Arti, Modena, v. 6, pp. 1–7.

[18] Abbate, E., 1969. Geologia delle Cinque Terre e dell'entroterra di Levanto (Liguria Orientale). Societa Geologica Italiana Memorie, v. 8, pp. 923–1014.

[19] Sagri, M., and Marri, C., 1980. Paleobatimetria e ambiente di deposizione delle unita torbiditiche Cretaceo–Superiori dell'Appennino settentrionale. Societa Geologica Italiana Memorie, v. 21, pp. 231–240.

CHAPTER 30

Hecho Turbidite System, Spain

Emiliano Mutti

Abstract

The Eocene Hecho Group submarine-fan and basin-plain turbidites fill an elongate basin in the south-central Pyrenees that was tectonically active during deposition. The total volume of these sediments is about 21,000 to 26,000 km^3. The bulk of the sand by-passed the fan-channel zone and was deposited in the lobe and fan-fringe environments. The stratigraphically lower part of the Hecho submarine fan was deposited during relative lowering of sea level.

Introduction

The Hecho Group turbidites offer a rather unique opportunity to study in detail a large deep-sea fan complex. The original depositional relations between these deep-marine sediments and their correlative shallow-marine and continental deposits in the general source area are preserved in outcrop. The facies characteristics and the depositional setting of the Hecho Group have been used for the formulation of general models [1], particularly those with "detached" lobes [2–4]. These models imply that the bulk of the sand by-passed the fan-channel mouths and was deposited in virtually flat outer-fan regions to form tabular bodies referred to as "outer-fan sandstone lobes" [2,5]. This paper summarizes the main, and still largely unpublished, results of recent work on the Hecho Group and focuses on the general depositional setting of these deposits.

General Geologic Setting

The Lower and Middle Eocene Hecho Group submarine turbidite system is a large clastic prism. It has a length of 175 km, a width of 40 to 50 km, a maximum thickness of about 3500 m, and a total volume of 21,000 to 26,000 km^3. It filled the western and deeper portion of the Tertiary South-Central Pyrenean Basin (for brevity referred to hereafter as TSCPB) that trended parallel to the Axial Zone of the Pyrenean tectonic province (Fig. 1) and opened toward the Atlantic [6,7]. The basin originated during the Late Paleocene and underwent synsedimentary structural deformation within an overall pattern of north-south compression related to the collisions of the Iberian and the southern-European margins along the north-Pyrenean fault zone [8,9]. The gradual uplift of the Axial Zone and the widespread occurrence of Triassic evaporites resulted in a general southward gravitational displacement of the Mesozoic and Tertiary cover of the Axial Zone contemporaneously with the infilling of the TSCPB [8]. Wrenching must also have played an important role, particularly along the Bigorre and the Segre faults (Fig. 1) that form the main present structural boundaries of the TSCPB to the west and east, respectively [9].

Depositional Setting

The depositional setting of the TSCPB immediately before the main phase of north-south compression that affected the Pyrenean domain during Late Eocene is shown in Figure 2. The vertical and lateral stratigraphic distribution of the main facies association delineates two principal depositional sectors within the TSCPB:

(1) The southeastern sector consists of at least four distinct depositional sequences whose boundaries are expressed either by stratigraphic unconformities or correlative abrupt facies changes. Except for the lowest sequence, the Ager, whose upper part consists of shal-

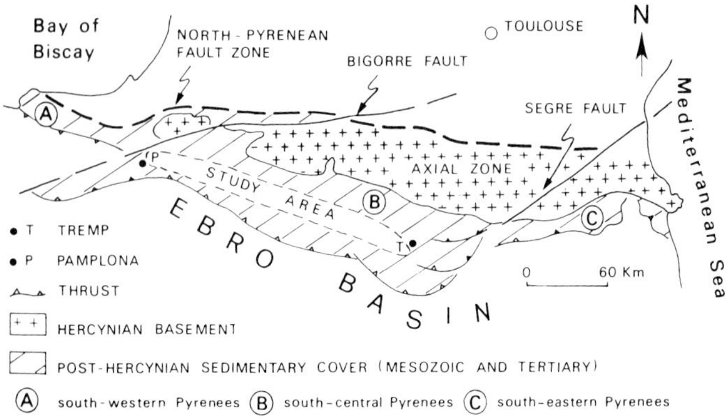

Figure 1. Simplified geologic map of the southern Pyrenees. Slightly modified after Souquet and others [9].

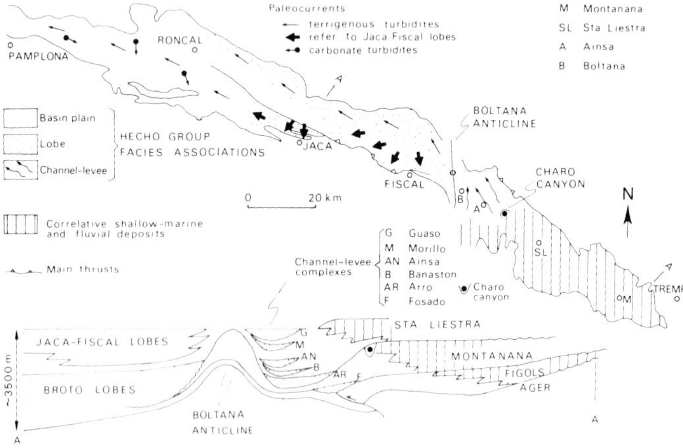

Figure 2. Paleogeographic sketch map and diagrammatic stratigraphic cross-section of the Hecho Group basin. Age of the different depositional sequences is the following: (1) Ager: Late Cretaceous to Paleocene; (2) Figols: Latest Paleocene–base of Early Eocene; (3) Montanana: Early Eocene; (4) Sta Liestra: Early to Middle Eocene. The scale of the cross-section is approximate.

low-marine carbonates marking a basinwide transgression, these sequences are essentially composed of fluviatile and deltaic deposits with a marked westward progradational character. Each sequence grades basinward into slope and basinal mudstone facies that enclose the different channel-levee complexes of the Hecho Group (Fig. 2).

(2) The western sector of the basin is characterized by a thick succession of turbidites referred to as Hecho Group [6]. The stratigraphic relations between the Hecho Group deposits and their correlative shallow-marine and continental strata of the eastern sector are diagrammatically shown in the cross-section of Figure 2.

The north-south-trending Boltana anticline is a prominent and largely synsedimentary structural feature in the eastern outcrop area of the Hecho Group (Fig. 2). It began to grow during the Early Eocene and is expressed by a thick succession of shallow-marine carbonates that formed contemporaneously with the fluvio-deltaic Montanana Sequence. The Boltana anticline separates the Hecho Group into two distinct portions. The eastern portion consists of six channel–levee complexes that were fed from southeast and south and are enclosed by predominant mudstone facies. These channel–levee complexes, except for the Fosado, which occurs at the very base of the Montanana Sequence, are shown in Figure 3. The Arro complex was probably fed by the laterally equivalent Charo canyon (Fig. 2), a feature some 200-m deep and 1000-m wide, which is cut into shelf sandstones and mudstones of the Montanana Sequence. The fill of this canyon is made up predominantly of chaotic mudstone facies and scattered lenses of coarse sandstone and blocky to cobbly conglomerate. Paleocurrent directions indicate northwesternward sediment transport.

The western portion of the Hecho Group outcrop area, extending from the Boltana anticline to the general region of Pamplona, consists of depositional sandstone lobes and associated fan-fringe deposits that grade downcurrent into a relatively sandstone-rich basin-plain facies association [2,10]. Numerous calcareous turbidites, mainly derived from north and northwest, are present in the basin–plain facies association (Fig. 2). Several of these turbidites are as thick as 100 m and voluminous as 50 km^3. They extend over a considerable distance, thus forming excellent stratigraphic markers across basin–plain and depositional lobe facies. These large-volume calcareous turbidites were derived from flanking carbonate platforms and are likely related to earthquakes along thrust faults [11].

West of the Boltana anticline, the lobe and the fan-fringe

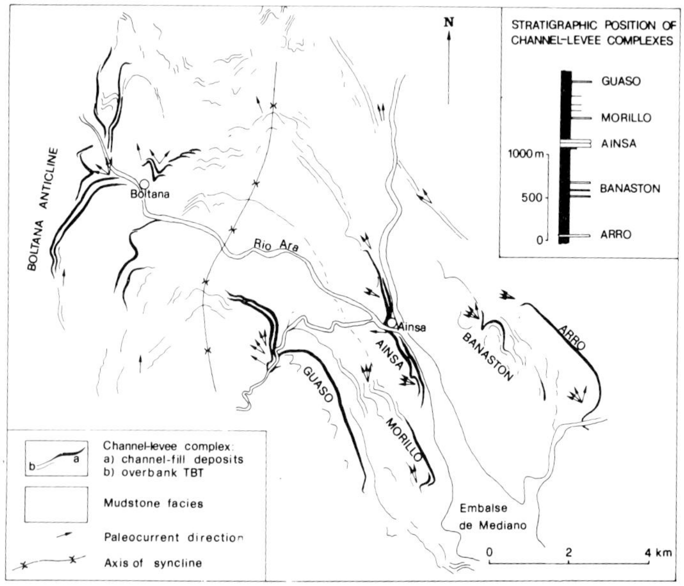

Figure 3. Geologic sketch map and paleocurrent directions of the Hecho Group channel–levee complexes in the Ainsa–Boltana region (modified after Mutti and other [15]). The individual channel–levee complexes have a thickness of 30 to 50 m, a width of 1 to 3 km, and a length of at least 7 km.

deposits are approximately 3500-m thick and can be split vertically into two main units that, in their eastern portions, are separated by a wedge of shalier deposits. The Broto lobes, the stratigraphically lower unit, form an elongate and 1000-m thick body that can be traced downcurrent to the Roncal region (Fig. 2), a distance of about 80 km. Paleocurrent directions that are consistently from the southeast together with gradual facies changes indicate that these sediments filled a long, narrow basin with a similar orientation (E. Remacha, personal communication, 1982). The overlying Jaca–Fiscal lobes have an aggregate thickness in excess of 1200 m and show a considerable spread in paleocurrent directions, as well as marked facies changes both laterally and vertically. The Jaca–Fiscal lobe complex is interpreted as a deposit made up of several coalescent lobe systems that were derived from the north and gradually deflected more westerly along the general basin axis [10]. Deflection and eventual pinching out into shale of individual lobe systems within the Jaca–Fiscal complex (details not shown in cross-section of Fig. 2) take place over distances of about 20 km.

The mudstone units that separate the different channel–levee complexes cannot be traced in outcrop across the Boltana anticline, making it difficult to establish detailed correlations between the channel–levee and the lobe deposits. However, biostratigraphic data and the fact that the bulk of the Broto lobes is younger than the Montanana Sequence indicate that most of the Broto lobes are roughly time-equivalent with the Arro, Banaston, and Ainsa channel–levee complexes. The Jaca–Fiscal lobes are correlative with the Morillo and the Guaso.

Both the channel–levee complexes and the lobe deposits onlap the shallow-marine carbonates of the Montanana Sequence on the opposite limbs of the Boltana anticline. These onlap relations are locally highly complicated both by large-scale submarine erosion prior to deposition of the Broto lobes on the western flank of the anticline and by patterns of divergent marine onlap (in the sense of Vail, Mitchum, and Thompson [12]), related to contemporaneous structural uplift, on the eastern one.

As indicated by the correlative unconformities in the easternmost part of the TSCPB, at least the Fosado and the Arro channel–levee complexes formed as a result of sudden phases of relative lowering of sea level. In the Arro complex, the lowering is also documented by the laterally equivalent Charo canyon, which is deeply incised into underlying shelf deposits. The Banaston, Ainsa, Morillo, and Guaso complexes, which are higher in the sequence, are probably related to periods of tectonic uplift during the deposition of the Sta Liestra Sequence. The correlative unconformities of these channel–levee complexes are not preserved or recognizable.

Relations Between Channels and Lobes

Channel–levee and sandstone lobe deposits occupy two geographically distinct areas within the original Hecho Group basin and lack direct vertical and lateral stratigraphic relations. The aggregate net sandstone thickness in the two different environments can be estimated as 300 to 500 m for the channel–levee complexes and 1500 to 2000 m in the lobes. The original transition, if any, between channel and lobes must have occurred in a very narrow zone across the Boltana anticline. These transitional sediments were probably removed by subsequent erosion.

Each channel–levee complex consists of intricately stacked or juxtaposed individual channel-fill facies and an abundance of associated and laterally equivalent thin-bedded, overbank turbidites (Fig. 3). With particular reference to the stratigraphically lower part of the Hecho Group sequence, the channel–levee complexes are characterized by a strong bimodality in both bed thickness and grain size. The channel–axis facies are composed of thick to very thick beds of facies A_1 and A_2 [1] conglomerate and pebbly sandstone. The margins of these channels and the laterally equivalent interchannel areas contain thin to very thin beds of very fine-grained facies D_1 and D_2 [1] sandstone. The abundance of these thin-bedded overbank deposits clearly indicates phases of prolonged channel activity associated with sand by-passing.

The sand that by-passed the channels was deposited as huge accumulations of laterally continuous and graded sandstone beds that comprise the lobe and the fan-fringe facies associations.

The lobe facies association consists of nonchannelized and laterally extensive bodies between 3- and 15-m thick that are made up of medium- to thick-bedded facies C [1] sandstone enclosed by roughly similar thicknesses of lobe-fringe thin-bedded facies D [1] turbidites. Commonly, these two types of deposits comprise thickening-upward facies cycles [2,10]. In the fan-fringe facies association, thick-bedded sandy turbidites are absent or rare and the thin-bedded turbidites form thickening-upward cycles considerably thinner than those of the lobe association. Both types of thickening-upward cycles probably reflect a process of compensation [13] related to the upbuilding of sand rather than a basinward progradation of the sandstone lobes.

Both the lower reaches of the channel–levee complexes and the most proximal sectors of the depositional sandstone lobes contain an abundance of coarse-grained and cross-stratified sandstone beds that are not describable in terms of the Bouma [14] Sequence. These beds, which are typically ungraded and relatively well sorted, form both thin, highly lenticular rippled units (facies E of Mutti and Ricci Lucchi [1]) or 10- to 30-cm thick, dune-shaped, and internally cross-stratified deposits (facies B_2 of Mutti and Ricci Lucchi [1]). These two types of beds are commonly intergradational over very short distances. In the Ainsa channel–levee complex, the B_2 beds are found within a channel–axis zone where they pinch out in a downcurrent direction in less than 10 m. In the Broto lobes, B_2 beds pass downcurrent into complete Bouma Sequence in less than 100 m. These coarse-grained and cross-stratified deposits probably formed as a result of flow expansion in the terminal zones of fan-channels and were deposited in a tractional regime by highly turbulent currents that were still transporting the bulk of their suspended sand load farther downfan. As indicated by the Broto lobes, *en-masse* deposition of Bouma A-divisions may follow immediately downcurrent from these cross-stratified sediments.

Acknowledgments

This summary is an outgrowth of several studies recently completed or still in progress as part of the Hecho Group Project (Universities of Barcelona, Montpellier, and Parma). Financial support was provided by CFP, Esso Exploration Inc., Petrobras, Shell Research B. V., and SNEA (P). Funds for the author were also provided by the CNR, Rome. Gratitude is particularly expressed to M. Seguret, J. Rosell, M. R. Estrada, E. Remacha, F. Fonnesu, G. Rampone, M. Sonnino, C. Treves, and A. Maymo. Constructive remarks from G. Allen, C. H. Nelson, T. H. Nilsen, E. G. Purdy, and P. R. Vail are also acknowledged.

References

[1] Mutti, E., and Ricci Lucchi, F., 1975. Turbidite facies and facies associations. In: E. Mutti and others (eds.), Examples of Turbidite Facies and Facies Associations from Selected Formations in the Northern Apennines. IX International Congress of Sedimentology, Nice-75, Field Trip A11, p. 21–36.

[2] Mutti, E., 1977. Distinctive thin-bedded turbidite facies and related depositional environments in the Eocene Hecho Group (south-central Pyrenees, Spain). Sedimentology, v. 24, p. 107–131.

[3] Mutti, E., 1979. Turbidites et cônes sous-marins profonds. In: P. Homewood (ed.), Sédimentation Détritique (Fluviatile, Littorale et Marine). Institut Géologique Université de Fribourg, Switzerland, p 353–419.

[4] Nilsen, T. H., 1980. Modern and ancient submarine fans: discussions of papers by R. G. Walker and W. R. Normark. American Association of Petroleum Geologists Bulletin, v. 54, p. 1094–1112.

[5] Mutti, E., Nilsen, T. H., and Ricci Lucchi, F., 1978. Outer fan depositional lobes of the Laga Formation (Upper Miocene and Lower Pliocene), East-Central Italy. In: D. J. Stanley and G. Kelling (eds.), Sedimentation in Submarine Canyons Fan and Trenches. Dowden, Hutchinson & Ross, Stroudsbourg, PA, p. 210–223.

[6] Mutti, E., and others, 1972. Schema stratigrafico e lineamenti di facies del Paleogene marino della zona centrale sudpirenaica tra Tremp (Catalogna) e Pamplona (Navarra). Memorie Società Geologica Italiana, v. 11, p. 391–416.

[7] Friend, P. F., and others, 1981. Fluvial sedimentology in the Tertiary South Pyrenean and Ebro Basins, Spain. In: T. Elliott (ed.), Field Guides to Modern and Ancient Fluvial Systems in Britain and Spain. University of Keele, p. 4.1–4.50.

[8] Seguret, M., 1972. Etude tectonique des Nappes et séries décollées de la partie Centrale du versant sud des Pyrénées. Ph. D. Thesis, University of Montpellier, Publications USTELA, Série Géologie Structurale, n. 2.

[9] Souquet, P., and others, 1977. La Chaîne Alpine des Pyrénées. Géologie alpine, v. 53, p. 193–216.

[10] Estrada, M. R., 1982. Lobulos deposicionales de la parte superior del Grupo de Hecho entre el anticlinal de Boltana y el Rio Aragon (Huesca). Ph. D. Thesis, Universidad Autonoma de Barcelona.

[11] Johns, D. R., and others, 1981. Origin of a thick redeposited carbonate bed in Eocene turbidites of the Hecho Group, south-central Pyrenees, Spain. Geology, v. 9, p. 161–164.

[12] Vail, P. R., Mitchum, Jr., R. M., and Thompson, III, S., 1977. Seismic stratigraphy and global changes of sea level, Part 3: Relative change of sea level from coastal onlap. In: C. E. Payton (ed.), Seismic Stratigraphy—Applications to Hydrocarbon Exploration. American Association of Petroleum Geologists Bulletin, Memoir 26, p. 63–81.

[13] Mutti, E., and Sonnino, M., 1981. Compensation cycles: a diagnostic feature of turbidite sandstone lobes. In: International Association of Sedimentologists; 2nd European Regional Meeting, Bologna, 1981, Abstracts, p. 120–123.

[14] Bouma, A. H., 1962. Sedimentology of Some Flysch Deposits. Elsevier, Amsterdam.

[15] Mutti, E., and others, 1981. Channel-fill and associated overbank deposits in the Eocene Hecho Group, Ainsa-Boltana region (South-Central Pyrenees). International Association of Sedimentology, 2nd European Regional Meeting, Bologna, 1981; Abstracts, p. 113–116.

CHAPTER 31

Marnoso-Arenacea Turbidite System, Italy

Franco Ricci Lucchi

Abstract

Submarine fans of different sizes, geometry, and petrology were built in the Marnoso-arenacea Basin, a migrating foredeep within an active continental margin. In an initial depositional stage, a well-developed basin plain received sediment from flows that by-passed restricted fan systems, now buried, located near the north end of an elongated basin. Minor fans grew near the steeper, tectonically deformed side of the basin. In the later stage, turbidite deposition was stopped in the former basin plain. Sediment sources and feeder channels shifted and fed fan lobes that prograded in a narrower trough and were distorted (choked). The tectonic control on development of megasequence and sand bodies is stressed here in contrast with previous emphasis on "inner" or "autocyclic" mechanisms.

Introduction

Deposits of the Marnoso-arenacea deep-sea fan system, ranging in size from tens to hundreds of meters in thickness, outcrop discontinuously in a northwest-southeast direction over a maximum length of 150 to 200 km in the northern Apennines. These strata were not deposited as a single fan system but as separate fill units within a tectonically active and migrating turbidite trough. As a result, multiple basin plains and fan bodies of different ages, locations, sizes, and composition can be recognized. One constant feature of the trough was its elongation parallel to the apenninic tectonic front. The width of the trough was narrow enough to cause axial constriction of both individual turbidite flows and resulting fan units.

Basin Setting

There are two separate types of basins in the Marnoso-arenacea deposits, or at least two distinct depositional stages and settings [1], with the transition recognized mainly in areas with a continuous record of sedimentation. The older main stage, or Inner Basin, is characterized by a single major basin with local topographic irregularities. The later or foredeep stage is split into separate minor basins located within the Apenninic-Padan-Adriatic Foretrough (Fig. 1). All of these active-margin (orogenic) basins fall in the category of foredeeps [2], that is, basins with asymmetrical profiles where tectonic influence on sedimentation is particularly strong on the "inner," steeper side (Fig. 2).

The main or Inner Basin reached a maximum length of 350 to 400 km in middle to late Miocene time. It is presently subdivided by a transversal tectonic line into a buried segment to the northwest (tectonic cover) and an outcropping segment to the southeast (Fig. 1A). The elongate basin was fed by both principal, longitudinal, and minor transverse sources, which built submarine fans of proportional sizes.

Principal sources fed fans that mainly occupy the buried portion of the basin (see [3]) and are therefore sedimentologically and geometrically poorly known. The clastic source for the fans was from the emergent Alps, i.e., from the far northern portion of an arcuate suture belt outside the still subaqueous Apenninic orogen. The apices of the principal fans switched from west to east along the foreland ramp. Although source areas changed, deep-sea fan systems persisted during both stages.

Minor fan systems were located on the orogenic flank of the basin and received their supply from a nearer, western part of the Alpine suture zone. The minor fans did not persist during basin evolution and did not coalesce or migrate.

Besides receiving sediment from multiple sources, other significant characteristics of the Inner Basin are: (1) basin-wide extent of individual turbidites, and (2) hemipelagic sed-

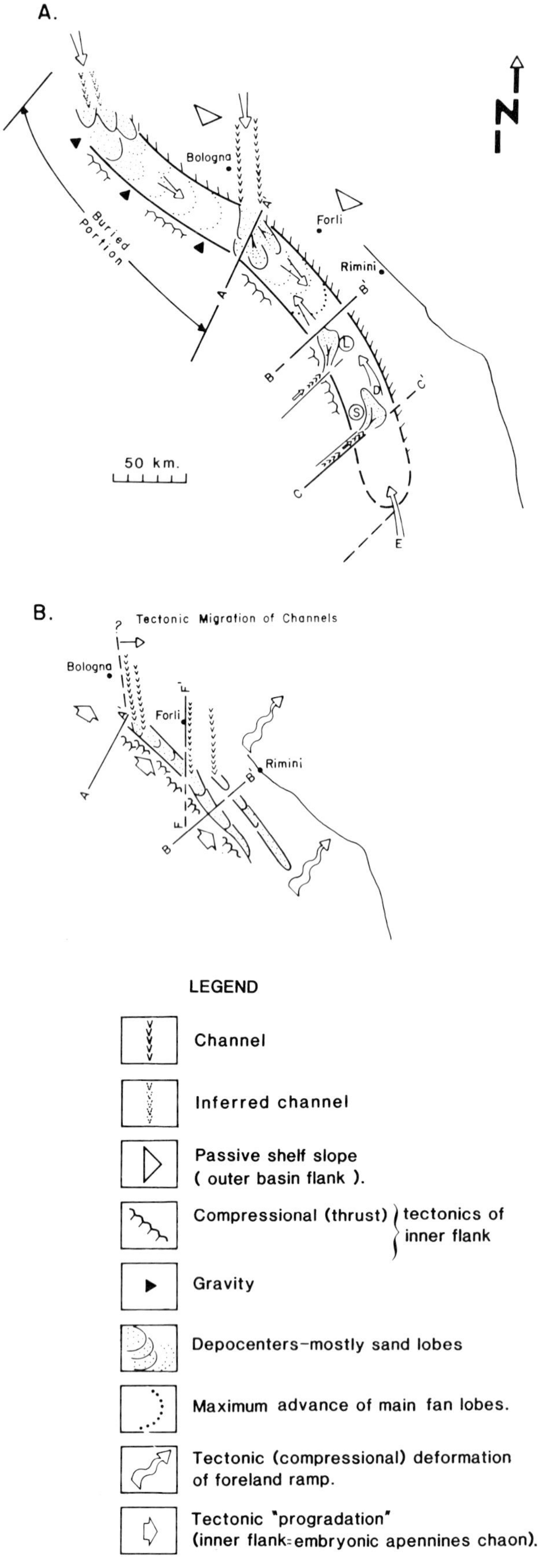

Figure 1. Depositional stages and dispersal patterns of Marnoso-arenacea. A. Main depositional stage (fan lobe and plain). B. Late depositional stage (fan fringe and lobes confined in narrow troughs; advancing channels, later abandoned). (A-A′) Sillaro tectonic line; (B-B′) Marecchia line; (C-C′) Perugia-Cagli line; (D) Contessa and some pre-Contessa flows unrelated to fan (L-S); (E) Calcareous turbidites from organic banks of Latium-Abruzzi platform; (F-F′) Forli line; Minor fans: L-Langhian, S-Serravallian.

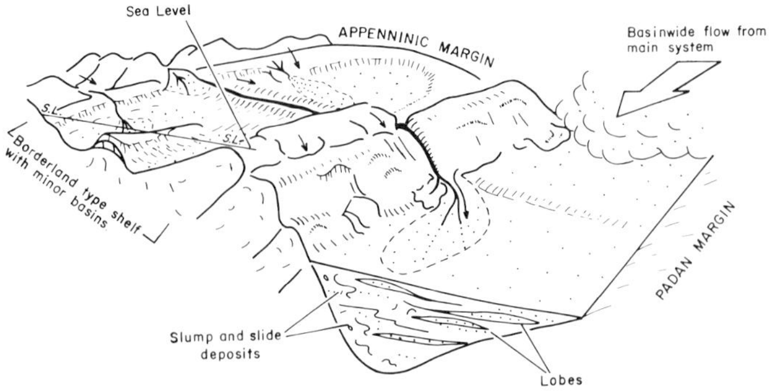

Figure 2. Schematic illustration of later stage deposition showing interaction of main basin wide flows with minor systems contributing to the main trough from borderland.

iments interbedded with turbidites. About 50% of the Marnoso-arenacea layers mantled the entire exposed basin plain (or most of it) [4] over a width of 20 to 30 km and a length of 90 to 170 km. The size and geometry of the basinal layers are independent of their composition and provenance; the first recognized examples were derived from minor sources and were used as stratigraphic markers. Prominent among these marker beds is the key layer Contessa (Figs. 1A and 3A). Hemipelagic beds a few centimeters thick (exceptionally 50–100 cm), which constitute 10% to 20% of the volume of the Marnoso-arenacea, are preserved mostly in the basin plain and slope associations.

The minor basins are basically structural channels ranging from a few kilometers to tens of kilometers long by a few kilometers wide, filled by sandy-muddy turbidite beds that, in some cases, exhibit scoured bases across their entire width. A typical basin-plain association is not recognizable except for the lowest part of the fill. The main clastic source and feeding system was from the anorogenic side of the basin while the orogenic side contributed slump, slide, olistostrome, and debris flow deposits (Fig. 1B).

Fan Model

Data from the Maronoso-arenacea formation were used extensively in the development of the early version of the Mutti and Ricci Lucchi model for fan facies development [5]. At that time, the Marnoso-arenacea was discussed in terms of a single depositional system rather than in terms of multiple sources and a two-stage development of the basin. This ini-

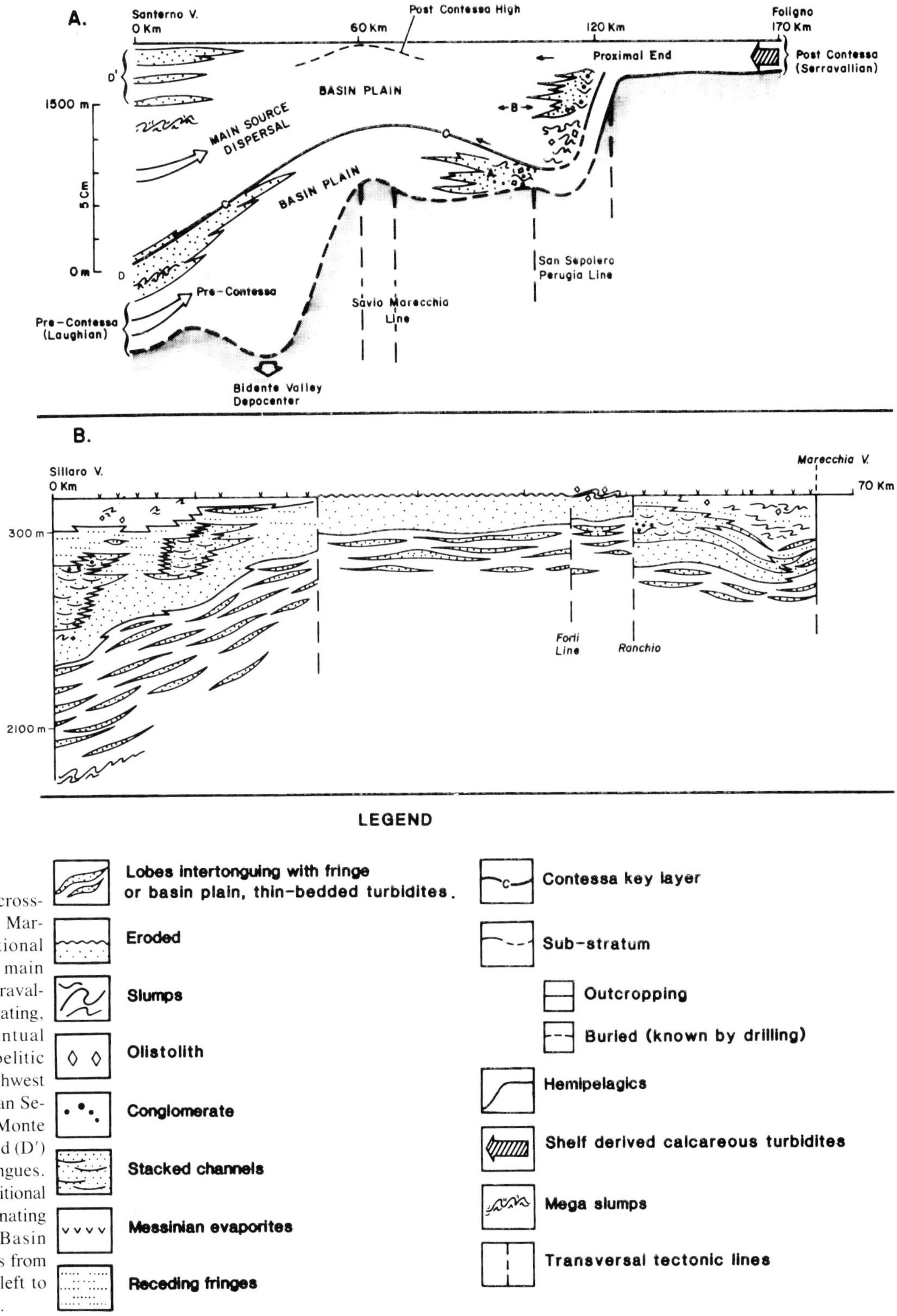

Figure 3. Longitudinal cross-sectional illustrations of Marnoso-arenacea depositional stages. A. Inner basin, main stage (Langhian and Serravallian), basin plain dominating, no vertical trend, eventual uplift with slides and pelitic draping. Basin strike northwest (left) to southeast. (A) San Sepoloro minor fan. (B) Monte Urbino minor fan. (D) and (D′) major outer fan, sand tongues. B. Foredeep (late) depositional stage, fan system dominating coarsening-up trend. Basin strike orientation changes from northwest-southeast on left to north-south at Forli line.

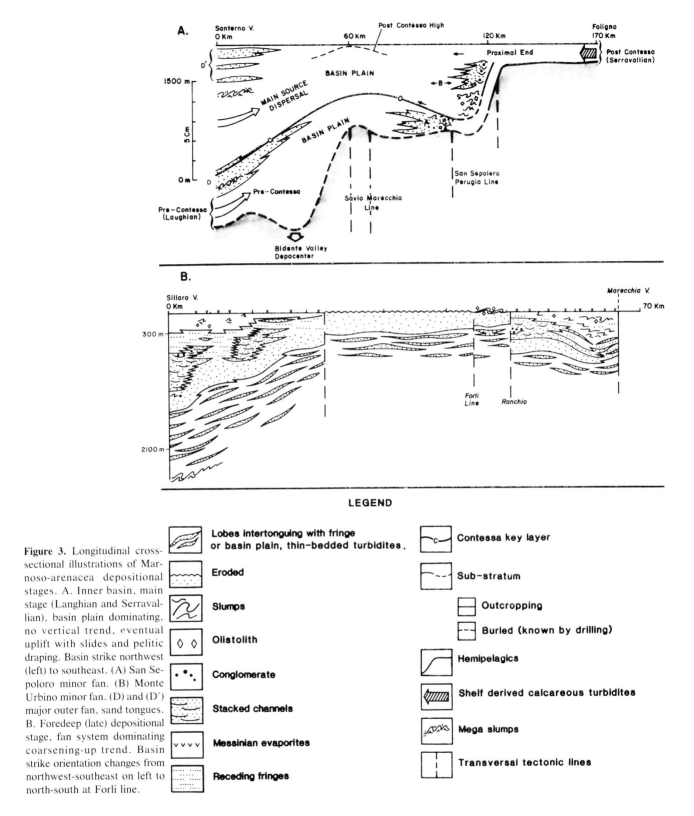

tial oversight is understandable because the geometry of channels and the inner portions of fans in the older, larger, and more persistent basin are almost unknown, whereas outer fan and basinal deposits are well exposed. In the latter, more confined stage, both channelized and nonchannelized fan sediments are prominent but the basin plain disappears or changes character substantially.

Information on Marnoso-arenacea fans is derived mostly

from outcrops and the distribution of sediments. The major recognized units of the fan systems can be summarized as follows:

1. Extensive nonchannelized sand bodies (Fig. 4) spreading out across and passing through the main system and interfingering with basin-plain deposits in the Inner Basin are interpreted as either outer fan depositional lobes [5] (equal to the high efficiency fan model of Mutti [6]) or as "special types" of basinal sand accumulation [7].
2. Nonchannelized sand bodies of undetermined length in minor troughs; same interpretation as 1, with more relief resulting from lateral confinement.
3. Channelized sand bodies with conglomerate lenses in minor basins connected with the main feeding system (Fig. 3B); interpreted as mid- and inner fan (in a three-zone model [5]) or as inner fan (in a two-zone model [8]).
4. Sand bodies of intermediate character in minor basins; possible channel lobe transition [9].
5. Minor fans of the Inner Basin that form isolated bodies with smaller channelized portions and more extensive nonchannelized bodies; size and geometry suggest an analogy with the modern Crati Fan [10,11]. However, the lobes of these minor fans are coarser grained than the Crati and would better fit a low-efficiency fan model [6,12] or in between high- and low-efficiency models and with a connection to a basin plain.

Until 1975, Marnoso-arenacea sand bodies were seen as elements of a depositional system dominated by its internal dynamics ("self-control" with uniform subsidence), with little consideration of any external geologic factors. More recently, evidence from detailed geometric analysis of these bodies [4,13,14] (Fig. 5), together with comparisons with a modern fan in a similar geodynamic setting [10,11] has led to a reevaluation of external controls, particularly tectonic activity.

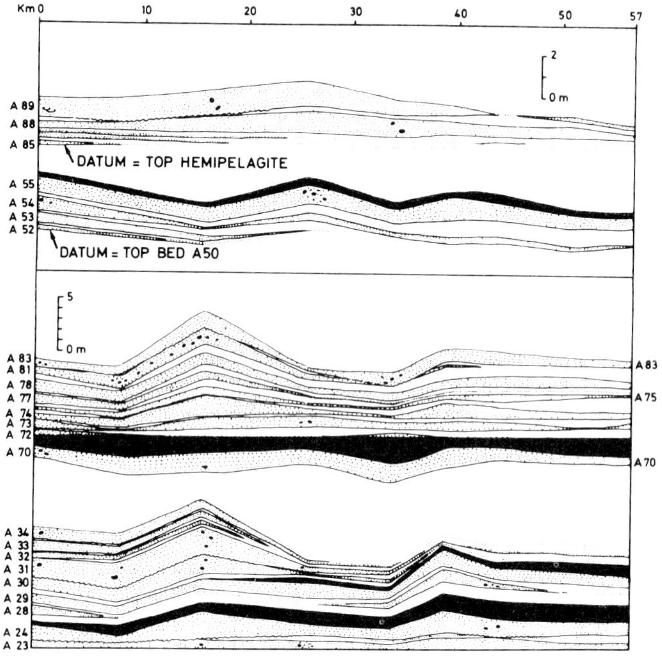

Figure 4. Longitudinal cross-sections of pre-Contessa sandstone bodies. Adapted from [14].

Fan Divisions

Channelized Sandstone Bodies in Foredeep Basins (Figs. 1B, 3B)

Stacked, channelized bodies reach up to 600 m in thickness; individual channel fills, recognizable in a few cases, are 10 to 50 m thick and 500 to 1500 m wide. The length cannot be determined for lack of downdip continuity of the outcrops. The stacked bodies are incised into nonchannelized ("lobe") sandstones. Large single beds are completely amalgamated (lack of abandonment and overbank facies) in the lower part of the sequence; they interfinger with or are interbedded with pelitic-sandy interchannel sediments in the upper part (Fig. 3B). On the basis of clast provenance and paleocurrents, a transport distance of about 100 km is estimated from the shoreline or inner shelf to the base of the slope.

The best channel exposures are ones that cut at right or oblique angles to the channel axes. They never display steep flanks or obvious levees and have depth-to-width ratios of 1:10 to 1:30. The channel fill is mostly aggradational thick-bedded sand, more evenly bedded toward the margins and more irregular and lenticular in some axial areas.

The above features might indicate that these deposits are from the terminal reaches of linear channels, probably aligned along discrete, shifting, or switching fractures and faults. Interpreted as distributaries of a midfan area in 1972 [5], they are now preferably seen, at least in part, as independent parallel channels organized into two main sets [10,11].

Nonchannelized Sandstone Bodies of the Main Basin (Figs. 1A, 3A, 4)

Nonchannelized sandstone bodies likewise vary in relationship to their basin location. In the Inner Basin, longitudinal sections of several bodies have been studied bed by bed. These have thicknesses of 3 to 10 m, unknown widths, lengths of 20 to 50 km, and are asymmetrical in shape. Grain size is mostly medium to fine sand (often present in the same bed: facies B and C, see Table 1). The sand bodies are vertically separated by fairly sharp contacts from under- and overlying basin-plain deposits (muddier turbidites, plus hemipelagites: facies association D_2, D_3, G). Typical "fan fringe" thin-bed-

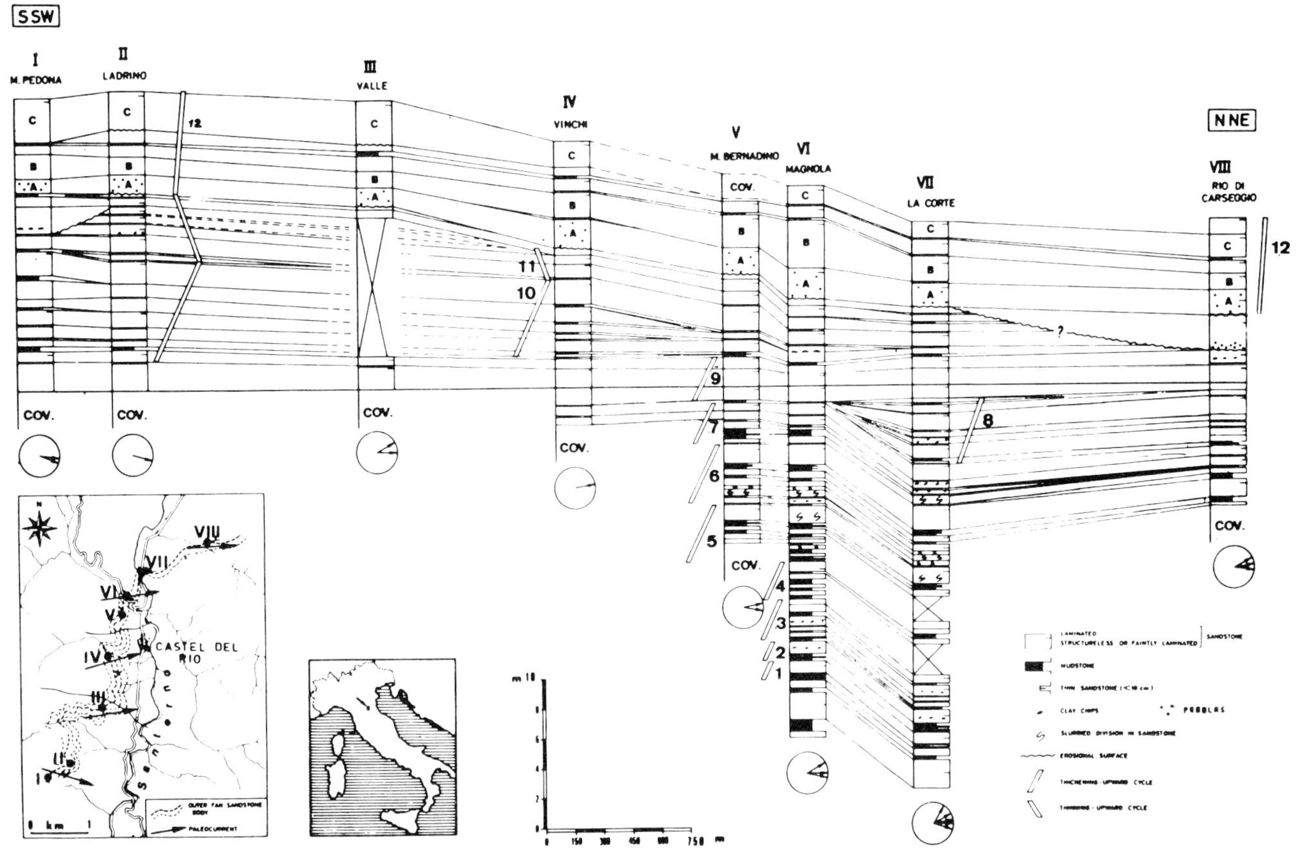

Figure 5. Detailed correlations of beds transversal to paleocurrent direction in a complex Foredeep lobe. From [13].

ded turbidites (thin-bedded and relatively sandy facies D_1) are absent or scarce. The upcurrent ends of sand bodies do not outcrop; consequently, the transition from the channels cannot be checked. Downcurrent, they merge into basin-plain deposits or abut against submarine highs (tilted segments of basin plain).

Nonchannelized bodies can be defined as simple or complex, on the basis of internal organization, and vertically ordered in terms of grain size and bed thickness (megasequences) or nonsequential. Changes occur from up- to downflow end [14]. A thickening-upward sequence, matched by an offlapping pattern, characterizes the progradational mode of growth, while the lack of a well-defined vertical order typifies the aggradational mode. The former growth pattern points to a mechanism of frontal accretion, while the latter may reflect lateral and vertical growth in the proximal part of a lobe and the occupation of a topographic depression (created by previous deposition plus compaction), or tectonic subsidence. Fining-upward megasequences statistically recorded by the author [7] should be included in the aggradational type.

Compared with suprafan lobes [15,16,17], the fan bodies of the main basin have conspicuously large sizes, in both relative and absolute terms. They seem more compatible with a high-efficiency dispersal system with extensive low-gradient fans emerging into vast basin plains.

Nonchannelized Sand Bodies of Foredeep

In the minor basins of the foredeep cross-current correlation is possible [9], but longitudinal sections are not available. Therefore, the width of the sand bodies can be estimated but their length remains undetermined. Vertically, sand bodies stand out in the lower fill, where they alternate with intervals of thinner bedded and muddier turbidites. The muddy turbidites thin and eventually disappear upwards and are overlain by an amalgamated sequence of stacked sand bodies. The thickness of individual bodies is 2 to 10 m (simple) or 7 to 45 m (composite). Widths in excess of 6.2 km have been determined in "minor" bodies (1.5–7.5 m thick) aggregated in a composite unit 45 m thick, that shows an overall thickening- and coarsening-upward trend. The extrapolated width of the composite "lobe" is 18 to 30 km [13].

The anatomy of a composite "lobe" can be determined (Fig. 5). The upper part of a "lobe" is formed by massive, amalgamated sandstone beds, locally pebbly, whose erosional bases cut through as much as 3 m of underlying de-

Table 1. Facies Associations and Sequences for Varied Fan-Systems in Marnoso-Arenacea Basin

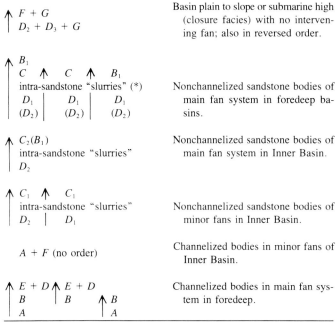

$\uparrow$ $F + G$ $D_2 + D_3 + G$	Basin plain to slope or submarine high (closure facies) with no intervening fan; also in reversed order.
$\uparrow$ B_1 C $\uparrow$ C $\uparrow$ B_1 intra-sandstone "slurries" (*) D_1 | D_1 | D_1 (D_2) | (D_2) | (D_2)	Nonchannelized sandstone bodies of main fan system in foredeep basins.
$\uparrow$ $C_2(B_1)$ intra-sandstone "slurries" D_2	Nonchannelized sandstone bodies of main fan system in Inner Basin.
$\uparrow$ C_1 $\uparrow$ C_1 intra-sandstone "slurries" D_2 | D_1	Nonchannelized sandstone bodies of minor fans in Inner Basin.
$A + F$ (no order)	Channelized bodies in minor fans of Inner Basin.
$\uparrow$ $E + D$ $\uparrow$ $E + D$ B | B $\uparrow$ B A | A	Channelized bodies in main fan system in foredeep.

*Facies unlabelled in Mutti and Ricci Lucchi classification, and consisting of a medium to thick-bedded sandstone-mudstone couplet with a chaotic division sandwiched between base and top (both flat) of the sandstone. The chaotic or "slurried" level is made of muddy sandstone, sandy mudstone or both, and shows a "whirly" or simply disordered structure with a variable amount of soft-cohesive inclusions (clay chips, silt-sandstone pillows, etc.). These intrabed chaotic deposits should be distinguished from unspecified chaotic bodies (meant as independent events) because they clearly occur within a turbidite. Their variety suggests different or combined origins [6].

posits. Most erosional surfaces are impressively flat and continuous, though punctuated by local irregular scours; they are regarded as the product of sheet flow erosion by single turbidity currents. Some others are more localized and possibly represent the shallow termini of midfan channels. This interpretation is significant in suggesting contiguity of lobes and channels, but it cannot be proven clearly because no major channels appear in the sequence.

Lateral thickness changes of individual massive beds are more apparent than in longitudinal sections of the Inner Basin lobes (5–100 cm km^{-1}). Successive beds tend to compensate mutually (i.e., the depoaxis of each bed is shifted from that of the previous one) resulting in an almost tabular body that gradually thins toward the north-northeast. Clearly, the topography created by the deposition of one bed influenced the deposition of the next.

Underlying the massive sandstones, medium-bedded turbidites with intrabed "slurries" are common. The abundant mud clasts are probably related to erosional surfaces described above: They were either lifted by eddies in turbidity currents or trapped within an accessory debris flow by a viscous flow within an overlying current flowing down the front or sides of lobes (see Table 1). Thin-bedded turbidites vertically associated within sandy sections are interpreted as lobe fringe or sandy basin-plain deposits. They belong to D_1 (dominant) and D_2 facies [8], lack a hemipelagic cover, and form successions up to 60 m thick punctuated by minor, thickening-upward or symmetrical cycles.

Minor Fans of the Inner Basin

In the minor Inner Basin fans (Fig. 1A), the thickness of the "lobe" bodies is more or less the same as in minor basins; the width is 4 to 8 km, and the length is 10 to 20 km. Fringe deposits are well developed in one case (Perugia-M. Urbino Fan), rare in the other (Sansepolcro Monte S. M. Tiberina Fan).

Relative proportions of channelized and nonchannelized sandstones vary from 1:2 to 1:3 in both large and small systems. As mentioned, no lateral physical relations can be documented between individual channelized and nonchannelized bodies. Vertically, channel sandstones occur on top of "lobe" sandstones. The contact is a large-scale erosional feature marking the acme of the regressive trend of the basin fill. It can be explained either as the aggregate product of lateral, time-transgressive migration of channels (Santerno-Sintria area, Fig. 3B) or as a broad, possibly simultaneous scour of a structural depression (up to 8 km wide in Savio Valley).

"Ambiguous" Sandstone Bodies

Tabular sandstone bodies can be observed in physical continuity (lateral or frontal) with channelized sandstones in the main system (Fig. 1B). These show well-defined base and top (sharp or rapidly transitional), fining-up or no trend internally, and dominant B_1 sandy facies (found in both channel fill and lobe top in more obvious cases). What is missing is the geometric evidence of channels, either stationary or migratory. These bodies are tentatively interpreted as channel-lobe transitions by assuming an attachment between lobes and channels [9].

Typical (Mega) Sequences

In terms of the Mutti-Ricci Lucchi facies scheme, the modal "themes" for the various parts of the Marnoso-arenacea are summarized in Table 1.

Sediment Dispersal Patterns

Transversal paleocurrent directions can be measured in a few outcrops of channelized bodies and lobes. The overall pattern of both the large and smaller basins is dominantly longitudinal, although transversal perturbations of axially directed flows have been recently reported [18,19]. This fact has several possible explanations: (1) reduced outcrop area of inner fans (sensu [8]) or (2) axial deflection and lateral confinement of both outer fan bodies and basinal turbidites resulting from the small basin width and area in comparison to the size of larger flows. The basic paleocurrent pattern shows longitudinal transport *after* introduction into the basin from entry points located both near one end and along sides. Moreover, the longitudinal pattern was bimodal in the Inner Basin but unimodal in the later trough, i.e., it was first multisourced and dominated by basin-plain deposits, then single-sourced and fan-dominated. The narrow basin-plain setting caused flow diversion toward either or both ends of the elongated basin depending on the location of the entry points and the orientation of the feeding channels.

Inner Basin

In the Inner Basin (Fig. 3A), an aggraditional, vertically monotonous succession more than 3000 m spans over 4.5 m.y. It can be split into two parts by a pronounced episode of fan growth and recession, with the start of the recessional stage marked by the key layer Contessa. The closure of the basin is tectonic, with draping and "slide" mudstones on top of turbidites. These closure facies are preserved rarely because of subsequent subaerial erosion. They point out, however, the absence of a coarse, progradational fill. Turbidite sedimentation was stopped by rising topographic barriers.

Foredeep Basins

In the foredeep basins (Fig. 3B), the trend is progradational (marginal fan facies replacing basinal ones) for the lower three-fourths of measured sections, then suddenly recessional. Typical basin-plain deposits are present only at the base.

Bounding Depositional Systems

Marginal sediments closer to the main source (the southern border of the Alps) are only locally preserved as fan-delta conglomerates or outer shelf to slope mudstones [3]. Marginal systems were involved in widespread subaerial erosion during the Messinian salinity crisis. The inner, Apenninic or orogenic margin (SW) provided limited land areas and probably had several submarine sills that trapped sediment in small shelf-slope basins upslope of the main turbidite trough of the Marnoso-arenacea. Overflow from silled basins occurred but was limited both in time (intermittent, with recurrence time in the order of 10^4–10^5 yr) and space. Marginal basins and the central turbidite basin were thus short circuited by tectonics and gravity flows. Not only are shelf biogenic limestones and clastics found in marginal basins but also in the deep-sea fan and basin-plain turbidites. These basins were built on thrust sheets, and their fills were displaced by tectonic remobilization of the sheets toward the migrating larger trough (position defined as semiallochthonous or "piggy back"). Remnants of marginal sequences are thus found in tectonic superposition on the main turbidite body. When subsidence shifted to the foredeep, emerging ridges in the orogen shed detritus to the various subbasins, by-passing the "dead" Inner Basin.

At the southeast or distal end, the main basin was fault-bounded. It was initially delimited by submarine ridges with hemipelagic deposition, but later deposition extended to the toe of a carbonate platform with rich benthic life and no terrigenous "pollution." After emplacement of the Contessa flow, this platform supplied more than 25 carbonate turbidite flows ("colombine"), one-third of which were of basinal extent. Influx of shelf detritus continued in the foredeep stage but could not reach the ends of the Marnoso-arenacea minor basins because they were captured by the opening Laga Basin [4,9].

The major depositional events of the Marnoso-arenacea, including opening and closing of the basins, correlate poorly with the global cycles of Vail [20]. The regional tectonic control was apparently dominant. One remarkable exception is the already quoted regressive-transgressive fluctuation in the Inner Basin centered on the Contessa marker (Langhian-Searravallian marker boundry). A widespread transgressive phase, marked by typical biogenic facies, characterizes various Mediterranean domains between 14 and 16 m.y. As for the foredeep, the receding trend of the fan system after the progradational acme in Tortonian time might correlate with the base of the short "Sahelian" sedimentary cycle (late Tortonian to early Messinian), which is developed in shallow-water areas resulting from relaxing after emplacement of nappes.

References

[1] Ricci Lucchi, F., 1975. Miocene paleogeography and basin analysis in Periadriatic Appennines. In: C. Squyres (ed.), Geology of Italy, v. 2, Castelfranco Veneto, Italy, pp. 129–236.

[2] Bally, A. W., and Snelson, S., 1980. Realms of subsidence. In: Memoir 6, Canadian Society of Petroleum Geologists, pp. 9–94.

[3] Cremonini, G., and Ricci Lucchi, F., 1982. Guida alla geologia del margine appenninico-padano. Societa Geologica Italiana, Guide Geologiche Regionali, Bologna, 248 pp.

[4] Ricci Lucchi, F., 1978. Turbidite dispersal in a Miocene deep-sea plain: the Marnoso-arenacea of the northern Apennines. Geologie en Mijnbouw, v. 57, pp. 559–576.

[5] Mutti, E., and Ricci Lucchi, F., 1972. Le torbiditi dell'Appennino settentrionale: introduzione all'analisi di facies. Societa Geologica Italiana Memorie, v. 11, pp. 161–199.

[6] Mutti, E., 1979. Turbidites et cônes sous-marins profonds. In: P. Homewood (ed.), Sédimentation détritique (fluviatile, littorale et marine), v. 1, Institut Géologique Université de Fribourg, Switzerland, pp. 353–419.

[7] Ricci Lucchi, F., 1975. Depositional cycles in two turbidite formations of northern Apennines. Journal of Sedimentary Petrology, v. 45, pp. 1–43.

[8] Mutti, E., and Ricci Lucchi, F., 1975. Turbidite facies and facies associations. In: Mutti, E., and others. Examples of Turbidite Facies and Facies Associations from Selected Formations of Northern Apennines. International Association of Sedimentologists, IX International Congress, Nice, Excursion Guidebook A-11, pp. 21–36.

[9] Ricci Lucchi, F., 1981. The Marnoso-arenacea: A migrating turbidite basin "oversupplied" by a highly efficient dispersal system. International Association of Sedimentologists, 2nd European Meeting, Bologna, Excursion Guidebook, pp. 231–275.

[10] Crati Group, 1981. The Crate Submarine Fan, Ionian Sea. A preliminary report. International Association of Sedimentologists, 2nd European Regional Meeting, Bologna. Abstracts, pp. 34–39.

[11] Ricci Lucchi, F., and others, 1983/84. The Crati Submarine Fan, Ionian Sea. Geo-Marine Letters, v. 3, pp. 203–210.

[12] Cazzola, C., and others, 1981. Geometry and facies of small, fault controlled deep-sea fan systems in a transgressive depositional setting (Tertiary Piedmont Basin, northwestern Italy). In: International Association of Sedimentologists, 2nd European Regional Meeting, Bologna, Excursion Guidebook, pp. 7–53.

[13] Ricci Lucchi, F., and Pignone, R., 1979. Ricostruzione geometrica parziale di un lobo di conoide sottomarina. Societa Geologica Italiana Memorie, v. 18, pp. 125–133.

[14] Ricci Lucchi, F., and Valmori, E., 1980. Basin-wide turbidites in a Miocene, oversupplied deep-sea plain: a geometrical analysis. Sedimentology, v. 27, pp. 241–270.

[15] Normark, W. R., 1970. Growth patterns of deep-sea fans. American Association of Petroleum Geologists, v. 54, pp. 2170–2195.

[16] Normark, W. R., 1978. Fan valleys, channels and depositional lobes on modern submarine fans; characters for recognition of sandy turbidite environments. American Association of Petroleum Geologists, v. 54, pp. 912–931.

[17] Normark, W. R., Piper, D.J.W., and Hess, G.R. 1979. Distributary channels and lobes, and mesotopography of Navy Submarine Fan, California Borderland, with applications to ancient fan sediments. Sedimentology, v. 26, pp. 749–774.

[18] Ellis, D., 1982. Palaeohydrodynamic and computer simulation of turbidites in the Marnoso-arenacea, northern Apennines, Italy. Unpublished Ph.D. Thesis, University of St. Andrews, 135 pp.

[19] Statera, I., and Ricci Lucchi, F., 1982. Caratteri sedimentologici e dispersione di alcune torbiditi di estensione bacinale nella Formazione Marnoso-arenacea. Mineralogica et Petrologica Acta, Bologna, v. 25, pp. 57–77.

[20] Vail, P. R., Mitchum, Jr., R. M., and Thomspson, III, S., 1977. Global cycles of relative change of sea level. In: C. E. Payton (ed.), Seismic Stratigraphy Applications to Hydrocarbon Exploration. American Association of Petroleum Geologists, Memoir 26, pp. 83–98.

CHAPTER 32

Peira-Cava Turbidite System, France

Arnold H. Bouma and James M. Coleman

Abstract

The Upper Eocene turbidites in the eastern French Maritime Alps can be thought of as consisting mainly of lateral migratory channel fills and overbank deposits in a middle and lower submarine fan setting. This interpretation is primarily based on: 1) the lenticular shapes of shale pebble nests and their lateral continuation as scattered pebbles along internal amalgamated contacts in the sandstone beds that are inferred to be lag deposits, 2) some large foreset bedding with preserved upper foreset and bottomset contacts overlain by finer material, 3) contacts tangential to the overall bedding within the sandstones, and 4) the variation in paleocurrent directions from sole markings on successive layers.

Introduction

The eastern part of the French Maritime Alps basin is bounded on the north by the Argentéra-Mercantour Massif and on the south by the Mediterranean Sea and that part of the French continent west of Nice [1–3]. The entire basin covered part of western Italy and extended westwardly to a NNW line between the Pelvoux and Estérel Massives. During uppermost Eocene and earliest Oligocene times, coarse-grained turbidites from three major source areas were deposited in various subbasins or depressions (Fig. 1).

The turbidites observed in the eastern part of the French Maritime Alps were deposited by density currents that were fed from a southern source area and are the subject of this review. Our interpretations draw on our experiences obtained during Deep Sea Drilling Project Leg 96 on the Mississippi Fan. Examples will be presented only from the Contes, Peira-Cava, and Tournairet areas because of our 1984 fieldwork in those locations. Emphasis is placed on the Peira-Cava area because of earlier studies [2]. The sandstones of these turbidites have been named Grès de Peira-Cava or Peira-Cava Sandstones by Bouma [2] and Grès d'Annot or Annot Sandstones by Stanley [1].

These Peira-Cava Sandstones have always been considered to be Flysch-type sediments. They consist of alternating sandstones and shales and were first interpreted as turbidites by Kuenen and others [4]. In the Peira-Cava area, they reach a total thickness of 550 to 650 m and show nearly continuous stratigraphic exposure in the several road cuts (Fig. 2).

The subbasins of the Eocene French Maritime Alps were bounded primarily by N-S trending highs of Jurassic limestones that were covered by Middle and Upper Cretaceous formations. The Upper Cretaceous formation consists either of medium-bedded, very hard and dense limestones with some irregular bedding and wedge-outs or of medium- to thick-bedded marly limestones that show some indistinct cut and fill structures. The series has been intensely folded and faulted.

Discordantly, the Upper Cretaceous formation in the Peira-Cava area is overlain by a 85-m thick brownish-colored nummulitic limestone. These limestones represent a northerly transgressive sequence over the Upper Cretaceous during early Upper Eocene. The brown limestone is overlain by blue marls (Marnes Priaboniennes); contacts normally are gradual rather than distinct. The marls range in thickness from 90 to 170 m and form badland topography. They are poorly bedded and show a barely perceivable grading in grain size which seldom exceeds 30 μ [2]. The carbonate content ranges from 54 to 60%. The bluish-gray marls gradually grade upward over about 10 m into the brownish-gray thin sandstone layers of the Peira-Cava Sandstone.

The Tertiary sediments have undergone at least three orogenic movements. The first occurred near the Middle to Upper Eocene boundary and was responsible for forming the N-S trending Jurassic ridges. The second orogenic movement took place during the Lower to Middle Oligocene, and the third occurred between the Upper Miocene

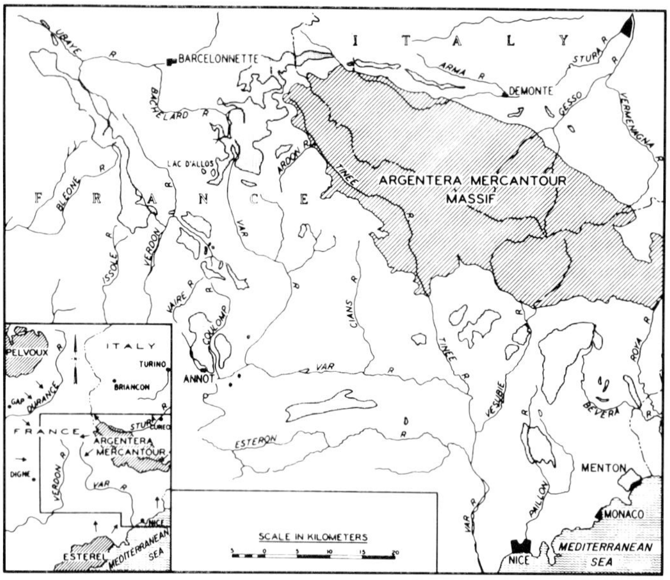

Figure 1. Map of southeastern France showing the turbidite areas (stippled patterns) and crystalline massifs (hachured pattern). Modified after [1].

and Lower Pliocene. Late Pliocene and Pleistocene uplift can be observed in the coastal area of the French Maritime Alps. Because no deposits of the late Tertiary age are found on top of the Peira-Cava Sandstone, the influence of such movements in the turbidite areas is difficult to establish.

Faure-Muret [5] established that the Mercantour Massif did not emerge prior to Oligocene time and that the petrography of the Massif rocks is not compatible with the turbidites of the Peira-Cava Sandstone. On the contrary, the petrography and paleocurrent directions indicate that the most likely source was the Estérel-Thyrenide chain, which may be extended into Corsica. The Mercantour Massive, however, formed a submerged topographic high and an effective northern boundary for the materials coming from the south.

Peira-Cava Sandstones

Although the French Maritime Alps basin covered a marine area in excess of 5000 km^2 in Upper Eocene time, only the subbasins with direct access to source material could be filled with submarine fan-type deposits. A complete reconstruction of those subbasins, their sizes and the possible interconnections is not possible because only part of the sediments have escaped later erosion. Based on earlier ideas that the entire Maritime Alps basin was one receiving area [4] and that turbidites are widespread deposits rather than deposited in smaller subbasins [2], we attempted to correlate individual layers between outcrops. Within the Peira-Cava area, such correlation of layer thickness and internal characteristics proved impossible [6]. Bouma [2] concluded that each deposit covered only a small portion of a subbasin and that the position of each subsequent deposit was governed by both the location of the source area and by the existing microtopography. In addition, each deposit would thin in lateral and downcurrent directions, resulting in more incomplete Bouma sequences.

About a decade later, the turbidite concept was enlarged to the submarine fan concept [7]. Geologists visiting the French Maritime Alps applied the Mutti and Ricci Lucchi model [8], emphasizing thinning-upward and thickening-upward sequences to identify channel fills and depositional lobes, respectively. A number of patterns can be distinguished, especially in the nearby continuous outcropping sections along the roads in the Peira-Cava area. The Mutti and Ricci Lucchi model [8], however, is not uniformly applied by different observers.

It is not the purpose of this review to introduce a new tabulation of the elements of that model, but rather to describe a number of observations that may shed a slightly different view on the mode of deposition. These new interpretations are based on similarities observed during Deep Sea Drilling Project Leg 96 on the Mississippi Fan in the Gulf of Mexico [Chapters 36 to 42]. The main comparative criteria are the recognition of channel migration with local erosion and lag deposits and of channel switching.

The Peira-Cava Sandstone layers were initially described in detail by Bouma [2] utilizing the concept of turbidity currents. A layer was defined as a sandstone bed with its overlying shale. Sometimes the contact between sandstone and its overlying shale is gradual; more often it is rather distinct. However, this contact normally is always less distinct than the contact at the bottom of the sandstone with the underlying shale.

The layers are basically parallel; the few layer terminations that could be observed take place over a very short distance. In the late 1950's, specific attention was paid to sedimentary structures, their vertical sequence (Bouma Sequence), and the position of the sandstone-overlying shale contact with regard to those sedimentary structures. Phenomena such as scouring, slumps, nests, and isolated shale and marl pebbles were described, but little attention was paid to potential process response. Later on, other investigators published more in-depth information about these phenomena [9].

Concentrations of shale and marl pebbles as well as occasional fragments of sandstone are typically arranged in semi-lenticularly shaped nests that show some asymmetry. Laterally, the pebbles commonly extend over some distance as isolated pebbles along an internal (amalgamated) contact in the sandstone. The pebble nests are either long and flat (Fig. 3) or narrow and thick (Fig. 4). This thick concentration was initially described as a large slump [2] of only local extension. Figure 4 shows two large pebble nests separated by a thin sandstone (arrow). The "massive" sandstone layer contains many internal contacts; some are

Figure 5. Sandstone bed with mega foreset bedding and lateral thinning overlain by fine-grained sandstone. The upper surface of the thinning sandstone has small ripples; the sandstone represents lateral migration of a channel fill with foresets and bottomsets. The sandstone with the mega foreset bedding is coarser and better-sorted than the normal turbidites in the section (for detail see Fig. 6). Peira-Cava area; for location see Figure 2.

direction. This is supported by the semi-lenticular shape of many of these pebble nests and their continuation and termination of isolated pebbles along an internal bedding plane within the sandstone layer. Drilling observations on the Mississippi Fan showed that channel shifting on the upper part of the lower fan (Sites 623 and 624) is common and that thin channel-fills alternate with overbank deposits.

Successive density currents moving through the channel will vary in size and intensity with time. When a major density current moves through a channel, a lateral migration can result, causing the outer channel side to be removed rapidly by rip-up scouring and slope failure. The resulting debris will be deposited immediately downstream as a lag deposit. Thus, the sandstones and their incorporated lag deposits can be considered as lateral accretion deposits.

Most layers containing pebble nests and isolated pebbles show a more general upward-fining and a more gradual sandstone-shale contact than do layers that have none or very few clasts. The latter type of deposit may fall more into the category of depositional lobes suggested by Mutti and Ricci Lucchi [8].

Conclusions

During the uppermost Eocene, the eastern part of the French Maritime Alps basin consisted of a number of subbasins, some or all of which were filled with submarine fan-type deposits.

The lack of correlation between layers within one turbidite area and, thus, between areas makes paleogeographic reconstructions rather incomplete. In general, it can be said that a few major complex channel-fills are present in the Menton and Contes areas, which likely places these deposits in the upper to middle-fan position. The Peira-Cava area then falls into the middle to lower-fan location. It is primarily characterized by lateral channel accretion and

Figure 6. Detail of blocks of sandstone on the left in Figure 5 (*) showing mega foreset bedding and curved erosional contact near the top (dashed line).

Figure 7. Thick sandstone layer with large and small dipping internal contacts that become tangential to the lower bedding plane or to internal amalgamated contacts. These "joints" are sedimentary in origin and indicate lateral sedimentation within a channel fill. Peira-Cava area; for location see Figure 2. See hammer at bottom (arrow) for scale.

overbank deposits. However, it is possible that part of the layers are sheet sands (depositional lobes) formed at the end of channels. The channel terminations probably moved upfan and downfan as well as laterally. The northern parts of the Peira-Cava and Tournairet areas are most distal, although some channel facies do occur.

A direct comparison of the drilling results from the Mississippi Fan is not justified. The Mississippi Fan was constructed on a passive margin in a large basin, and had a source area with a low sand/clay ratio. The Maritime Alps subbasins were small, their source area had a much higher sand/clay ratio, and the basin slopes probably had steeper gradients. The northern terminations of the Peira-Cava and Tournairet subbasins were dipping up onto the still submerged Argentéra-Mercantour Massif. These tectonic settings may have resulted in faster and stronger density flows that lost their momentum on the northern upslope. Several source points may have been active to supply sediment to the individual subbasins, of which some must have been connected longitudinally.

Nevertheless, it is striking that the concept of migratory channels and the frequent switching of channels in the downfan area can apparently be used on a large submarine fan in a passive margin setting as well as on small subbasins in an active margin setting.

References

[1] Stanley, D. J., 1961. Etudes sédimentologiques des Gres d'Annot et de leurs équivalents lateraux. Thesis, University de Grenoble, Editions Technip, Paris, 158 p.

[2] Bouma, A. H., 1962. Sedimentology of Some Flysch Deposits: A Graphic Approach to Facies Interpretation. Elsevier, Amsterdam, 168 p.

[3] Stanley, D. J., and Bouma, A. H., 1964. Methodology and paleogeographic interpretation of Flysch formations: a summary of studies in the Maritime Alps. In: A. H. Bouma and A. Brouwer (eds.), Turbidites. Developments in Sedimentology 3. Elsevier, Amsterdam, pp. 34–64.

[4] Kuenen, Ph. H., and others, 1957. Observations sur les Flysch des Alpes Maritimes Françaises et Italiennes. Bulletin Société Géologique de France, 6 ême Série, v. 7, pp. 11–26.

[5] Faure-Muret, A., 1955. Etudes géologiques sur le massif de l'Argentéra-Mercantour. Mémoir Carte Géologique de France, 366 p.

[6] Bouma, A. H., 1959. Some data on turbidites from the Alpes Maritimes (France). Geologie en Mijnbouw, v. 21, pp. 223–227.

[7] Normark, W. R., 1970. Growth patterns of deep-sea fans. American Association Petroleum Geologists Bulletin, v. 54, pp. 2170–2195.

[8] Mutti, E., and Ricci Lucchi, F., 1972. Le torbiditi dell'Appennino settentrionale: introducione all'analisi di facies. Memorie Societa Geologia Italiana, v. 11, pp. 161–199.

[9] Mutti, E., and Nilsen, T. H., 1981. Significance of intraformational rip-up clasts in deep-sea fan deposits. International Association of Sedimentologists, 2nd European Regional Meeting–Abstracts, Bologna, Italy, pp. 117–119.

CHAPTER 33

Torlesse Turbidite System, New Zealand

T. C. MacKinnon and D. G. Howell

Abstract

The Torlesse terrane of New Zealand is an ancient subduction complex consisting of deformed turbidite-facies rocks. These are mainly thick-bedded sandstone (facies B and C) with subordinate mudstone (facies D and E), comparable to inner- and middle-fan deposits of a submarine fan. Strata were deposited in trench-floor and trench-slope settings that received sandy sediment from slope-cutting submarine canyons. The dominance of sandstone suggests that some mudstone may have been selectively subducted. Construction of a detailed sediment dispersal model is not possible because tectonic deformation has largely destroyed original facies relationships and paleocurrent patterns.

Introduction

Complexly deformed turbidite sequences, inferred to have originated in ancient subduction complex settings, are particularly common in coastal ranges around the margins of the Pacific Ocean. The Torlesse terrane of New Zealand is such a deposit [1,2]; other well-known examples include the Franciscan assemblage of California [3], part of the Chugach terrane in Alaska [4], and the Shimanto terrane of Japan [5].

Turbidite sedimentation in ancient subduction complexes is poorly understood in comparison with that of submarine fans deposited in other settings because deformation associated with subduction has destroyed most of the original stratigraphic relationships, including basin geometry [6]. The study of modern subduction zones, mainly through the Deep Sea Drilling Project and seismic-reflection studies, has provided some constraints on basin geometry and sediment dispersal [7,8] and holds great promise for future work. But as yet, it has been difficult to integrate the modern data with the ancient, except in the most general way.

The Torlesse terrane is a good example of turbidite facies deposition in an ancient subduction complex characterized by extensive accretion. Exposures are excellent over wide areas of the terrane, particularly in the gorges and high Southern Alps of the South Island (Fig. 1). In addition, there is sufficient age control to allow reasonable inferences on the accretionary history of the terrane, a critical factor to sedimentologic interpretation.

In this paper, we describe turbidite facies of the Torlesse terrane on the South Island and possible sedimentation models. Also, the limitations of stratigraphic analysis imposed by structural deformation are evaluated.

Geologic Setting

The Carboniferous to Lower Cretaceous Torlesse terrane is the so-called basement rock for large areas of the North and South Islands of New Zealand (Fig. 1) and makes up a large portion of the Chatham Rise to the east. The extent of the Torlesse that is covered by the sea or that has been tectonically displaced is unknown but must be considerable.

Rock types within the Torlesse consist primarily of quartzofeldspathic sandstone and subordinate mudstone and conglomerate. Tectonically interleaved are sequences of volcanic rocks with associated chert and limestone. Almost all of the clastic rocks were deposited by sediment gravity flows in a deep-marine environment [2,9]. Shallow-marine and terrestrial deposits are known from a few small areas [2].

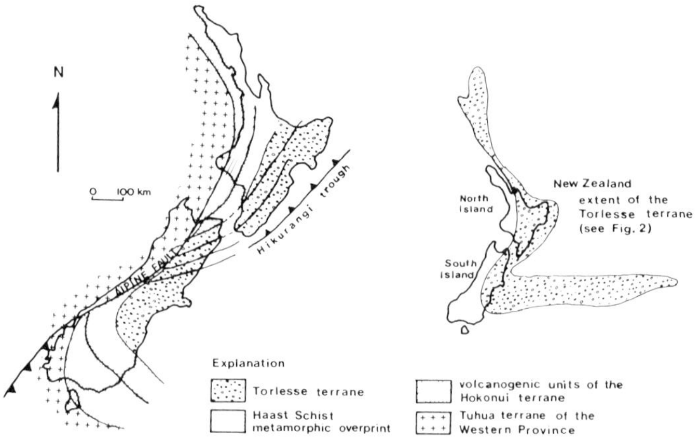

Figure 1. Distribution of Torlesse graywacke and associated strata in New Zealand and adjoining regions.

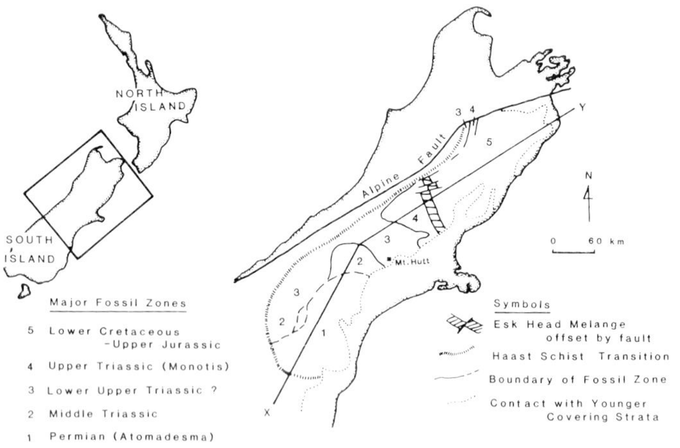

Figure 2. Distribution of biostratigraphic units within the Torlesse terrane of the South Island, New Zealand.

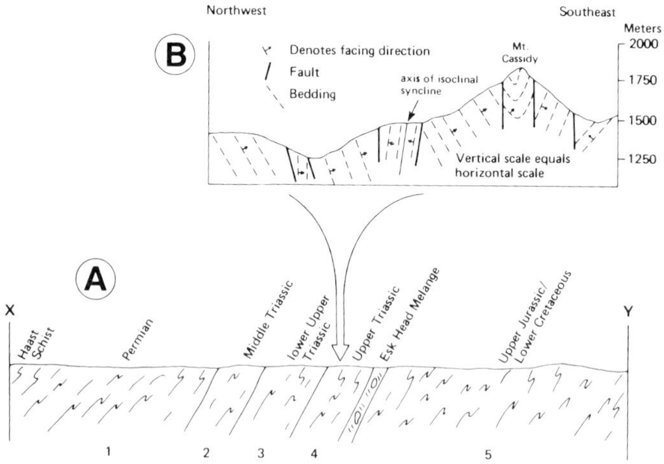

Figure 3. (A) Schematic cross section of the Torlesse terrane. See Fig. 2 for location. The stratigraphic arrangement demonstrates the accretionary outbuilding of this graywacke terrane. (B) Detailed cross section from near the crest of the central Southern Alps [10]. The faults are pre- or synmetamorphic in origin and may have formed during accretion. Strata cannot be correlated across fault boundaries.

The interpretation that the Torlesse is a subduction-related Jurassic-Lower Cretaceous deposit rests largely on structural and metamorphic features. Beds are nearly everywhere steeply dipping, and several periods of deformation are recognized. The first phase of deformation apparently began shortly after deposition. Isoclinal, recumbent, and steeply plunging folds are common, and areas of melange and broken formation are present. Metamorphism is zeolite to prehnite-pumpellyite grade in the Torlesse and increases to greenschist facies in the west, where rocks of Torlesse parentage are known as the Haast Schist. Despite pervasive deformation, much of the Torlesse appears fairly unaltered, and sedimentary structures are generally well preserved.

Distribution of fossils is a key factor in unraveling accretionary history. Fossil control in the Torlesse indicates five mutually exclusive stratigraphic zones: Permian, Middle Triassic, lower Upper Triassic, Upper Triassic, and Jurassic-Lower Cretaceous. Although somewhat structurally complicated, the zones are progressively younger toward the northeast. Systematic outbuilding of an accretionary prism may account for this sequence (Figs. 2 and 3).

The ultimate origin of the Torlesse and its relation to coeval terranes in New Zealand are complex and controversial. Immediately to the south and west of the Torlesse lies a series of volcanogenic units thought to represent an island arc-trench complex (Fig. 1). The Torlesse terrane apparently was rafted from some unknown location into the subduction zone associated with this volcanogenic complex. This collision and ultimate suturing event may account for the metamorphism represented by Haast Schist strata. On the south and west side of the volcanogenic units (Hokonui terrane of Howell [1]), lies the Western Province or Tuhua terrane, once part of the Gondwana continental mass. The Western Province and previously adjacent terranes traditionally have been considered a suitable provenance for the quartz-feldspathic strata of the Torlesse terrane; however, the presence of the intervening volcanogenic Hokonui terrane complicates reconstructions. Alternate plate-tectonic reconstructions of the Torlesse and coeval terranes in New Zealand are given by Howell [1] and MacKinnon [2].

Limitations Imposed on Fan Models by Structural Complications

Regional and detailed mapping indicate that Torlesse strata are broken into numerous steeply dipping fault-bounded blocks or slabs that may have formed during the accretionary process. Facies relationships are generally intact within individual slabs, but stratigraphic relations between adjacent slabs are unknown. Slab thickness is generally not great, perhaps less than 1 km in most cases. As an example, detailed mapping of a 6 km^2 glacially scoured area near the crest of the Southern Alps shows that maximum unbroken stratigraphic

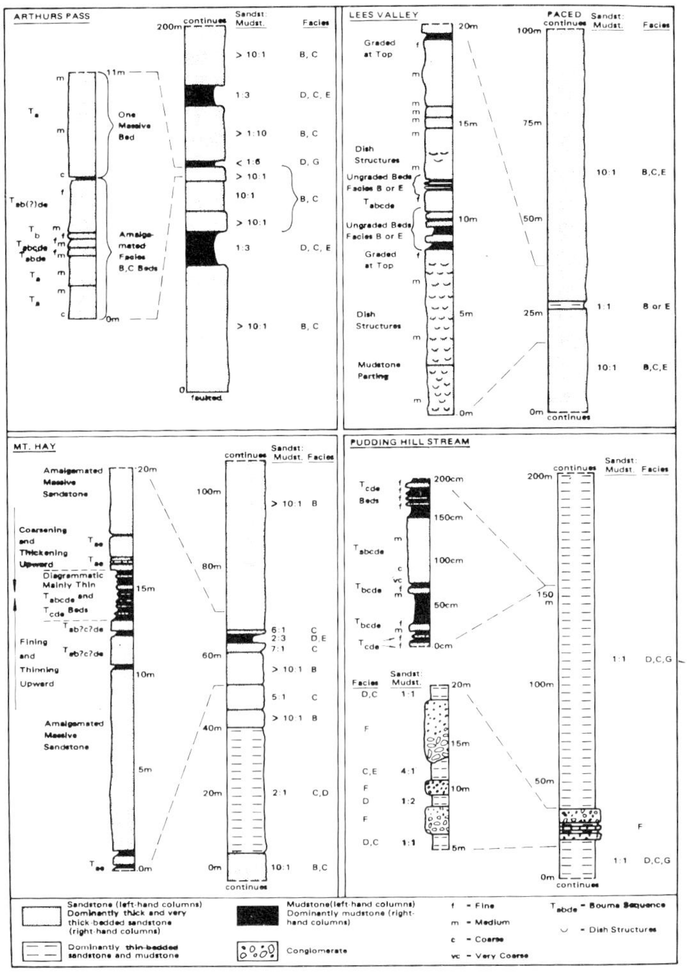

Figure 4. Selected stratigraphic columns of the Torlesse terrane showing typical lithofacies relations. Modified from MacKinnon [8].

thickness between faults is 600 m [10]. Lateral continuity is severly limited as well because beds cannot be walked out more than 200 m before they are truncated by faults or obscured by other structural complications.

The complex structure hampers further the application of fan models because it precludes regional paleocurrent analyses. Internally consistent paleocurrent data have been reported from several areas [11], but no consistent regional pattern is apparent. The reconstruction of fold history necessary for paleocurrent analyses is typically ambiguous; in many cases the presence of steeply plunging folds and areas of melange allow for more than one solution to the unfolding history.

In light of these kinds of structural complexities, what can be done to evaluate the depositional patterns of the Torlesse and other similar deposits? Diagnostic characters unaffected by structural disruption include the relative abundance of facies and their distribution in unfaulted vertical sequences (generally <1000-m thick). These characters can be documented and incorporated in dispersal models that must of necessity rely heavily on modern analogs.

Figure 5. Panoramic cross-sectional view of Torlesse Strata, Rugged Range, Southern Alps. This is a typical Torlesse sequence of noncyclic, thick-bedded sandstone (facies B and C) interbedded with bundles of mudstone (facies D and E).

Facies Associations and Fan Facies

Though submarine fan models cannot be applied blindly to subduction complex settings, facies associations within vertical sequences should be similar and have broadly similar meanings. Therefore, in this discussion the Mutti and Ricci Lucchi facies [12] are used and reference is made to fan subenvironments with their limitations in mind.

The bulk of the Torlesse consists of sandstone-dominated "packets" of facies B and C, alternating with much thinner mudstone-dominated packets described largely by facies D and E (Figs. 4, 5, and 6). The sandstone-dominated packets range from several meters to over 100-m thick whereas the mudstone-dominated packets are generally less than 10- to

Figure 6. Cross-sectional view of ~45 m of Torlesse Strata, Ohau Ski Basin, Southern Alps. Interbeds of thick-bedded sandstone (principally facies B) and mudstone (principally facies E); stratigraphic tops to the right.

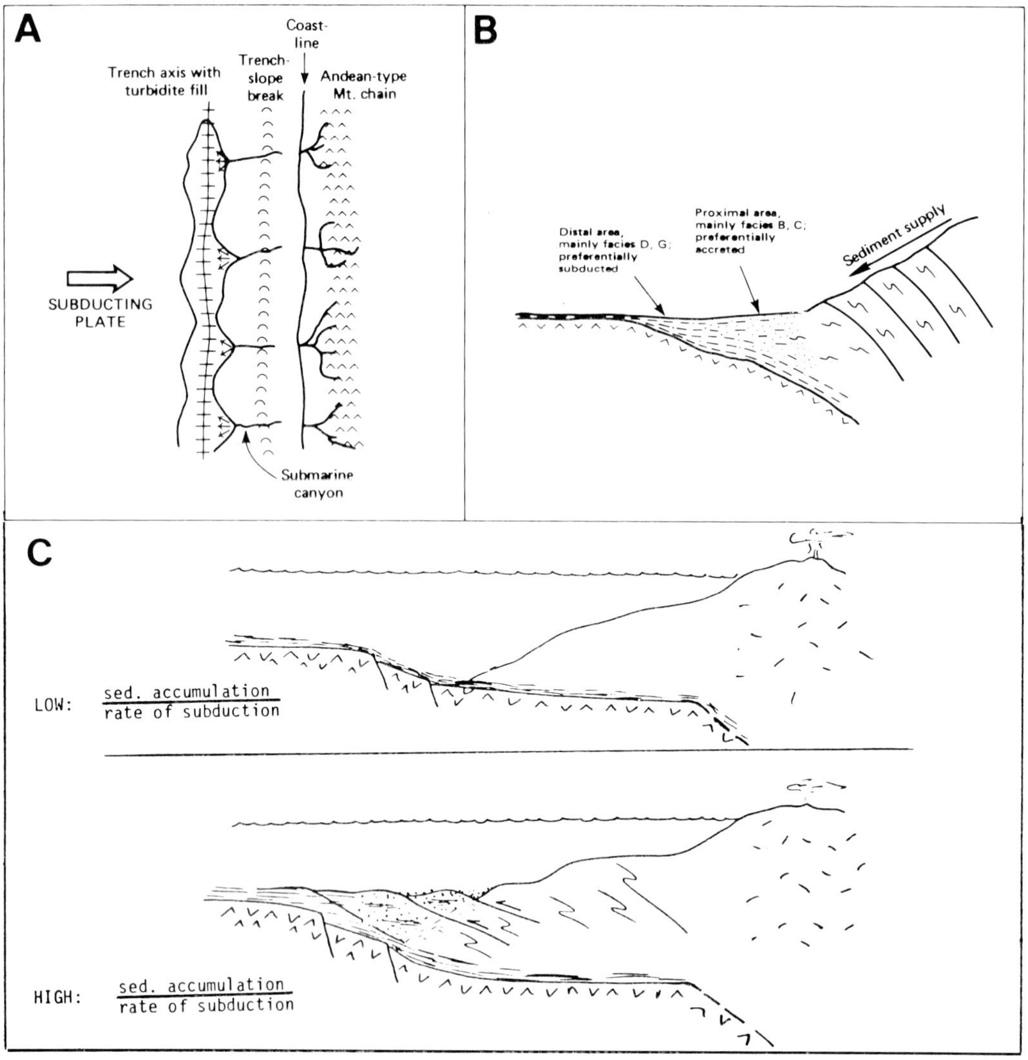

Figure 7. Possible tectonosedimentary setting for the Torlesse. Map view (A) shows Andean-type arc drained by rivers spaced at regular intervals along the coast (e.g., southern Chile today). Sediment bypasses shelf and slope areas via submarine canyons leading directly to the trench floor. Cross section (B) shows how fine-grained, thin-bedded facies deposited seaward of the trench axis might be preferentially subducted whereas sandy facies are accreted. (C) Possible tectonostratigraphic effects in a subduction complex as a consequence of varying rates of subduction vs. sediment accumulation.

20-m thick. The other facies, A, F, and G, are distinctly subordinate. The sandstone/mudstone ratio is notably high, at approximately 2.5:1.

Sandstone-dominated packets typically have a sandstone/mudstone ratio of 10:1 or greater. Sandstone beds are generally thick (>60 cm) to very thick (>120 cm). Amalgamated beds are common. Typical amalgamated layers are 0.5- to 3-m thick (Fig. 4), and composite beds may be as much as 30-m thick. Other common features include plane parallel laminations, dish structures, rip-up clasts, and load structures.

Conglomerate and coarse-grained sandstone (facies A), where present, are usually associated with facies B sandstone. The conglomerate is typically pebbly, but boulder-size clasts also occur. Most conglomerate beds are <20-m thick but may reach over 150 m in thickness.

In mudstone-dominated packets turbidite beds are either planar (facies D) or wavy to lens-shaped (facies E). The sandstone/mudstone ratio is typically 1:1 to 1:2, with sandstone beds usually less than 30-cm thick and containing T_{c-e} Bouma divisions. However, lithology of mudstone packets is highly variable, and facies F and G are commonly present also.

In terms of published submarine fan models, the typical Torlesse-facies association of B and C facies with subordinate D and E facies is inferred to be characteristic of deposition in middle-fan or suprafan areas or possibly in broad interfan channels. In most areas it is difficult to find a mudstone-dominated interval greater than 20-m thick, and those over 100-m thick are rare. Most mudstone-dominated packets can be interpreted as middle-fan associations, e.g., interchannel or interlobe deposits, or deposits in inactive areas of a broad interfan channel.

Surprising is the lack of thick sequences of thin-bedded turbidite facies representative of deposition in outer-fan basin–plain environments. The one major exception on the South Island includes most of the Lower Cretaceous strata. These strata are classic outer-fan and basin–plain deposits, mainly facies D and G with a sandstone/mudstone ratio of 1:1 or less.

Two other large areas containing predominantly mudstone are known, but these are apparently slope deposits rather than outer-fan basin-plain deposits. These are the Esk Head Melange of probable Late Jurassic age, and a roughly 50-km^2 area of lower Upper Triassic strata near Mount Hutt (Fig. 2).

The Esk Head Melange is composed primarily of disrupted mudstone containing "floating" blocks of sandstone, volcanics, limestone, and chert. Facies and structural considerations suggest that the mudstone originally consisted of facies G or F slope deposits that were subsequently tectonically disrupted.

The strata near Mount Hutt are composed primarily of facies D and C turbidites with a sandstone/shale ratio of approximately 1:1. These strata are unusual in that sandstone beds are much coarser grained than typical facies D and C beds elsewhere in the Torlesse; e.g., some beds as thin as 10-cm thick contain small pebbles and granules at their bases. Also present are rare but significant boulder-conglomerate beds up to 5-m thick containing clasts up to 1.5 m in diameter in a mudstone matrix. These debris-flow strata were most likely deposited in a slope or base-of-slope setting and may represent an area of significant sand by-pass.

In summary, turbidite facies in the Torlesse are dominated by thick to very thick bedded sandstone (faces B and C) with subordinate thin-bedded sandstone and mudstone (facies D and E). The overall sandstone/shale ratio is high at 2.5:1. The facies associations resemble those inferred for middle-fan channel and lobe deposits and perhaps deposits of broad interfan channels. Slope deposits have also been recognized in two areas. Outer-fan basin-plain deposits are rare except in Lower Cretaceous strata.

Depositional Setting

Included as depositional sites within subduction complexes are trench floors and trench slopes, which may include distinct trench-slope basins [7]. Sediment may be dispersed either by axial flow along the trench axis or by dispersal perpendicular to the axis down the trench slope. Geometry and facies distribution are dependent on sediment supply and rate of deformation, determined in part by the rate of subduction.

In the Torlesse, the high sandstone/mudstone ratio and virtual lack of fan facies other than those reflecting sandy middle-fan or inner-fan facies are most critical to any interpretation. These features seem to rule out sediment dispersal by long-distance axial flow along a trench axis. If this were the case, source proximal-distal relations for coeval strata should be apparent on a regional scale along strike (e.g., Chugach terrane in Alaska [4]), yet none are recognized in the Torlesse.

More likely, sediment entered the trench at various points along the trench axis via submarine canyons (Fig. 7). The persistent sandy nature of the Torlesse could be explained if submarine canyons delivered sediment directly from a rugged source terrane to the trench floor and perhaps trench-slope basins, effectively by-passing shelf or forearc basin areas. Furthermore, the high sandstone/mudstone ratios and the paucity of thin-bedded facies that we now see could be better explained if some of the muddy detritus traveled across the trench axis out onto the subducting plate where it might eventually be selectively subducted.

A possible modern analog is part of the Middle American Trench off Mexico as documented by Moore and others [13]. The system is characterized by a sandy and voluminous sediment supply, narrow shelf (no forearc basin), and a steep trench slope cut by submarine canyons. Sandy sediments are accumulating on the trench floor at rates calculated to be 7 to over 100 times faster than in slope and shelf areas.

It is important to note that the Torlesse and the Middle American Trench analog are not typical of modern subduction complexes. In most modern subduction complexes, including part of the Middle American Trench off Guatemala, little sediment reaches the trench floor and sediment subduction prevails (DSDP results).

This paucity of sediment in modern trenches has been cited before as an argument against trench-floor deposition for complexly deformed flysch deposits like the Torlesse that are so common in the Upper Paleozoic-Mesozoic-Lower Cenozoic rock record of the circum-Pacific region [14]. Possibly this disparity in the modern versus the ancient record is one of style and preservation. In the past, more of the Pacific margin may have been like the Middle America Trench off Mexico than it is today. Furthermore, whenever sediment supply was low, sediment subduction would likely prevail, resulting in a lack of rock record during these time periods.

One of the purposes of this paper is to outline the difficulties in determining sediment-dispersal patterns for ancient subduction complexes. Because of structural complications, interpretations of Torlesse sedimentary patterns have been based mainly on overall sandstone/mudstone ratio and facies proportions rather than their distribution, coupled with a heavy reliance on comparisons with subduction complex dispersal models based on modern examples. The possibilities of selective subduction further complicates reconstructions. A more detailed regional picture of original sediment dispersal patterns such as can be gleaned from submarine fans in passive settings (e.g., most other papers in this volume) probably cannot be obtained from deposits like the Torlesse.

References

[1] Howell, D. G., 1980. Mesozoic accretion of exotic terranes along the New Zealand segment of Gondwanaland. Geology, v. 8, pp. 487–491.

[2] MacKinnon, T. C., 1983. Origin of the Torlesse terrane and coeval rocks, South Island, New Zealand. Geological Society of America Bulletin, v. 94, pp. 967–985.

[3] Blake, M. C., and Jones, D. L., 1974. Origin of Franciscan melange in Northern California. In: R. H. Dott and R. H. Shaver, (eds.), Modern and Ancient Geosynclinal Sedimentation. Society of Economic Paleontologists and Mineralogists Special Publication 19, pp. 345–358.

[4] Nilsen, T. H., and Zuffa, G. G. 1982. The Chugach terrane—A Cretaceous trench-fill deposit, southern Alaska. In: J. K. Leggett, (ed.), Trench-Forearc Geology; Sedimentation and Tectonics in Modern and Ancient Active Plate Margins. Geologic Society of London Special Publication 10, pp. 263–327.

[5] Taira, A., Okeda, H., Whitaker, J. H. McD., and Smith, A. J., 1982. The Shimanto Belt of Japan: Cretaceous–lower Miocene active margin sedimentation. In: J. K. Leggett (ed.), Trench-Forearc Geology; Sedimentation and Tectonics in Modern and Ancient Active Plate Margins. Geologic Society of London Special Publication 10, pp. 5–26.

[6] Scholle, D. W., von Huene, R., Vallier, T. L., and Howell, D. G., 1980. Sedimentary masses and concepts about tectonic processes at underthrust ocean margins. Geology, v. 8, pp. 564–568.

[7] Underwood, M. G., and Bachman, S. B., 1982. Sedimentary facies associations within subduction complexes. In: J. K. Leggett (ed.), Trench-Forearc Geology; Sedimentation and Tectonics in Modern and Ancient Active Plate Margins. Geologic Society of London Special Publication 10, pp. 537–550.

[8] Schweller, W. J., and Kulm, L. D., 1978. Depositional patterns and channelized sedimentation in active eastern Pacific trenches. In: D. J. Stanley and G. Kelling (eds.), Sedimentation in Submarine Canyons, Fans, and Trenches. Dowden, Hutchinson, and Ross, Stroudsberg, PA, pp. 311–325.

[9] Howell, D. G., 1981. Submarine fan facies in the Torlesse terrane, New Zealand. Journal of the Royal Society of New Zealand, v. 11, pp. 113–122.

[10] MacKinnon, T. C., 1980. Geology of *Monotis*-bearing Torlesse rocks in Temple Basin near Arthurs Pass, South Island, New Zealand. New Zealand Journal of Geology and Geophysics, v. 28, pp. 68–81.

[11] MacKinnon, T. C., 1980. Sedimentologic, petrographic, and tectonic aspects of Torlesse and related rocks, South Island, New Zealand. Ph.D. thesis, University of Otago, Dunedin.

[12] Mutti, E., and Ricci Lucchi, F., 1972. Le torbiditi dell'Apennino settentrionale: introduzione all'analisi di facies. Society Geologia Italiana Memorie, v. 11, pp. 161–199. (Translated by Tor Nilsen into English, 1978, Turbidites of the Northern Apennines, Introduction to facies analysis. International Geology Review, v. 20, pp. 125–166.)

[13] Moore, J. C., Watkins, J. S., and others, 1982. Facies belts of the Middle America Trench and forearc region, southern Mexico: results from Leg 66 DSDP. In: J. K. Leggett (ed.), Trench-Forearc Geology; Sedimentation and Tectonics in Modern and Ancient Active Plate Margins. Geologic Society of London Special Publication 10, pp. 537–550.

[14] Scholl, D. W., and Marlow, M. S., 1974. Sedimentary sequences in modern Pacific trenches and the deformed circum-Pacific eugeosyncline. In: R. H. Dott and R. H. Shaver, (eds.), Modern and Ancient Geosynclinal Sedimentation. Society of Economic Paleontologists and Mineralogists Special Publication 19, pp. 193–211.

V

Ancient Turbidite Systems

Passive Margin Setting

CHAPTER 34

Brae Oilfield Turbidite System, North Sea

Dorrik A. V. Stow

Abstract

The Brae oilfield reservoir in the North Sea comprises Upper Jurassic resedimented conglomerates and sandstones interbedded with organic-rich silstone and mudstone thin-bedded turbidites. The system represents a series of small overlapping fans that form a thick (300 m) slope-apron accumulation of sediments deposited in a narrow (<10 km wide) belt along an active fault zone. The complex lateral and vertical distribution of facies was due mainly to variable tectonic activity, and partly also to sediment supply and sea-level changes.

Introduction

The Brae oilfield in the North Sea was discovered by the Pan Ocean Group in 1975. Since the discovery well, 14 more wells have been drilled on License Block 16/7a and over 2500 m of core have been recovered. The field is structurally complex and can be divided into northern, central, and southern parts. Together these make up the Brae slope-apron system. The operator and her partners have released a significant amount of structural and sedimentological data in papers by Harms and others [1] and Stow and others [2]. These data, including core descriptions and analysis, electric logs, dipmeter logs, and seismic profiles, form the basis of this contribution. It should be noted that the former interpretation [1] as partly subaerial fan deltas differs significantly from the latter [2] suggesting overlapping submarine fans that formed a slope-apron system along a submarine fault scarp.

Geological Setting

The Brae slope-apron was developed on the western margin of the South Viking Graben adjacent to a major fault escarpment at the edge of the Fladen Ground Spur (Figure 1). The Graben is part of an elongate central rift system in the North Sea Basin that was initiated in the Late Triassic and has since been periodically reactivated as a major depocenter [3]. An intense phase of tectonic activity in the Late Jurassic caused uplift of the Fladen Ground Spur along a series of north to northeast-trending en-echelon faults that dip about 60° to 80° eastwards and became progressively younger to the west. Many of the fault blocks are antithetically rotated so that the deepest part of the downthrown fault block is adjacent to the Spur. Smaller antithetic faults downthrowing to the west within the Graben occur subparallel to the main fault-zone trend.

Rapid erosion of the newly uplifted terrain resulted in deposition of a thick sequence of coarse clastic sediments in a relatively narrow (~5 km) elongated zone extending for at least 15 to 20 km along the faulted margin. These Upper Jurassic (mainly Kimmeridgian to Volgian) sandstones and conglomerates, forming the Brae field reservoir, interdigitate basinwards along-strike and up-section with organic carbon-rich shales (the Kimmeridge Clay Formation) that provide the hydrocarbon source for the Brae field and associated plays.

The Brae slope-apron system overlies Middle Jurassic sandstones, shales, and limestones and is juxtaposed against impermeable Devonian sandstones and conglomerates to the

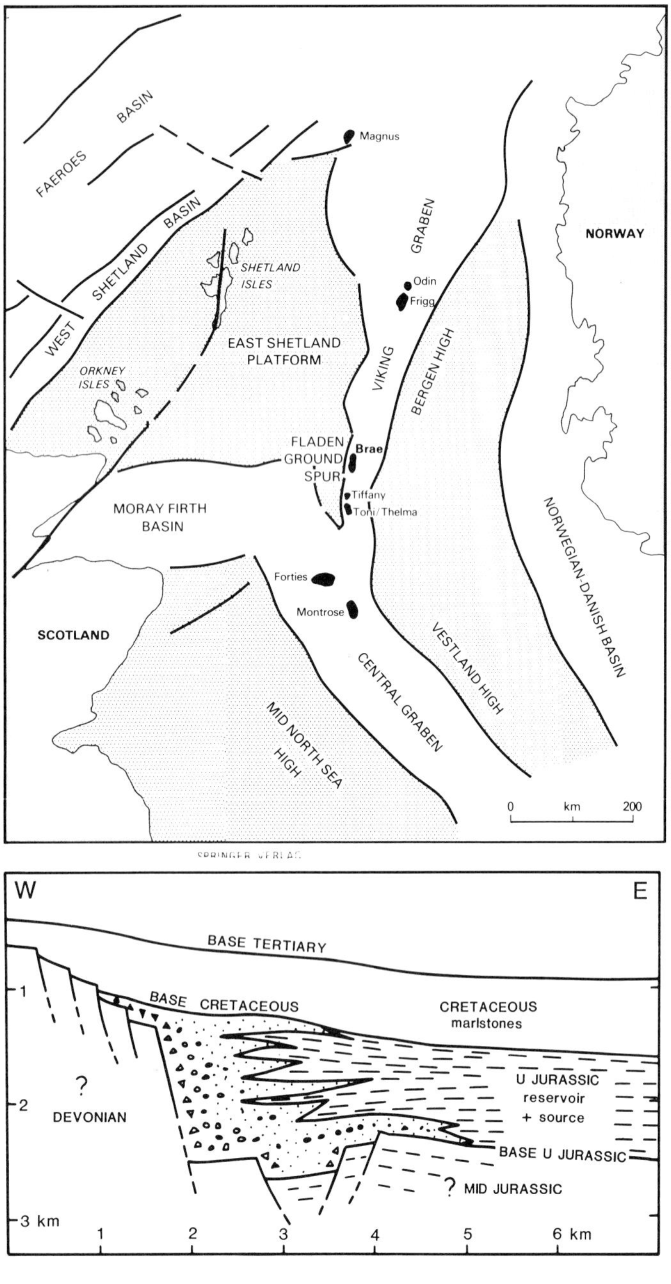

Figure 1. Location map (upper) Brae field, North Sea. Other oil and gas fields shown have also been interpreted as submarine-fan reservoirs. Cross-section runs across Brae field from Fladen Ground Spur to Viking Graben.

west. Latest Jurassic shales, followed unconformably by Lower Cretaceous marlstones and younger strata overlie the slope-apron sediments and onlap onto presumed Devonian basement to the west. Tectonic activity diminished through the Kimmeridgian, and subsequent minor tectonism has resulted in a series of gentle anticlines parallel to the Graben margin that now form the structural hydrocarbon trap.

Sediments

Four main facies groups are present in the Brae cores: mudstones, sandstones, conglomerates, and slumps (Figure 2), the first three of which can be further subdivided into separate facies. The *mudstone group* comprises interlaminated dark-grey micaceous mudstone and light-grey siltstones or fine-grained sandstones, and shows complete gradation from dominantly mudstone to dominantly sandstone facies. A range of microstructures, including basal scouring and mud injection, grading, fading ripples, and climbing low-amplitude ripples, indicate deposition from turbidity currents.

The *sandstone group* includes thin-bedded (1 to 10 cm) sandstones with internal grading, parallel and cross-lamination, and medium- to thick-bedded (>10 cm to over 40 cm) sandstones and pebbly sandstones that are commonly massive or with slight positive grading. They are interpreted as the deposits of higher-concentration turbidity currents and associated flows. They are quartz-rich with minor feldspar, mica, and other minerals, and variable amounts of carbonaceous debris, mudstone chips, and shell material. Porosity and permeability characteristics are commonly good, making these the main reservoir facies, but silica and calcite cementation are locally important. In addition, authigenic illite may also be present.

The *conglomerate group* is quite varied, including breccias, pebbly sandstones, pebbly mudstones, and possible tectonic breccia. Both graded-stratified beds 20 cm to 200 cm thick and massive matrix- or clast-supported conglomerates of indeterminate bed thickness are present. These probably result from rock-fall, debris flow, and other mass-flow processes. Clasts are very variable in size (1 cm to 150 cm), from angular to rounded in shape, and comprise presumed Devonian sandstones, quartz pebbles, dark-grey (? Jurassic) mudstones, dolomite, shell fragments, and carbonaceous debris. The sandstone matrix in some wells has good porosity and permeability, whereas others are very tightly cemented with carbonate.

Distinct *slump units* of variable thickness occur throughout, showing convoluted, contorted, and overturned laminae, small-scale faulting, steeply inclined lamination, and chaotic mudstone-sandstone mixes. They are indicative of deposition on a slope with periodic tectonic activity and/or rapid sediment build-up.

The different well sections have very different proportions of these facies groups (Figure 3). Overall, the mudstone facies are slightly more common (about 45%) than the conglomerate facies (about 35%), with sandstones being less abundant (about 20%) and *recognizable* slump units relatively unimportant (<3%). Three scales of vertical sequences are recognized: thinning- and thickening-upwards sequences over 5 to 20 m, mainly thinning/fining-upwards megasequences over 50 to 150 m, and gradual basin-fill fining-up-

Figure 2. Photographs of typical resedimented facies from Brae field slope-apron system. From left to right, mudstone-group, sandstone-group, conglomerate-group, and slump facies. Core widths about 5 cm.

ward over the complete 300 to 600 m of section (Figures 3 and 4).

Horizontal variability is very marked. The basin-fill sequence is the only one that can be correlated with certainty between all wells. From three to six megasequences occur in a number of the wells, but the lateral variation over even short distances in both north-south and east-west directions makes correlation very tenuous. There is a general trend from more conglomeratic close to the main fault zone to more mudstone-rich in the east or basinwards. Compositional differences are also noted: shell debris and glauconite are most common in wells D, G, and H, but rare in the southern group of wells, J, K, L, M, and N. Well B appears to be mainly thick, clean sandstones and well J mainly mudstones, but neither were cored intensively so that their compositional characters are difficult to ascertain. Wells C, E, and F are on higher fault blocks or on the Fladen Ground Spur and comprise fractured and faulted conglomerates over Devonian basement.

The inferred paleogeography [3], the composition of clasts, and the overall fining of sediments to the east all imply a sediment source to the west. Dipmeter logs from the mudstone facies of several wells show low easterly dips after removal of a small structural component, commonly with slight vertical oscillations (2° to 4°) over 3 to 10 m of section. These have been interpreted tentatively as resulting from prograding mud lobes. The sandstone facies show less regular but mainly easterly dips, sometimes with irregular upward shallowing and steepening trends, whereas the conglomerates often have a random bag-o'-nails dipmeter pattern. These data certainly corroborate the derivation of sediment from the west but are not adequate to define radial or other paleocurrent trends.

Flora and Fauna

Marine microplankton together with rare marine macrofossils, including ammonites, belemnites, and fragments of shallow-water bivalves are found throughout the Brae wells in the mudstone facies. These are mixed with terrestrial spores, pollen, and woody debris that are mostly well dispersed and rarely occur as thin carbonaceous horizons. The same mixed marine and terrestrial biogenic assemblage is restricted but still evident in many of the interbedded sandstones and, more rarely, the conglomerates. Clearly, the environment was marine but with a significant and probably local supply of terrestrial material.

Shelly debris is present mainly in the three wells in the central Brae area, and this may imply local development of

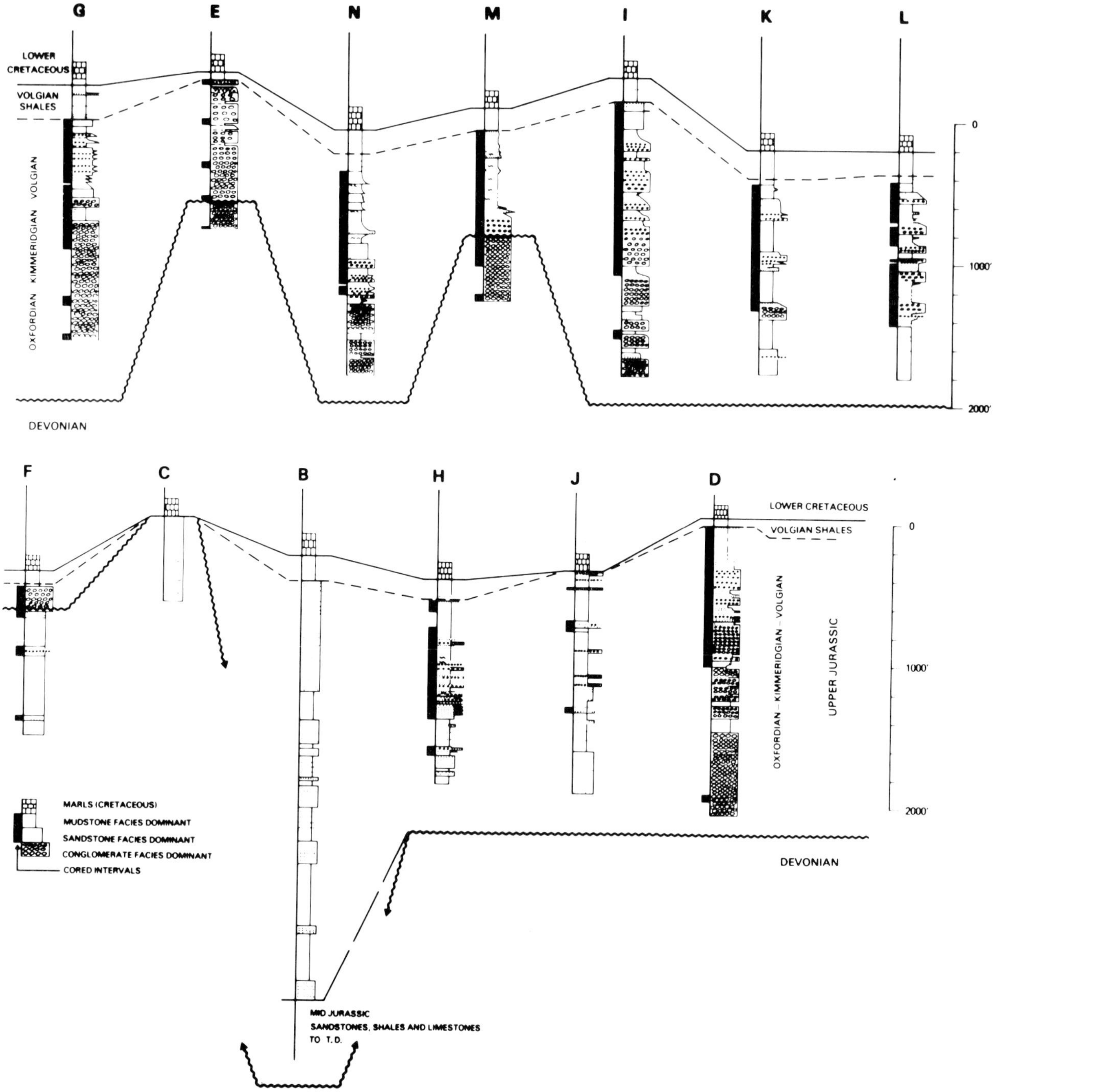

Figure 3. Summary well logs from Brae field slope-apron system, showing distribution of main facies groups and cored intervals. Well locations shown on Figure 4 (after [2]).

a narrow shelf on which a shallow-water benthic community existed and was periodically redeposited by turbidity currents and associated downslope flows. Bioturbation is almost completely absent even within the finely laminated mudstones, although the relatively high organic-carbon contents should have attracted a vigorous benthic infauna. It appears that conditions were unfavorable either because of anoxic/near-anoxic bottom waters and/or a very rapid sediment input.

Discussion

The Brae field wells were drilled through a series of small (<10-km radius) overlapping submarine fans that form a complex sediment apron along the faulted scarp margin of the Viking Graben (Figure 5). Distinctive fan geometry and morphology are not well established so that the system is better designated as a slope-apron, deposited for the most

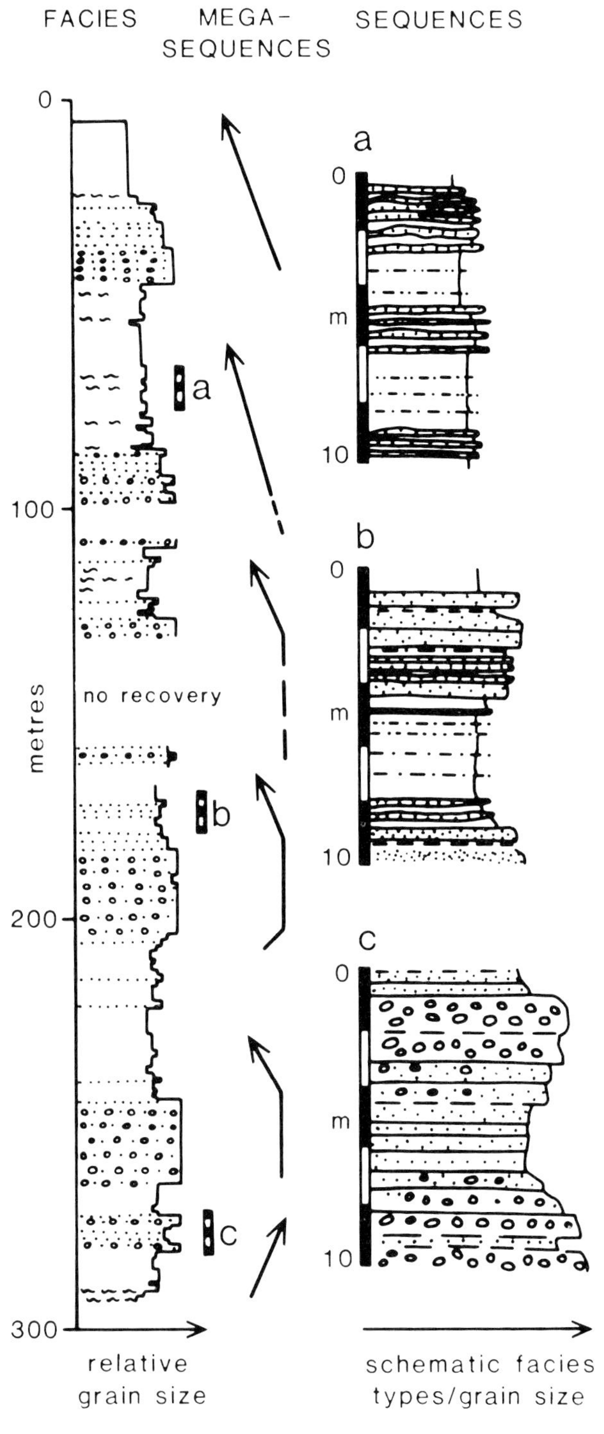

Figure 4. Detail of facies and vertical sequences through well L, southern Brae oilfield.

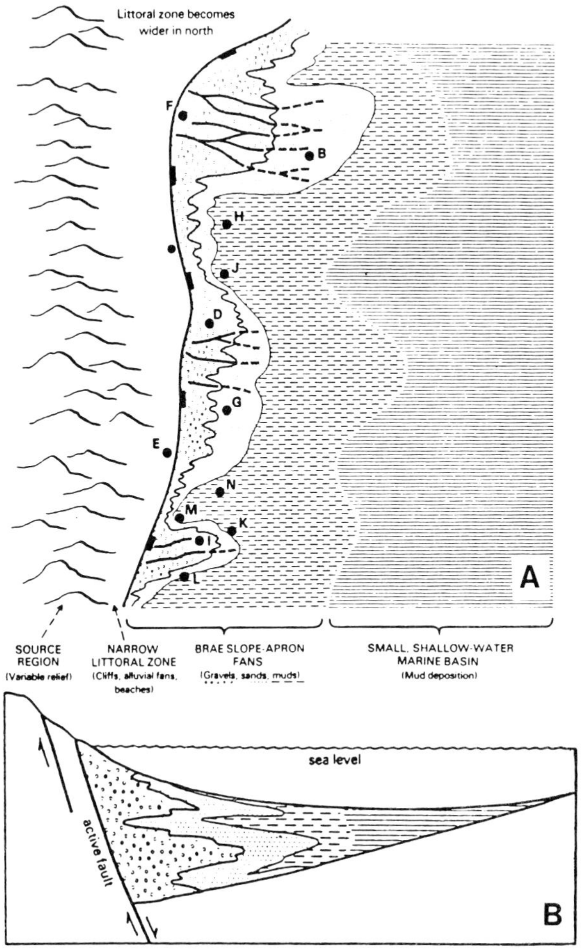

Figure 5. Schematic facies distribution for Brae field slope-apron system, North Sea, Upper Jurassic. A. Plan view showing boundary fault and well locations; facies indicated in solid ornament; solid and dashed lines represent probable submarine braided-channel system; B. Cross section showing schematic slope-apron sequence (after [2]).

part below wave base in a shallow marine basin. The three main facies groups occur in an irregular slope-parallel arrangement: a base of fault-scarp breccia-conglomerate association passing basinwards through pebbly sandstones and sandstones to a progressively more mudstone-dominated association. However, there is a complex interdigitation of these facies, with marked lateral and vertical facies changes which indicate that some channeling of the coarser facies occurred during deposition.

The main control on the development of this system appears to have been tectonic, but with important secondary controls being sediment supply and sea-level changes. Major fault movements and basin subsidence in the Oxfordian produced the first influx of coarse clastic sediments to the basin margin. Several subsequent episodes of active tectonism followed by relative dormancy probably resulted in up to six fining-upward megasequences through the Oxfordian-Kimmeridgian. Subsidence was most pronounced adjacent in the main fault zone so that a relatively thick but narrow wedge of sediment was developed. Faulting was not uniform along the margin either in time or place, so that the resulting separate redeposited systems are not readily correlated. "Piano-key" tectonics, causing differential uplift and subsidence, and transcurrent (east-west) faults, both acted to complicate the pattern of sedimentation. There is some evidence to suggest that these en-eschelon offset faults may have served as structurally controlled conduits to funnel sandy sediments beyond the slope-apron system into the Central Graben.

At times and along some parts of the margin, subaerial alluvial fans probably fed directly into the sea. In other parts, there was a narrow littoral zone in which pebbles were rounded, sands sorted, and both were mixed with broken shell fragments. This coastal zone was wider to the north, such that more mature dominantly sandy sediments were reworked downslope. Sediment supply may have been greater to the north because the platform was larger. A general decrease in fault activity, sediment supply and/or a relative rise in sea level through the Late Jurassic led to an overall fining-upward basin-fill sequence culminating in the Volgian black-shale transgression.

Similar slope-apron accumulations of turbidites and associated facies have been described in detail from both the Jurassic of east Greenland [4] and the Devonian-Carboniferous of the Polish Carpathians [5]. Recent work has shown that the South Arabian margin provides a modern example of this type of sedimentation (J. C. Faugeres, personal communication, 1983). Although not yet widely recognized, slope-apron systems are probably at least as common in the ancient record as the more classical submarine fan sequences.

Acknowledgments

The author acknowledges financial support from the Royal Society of Edinburgh, secretarial and technical assistance from the Grant Institute of Geology, and former colleagues in BRITOIL for many fruitful discussions on the sedimentological problems of the Brae field. Drs. John Damuth and Paul Griffiths reviewed an earlier version of the manuscript.

References

[1] Harms, J. C., and others, 1981. Brae field area. In: L. V. Illing and G. D. Hobson (eds.), Petroleum Geology of the Continental Shelf of Northwest Europe. Heyden, London, pp. 352–357.

[2] Stow, D. A. V., Bishop, C. D., and Mills, S. J., 1982. Sedimentology of the Brae Oilfield, North Sea: fan models and controls. Journal of Petroleum Geology, v.5, pp. 129–148.

[3] Ziegler, P. A., 1981. Evolution of sedimentary basins in northwest Europe. In: L. V. Illing and G. D. Hobson (eds.), Petroleum Geology of the Continental Shelf of Northwest Europe. Heyden, London, pp. 3–39.

[4] Surlyk, F., 1978. Submarine fan sedimentation along fault scarps on tilted fault blocks (Jurassic/Cretaceous boundary, East Greenland). Bulletin Grølands Geologiske Undersgelse, v.128, 108 pp.

[5] Nemec, W., Porebski, S. J., and Steel, R. J., 1980. Texture and structure of resedimented conglomerates: examples from Ksiaz Formation (Famennian–Tournaisian), southwestern Poland. Sedimentology, v.27, pp. 519–538.

CHAPTER 35

Kongsfjord Turbidite System, Norway

Kevin T. Pickering

Abstract

The late Precambrian Kongsfjord Formation submarine fan is as much as 3200-m thick and contains inner, middle, outer, and transitional fan environments such as a fan lateral margin. It forms the oldest exposed part of a fan-slope-delta system believed to be comparable in size with modern medium-sized fans and deposited along either a passive "Atlantic-type" continental margin or within an aulacogen. Palaeocurrent data suggest that the fan-slope-delta system prograded toward the east-northeast, and the petrography is typical of a stable low/medium-grade metamorphic and intrusive acid-igneous source area.

Introduction

The late Precambrian Kongsfjord Formation is a succession of turbidite and other sediment gravity-flow deposits up to 3.2-km thick that crops out discontinuously over a northwest-southeast distance of about 90 km on the northeast of Varanger Peninsula, N. Norway (Fig. 1). The formation only occurs northeast of the NW-SE oriented Trollfjord-Komagelv Fault that is interpreted as a palaeotransform fault with dextral displacement between 500 to 1000 km which occurred post-640 and pre-500 m.y. BP [1].

Based on Rb-Sr whole-rock isotopic analyses of cleaved mudstones, the Kongsfjord Formation has undergone one major syn-metamorphic fold deformation at about 520 ± 47 m.y. BP [2]. Although deformation intensity increases westward, the major tectonic style is that of relatively open southwest-plunging folds. Metamorphism was up to the lower greenschist facies and, locally, dolerite dykes are abundant, especially in some of the finer-grained lithologies. The excellent coastal exposures, together with the inland stream sections and cliffs, allow detailed sedimentologic mapping despite the age and deformation history of the Kongsfjord Formation.

Stratigraphy and Regional Setting

The Kongsfjord Formation is the oldest exposed part of the approximately 9-km thick late Precambrian Barents Sea Group (Fig. 1). The group forms an overall regressive sequence from deep-water submarine fan (Kongsfjord Formation), through basin slope and fluvio-deltaics (Båsnaering Formation, 2500 to 3500-m thick), to shallow marine and intertidal/supratidal carbonates/clastics (Båtsfjord Formation, 1500-m thick) and shallow marine and fluviatile deposits (Tyvofjell Formation, 1500-m thick). In the absence of reliable chronostratigraphic horizons, the boundaries/transition between these formations provide useful lithostratigraphic markers.

Because the transition from the Kongsfjord Formation is lithologically similar across the Varanger Peninsula, this has been used as a datum line to correlate the two principal sections (Fig. 2). While it is possible that the transition is highly diachronous across the peninsula, it has been tentatively assumed that the Kongsfjord/Båsnaering Formation transition, to a first approximation, is a chronostratigraphic horizon.

Fan Definition and Fan Divisions

The main reasons for considering the Kongsfjord Formation to be an ancient deep-water submarine fan are: 1) the formation comprises more than 3.2 km of turbidites and other sediment gravity-flow deposits; 2) there is no evidence of deposition in water above wave-base, while symmetric/asymmetric wave-ripples are evident upward from the base of the immediately overlying Båsnaering Formation [3,4]; 3) using palaeocurrent data solely from sandstone packets (considered to be the most reliable slope indicators), there

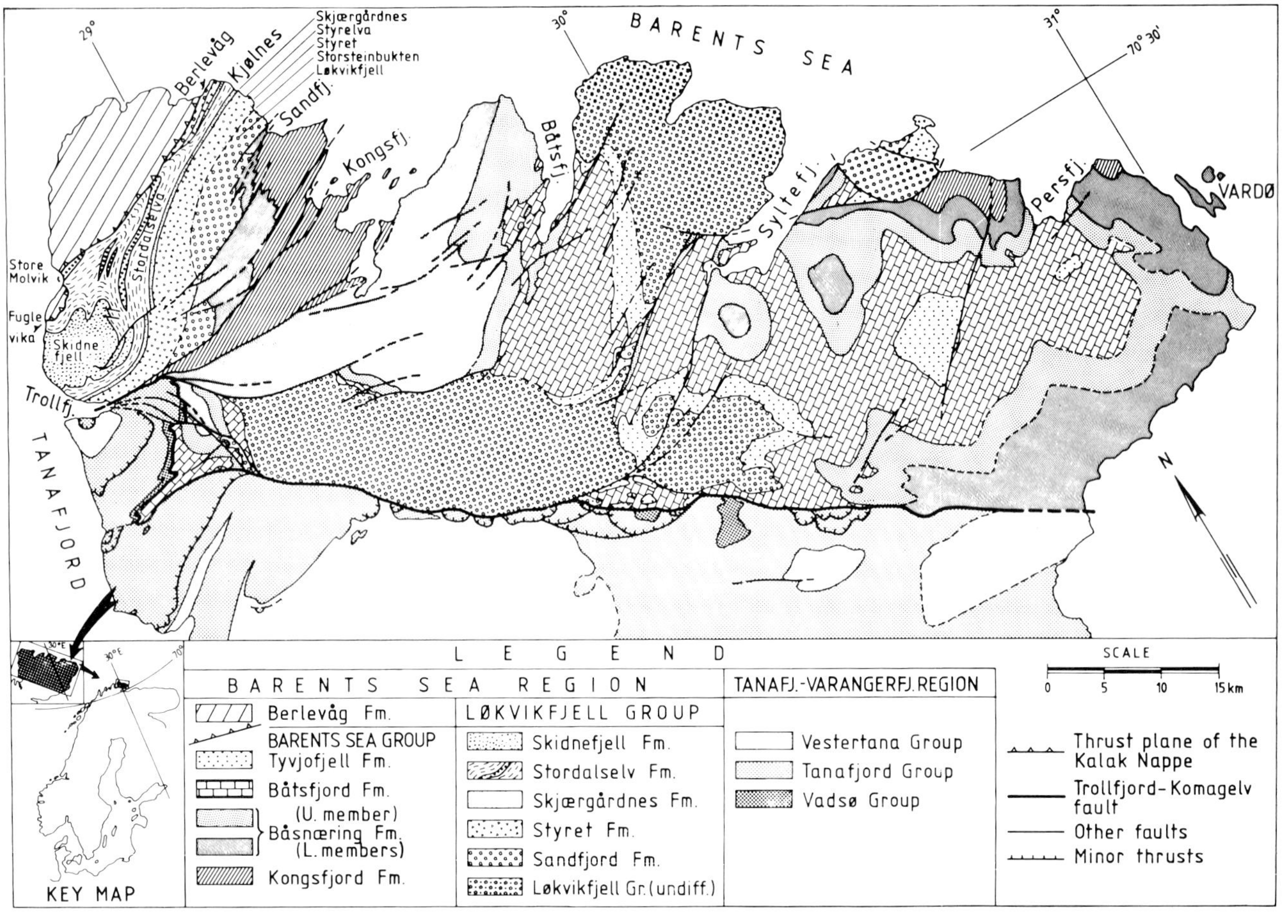

Figure 1. Geologic map of Barents Sea region, North Norway, showing the Kongsfjord Formation and the immediately overlying Båsnaering Formation. The stratigraphy of the Barents Sea Group is shown toward the bottom left of the legend. Map from Siedlecka and Edwards [3].

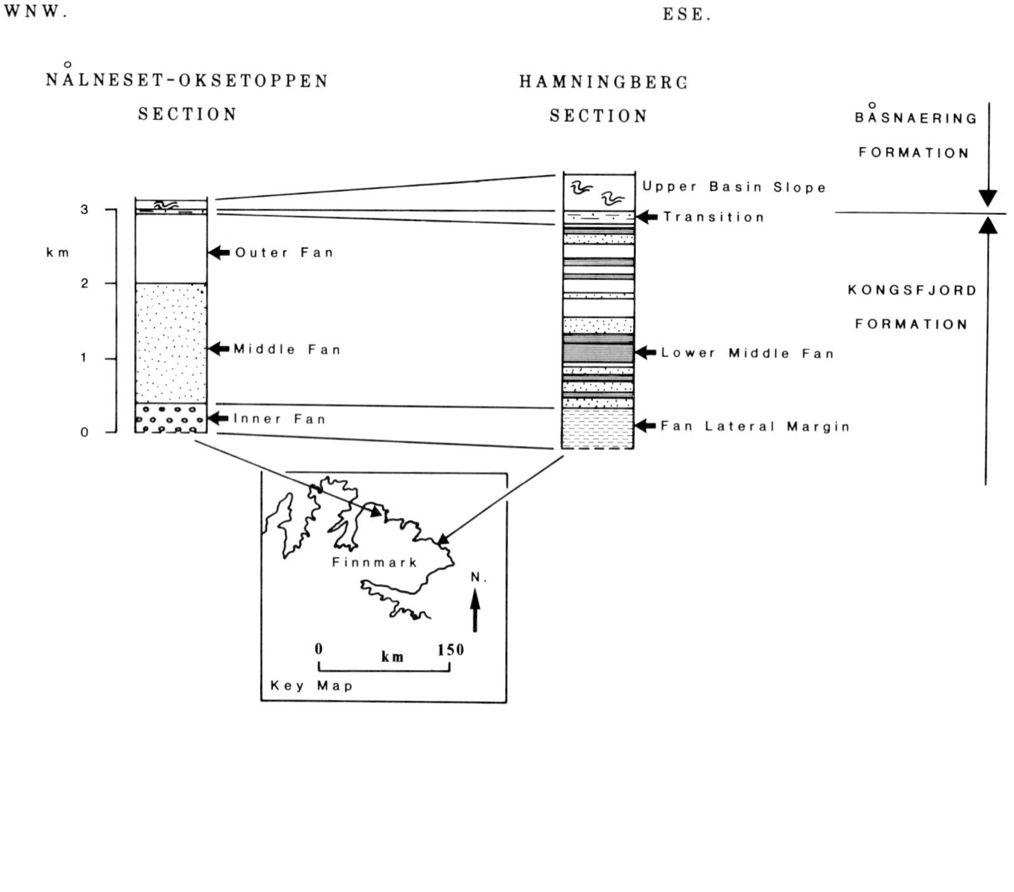

Figure 2. Depositional environments of the Kongsfjord Formation in the two principal sections. A lithostratigraphic correlation is also shown.

is a radial distribution about an easterly azimuth, with many showing an east-northeast sense of flow (Fig. 3); and 4) it is possible to reconstruct a submarine fan model with inner, middle, outer, and transitional fan environments based on the intrinsic sedimentary features of the formation (Table 1) and by analogy with modern and ancient submarine fans (Fig. 4).

Inner-fan mainly sandstone conglomerates occur in packets that vary from 70 to 200-m thick, separated by siltstone/mudstone packets from 25- to 45-m thick [5]. Within the sandstone conglomerate packets, there are typically 10 to 30 m-thick units of amalgamated, mainly coarse-grained sandstones commonly showing an erosional base to the unit. Pebbly mudstones up to 4.25-m thick, interpreted as debris flow deposits, and reworked, relatively "clean" cross-stratified sandstones are common in these deposits compared with the rest of the formation. By analogy with the Lago Sofia conglomerate and sandstone lenses in the Upper Cretaceous Cerro Toro Formation in southern Chile [6], the Kongsfjord Formation inner-fan deposits are thought to be probably lateral from the main channel axes (Fig. 5).

Mid-fan channel, mainly sandstone-rich packets typically vary from 10 to 30-m thick, with individual beds or packets of beds, or both, that wedge out and cut down up to 11 m over a lateral distance of 150 m. Such deposits are separated by siltstone/mudstone intervals up to 25-m thick, showing diverse palaeocurrents compared with most of the channel deposits [7]. Based on detailed facies, facies-association, and sequence analyses, a depositional model for mid-fan sedimentation (Fig. 6) was developed [7] and includes many of the features of the mid-fan model proposed by Mutti [8].

Outer-fan sections are characterized by 2 to 15-m thick lobe and lobe-fringe deposits separated by fan fringe deposits up to 70-m thick [9]. Type I and Type II lobe sequences are also observed [9] (Fig. 7). The distinction primarily is based on the vertical frequency of lobes within fan-fringe deposits. Type I lobes plus lobe-fringe deposits comprise 60% of the section compared with 22% of a continuous section for Type II lobe sequences. Type I lobes are separated by lobe fringe plus fan-fringe deposits of similar thicknesses, such that in vertical sections the lobes are regularly spaced. Type II lobes occur irregularly within sections where lobe fringe plus fan-fringe deposits vary up to 70-m in thickness. Type I lobe sequences are thought to be the result of regular lobe switching immediately downfan from mid-fan channels within a relatively proximal outer-fan environment. Type II lobe sequences are believed to be the consequence of sporadic (erratic) lobe progradation in a relatively distal region that usually received fan-fringe sedimentation. Type I lobe sequences represent regular and stable deposition patterns, probably controlled by intra-fan (intra-basinal) processes, whereas Type II lobe

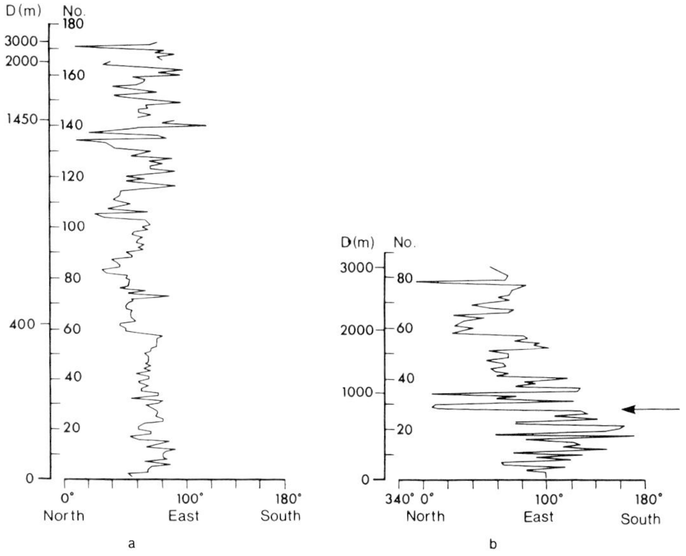

Figure 3. Palaeocurrents from (a) Nålneset-Øksetoppen section, and (b) Hamningberg section. Data from flute casts within packets of amalgamated sandstones, which are thought to be the most likely slope indicators for basin and overall fan surface. "D" is distance from base of formation (meters); "No." is the nth reading from base of formation. At about 800 m from the base in Figure 3b, there is a swing from an east to south-easterly direction, to a more easterly and north-easterly direction as shown by the position of the arrow. See text for explanation.

sequences developed due to relatively catastrophic intra-fan (e.g., major channel relocation) or extra-fan/extra-basinal (source control) processes [9].

"Transitional fan environments" are recognized [10], with attributes that are intermediate between the more typical fan environments listed in Table 1. The three transitional fan environments described by Pickering [10] are: middle to outer-fan; fan lateral margin, and fan to upper-basin-slope/prodelta.

Petrography

Thirty-one representative sandstone thin-sections, with over 1000 points per slide, were counted using 30 μ as the upper limit to define matrix. Matrix content ranges from 12.5 to 50.4%, with an average of 28.3%. Excluding matrix, the three main components are quartz, feldspar, and polymineralic rock fragments (Fig. 8). Figure 8a emphasizes the provenance, whereas Figure 8b shows the compositional maturity.

Feldspar is relatively common and, in many cases, constitutes more than 15% of the sandstone. Therefore, many of the rocks are arkosic graywackes, with the proportion of rock fragments being less than that of the feldspars. Alkali feldspars are the most abundant type, with subordinate,

Figure 4. (A) Inner-fan channel granule/pebble sandstones in packet of amalgamated beds dipping steeply and younging to the right. Location about 120 m from base of section at Nålneset. Human scale. (B) Mid-fan channel deposits, SE Kongsfjord. Thinning and fining-upward sequence is 15-m thick (left, shown by arrow). "Normal" bedding is horizontal bedding contacts on the right with successive excavation surfaces dipping to the left, suggesting lateral migration of the channel margin. (C) Mid-fan channel margin showing slide immediately above very thick basal sandstone bed (arrow). The slide thickens to channel axis on right. Overlying "normal" bedding is toward the top of the plate. Human scale bottom right (boxed). Nålneset-Øksetoppen section. (D) Fan lateral margin deposits toward the base of Hamningberg section, showing large clastic dyke-cutting thin-bedded siltstones/mudstones that thin toward the top of plate. (E) Outer-fan (mainly lobe plus lobe fringe) deposits younging to the left. About 150 m of section. Lobes are light-colored amalgamated sandstones (three distinct packets); very dark-colored fine-grained beds are fan fringe, and intermediate-colored packets are lobe fringe. (F) Slide folds in upper basin-slope deposits. Plate width about 1.5 m of deposits from SE Kongsfjord. (G) Typical thin-bedded siltstone turbidites forming the bulk of the upper basin-slope deposits. Hamningberg section.

Table 1. Intrinsic Sedimentary Features of Submarine Fan Formation

Inner-Fan Deposits		
Interpretation	Channel Axis/Channel Margin*	Interchannel/Levee†
Definition	Very thick to very thin-bedded turbidite and other mass flow deposits (debris flow deposits common), thalweg channels within larger-scale channels	Thin to very thin-bedded trubidites
Bedding pattern	Irregular, wedging	Sheet-like
Common internal sedimentary structures	Ta–Tc: traction cross-bedding, poor to well organized stratified and graded beds	Td–Te; hemipelagic (?) structureless mudstone
Typical grain sizes	Small pebble to medium-grained sandstone	Siltstone to mudstone
Estimated % sandstone	>50%	<10%
Amalgamation	Very common	Absent
Palaeoflow relative to adjacent environments in *this* table	Variable	Variable
Other features	Local deep and narrow scour and fills, and small-scale channelling; thickening, coarsening, thinning, and fining-upwards sequences	—

*By analogy with the Lago Sofia conglomerates (Winn and Edwards [6]), these deposits are interpreted as having accumulated in the marginal parts of inner-fan channels. Thicker, coarser-grained amalgamated conglomerates and sandstones may represent thalweg channels.
†It was impossible to distinguish levee and interchannel deposits.

Middle-Fan Deposits				
Interpretation	Channel Fill	Channel Margin	Levee	Interchannel
Definition	Very thick to medium-bedded turbidites and other mass flow beds filling main parts of sediment conduit	Thick to thin-bedded turbidites laterally between channel fill and levee deposits	Medium to thin-bedded turbidites lateral and proximal to channel fill deposits	Thin to very thin-bedded turbidites lateral to channel fill deposits; distal overspill from channel
Bedding pattern	Irregular, wedging	Irregular, wedging	Irregular, wedging	Sheet-like
Common internal sedimentary structures	Ta–Tc; traction cross-bedding	Tb–Td	Tc–Td	Tc–Te; structureless mudstone
Typical grain sizes	Granules to medium-grained sandstone	Medium to fine-grained sandstone	Fine-grained sandstone to siltstone	Very fine-grained sandstone to mudstone
Estimated % sandstone	>90%	60–90%	60–20%	<20%
Amalgamation	Very common	Uncommon	Absent	Absent
Palaeoflow relative to adjacent environment in *this* table	Variable	Variable	Variable	Variable
Other features	Often thinning and fining-upward sequence on scale of meters	Slides; displacement on scale of meters	Packets of beds divided by eroso-depositional surfaces; soft-sediment deformation	Some sheet-like, and channelled sandstone packets less than 2 m thick-crevasse deposits?

Outer-Fan Deposits			
Interpretation	Lobe	Lobe Fringe	Fan Fringe
Definition	Very thick to medium-bedded turbidites as sheet-like beds forming topographic highs immediately downfan from a channel mouth	Medium to thin-bedded turbidites peripheral to lobe deposits, as distal equivalents	Thin to very thin-bedded turbidites in regularly bedded packets representing the most distal submarine fan deposits
Bedding pattern	Sheet-like; localized scour-and-fill; may cut down as packet over hundreds of metres	Sheet-like	Sheet-like
Common internal sedimentary structures	Ta–Tc	Tb–Te	Tb & Tc–Te; hemipelagic (?) structureless mudstone
Typical grain sizes	Very coarse to medium-grained	Fine-grained	Very fine-grained
Estimated % sandstone	>80%	80–40%	<40%
Amalgamation	Very common	Rare	Absent
Palaeoflow relative to adjacent environments in *this* table	Similar	Similar	Similar
Other features	Thickening and/or coarsening-, and thinning and/or fining-upward sequences	As for lobes	Impressive regularity of bedding vertically and laterally; no sequences

Table 1. *continued*

	Upper Basin-Slope Deposits
Definition	Thin to very thin-bedded turbidites, wave-generated and hemipelagic (?) deposits, slides, deposits sandwiched between deep water and shallow water, shelf deposits
Bedding pattern	Sheet-like; wedging; irregular
Common internal sedimentary structures	Tc–Te; hemipelagic (?) structureless mudstone; bidirectional ripples
Typical grain sizes	Very fine-grained sandstone to mudstone
Estimated % sandstone	<20%
Amalgamation	Absent
Palaeoflow relative to adjacent packets of beds in *this* environment	Similar or opposing direction
Other features	Slide, slump, and other soft-sediment deformation common; no sequences; no coarser grained, thicker bedded channel fills

variously twinned plagioclase feldspars. Rock fragments generally constitute a small proportion and include polycrystalline quartz, acid-igneous feldspar-quartz intergrowths, intraformational mudstones, chert, and detrital mica. Most of the polycrystalline quartz is polygonized or elongated, characteristic of low/medium-grade metamorphic rocks. The quartz-perthite intergrowths probably were derived from acid-igneous intrusive and gneissic rocks.

Thus, the Kongsfjord Formation quartzo-feldspathic graywackes are believed to have been derived from a relatively stable source area of low/medium-grade metamorphic and acid-igneous intrusive rocks.

Palaeogeography

Correlations between sections across the Varanger Peninsula are based on lithostratigraphy. By comparing principal sections, a tentative palaeogeography for the depositional history of the Kongsfjord Formation can be shown (Fig. 9). Although the greatest separation of sections today is about 100 km (probably, they were considerably more separate if tectonic folding is taken into account), palaeocurrent data suggest that the downslope separation, perpendicular to the strike of the putative basin margin, was on the order of 20 to 30 km.

In the principal sections, the Kongsfjord Formation shows an overall 3000-m thick thinning and fining-upward sequence from inner, through middle, to outer-fan deposits. There is a relatively thin (up to 100 m-thick) section mainly

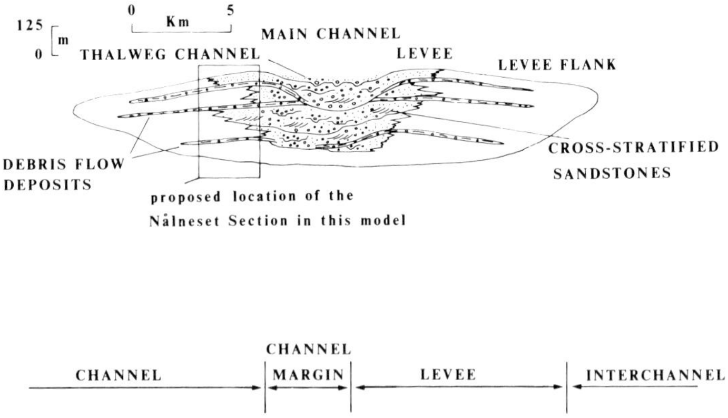

Figure 5. Model of inner-fan sedimentation for the Kongsfjord Formation (modified after Winn and Dott [6]). Thin sandstones from overbanking flows wedge out from the levees into overbank areas, together with debris flows from the channel. Conglomerates and sandstones are confined mainly to the channel, with the coarsest-grained sediments in thalweg channels.

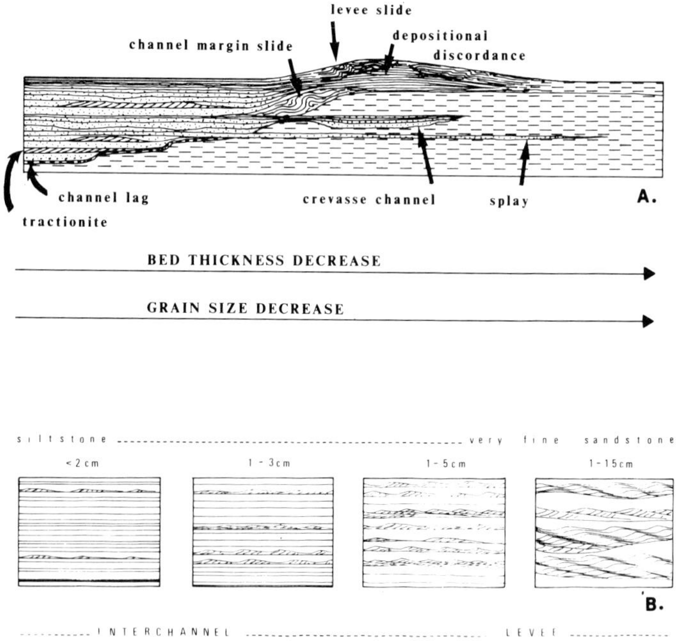

Figure 6. (A) Depositional model for mid-fan sedimentation in the Kongsfjord Formation, as compared with that by Mutti [8]. From Pickering [4]. (B) Model showing the transition from levee to overbank and interchannel deposits in the Kongsfjord Formation. Only ripple-lamination and bed shape are shown. Each bed is graded, and the blank part represents parallel lamination. Typical bed thicknesses are shown, and in the Bouma T_c division of the levee deposits, soft-sediment deformation is common. From Pickering [4].

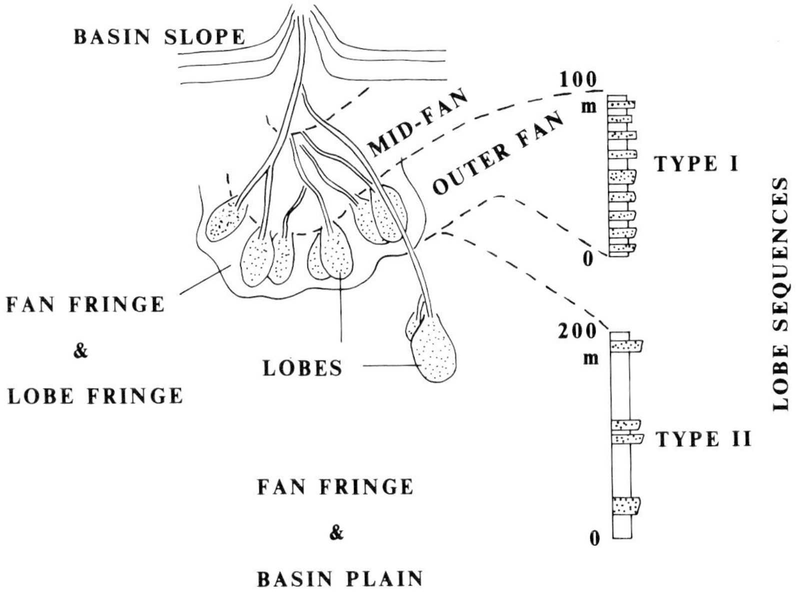

Figure 7. Depositional model for Type I and Type II lobe sequences in the Kongsfjord Formation. Lobes developed by continual switching and stacking in the region immediately downfan from mid-fan channels give Type I sequences. Lobes that abruptly prograde the fan onto the basin plain and/or fan fringe, followed by a return to more stable lobe accretion toward the source of the fan fringe, generate Type II lobe sequences. See text for explanation. From Pickering [5].

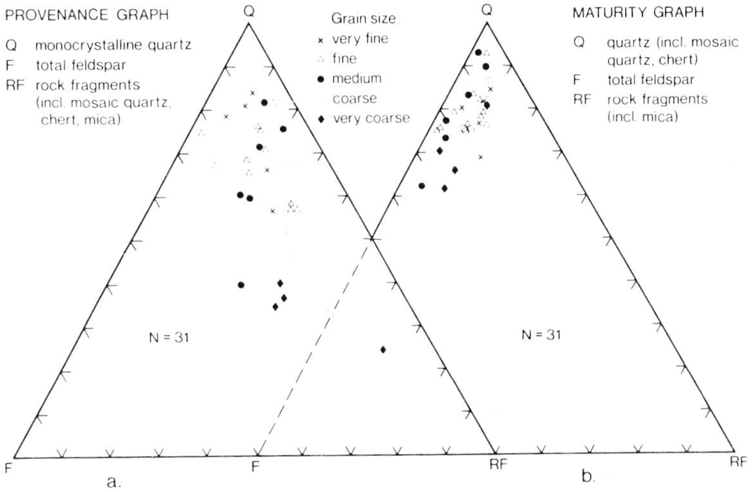

Figure 8. Triangular graphs showing provenance and compositional maturity of 31 representative thin-sections from the Kongsfjord Formation Sandstones. Note the effect of grain size on composition.

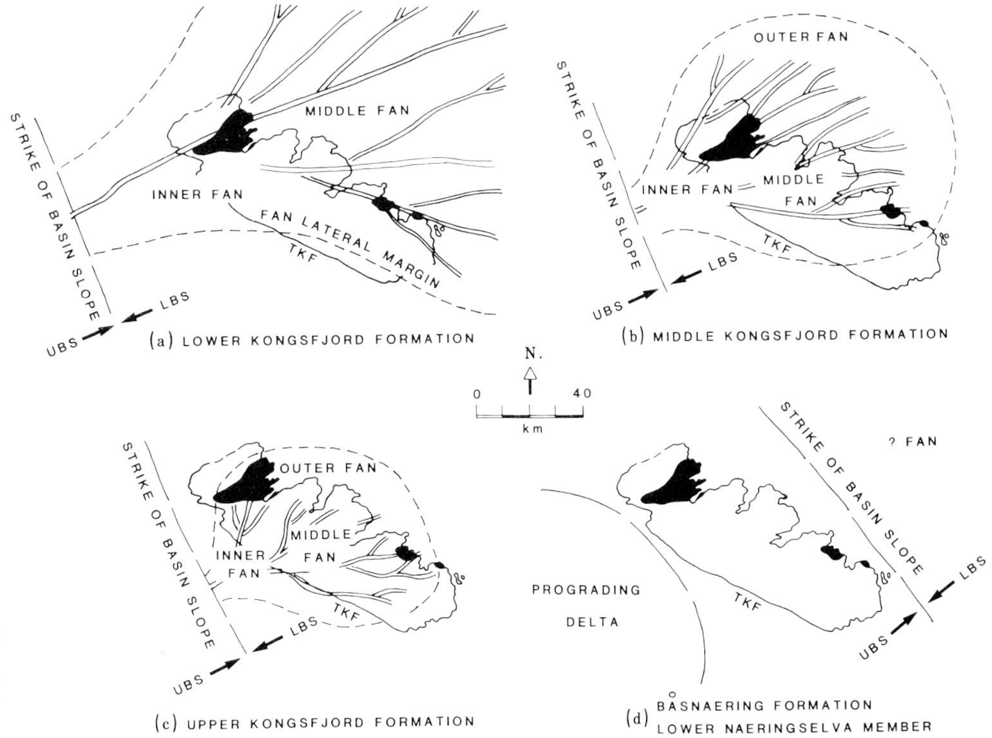

Figure 9. Depositional history of the Kongsfjord Formation and Båsnaering Formation fan-slope-delta system. Outline of Varanger Peninsula north of the Trollfjord-Komagelv Fault (minus the extreme western Caledonian nappe) and the Kongsfjord Formation are shown (dashed lines). Clearly, the fan shape is unknown, but tentative boundaries are shown on the basis of lithostratigraphic correlations and palaeocurrent patterns. See text for explanation.

consisting of channel and related deposits immediately below the upper basin-slope/prodelta (Fig. 2). The lithostratigraphy and environmental interpretations suggest a limited east-southeastward shift in the locus of fan sedimentation with time (Fig. 9). Palaeocurrent data corroborate this, since the change in direction in the lower part of the Hamningberg Section (Fig. 3) is believed to be a change from fan lateral margin to more axially deposited sediments. However, the fan remained in roughly the same area of the depositional basin, and the major "retrogradational" sequence may reflect an overall decrease in the rate of submarine fan growth, possibly because of an increase in the rate of basin subsidence and/or enhanced entraiment of coarser-grained sediments in the north-easterly prograding fluvio-deltaic and slope system that formed the overlying Båsnaering Formation.

Conclusions

The late Precambrian fan probably was comparable in size with modern medium to large-radius submarine fans. The great thickness (about 9 km) of the Barents Sea Group is probably fully marine, and the lack of contemporaneous volcanism, the stable source area, the low grade of metamorphism, the lack of complex polyphase deformation history, and the overall regressive sequence generated by the Barents Sea Group Formations all suggest similarity to a passive "Atlantic-type" continental margin or an aulacogen. Although the original radius of the fan is unknown, perhaps the best modern analogues for the Kongsfjord/Båsnaering Formation fan-slope-delta system are the Mississippi, Niger, and Rhone submarine fan systems.

References

[1] Kjøde, J., and others, 1978. Palaeomagnetic evidence for large-scale dextral displacement along the Trollfjord-Komagelv Fault, Finnmark, North Norway. Physics of Earth and Planetary Interiors, v. 16, pp. 132–144.

[2] Taylor, P. N., and Pickering, K. T., 1981. Rb-Sr isotopic age determination on the late Precambrian Kongsfjord formation, and the timing of compressional deformation in the Barents Sea Group, East Finnmark. Norges Geologiske Undersøkelse, v. 367, pp. 105–110.

[3] Siedlecka, A., and Edwards, M. B., 1980. Lithostratigraphy and sedimentation of the Riphean Båsnaering formation, Varanger peninsula, North Norway. Norges Geologiske Undersøkelse, v. 355, pp. 27–47.

[4] Pickering, K. T., 1982. A Precambrian upper basin-slope and prodelta in northeast Finnmark, North Norway—a possible ancient upper continental slope. Journal of Sedimentary Petrology, v. 52, pp. 171–186.

[5] Pickering, K. T., 1981. The Kongsfjord Formation–a late Precambrian submarine fan in north-east Finnmark, North Norway. Norges Geologiske Undersøkelse, v. 367, pp. 77–104.

[6] Winn, R. D., and Dott, R. H. Jr., 1979. Deep-water fan-channel conglomerates of late Cretaceous age, southern Chile. Sedimentology, v. 26, pp. 203–228.

[7] Pickering, K. T., 1982. Middle-fan deposits from the late Precambrian Kongsfjord formation submarine fan, northeast Finnmark, northern Norway. Sedimentary Geology, v. 33, pp. 79–110.

[8] Mutti, E., 1977. Distinctive thin-bedded turbidite facies and related depositional environments in the Eocene Hecho group (south-central Pyrenees, Spain). Sedimentology, v. 24, pp. 107–131.

[9] Pickering, K. T., 1981. Two types of outer fan lobe sequence, from the late Precambrian Kongsfjord formation submarine fan, Finnmark, North Norway. Journal of Sedimentary Petrology, v. 51, pp. 1277–1286.

[10] Pickering, K. T., 1983. Transitional submarine fan deposits from the late Precambrian Kongsfjord formation submarine fan, NE Finnmark, N. Norway. Sedimentology, v. 30, pp. 181–199.

VI

Mississippi Fan, DSDP Leg 96

Seismic Surveys and Drilling Results

CHAPTER 36

Mississippi Fan: Leg 96 Program and Principal Results

Arnold H. Bouma, James M. Coleman, and DSDP Leg 96 Shipboard Scientists

Abstract

Nine sites were drilled during Deep Sea Drilling Project Leg 96 on the Mississippi Fan, four on the middle fan, four on the outer fan, and one through a slump. The drilling established lateral and upward channel migration in the middle fan, frequent channel shifting of the lower-fan channel, and sheet sands at the channel terminus. Sediments deposited in the middle-fan channel fine upward, starting with gravel. The sheet sands in the upper two fanlobes have 41% and 64% net sand based on gamma-ray logs. They were deposited during the upper and lower late glacial stages of the Wisconsin. A 29-m thick carbonate debris flow underlies those sands.

Introduction

The Mississippi Fan is a broad arcuate accumulation of Pleistocene deposits (Fig. 1). Examination of a large number of seismic reflection profiles enabled us to establish eight distinct regional acoustic reflectors [1; Chapter 21, Fig. 3). Construction of structural contour and isopach maps shows that the Mississippi Fan actually consists of many elongated bodies called fanlobes [1, Fig. 2; Chapter 21]. The presentations of the maps in Figure 3 of Chapter 21 indicate a number of important aspects: 1) each fanlobe is an elongated sediment body with its main thickness coincidental with the longitudinal axis and its thickest accumulation on the middle fan; 2) the juxtaposition of fanlobes is controlled by the paleo-relief of the older ones and by the location of the sediment source; 3) through the Pleistocene, the fanlobes shifted from west to east and toward deeper water; and 4) each structure and isopach map reveals more complexity than would be expected from a single fanlobe. The resolution and density of many seismic reflection profiles are insufficient at this time to break out more detail and plot smaller fanlobes in a consistent manner.

The general movement of fanlobes from west to east cannot be tied as yet to canyons incised into the shelf because of correlation problems caused by the high density of salt diapirs on the continental slope. It should be sufficient to indicate that many completely filled canyons have been observed near the shelf break.

Parts of the youngest fanlobe have been surveyed in detail by medium resolution seismic reflection systems and by side-scan sonar (GLORIA, Sea MARC I, and EDO). The results show a sinuous channel on the middle fan, with adjacent overbank deposits and patterns that may be analogous to crevasse splays. Although a migratory pattern is suggested [2,3], proof cannot be substantiated by such surveys alone. The drilling provided a much more conclusive answer. The Sea MARC data over the lower fan indicate a central channel, slightly sinuous, flanked by linear imprints that suggest former channel courses. Bifurcation is often observed near the end of the lower-fan channel, downfan of which no observable channel pattern can be found.

Drilling Program

Leg 96 of the Deep Sea Drilling Program ran from September 29 to November 8, 1983 using the D/V Glomar Challenger as drilling platform. Because the leg commenced in Fort Lauderdale, Florida, drilling started on the lower fan (Sites 614 and 615) and then continued on the middle fan (Sites 616 and 617). Weather and currents forced us to abandon the area and go to the intraslope basins selected for drilling on the continental slope off Louisiana (Sites 618 and 619; Fig. 1, Table 1). From there, the drilling vessel returned to the middle fan and drilled in overbank deposits (Site 620) and the channel (Site 621: axis of present channel; Site 622: inner bend or "pointbar" of present channel).

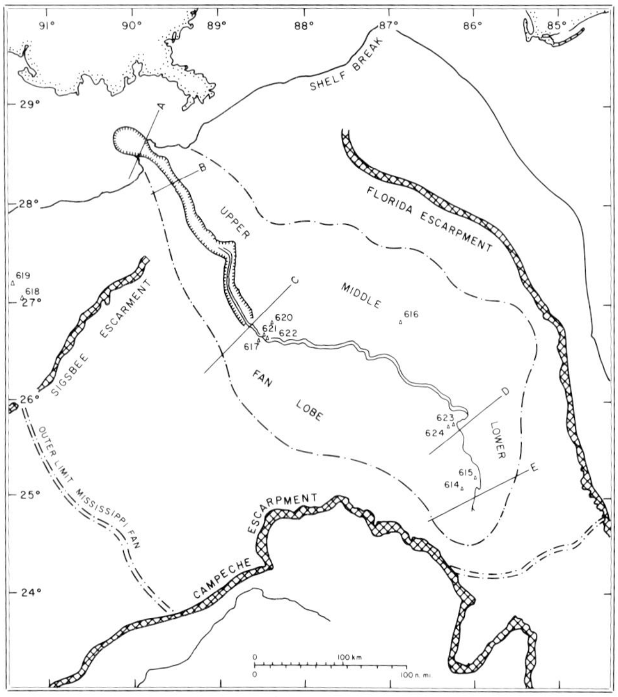

Figure 1. General outline of the Mississippi Fan, Gulf of Mexico, with generalized position of the youngest fanlobe. Locations of drill sites and positions of cross-sections (Fig. 2) are presented.

Because insufficient time was left to return to the area of Site 614 and drill a second deep hole, instead we drilled two 200-m (maximum permitted subbottom depth) holes in the area of distinct channel switching (Sites 623 and 624).

Several wire line coring systems were used on Leg 96. The Advanced Hydraulic Piston Corer and the Extended Core Barrel were used most frequently. The rotary core was used at Site 620 because deep penetration was necessary at that site.

Some of the holes could not be drilled to the approved depth of penetration because of downhole drilling problems. At Site 616, the lower drill collar became totally stuck and had to be severed. For safety reasons, the presence of gas was monitored continuously. Contrary to expectations, only traces of gas were encountered.

Core recovery typically was high (70 to 100%) for the upper 80 to 90 m of the sediment column at each site. Below that depth, recovery dropped sharply and was irregular. Sands thicker than 20 to 25 cm could not be recovered without disturbance; they were either completely fluidized or could not be retained by the core catcher assembly and, therefore, were not recovered at all.

Though we initially planned to log only the two deep holes (one on each of the lower- and middle-fan regions), we soon decided to log as many holes as possible to offset the generally poor core recovery at depth.

Seismic Characteristics of a Fanlobe

A fanlobe may be seismically defined as a depositional unit bounded by laterally continuous seismic reflectors, display mappable internal seismic facies, and be formed in a relatively short geological time. A number of fanlobes make up a submarine fan; fanlobes are separated from each other by fine-grained sediments and/or pelagic oozes deposited during periods of low activity of fan development (for details see Feeley and others, Chapter 37).

A fanlobe can be described as a channel-overbank complex in which internal seismic facies change both laterally and in a downfan direction. It is elongated in shape and its width is dependent on the preexisting topography, the amount of sediment available, and the size of the transport mechanisms with regard to the dimensions of the channel. Seismically, it is often difficult to determine fanlobe boundaries because the parallel reflector patterns of the flank sediments may not differ substantially from underlying and overlying deposits. The thickness of a fanlobe varies and is primarily dependent on the total amount of material delivered from the source area during the period of active fan formation.

Drilling during Leg 96 was concentrated on the youngest or modern fanlobe as determined from seismic and side-scan sonar surveys. The drilling program (Table 1) called for a few deep sites to obtain a time-stratigraphic framework of the upper fanlobes, while the other sites were more directed at obtaining sedimentological, paleontological, chemical, and geotechnical information to typify fanlobe subenvironments.

Assuming that the modern fanlobe and its overall characteristics typify the older fanlobes of the Mississippi Fan, it behooves us to discuss it in some detail. The modern fanlobe starts on the outer shelf, crosses the continental slope and rise, and terminates in the abyssal plain (Fig. 1). The fanlobe can be divided into four regions, each having certain typical characteristics that can be seen on seismic reflection profiles. The four regions are: 1) an upslope erosional canyon, 2) an upper fan that terminates at or slightly beyond the base of slope, 3) an aggradational middle fan with a sinuous axial channel, and 4) an aggradational lower fan with many channels of which only one is active at a given time. Schematic seismic sections across these regions (Fig. 2) are based on an evaluation of a large number of seismic records (primarily strike-oriented) and emphasize certain acoustical aspects. The general characteristics typical of each of these regions are discussed below.

Mississippi Canyon

A brief description and age history of the Mississippi Canyon is given in Chapter 21 and in [1,4]. Seismically, one

Table 1. Drilling Sites and General Core Information, Leg 96

Site No.	Latitude	Longitude	Water Depth (m)	Penetration Depth (m)	Meters Cored	% Core Recovery	Type of Cores	Type of Well Logs	Principal Lithologies
614[1]	25° 04.08′N	86° 08.21′W	3314	37.0	37.0	100	HPC/XCB	—	—
614A	25° 04.08′N	86° 08.21′W	3314	150.3	75.0	75	HPC/XCB	—	M-s, S
615	25° 13.34′N	85° 59.53′W	3284	523.2	419.3	42	HPC/XCB	DIL,LSS,GR FDC,CNL,GR	S, M-s
615A[2]	25° 13.35′N	85° 59.55′W	3284	208.5	74.5	70	HPC/XCB		—
616[3]	26° 48.67′N	86° 52.83′W	2999	371.0	307.8	47	HPC/XCB	FDC,CNL,GR	M-s, S
616A	26° 48.65′N	86° 52.86′W	2999	132.4	38.4	63	HPC/XCB		—
616B[2]	26° 48.66′N	86° 52.85′W	2984	204.3	143.2	79	HPC/XCB		—
617	26° 41.93′N	88° 31.67′W	2468	191.2	130.1	86	HPC/XCB		M-s, M
617A[2]	26° 41.93′N	88° 31.67′W	2467	73.0	73.9	77	HPC/XCB		—
618	27° 00.68′N	91° 15.73′W	2422	92.5	78.0	87	HPC/XCB		M
618A[4]	27° 00.68′N	91° 15.73′W	2422	47.6	28.7	65	HPC/XCB		—
619	27° 11.61′N	91° 24.54′W	2273	208.7	134.4	83	HPC/XCB		M, M-s
619A[5]	27° 11.61′N	91° 24.54′W	2273	5.3	5.3	100	HPC/XCB		0
620	26° 50.12′N	88° 22.25′W	2612	422.7	421.3	47	F93 CK	DIL,LSS,GR	M, M-s
621	26° 43.86′N	88° 29.76′W	2485	214.8	157.3	87	HPC/XCB	LSS,GR,CAL FDC,CNL,GR	M-s, M, S, G
622	26° 41.41′N	88° 28.82′W	2495	208.0	132.7	75	HPC/XCB	LSS,GR,CAL	M-s, M, S, G
622A[5]	26° 41.41′N	88° 28.82′W	2495	5.6	5.6	99	HPC/XCB		0
623	25° 46.09′N	86° 13.84′W	3188	202.2	110.2	81	HPC/XCB	LSS,GR,CAL	M-s, S
624	25° 45.24′N	86° 16.63′W	3198	199.9	109.8	69	HPC/XCB		M-s
624A[2]	25° 45.24′N	86° 16.63′W	3198	207.6	103.7	84	HPC/XCB	DIL,LSS,GR FDC,CNL,GR	—

[1]Core barrel separated, round trip required.
[2]Extra continuous core for shore-based geotechnical studies.
[3]Pipe stuck, severing required, 616A got stuck at 132.4 m.
[4]Extra cores for geochemical studies.
[5]Extra cores for paleontological studies.
HPC = hydraulic piston core, includes advanced piston core and variable length hydraulic piston core; XCB = extended core barrel system, used below 100 to 130 m; F93 CK = standard rotary coring bit; LSS = long-spaced sonic log; GR = gamma-ray log; CNL = compensated neutron log; FDC = formation density log; DIL = dual-induction lateral log; CAL = caliper log; M-s = mud with thin and thick silt laminae; M = mud; S = sand; G = gravel; 0 = foraminifera and/or nannoplankton ooze.

observes shelf deposits with a few distinct horizontal reflectors separated by wider zones with either semiparallel and discontinuous or oblique reflectors (Fig. 2A). A major erosional structure that is partly filled represents the actual canyon. The bottom of the fill shows hyperbolics that are overlain by a discontinuous, parallel to slightly irregular, reflector pattern. The shallow erosional cutting adjacent to the canyon is a slide scar that was filled later by fine-grained sediments.

We suggest that development of this canyon initially began about 50,000 to 55,000 years ago on the middle slope and retrogressed onto the shelf by 25,000 to 27,000 years B.P. by large-scale slumping on an unstable shelf-slope area during a low stand of sea level or its initial rise [4]. Retrogressive slumping lengthened and widened the canyon further upshelf.

Upper Fan

The fan area from the Mississippi Canyon at approximately 1200-m water depth to the base of slope at about 2000 m is called the upper fan [1; Chapter 21]. It is characterized by a large cut and fill structure with a slight convex upper surface (Fig. 2B). The area has acted as the conduit to transport the source materials downfan. Seismically, a fill pattern similar to the canyon is observed, but with a larger width to depth ratio and with the presence of a second smaller cut and fill structure in the larger one. The smaller channel is not completely filled, but is flanked by laterally discontinuous reflectors that are inferred to represent overbank deposits. The hyperbolic acoustical patterns at the base of each channel may represent slump materials or homogeneous sandy deposits, such as from debris flows.

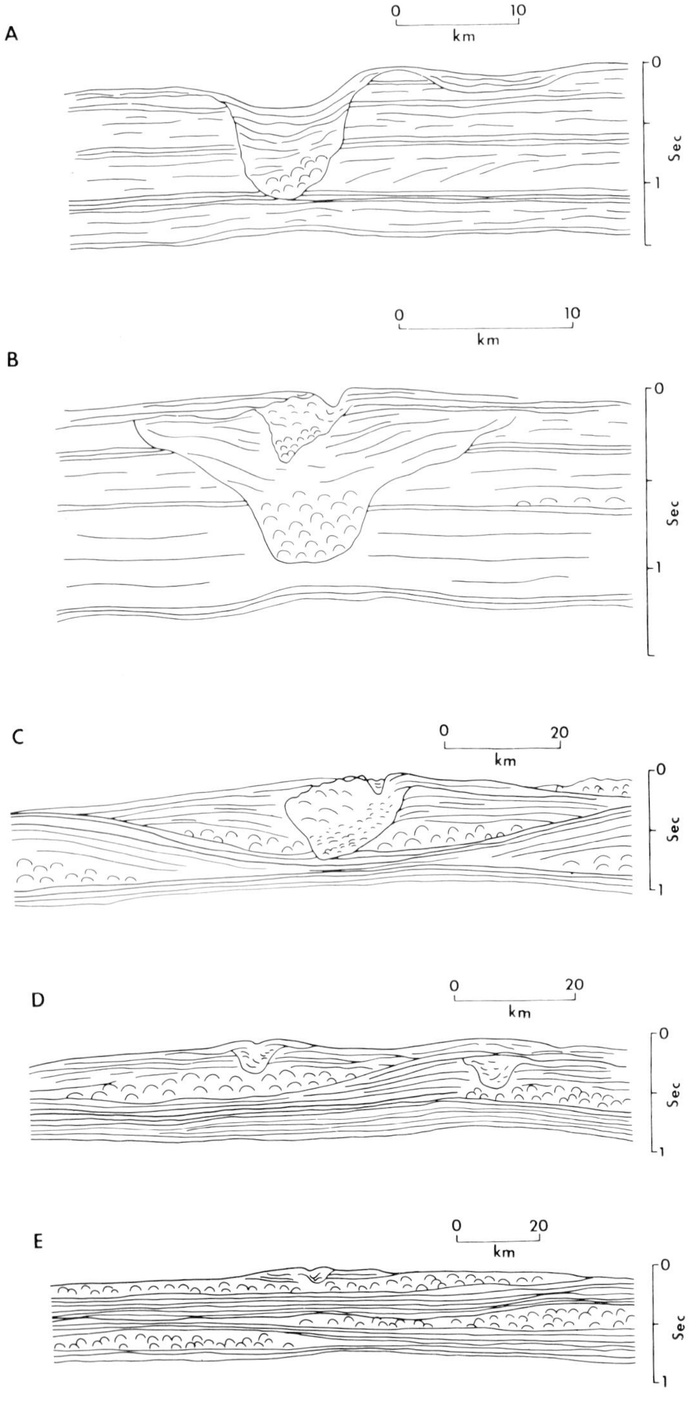

Figure 2. Schematic cross-sections across the youngest fanlobe of Mississippi Fan. These sections are based on seismic reflection profiles. For location, see Figure 1. For discussion, see text.

Middle Fan

The thickest accumulation of sediment occurs in the middle part of the fanlobe. Figure 2C shows the lenticular nature and the change in acoustical patterns of the middle-fan region. An aggradational channel complex is located along the apex. This 2 to 4-km wide channel is migratory in nature and appears on side-scan sonar images as a leveed, sinuous channel-levee complex [1–3]. Seismically, one observes an asymmetric channel profile with an acoustical, high amplitude zone angling laterally from the base of the channel toward the present thalweg. Drilling results from Sites 621 and 622 indicate that this high amplitude zone consists of coarse material (gravel and sand) and likely is interlayered with pebbly mudstones and muds. The migratory characteristics of the middle-fan channel are discussed in Chapter 41; a more detailed description of the middle fan is presented in Chapter 40. The discontinuous reflectors that flank the channel complex represent primarily overbank deposition.

Lower Fan

The area between approximately 3100-m water depth on the north and 3350-m water depth near the Florida Straits in the south is covered by the lower fan (see also Chapter 42). The sinuosity and dimensions of the channel have decreased gradually from that typical of the upper middle-fan channel. Side-scan sonar images show a single distinct channel and numerous indistinct linear images on both sides of the channel that are interpreted as abandoned channels. Drilling results at Sites 623 and 624 support the idea of frequent channel switching, which implies that a given channel is active for a rather "short" time and then switches position. Figure 2D shows the seismic characteristics for the major upfan area of the lower fan. The seismic reflectors are more continuous than in the middle-fan region and are interspersed by acoustically slightly semitransparent to chaotic flat lenses and small channel structures.

Near the end of the lower fan, the seismic pattern becomes more regular and more parallel. If any channels were present, they were too small to identify by high resolution seismic techniques (Fig. 2E). Drilling at Sites 614 and 615 recovered sheet sands that may be equivalent to depositional lobes described from ancient turbidite sequences. The sand content is very high based on gamma-ray well log data. The youngest fanlobe (late Wisconsin in age) has a net sand content of 41%, while the underlying Early Wisconsin fanlobe contains 65% net sand (see also Chapter 39).

Summary of Main Drilling Results

The Leg 96 drilling of four sites on the middle fan in and near the sinuous channel, of four sites on the lower fan, and of one site in the slump area described by Walker and Massingill [5] has produced some very important information, a summary of which follows. The remaining chapters in this section provide additional detail and interpretation. The chapter by Feeley, Buffler, and Bryant (Chapter

37) is not related to the drilling program. However, it provides a more detailed interpretation of the seismic stratigraphy and seismic facies of the entire Mississippi Fan.

1) The Mississippi Fan consists of several fanlobes, each having an elongated shape. Each fanlobe is connected to a submarine canyon that is incised into the outer continental shelf and upper slope. During the course of the Pleistocene, a general migration of fanlobe deposition took place from west to east and toward deeper water.

2) A fanlobe is basically a channel-overbank complex that can be divided into four regions: 1) A canyon, most likely formed by massive slope failure, and considered to be the main intermediate source area. 2) The upper fan, which terminates near the base of slope and acted as a conduit for transport of sediment downfan. It most likely contains a lag deposit, while the remainder of the fill consists of fine-grained sediments deposited after major transport ceased. 3) The middle fan, which is an aggradational unit with a convex upper surface and a large sinuous migratory and aggradational channel running along its apex. 4) The lower fan, which is also aggradational in nature and which has several more or less parallel channels, of which only one is active at a given time. Near the end of the channels, bifurcation may take place and sand sheets are deposited at the terminus of each channel.

3) The single sinuous channel on the middle fan is migratory and aggradational in nature. It has a general similarity to migratory fluvial systems. The channel fill is a fining-upward sequence starting with gravelly deposits. The overlying sandy section is thicker in the "pointbar" area than in the "thalweg" area. The basal gravelliferous and sandy sediments represent channel lag and lateral accretion deposits that are characterized acoustically by a high amplitude zone that shifts laterally and climbs stratigraphically.

The "passive" channel fill is characterized by a lack of indigenous fauna and contains only a sparse, predominantly reworked upper and middle neritic microfauna. The overbank deposits contain some indigenous fauna in addition to the reworked species.

4) Coarser-grained sediments, including the majority of the upper and middle neritic foraminifera, must have been confined primarily to the channel, and only fine-grained materials were able to flow over the levees onto the overbank areas. During a time of active transport, the channel was likely not wider than about 4 to 6 km and not deeper than 100 to 140 m near Sites 621 and 622. To allow for the rapid deposition of thick overbank deposits, the sediment volumes must have been very large and density flows high compared with the channel dimensions.

5) In contrast to the single channel on the middle fan, the lower fan contains a number of more or less parallel channels, only one of which seems to be active at any given time. This causes deposition of a semi interleaved stacking of thin channel deposits that are bounded on top and bottom and laterally by fine-grained overbank deposits, significantly different from the thick confined channel deposits characteristic of the middle-fan region.

6) Significant amounts of sand, derived from a source area with a low sand/clay ratio, were moved across the fan to its distal area, a transport distance in excess of 600 km.

7) The upper two fanlobes (drilled at Site 615) consist of a late Wisconsin fanlobe and an underlying early Wisconsin fanlobe. The deposits are very sandy and contain 41% and 65% net sand, respectively. Each sequence starts with a coarsening-upward sand to mud unit, overlain by bedded sands and topped by a thinner fining-upward unit.

8) The generally low recovery of sands and the fluidization of most of those recovered makes comparisons with reports on ancient turbidite sequences concerning internal sedimentary structures and vertical sequence trends impossible. This makes it difficult to evaluate such concepts as depositional lobes and compensation-cycles.

9) Underlying seismic Horizon 30 [Chapter 21] at Site 615, a 29-m thick carbonate deposit was recovered that shows an upward-fining sequence from pea-gravel-sized fragments of limestones to a foram-rich nannoplankton ooze to a nannoplankton ooze. Shallow water benthic species were also observed. It is not yet certain whether this is one unit or multiple units. Tentatively, based on sparse seismic data, we consider it to be a single debris flow deposit that may have been derived from the central Florida Platform-Escarpment to the northeast.

10) Sedimentation rates were extremely high on the Mississippi Fan during the Late Pleistocene. For the youngest fanlobe, deposited during the late Wisconsin glacial stage, accumulation rates ranged from nearly 12 m/1000 yr for the channel fill and nearly 11 m/1000 yr for the overbank areas on the middle fan to about 5 to 6 m/1000 yr for the lower fan. These accumulation rates are averages over the vertical column of the drill sites. The volume of sediments deposited during the late Wisconsin glacial stage has not yet been calculated. In the late Pleistocene, sedimentation rates were high during periods of lowered sea level and were relatively lower during periods of high sea-level stands.

11) Drilling at Site 616, in the area reported by Walker and Massingill [5] to be an extensive surficial slump, could not accurately establish the thickness of the disturbed zone (95 to 105 m). The disturbed zone is underlain by bedded silts and clays with some sand layers.

The slump includes packages of laminated to thin-bedded muddy series, ranging in thickness from a few meters to more than 10 m. Within each package the laminae have consistent dips (up to 65°), but no two packages have the same dip, nor does there seem to be any increasing or decreasing dip sequence downcore. Either heavily disturbed zones or normally bedded series were found between the packages. We interpret this "slump" to consist of a series of slides. How large the individual slides are and where they originated from are not known.

References

[1] Bouma, A. H., Stelting, C. E., and Coleman, J. M., 1983/84. Mississippi Fan: internal structure and depositional processes. Geo-Marine Letters, v. 3, pp. 147–153.

[2] Garrison, L. E., Kenyon, N. H., and Bouma, A. H., 1982. Channel systems and lobe construction in the Mississippi Fan. Geo-Marine Letters, v. 2, pp. 31–39.

[3] Kastens, K. A., and Shor, A. N., 1985. Depositional processes of a meandering channel on the Mississippi Fan. American Association of Petroleum Geologists Bulletin, v. 69, pp. 190–220.

[4] Coleman, J. M., Prior, D. B., and Lindsay, J. F., 1983. Deltaic influences on shelf edge instability processes. In: D. J. Stanley and G. T. Moore (eds.), The Shelf Break, Critical Interface on Continental Margins. Society of Economic Paleontologists and Mineralogists Special Publication 33, pp. 121–137.

[5] Walker, J. R., and Massingill, J. V., 1970. Slump features on the Mississippi Fan, northeastern Gulf of Mexico. Geological Society of America Bulletin, v. 81, pp. 3101–3108.

CHAPTER 37

Depositional Units and Growth Pattern of the Mississippi Fan

Mary H. Feeley, Richard T. Buffler, and William R. Bryant

Abstract

Basin-wide unconformities within the Plio-Pleistocene section of the Mississippi Fan define eight major sequences. Isopach maps of the sequences reveal that the fan depocenter has migrated both eastward and seaward through its development. Definition and interpretation of seismic facies suggest mass transport, with channelized/unchannelized turbidite deposition were the important depositional mechanisms during fan growth. Although each sequence has a unique history, certain characteristics, particularly the vertical and lateral succession of seismic facies, are common to the majority of sequences. This succession suggests a dominant controlling mechanism on fan development, possibly sea-level fluctuations, with secondary influences of salt tectonics and sediment supply.

Introduction

The Mississippi Fan is a large deep-sea fan in the eastern Gulf of Mexico, consisting of a broad arcuate accumulation of predominantly Pliocene and Pleistocene sediments (Fig. 1) [1]. The fan is flanked on the east by the West Florida carbonate platform and on the north and west by the Texas-Louisiana Continental Slope. The deeper parts of the fan merge with the Florida Plain to the southeast and with the Sigsbee Plain to the southwest.

The Mississippi Fan has been studied by several investigators [2–6]. These early studies were largely limited by the vast size of the fan to only defining basic physiography and morphology. More recent work on the fan has concentrated on the younger fan deposits. Garrison and others [7] identified a sinuous channel along the apex of the most modern fanlobe. The DSDP Leg 96 drilling strategy emphasized the more recent fan deposits and concentrated on the central channel and adjacent deposits and their role in the formation of the most recent fanlobe(s). Preliminary results from the analysis of the site survey and core information have greatly increased our knowledge of the depositional mechanisms active in deep water [1,8,9].

This chapter briefly presents some of the results of a seismic stratigraphic analysis of single channel and multichannel seismic profiles collected across the Mississippi Fan during the last 15 years [1]. First, we describe the major depositional units and seismic facies. This is followed by a discussion of the general growth pattern of fan development, integrating the interpretation of transport/deposition mechanisms with the evolution of individual fanlobes.

Data Base

Two main data sets were used for the bulk of the analyses: approximately 6000 km of 12-fold multichannel data collected by the University of Texas Institute for Geophysics during 1976 through 1978, and approximately 3000 km of single channel reflection profiles obtained in 1969 during USNS Kane cruises [1]. Additional high resolution data, collected by the U. S. Geological Survey and Lamont-Doherty Geological Observatory, supplemented these principal data sets.

Seismic Stratigraphy

Depositional Packages

The Mississippi Fan consists of a broad arcuate accumulation of predominantly clastic detritus that has accumulated during at least the last 3 m.y. [6]. The fan reaches a maximum thickness of over 3 km in its central part. The axis of maximum thickness is centered more or less coincident with the axis of the Mississippi Canyon to the north (Fig. 1). The fan thins to the southeast and southwest, reaching minimum thicknesses of 1700 and 1600 m, respectively, beneath the Sigsbee Plain and the Florida Plain (Fig. 1) [1].

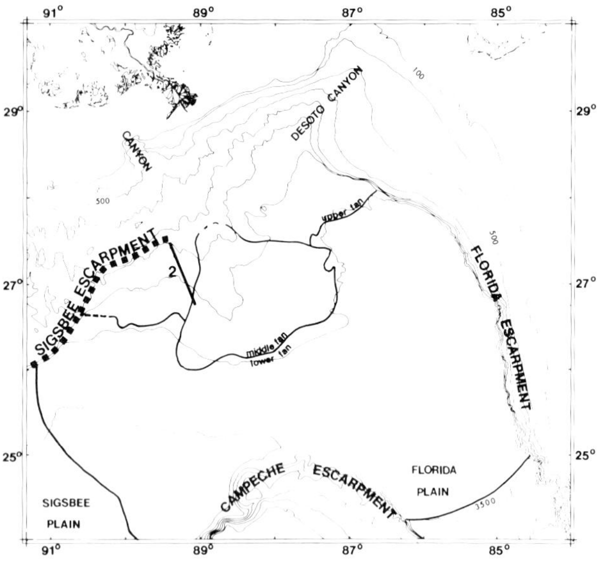

Figure 1. Bathymetry of the eastern Gulf of Mexico, with upper and middle Mississippi Fan subdivisions as defined by Moore and others [6] and outer-fan limit as defined by Bouma and others [9]. Contour interval = 500 m. Heavy line indicates location of Figure 2.

Basin-wide unconformities within the Mississippi Fan define eight seismic depositional packages or sequences (sequences I through VIII, Fig. 2). The base of each of the sequences (i.e., fanlobe) is, in general, erosional updip, becoming more conformable downdip. These packages represent major growth phases in the development of the Mississippi Fan complex. The base and thickness of each sequence were mapped to examine the gross evolutionary history of the fan complex [1]. In general, each of the sequences is lenticular in cross-section and thins laterally from a central depositional ridge. Analysis of the isopach maps reveals that the maximum accumulations of fan deposits for each sequence have migrated generally to the east and seaward through time. This shift is illustrated in Figure 3, which shows the updip position and orientation of the axis of maximum thickness for each of the sequences [1]. Several possible factors may have influenced this shift in the depocenter including: 1) an eastward shift of the source area, 2) movement within the Texas-Louisiana salt dome province and along the Sigsbee Escarpment (Fig. 1), and/or 3) an affect of pre-existing topography controlling the position of subsequent deposition.

Changes in the position of the source area can only be inferred because of the complexities involved with salt diapirism on the outer shelf and slope in the northern Gulf of Mexico. However, major shifts in the position of the Mississippi River since 50,000 years B.P. have been documented [2]. Similar shifts during various stages of the Pleistocene possibly played a critical role in the migration of the fan depocenter. In addition, examination of seismic profiles and isopach maps of each of the sequences indicates contributions to the fan from sources other than the Mississippi River Embayment, particularly from the Campeche Escarpment and the Florida Escarpment [1,10–12]. This observation is illustrated in Figure 3, in which a general migration of the fan depocenter from west to east for sequences I through V is shown. This was followed by a significant shift in the axis of maximum thickness further eastward (sequence VI), suggesting major influxes of sediment from the northeast Gulf of Mexico, possibly the DeSoto Canyon (Fig. 1). Sediment contributions from the adjacent margins have apparently added large volumes of clastic detritus to the outer areas of the fan.

Salt tectonics may have played an important role in the position and orientation of deposition of each successive fan sequence. For example, thickness relationships for sequences I and II suggest that the maximum accumulations of these older fan deposits are now shelfward of the present salt front. During or after deposition of these units, the front shifted to the southeast, disrupting the fan deposits, and making correlation across the escarpment difficult. By the time of deposition of sequence IV, the deep-water depocenter had shifted well beyond the Sigsbee Escarpment (Fig. 2).

Seismic Facies

In addition to the eight major depositional units just described, seven seismic facies have been identified within the Mississippi Fan based on reflection character and configuration, external geometry, and lateral facies association [1]. The facies are: 1) mounded chaotic, 2) high amplitude/low continuity (channel), 3) high continuity/low amplitude (overbank), 4) onlapping fill, 5) parallel layered, 6) progradation, and 7) conformable drape. Interpretation of these seismic facies suggests that various sediment gravity flows, together with pelagic and hemipelagic deposition, are the basic sedimentary processes operative in this deep-water environment.

For the interpretation of the growth pattern of each of the depositional sequences, the important facies are mounded chaotic, channel, overbank, onlapping fill, and conformable drape facies. The mounded chaotic facies have a distinct mounded external form with an internal hummocky or chaotic reflection pattern (Figs. 2a and b). The base of these mounds may be erosional, and diffractions may be associated with their upper surface. In places, these mounds are onlapped by overlying material. Up to 30% of an individual fan sequence (sequence V) is comprised of this mounded chaotic facies. The transport/deposition mechanism envisioned is some type of mass transport process, either slumping or debris flows.

The interpreted channel (high amplitude/low continuity) and overbank (high continuity/low amplitude) seismic facies are usually found associated with each other and form a

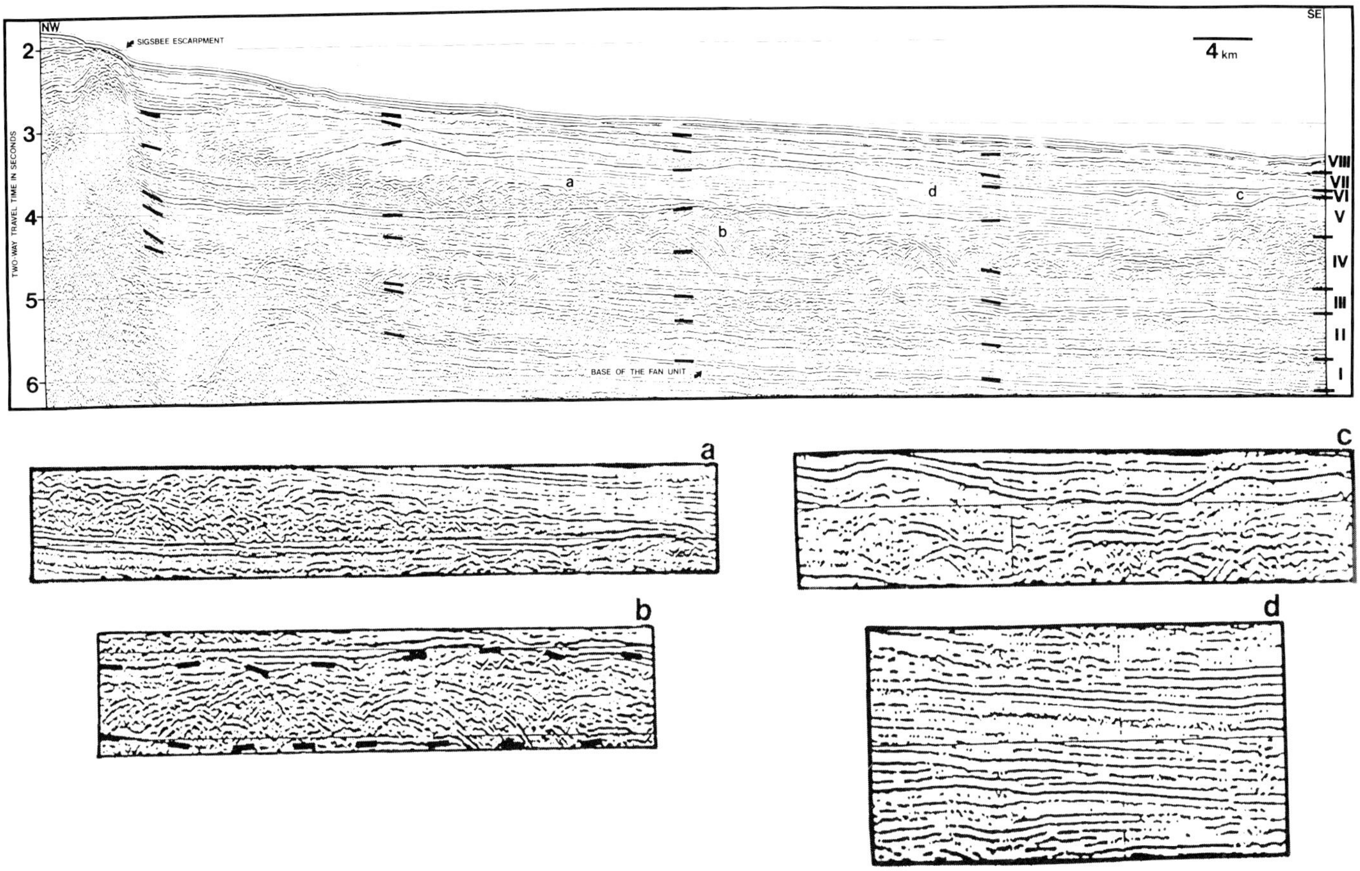

Figure 2. A 12-fold multichannel seismic line across the Mississippi Fan. Major sequences are indicated. Enlarged examples of mounded chaotic facies (a and b) and channelized lobe facies (c and d) are presented. For location, see Figure 1.

"channelized fan lobe" (Figs. 2c and d). The external geometry of the deposit is lenticular, with the channel facies associated with the thickest areas of each individual channelized lobe (Fig. 2d). The mechanism for the deposition of these two seismic facies may be visualized as a series of debris flows and/or turbidity flows flushed down an established channelized system (see Stelting and others, Chapter 41). Coarse material is confined to the channel axis. If the magnitude of individual flows is too great for the capacity of the channel, overflow of the channel banks may take place and sheet flow-type overbank material is deposited laterally.

High amplitude, moderate continuity, parallel to subparallel reflection patterns characterize the onlapping fill facies. It occurs predominantly as a fill between the mounded chaotic facies (Fig. 2b) on the upper and middle fan and tends to fill downdip of pre-existing topography on the middle to lower fan. The facies possibly results from turbidity current deposition.

The conformable drape facies is mainly associated with the unit boundaries (Fig. 2c). High continuity and amplitude reflectors are characteristic, and individual cycles can be traced over hundreds of kilometers. Changes in thickness are very gradual. Similar seismic facies have been associated with interglacial and late glacial age faunas, and are formed primarily during high sea-level stands by pelagic and hemipelagic deposition [5].

Growth Pattern

Lateral and Vertical Succession of Seismic Facies

Each of the identified seismic sequences is basically a separate fan complex with distinct physiographic, morphologic, and depositional zones [1]. Therefore, each sequence has a unique history, reflecting the complex interaction of three primary factors: sea-level fluctuations, salt tectonics, and the magnitude and position of sediment input. However, certain characteristics, particularly the order of depositional events and the distribution of seismic facies (both vertically and laterally) are common to the majority of the sequences. Table 1 summarizes this general vertical sequential succession of seismic facies. These are: 1) initial deposition of regionally extensive, mounded chaotic units; 2) deposition of an onlapping fill facies, which covers the mounded cha-

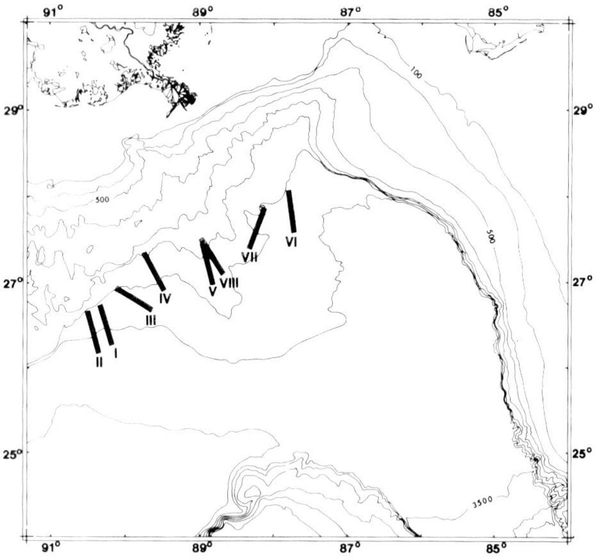

Figure 3. Map of updip position and orientation of the axis of maximum thickness for each of the eight fan sequences.

Table 1. General Vertical Succession of Seismic Facies Within an Individual Fan Sequence

Conformable drape
Channelized lobe (channel and overbank)
Onlapping fill
Mounded chaotic

otic units, thickening into topographic lows, and filling the lows formed downdip of pre-existing topography; 3) channelized lobe development with a dominance of channel and overbank deposits; and 4) blanket or conformable drape deposition of predominantly hemipelagic and pelagic sediments. The relative importance of each stage in this succession varies between sequences, but is best developed in the lower sequences (sequences I through VI).

In addition to the vertical succession of seismic facies, there are distinct transitions in the facies and, therefore, the inferred transport/deposition mechanisms downfan. Table 2 summarizes these changes of interpreted mechanisms from the upper fan to the lower fan for each of the seismic sequences. The upper-fan area is dominated by slump-type, thick mass-transport deposits. In addition, channelized turbidity currents and/or debris flows possibly form thick accumulations of channel and overbank deposits. In the middle-fan area, there is an increase in debris flow-type mass-transport facies. Also, as the channel narrows, turbidity currents are possibly less confined and thick accumulations of unchannelized turbidity flows may form on the outer flanks of the middle fan. Unchannelized density-flow deposits may comprise the majority of the lower-fan sediments. Channelized turbidity currents are not easily identified on the deeper-penetration seismic records. Debris flow deposits dominate the mass-transport regime as unstable sediment failed off adjacent margins, particularly the Campeche Escarpment.

Variations in Seismic Facies Distributions

Although the general growth pattern for fan development appears to be valid for the majority of the sequences, some changes in facies distributions are observed for the younger fan sequences. These changes are: 1) a general reduction in the measured surface areas and calculated volumes of progressively younger sequences, and 2) a general decrease in the volume of mounded chaotic facies, coupled with a general increase in the volume of the channelized lobe facies (particularly sequences VII and VIII).

These observations suggest a change in the depositional regime on the fan during Late Pleistocene. The decrease in inferred mass-transport deposits may be the result of either reduced direct effects of salt tectonics as the fan depocenter migrated beyond the Sigsbee Escarpment and/or a decrease in sediment contributions from escarpments bordering the lower-fan area. Although the mounded chaotic seismic facies are interbedded at the base of the escarpments, seismic patterns suggest channelized and unchannelized flows dominated the construction of the younger fan sequences.

Evolution of an Individual Sequence

The general cyclic vertical succession of seismic facies described earlier suggests the possibility of a common controlling mechanism on fan development. Stow and others [13] outlined three primary controls on fan development and deep-sea sedimentation in general. These are: 1) sediment type and supply, 2) tectonic setting and activity, and 3) sea-level fluctuations. Of these controls, sea-level fluctuations appear to be the dominant factor in individual sequence development on the Mississippi Fan [1].

The control on fan development by sea-level fluctuations suggests the following possible ordering of events. First, initial deposition within a sequence is possibly triggered by the initiation of a fall in sea-level. Progradation across the shelf-edge occurs as deltaic and nearshore deposits progressively shift seaward. Rapid deposition on the outer shelf and upper slope results in large-scale sediment failure and the accumulation of thick mounded chaotic units in the deep basin. Initiation of canyon development and possible erosion in the deep basin may take place during this stage.

During the late stages of the fall in sea-level, progradation of the shelf-edge continues. However, the rate of fall of sea-level has declined and there is a general reduction in the volume of material reaching the deep basin from along the margins. Sediment reaching the fan is deposited as a fill, thickening into topographic lows adjacent to the mounded chaotic facies. Next, with the rise in sea-level,

Table 2. Summary of the Changes in Dominant Transport/Deposition Mechanisms Downfan

SEDIMENTARY PROCESS		
Turbidity Currents	Mass Transport	
channelized	slumping	UPPER FAN
↓	slumping and debris flows	MIDDLE FAN
unchannelized	debris flows	LOWER FAN

channelized lobe development occurs. Other sediment sources to the fan are cut off, and deposition is apparently confined to a major canyon and an associated channel of the fan itself. Channelized density flows dominate the depositional regime. During the last stages of the rise and a relative highstand of sea-level, hemipelagic and pelagic deposition dominates, draping the deep-water fan deposits. Mass-transport events may occur, but their importance during this stage is minor.

Conclusions

1. Eight major seismic sequences comprise the Plio/Pleistocene section of the Mississippi Fan. These sequences are bounded by basin-wide unconformities.

2. Isopach maps of each sequence suggest that the axis of maximum accumulation for each sequence has migrated both eastward and seaward through time.

3. Significant contributions to the fan have come from sources other than the Mississippi Embayment, particularly from the Campeche Escarpment, the Florida Escarpment, and the DeSoto Canyon.

4. The basic transport/deposition mechanisms that are believed to be significant are: 1) mass transport (slump, debris flow), 2) turbidity current flow (channelized and unchannelized), and 3) pelagic/hemipelagic deposition.

5. Each sequence is basically a separate fan complex with a unique history and distinct physiographic, morphologic, and depositional zones.

6. Mass transport appears to be a dominant process for deposition in the submarine environment. Interpreted mass-transport deposits may comprise up to 30% of an individual sequence.

7. Channelized lobe development (channel and overbank deposits) apparently occurs late in the evolution of the majority of the sequences and may be associated with a rise in sea level.

8. The following factors appear to have exercised the greatest control on the deposition of the fan sequences: 1) the rate and magnitude of sea-level fluctuation; 2) the amount and location of unstable sediment deposited on the outer shelf and upper slope; 3) the position of submarine canyon development and its association with a source of sediment on the shelf; 4) salt tectonics; and 5) the position of pre-existing fan deposits.

References

[1] Feeley, M. H., 1984. Seismic stratigraphic analysis of the Mississippi Fan. Ph.D. dissertation, Texas A&M University, College Station, Texas, 209 p.

[2] Fisk, H. N., and McFarlan, E., 1955. Late Quaternary deltaic deposits Mississippi River. In: A. Poldervaart (ed.), Crust of the Earth. Geological Society of America Special Paper 62, pp. 279–302.

[3] Walker, J. R., and Massingill, J. L., 1970. Slump features on the Mississippi Fan. Geological Society of America Bulletin, v. 81, p. 3101–3108.

[4] Huang, T. C., and Goodell, H. G., 1970. Sediments and sedimentary processes on the eastern Mississippi Cone, Gulf of Mexico. American Association of Petroleum Geologists Bulletin, v. 54, pp. 2070–2100.

[5] Sangree, J. B., and Widmier, J. M., 1977. Seismic interpretation of clastic depositional facies. American Association of Petroleum Geologists Memoir 26, pp. 165–184.

[6] Moore, G. T., and others, 1978. Mississippi Fan, Gulf of Mexico—physiography, stratigraphy and sedimentation patterns. In: A. H. Bouma, G. T. Moore, and J. M. Coleman (eds.), Framework, Facies, and Oil-trapping Characteristics on Upper Continental Margin. American Association of Petroleum Geologists Studies in Geology 7, pp. 155–191.

[7] Garrison, L. E., Kenyon, N. H., and Bouma, A. H., 1982. Channel systems and lobe construction of the eastern Mississippi Fan lobe. Geo-Marine Letters. v. 2, pp. 31–39.

[8] O'Connell, S., 1983. Lower Mississippi Fan depositional processes (abstract). EOS (Transactions, American Geophysical Union), v. 64, p. 241.

[9] Bouma, A. H., Stelting, C. E., and Coleman, J. M., 1983/84. Mississippi Fan: internal structure and depositional processes. Geo-Marine Letters, v. 3, p. 147–153.

[10] Addy, S. K., and Buffler, R. T., 1984. Seismic stratigraphy of the shelf and slope, northeastern Gulf of Mexico. American Association of Petroleum Geologists Bulletin, v. 68, pp. 1782–1789.

[11] Mitchum, R. M., Jr., 1978. Seismic stratigraphic investigation of the west Florida Slope, Gulf of Mexico. In: A. H. Bouma, G. T. Moore, and J. M. Coleman (eds.), Framework, Facies and Oil-trapping Characteristics on Upper Continental Margin. American Association of Petroleum Geologists Studies in Geology 7, pp. 193–223.

[12] Lindsay, J. F., Shipley, T. H., and Worzel, J. L., 1975. The role of canyons in the growth of the Campeche Escarpment. Geology, v. 3, pp. 533–536.

[13] Stow, D. A. V., Howell, D. G., and Nelson, C. H., 1983/84. Sedimentary, tectonic, and sea-level controls on submarine fan and slope-apron turbidite systems. Geo-Marine Letters, v. 3, pp. 57–64.

CHAPTER 38

Mississippi Fan Sedimentary Facies, Composition, and Texture

Dorrik A. V. Stow, Michel Cremer, Laurence Droz, William R. Normark, Suzanne O'Connell, Kevin T. Pickering, Charles E. Stelting, Audrey A. Meyer-Wright, and DSDP Leg 96 Shipboard Scientists

Abstract

Eight different sedimentary facies recognized in the Mississippi Fan sediments drilled during DSDP Leg 96 are defined on the basis of lithology, sedimentary structures, composition, and texture. Pelagic biogenic sediments are of minor importance volumetrically compared with the dominant resedimented terrigenous facies. Clays, muds, and silts are most abundant at all sites, with some sands and gravels within the mid-fan channel fill and an abundance of sand on the lower fanlobe. Facies distribution and vertical sequences reflect the importance of sediment type and supply in controlling fan development.

Introduction

This contribution documents the sedimentary facies recovered at the nine Mississippi Fan Sites during DSDP Leg 96 and summarizes the preliminary results of our sedimentological analyses.

Other chapters in this publication outline the general geological framework (Chapter 21) and certain specific attributes of the fan (Chapters 40, 42). The following points, however, are worth emphasizing for this facies analysis. The Mississippi Fan is a relatively large, mud-dominated, elongate-type fan [1,2]. The immediate source area is a major delta and prograding shelf with shifting submarine canyons. Very rapid fan construction occurred throughout the Pleistocene, with a complex interplay of sedimentary, sea-level, and tectonic controls on fan development.

The nine fan sites were drilled to depths of between about 150 and 525-m subbottom depth and were cored continuously. Core recovery was best in the top 80 to 100 m although, in most cases, a good suite of wire line logs has enabled us to interpret lithologies in the deeper parts of wells where core recovery was lower.

Sedimentary Facies

Eight sedimentary facies are recognized in Mississippi Fan sediments on the basis of lithology, sedimentary structures, composition, and texture (Fig. 1) (see also Chapters 40, 42, 44, 46). Calcareous biogenic sediments are volumetrically minor, but significant at certain horizons. They can be divided into two facies on the basis of the carbonate content. Terrigenous sediments are dominant and can be divided into six distinct facies, ranging from the finest-grained clays and muds to coarser-grained pebbly muds and gravels. There is some gradation between facies, and locally all occur intermixed within disturbed units.

1. Oozes and Muddy Oozes

The biogenic sediments are a minor but ubiquitous facies that were recovered as a relatively thin unit (5 to 50 cm) at the surface of most sites. Staining with Rodamin-B dye commonly showed that none of the organisms recovered were living, so that the actual thickness of the unit on the seafloor is probably slightly greater than the core interval obtained. The facies also occurs as a thick unit (about 30 m) at the base of the deepest hole penetrated on the lower fan (Site 615).

In the surficial biogenic layer, there is no internal bedding or other primary sedimentary structure visible. The sediment appears homogeneous and is probably thoroughly bioturbated. It is very poorly sorted, with a fine-sand to silt-size grade for the biogenic component and a variable fine-silt to clay grade, terrigenous admixture. It is a yellowish brown, marly calcareous ooze in which planktonic foraminifers are dominant, nannofossils and siliceous or-

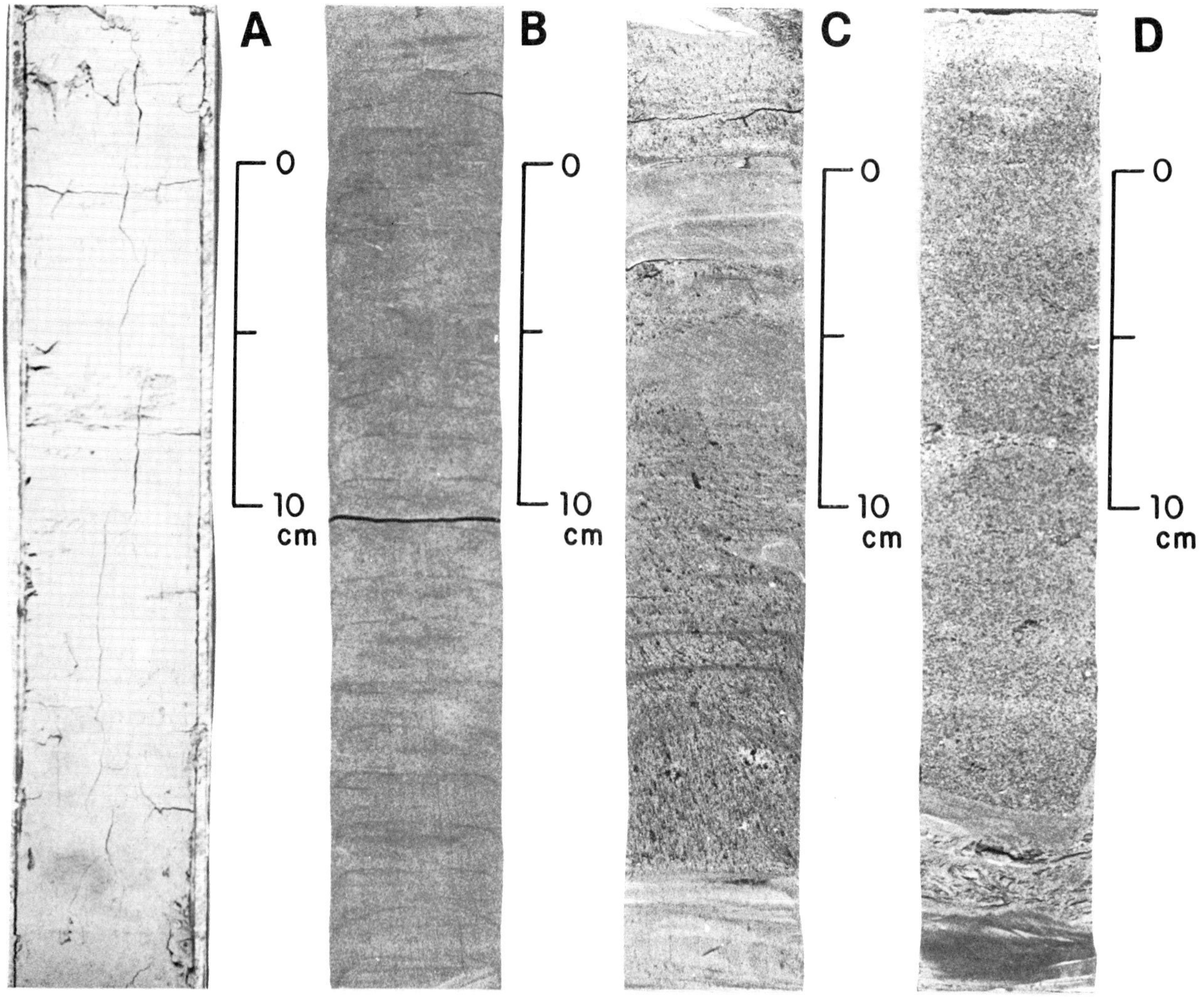

Figure 1. Photographs of seven of the sedimentary facies from Mississippi Fan cores (subbottom depths correspond to top of each photograph). (A) Ooze, very fine-grained, nannofossil dominant (Site 615; 480.9 m). (B) Mud with very thin, dark mud laminae (Site 616; 306.0 m). (C) Silty mud, poorly-sorted and carbonaceous near base grading up to fine mud (Site 615; 192.4 m). (D) Sand, lower part of medium-thick graded sand bed (Site 623; 65.9 m). (E) Silt-laminated mud, occurring as probable graded laminated units (Site 621; 158.6 m). (F) Pebbly mud (Site 621; 195.5 m). (G) Gravel (Site 621; 214.3 m).

ganisms form less than 10% of the sediment, and terrigenous material comprises up to 25% of the sediment. Rare, black, authigenic iron-sulphide-rich mottles are present.

The light bluish-gray to yellowish gray oozes recovered near the base of Site 615 are also relatively homogeneous and structureless when observed visually (Fig. 1A). However, there are subtle grain size variations within an overall normally graded sequence that extends through the upper 28 m of recovered section. This ooze grades from a thin (10 cm) coarse gravelly layer at the base, with chalk and shelf-depth bioclastic debris up to 15 mm in size, through a shelly, foraminferal-rich nannofossil ooze to a very fine-grained, pure nannofossil ooze in the top several meters. The biogenic material consists of a high percentage of reworked Cretaceous, Pliocene, and Pleistocene forms as well as some contemporary Pleistocene planktonics. This sequence overlies approximately 1 m of very fine-grained Pleistocene pelagic nannofossil ooze and calcareous mud without reworked fauna.

2. *Calcareous Muds*

There is a complete gradation between the biogenic oozes and calcareous mud facies, the distinction being made on the basis of carbonate percent. At some sites, the surficial biogenic-rich layer contains less than 50% $CaCO_3$ and is more properly termed "calcareous mud." It is structure-

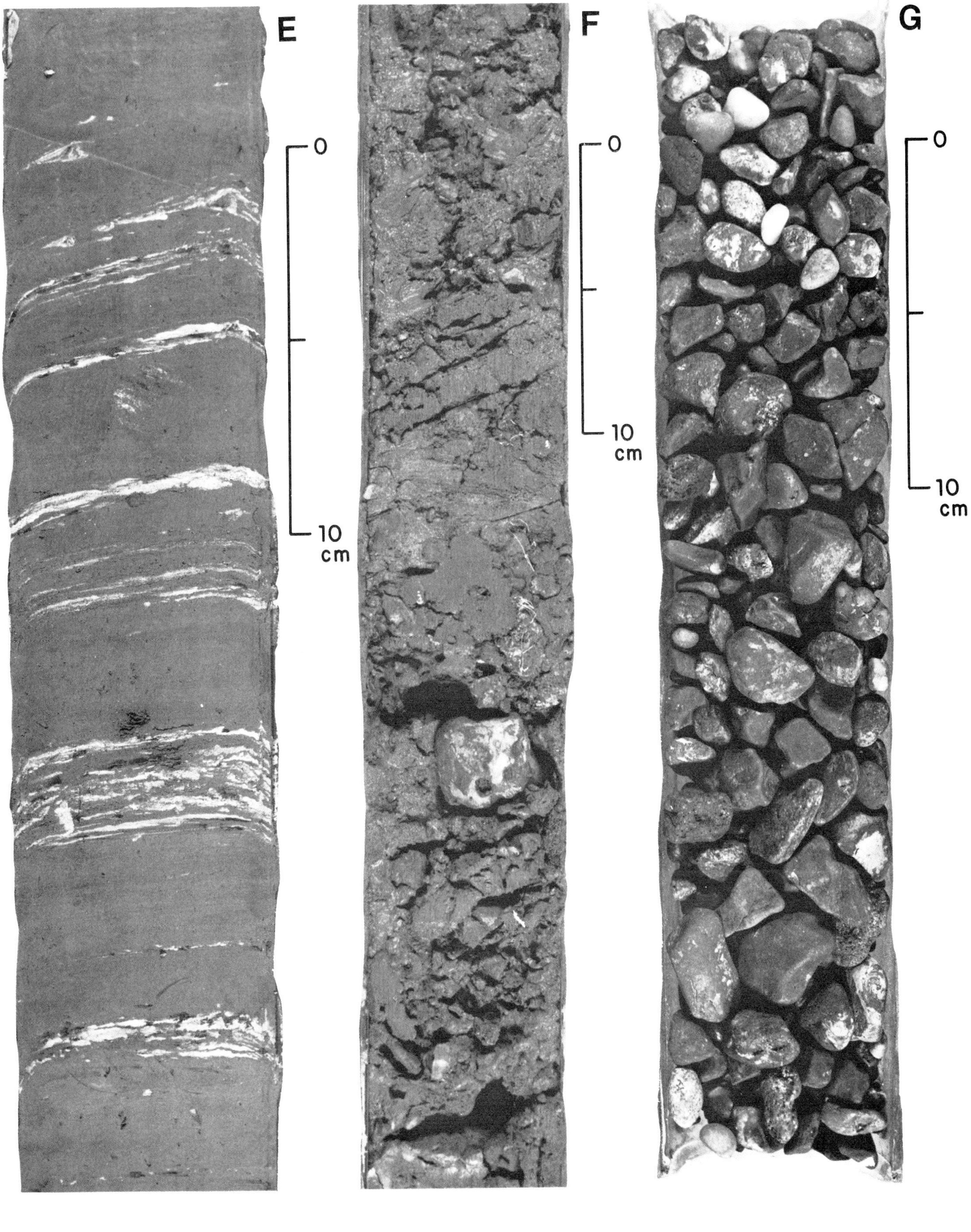
E
0
10
cm
F
0
10
cm
G
0
10
cm

less, fine-grained, and has a poorly-sorted admixture of sand-sized planktonic foraminifers, calcareous nannofossils, rare siliceous biogenics, and terrigenous silt and mud.

The very bottom 50 cm of recovered section at Site 615 is a brownish-colored calcareous mud with up to 15% foraminifers and nannofossils that underlies the nannofossil ooze. It is mainly structureless or, in part, indistinctly laminated.

3. Clays and Muds

This sedimentary facies represents the very finest-grained terrigenous sediments recovered, including the fine muds and true clays, having between 60 and 90% clay-size fraction and generally less than 0.5% sand-size material. These sediments occur in thin to very thick units, commonly without any clear bedding or primary sedimentary structures. Probable bioturbational mottling, however, is rare so that the homogeneity appears primary. In other cases, there are rare, very thin, silt laminae or a distinct color banding, commonly accentuated by dark-colored, iron-sulphide-rich bioturbationally mottled layers. Much of the apparently structureless muds have a very subtle, regular banding (Fig. 1B) only evident on close inspection or on X-radiographs (Chapter 44).

These clays and muds are dominantly terrigenous (quartz, feldspar, and clay minerals), with a small (< 5%) percentage of calcareous nannofossils, including both contemporary Pleistocene and reworked Pliocene forms. In the upper parts of the mid-fan channel sites, they locally occur as dark-colored gas-disrupted muds.

4. Silty Muds and Muddy Silts

The coarser-grained muds and poorly sorted silts form a facies gradational with the finer clays and muds. They contain between 10 and 60% clay and up to approximately 5% sand. This sedimentary facies forms beds from about 5 cm to 1m or more in thickness, or occurs as very thick, essentially unbedded, visually structureless intervals. Silt-sized quartz and clay minerals are the dominant components, with minor feldspar, carbonate grains, micas, lignite, and heavy minerals. Many of the grains appear to be partially altered or coated with iron-oxides.

This facies also includes distinctive dark-colored lignite-bearing, silty mud beds, ranging from about 5 to 50 cm in thickness. These lignite-bearing beds occur in three main types: 1) those that are completely structureless with gradational contacts; 2) those that are organized into distinct beds, in some cases with indistinct normal grading and floating mud clasts; and 3) those that occur as clearly graded beds, commonly forming part of a thicker graded bed from laminated silt or sand at the base to fine-grained homogeneous mud or clay at the top (Fig. 1C).

5. Silt-Laminated Muds

The most common sediments at many of the sites are silt-laminated muds, occurring over intervals of a few centimeters to a few tens of meters in thickness. This sedimentary facies ranges from uniform muds with only 5 to 10% thin silt laminae to muds with over 50% silt laminae and thin silt beds (Fig. 1E). Visually observable laminae frequency may reach 400 to 500 per meter of section. However, the very thin silt laminae are difficult to resolve visually and X-radiographs show a still greater abundance in parts. The thicker laminae commonly show internal parallel lamination or micro-cross lamination and slight normal grading. The bases are commonly sharp, locally scoured, loaded, and with flame structures; the tops may be sharp or gradational.

In many cases, the laminae are more or less regularly spaced and apparently ungrouped. However, at least three types of groupings or graded laminated units are recognized, each ranging from about 3 to 10 cm in thickness: 1) units of up to 10 to 15 laminae that show a regular upward decrease in thickness and grain size of laminae; 2) units with fewer silt laminae that grade upwards through gray, reddish, and gray-black mottled mud, and 3) more irregular units with discontinuous and lenticular laminae showing load, flame, and micro-slump structures indicative of very rapid deposition.

This facies is compositionally and texturally very similar to the silty mud and muddy silt facies, being fine-grained and dominantly terrigenous, but with a much better sorting in terms of separation of the silt and clay fractions. The silts locally include significant angular detrital carbonate and, more rarely, volcanic ash.

6. Silts and Sands

Silts and sands are a common facies in parts of the fan, and occur in intervals from less than 10 cm to over 10 m in thickness. Sand loss by wash out and section increase by flow-in during the coring process mean that some of the thickest (1.5 to 10 m) sandy intervals recovered probably do not represent single bed thicknesses (see Chapter 36).

The thicker beds commonly appear to be structureless, whereas most of the thinner sand and silt beds show some internal sedimentary structures. Many of the beds show clear positive grading (Fig. 1D). These are commonly organized in partial Bouma T_a to T_b sequences with massive, parallel, and cross-laminated divisions. The bottom contacts are invariably sharp and commonly loaded or scoured; the upper contacts are either sharp or gradational.

Grain size varies both within and between beds. The maximum size at the base of the thicker beds is as much as 5 mm (pebble-sized). The mean size is most commonly fine to medium sand (125 to 250μm), and there is a high proportion of silt. The larger grains are commonly well-rounded, spherical or elongate, and highly polished. The thinner beds tend to be better-sorted, medium to coarse silt-sized (16 to 63μm), and with a maximum size rarely exceeding 150μm (fine sand). The finer grains are often highly angular and irregular in shape. There are rare medium- to coarse-grained thin sand beds.

The sands and silts are dominantly terrigenous and quartzose with minor biogenic material.

7. Muddy Gravels and Pebbly Muds

This is a relatively rare facies encountered only at the two mid-fan channel sites in intervals up to 4-m thick (Chapter 40). Pebbles are as much as several centimeters in diameter, very poorly sorted, and set in a clay-silt-sand matrix (Fig. 1F). There are no bedding or internal structures evident. Clasts include chertz, quartz, jasper, mudstones, and shell fragments.

8. Gravels

True clast-supported gravel was recorded only in a 60 cm-thick section near the base of Site 621 in the channel thalweg (Fig. 1G). Clasts range up to 3 cm in size, are very poorly sorted, and have a composition similar to that of the pebbly mud facies. The clasts are mostly rounded to subrounded in shape, and show an abrupt grading over a few centimeters into overlying medium-grained sands. The coring process may have washed out any fine-grained matrix and disturbed any original structure that might have been present.

Sediment Composition

There is a broad compositional similarity of sediments within any one facies, as well as between many of the facies. This uniformity is reflected in the sand and silt mineralogy (thin-section and grain-mount data), clay mineralogy (X-ray diffraction analyses), inorganic geochemistry (X-ray fluorescence spectrometry), and carbonate content (bomb analyses). Standard analytical techniques were used in each case.

1. Sand and Silt Mineralogy

The sand and sandy silt beds are uniformly terrigenous (95 to 98%). Quartz is the dominant mineral, with secondary feldspars, micas, and carbonates, and accessory heavy minerals, glauconite, and lithic fragments. The heavy mineral suite commonly includes amphiboles, pyroxenes, epidote, zircon, tourmaline, and opaque grains. The small biogenic fraction (2 to 5%) comprises foraminifers, shallow-water shell debris, and lignitic material.

The generally finer-grained, thin silt laminae show a similar composition to the thicker sand beds but commonly have, in addition, a variable and significant proportion (10 to 25%) of clastic carbonate material of undetermined origin. The silt laminae as well as the dispersed silt fraction of the silty mud and muddy silt facies appears relatively richer in altered or iron-stained grains of indeterminate composition. Volcanic ash is locally important.

2. Clay Mineralogy and Inorganic Geochemistry

The less than 4-μm size fraction from all eight facies types (120 samples) was analyzed by X-ray diffraction and semi-quantitative estimates of mineral abundances made from peak-height and peak area measurements. A generalized "average" value shows that the four main clay minerals identified (kaolinite, chlorite, illite, and smectite) are each present in comparable proportions, although smectite is locally more abundant, quartz and feldspars occur in lesser amounts (about 8 and 5%, respectively), and calcite, dolomite, and aragonite are variably present in minor quantities.

In fact, the ranges of clay mineral abundances are quite large, although it is difficult to correlate this variability with differences in facies or location on the fan. The only apparent facies differences occur in the oozes and muddy oozes, which show relatively greater calcite and aragonite, and in the thick sands, which have relatively greater quartz and feldspar, a higher chlorite to kaolinite ratio, and a generally smaller clay-size fraction. It is partly these facies differences that are also reflected in the observed regional variations that show the lower-fan sites to have relatively more quartz and feldspar than the mid-fan sites, and the mid-fan overbank sites to have relatively more carbonate than the mid-fan channels or the lower-fan sites.

Inorganic geochemical data from some 150 ground whole-rock samples have still to be analyzed in detail. The generalized "average" composition of major element oxides is: SiO_2 (50 to 60%), Al_2O_3 (10 to 15%), FeO/Fe_2O_3 (4 to 6%), MgO (2 to 3%), CaO (2 to 4%), Na_2O (2%), and K_2O (2 to 5%) with minor amounts of MnO, TiO_2 and P_2O_5. The SiO_2 abundance varies more widely than this range, but in an inverse relation with CaO and Al_2O_3. Trace element abundances measured are all relatively low to average compared with data from other deep-sea sediments.

3. Carbonate Content

The percentage of carbonate was measured for over 200 samples and shows wide variation from 0 to more than 80%. The true oozes contain over 75% carbonate, the muddy oozes have an admixture of up to 50% terrigenous material, and the calcareous muds range from 10 to 50% carbonate. In each of these facies, the carbonate is dominantly pelagic foraminifers and nannofossils; however, in the resedimented oozes at the base of Site 615, benthic foraminifers and shallow-water shell debris are also present.

The terrigenous facies mostly contain less than 10% carbonate (rarely up to 18%), and this component is a mixture of mainly reworked pelagic biogenics and carbonate silt of indeterminate origin. The lower-fan sites (Sites 614 and 615) average 2.8% and the mid-fan channel sites average 3.7% carbonate, whereas the overbank sites on both the mid-fan (Sites 616, 617, and 620) and lower-fan sites (Sites 623 and 624) average around 8% carbonate. These apparent regional differences may be related to facies differences, because there is less carbonate in both the silt-sand and clay-mud facies than in the silt-laminated mud facies.

Sediment Texture

The grain size characteristics just described for each of the separate facies were determined from some 120 granulometric analyses using the sieve and pipette method. The differences between sedimentary facies are clearly distinguished using either a triangular plot of sand-silt-clay percentage (Fig. 2a) or typical cumulative frequency curves (Fig. 2b). The thick-bedded coarser-grained sands and finest-grained clays are both relatively well-sorted, but with a distinct fine tail (hyperbolic curve). The silt-laminated muds appear less well-sorted with a broad fine tail (hyperbolic-logarithmic curve), although individual silt and mud laminae show much better sorting when analyzed separately. The silty mud facies are poorly sorted with a broad coarse tail (parabolic-logarithmic curve). Only a few analyses are presently available for the ooze and calcareous mud facies, and these generally show an irregular very poorly sorted distribution (logarithmic-tending curve).

All the sediments drilled are unconsolidated with moderate to very high water contents and porosity values. Physical property measurements are reported elsewhere (Chapter 43) and grain shape and surface texture analyses have not yet been completed.

Discussion

1. Facies Interpretation

Apart from the thin surface layer of calcareous muds and oozes slowly deposited by pelagic or hemipelagic settling, most of the sediment recovered shows evidence of resedimentation from shallower water. This evidence includes: 1) the very rapid rates of sedimentation (6 to 12 m/1000 yr) (Chapter 39); 2) the dominant terrigenous composition with land-derived plant material and a sparcity of contemporary planktonic tests (Chapters 39, 44); 3) the abundance of primary sedimentary structures suggesting deposition from turbulent suspension, debris flows, or sediment slides (Chapters 44, 45); and 4) the almost complete absence of secondary biogenic structures.

In detail, however, there are certain aspects of these resedimented facies that still require interpretation. The clays and muds, in particular, although very rapidly deposited, are commonly structureless with little evidence of the type of flow from which they were deposited. In some cases, barely perceptible, very thin, darker-colored, mud laminae occur at a spacing of 1 to 3 cm through several meters of section. These appear to be primary in origin, rather than caused by coring disturbance, and are similar to the thick-bedded unifite muds studied by Stanley [3], perhaps representing deposition from very large, slump-derived, muddy turbidity currents. In other cases, there is color banding on a centimeter to decimeter scale, with irregular dark iron-sulphide mottling that suggests bioturbational activity. The sediments may have been deposited partly as thin-bedded mud turbidites and partly from relatively concentrated hemipelagic suspensions.

The silty mud and muddy silt facies are also enigmatic in their origin. Where graded, they are similar to the disorganized turbidites studied by Stow and Piper [4], perhaps resulting from very rapid deposition or from a poorly developed turbidity current. Where structureless, they might be better interpreted as having settled out of concentrated hemipelagic suspensions. The silt-laminated muds, in contrast, show clear evidence of deposition from normal, low concentration, turbidity currents, having many of the characteristics of fine-grained turbidites [4–6].

The thin- to very thick-bedded silts and sands commonly show evidence of deposition from high concentration turbidity currents [7], the thicker beds perhaps having been influenced by grain flow or fluidized flow in the final stages of deposition [8]. The structureless aspect of many of these thick sands and the possibility of coring disturbance make firm interpretation difficult. The apparent graded top of the lone gravel unit from the channel thalweg site also suggests turbidity current transport, whereas the pebbly muds presumably result from debris flows.

Mass movement is exhibited as local, small-scale, overturned folds and microfaults most commonly in the mid-fan levee site (Site 617) and less frequently within the channel and the lower-fan sites. The top 90 m of the section at Site 616 has undergone mass movement, as indicated by highly inclined lamination, overturned folds, and possible repeat sections. It is not yet clear whether this is indeed a single far-travelled megaslide, as proposed by Walker and Massingill on the basis of seismic evidence [9], or a series of large (10 to 15-m thick) slide blocks of perhaps more local origin.

2. Facies Distribution and Sequences

The percentage of different sedimentary facies present in the recovered section at each site is shown in Table 1. The

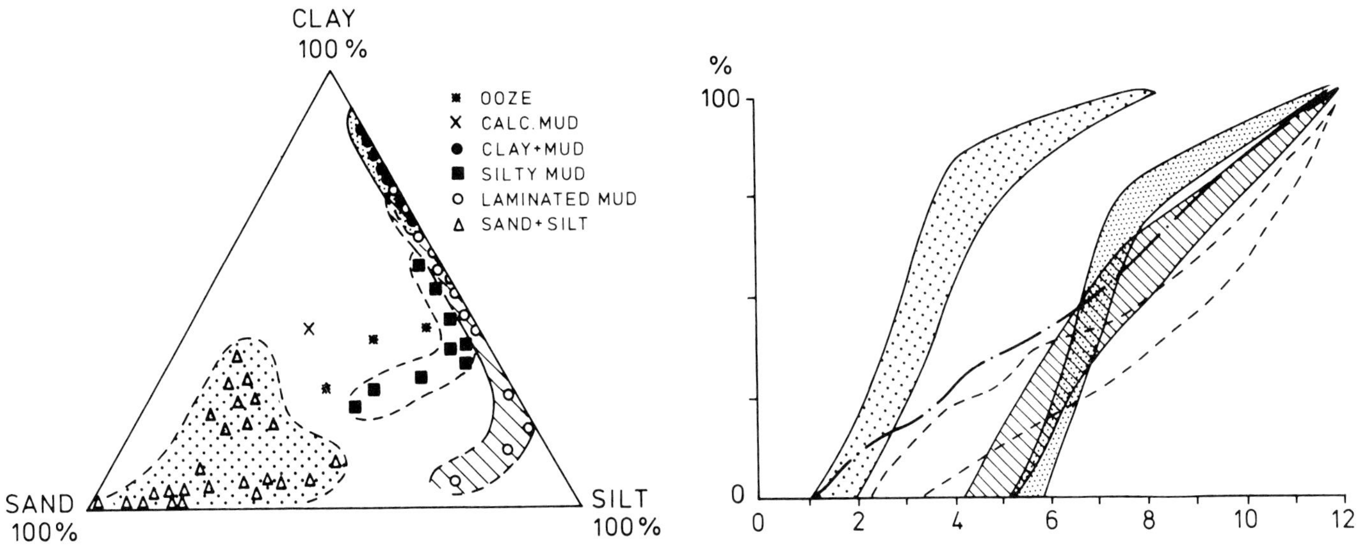

Figure 2. Grain size characteristics of Mississippi Fan sediments. (A) Triangular plot of sand-silt-clay percentages. (B) Typical cumulative frequency curves for different sedimentary facies.

biogenic facies are of minor importance, occurring only as a thin surface veneer over the fan and at the base of the deepest site. The fine-grained terrigenous facies are dominant everywhere. Silt-laminated muds are most abundant close to the channel on both the mid- and lower-fan, and also at Site 616 where they probably represent overbank deposits adjacent to a former channel. The clay and mud facies are more abundant both away from the channel and as thick, passive channel-fill deposits. The coarser-grained silt and sand facies are most abundant in mid-fan channel sites. Wire line logs suggest that core recovery was more complete in the fine-grained facies, so that actual percentages of sands and silts are relatively higher than recorded and might be as much as 60 to 70% on the lobes and 20% in the channels. Pebbly muds and gravels are a very minor facies, recovered only in the channels.

The vertical sequences in which the various facies occur are described in more detail in papers in this volume dealing specifically with the mid-fan and lower fan (Chapters 40, 42). We simply note here that although clear trends of grain size and bed thickness are observed, vertical sequences are rather more variable than those classically related to fan deposits [8]. In particular: 1) the lower-fan lobe sites show coarsening-upward, fining-upward, blocky, and irregular sequences over tens of meters of section; 2) the mid-fan channel sites show somewhat irregular fining-upward sequences and a monotonous mud fill, and 3) the mid-fan levee and overbank sites show coarsening-upward, fining-upward, and symmetric sequences. Smaller-scale sequences over 2 to 10 m of section in the lower fan lobe sites might be considered similar in origin to the compensation cycles described by Mutti and Sonnino [10], although coarsening, fining and symmetrical sequences are all present.

References

[1] Bouma, A. H., Stelting, C. E., and Coleman, J. M., 1983/84. Mississippi Fan: internal structure and depositional processes. Geo-Marine Letters, v. 3, pp. 147–154.

Table 1. Percentages of Different Sedimentary Facies in Recovered section at Each Mississippi Fan Site

	Middle Fan					Lower Fan			
	Overbank		Channel		Margin	Lobe		Channel/Levee	
Facies/Sites	617	620	621	622	616	614	615	623	624
1. Oozes and muddy oozes	0.1	0.1	0.2	0.1	0.1	1	17	0	<0.1
2. Calcareous muds	0.1	0	0	0	0	0.1	0.3	0	0
3. Clays and muds	16	72	68	50	22	14	10	28	38
4. Silty muds and muddy silts	2	8	10	4	10	23	10	10	5
5. Silt-laminated muds	82	20	14	33	65	14	25	56	56
6. Silts and sands	0	0	4	12	3	48	38	6	1
7. Muddy gravels and pebbly muds	0	0	3	0.5	0	0	0	0	0
8. Gravels	0	0	0.4	0	0	0	0	0	0

[2] Stow, D. A. V., Howell, D. G., and Nelson, C. H., 1983/84. Sedimentary tectonic and sea-level controls on submarine fan and slope-apron turbidite systems. Geo-Marine Letters, v. 3, pp. 57–64.

[3] Stanley, D. J., 1981. Unifites: structureless muds of gravity-flow origin in Mediterranean basins. Geo-Marine Letters, v. 1, pp. 77–84.

[4] Stow, D. A. V., and Piper, D. J. W. 1984. Deep-water fine-grained sediments: facies models. In: D. A. V. Stow and D. J. W. Piper (eds.), Fine-Grained Sediments: Deep-Water Processes and Facies. Geological Society London Special Publication 15, Blackwell Scientific Publications, Oxford, pp. 611–645.

[5] Stow, D. A. V., and Shanmugam, G., 1980. Sequence of structures in fine-grained turbidites: comparison of recent deep-sea and ancient flysch sediments. Sedimentary Geology, v. 25, pp. 23–42.

[6] Piper, D. J. W., 1978. Turbidite muds and silts on deep-sea fans and abyssal plains. In: D. J. Stanley and G. Kelling (eds.), Sedimentation in Submarine Canyons, Fans and Trenches. Dowden, Hutchinson and Ross, Stroudsburg, PA, pp. 163–176.

[7] Bouma, A. H., 1962. Sedimentology of Some Flysch Deposits. Elsevier, Amsterdam, 168 pp.

[8] Walker, R. G., 1978. Deep water sandstone facies and ancient submarine fans: models for exploration for stratigraphic traps. American Association of Petroleum Geologists Bulletin, v. 62, pp. 932–966.

[9] Walker, J. R., and Massingill, J. V., 1970. Slump features on the Mississippi Fan, NE Gulf of Mexico. Geological Society of America Bulletin, v. 81, pp. 3101–3108.

[10] Mutti, E., and Sonnino, M., 1981. Compensation cycles: a diagnostic feature of turbidite sandstone lobes. International Association Sedimentologists, 2nd European Regional Meeting, Bologna, Abstracts, pp. 120–123.

CHAPTER 39

Biostratigraphy and Sedimentation Rates of the Mississippi Fan

Barry Kohl and DSDP Leg 96 Shipboard Scientists

Abstract

Paleontologic studies of DSDP Leg 96 cores support several mechanisms for deposition of Quaternary sediments on the Mississippi Fan: 1) density flow, 2) debris flow, and 3) suspension load. The constituents of the faunal assemblages suggest that the sediments were derived from: 1) upper Mississippi Valley—reworked Cretaceous foraminifers and radiolarians, 2) Louisiana continental shelf—displaced shallow water benthic foraminifers, and 3) Florida continental shelf—displaced carbonate bank assemblage.

Calculated sedimentation rates for the fan are 2 to 13 cm/1000 yr for the Holocene, 600 to 1100 cm/1000 yr for the late Wisconsin glacial, and 71 cm/1000 yr for the Wisconsin interstadial.

Introduction

Nine sites penetrating Wisconsin glacial sediments were cored in the Mississippi Fan in water depths ranging from 2400 to 3300 m. Five of these sites are located on the middle fan and four are on the lower fan. They are located between 300 to 550 km from the present Mississippi River Delta (Fig. 1).

The paleontologic problems encountered on Leg 96 are rather unique for the Deep Sea Drilling Project. High sedimentation rates (11 m/1000 yr for mid-fan and 6 m/1000 yr for lower-fan sites) caused a significant dilution of planktonic and benthic foraminiferal assemblages. Calcareous nannoplankton floras in the glacial sediments are dominated by reworked Cretaceous forms. The strong regional reflectors seen on reconnaissance seismic surveys were initially considered to be hemipelagites. But the only deep-sea planktonic oozes encountered were a thin (< 2 m) Holocene veneer on the present sea floor and a foraminifer-calcareous nannofossil-rich Wisconsin interstadial debris flow (Horizon 30) at Site 615 on the lower fan.

Biostratigraphy

A biostratigraphic framework for the Gulf of Mexico proposed by Ericson and Wollin [1] and expanded by Kennett and Huddleston [2] is used to subdivide the late Quaternary. In addition to foraminifers and calcareous nannoplankton commonly used in the Gulf of Mexico, tephrochronology, oxygen isotope stratigraphy, and paleomagnetics are also employed to date the Pleistocene sediments of Leg 96.

Figure 2 summarizes the zonation for the late Quaternary and includes the oldest age material penetrated (Zone W, early Wisconsin glacial). Older sediments were not encountered because of the high sedimentation rates and shallow penetration of the coreholes, which normally did not exceed 200 m.

Criteria for Recognition of Ericson Zone Boundaries

Zone Z: The top of the Holocene is defined by the sediment-water interface (see additional discussion under Sedimentation Rates). The Z/Y boundary or base of the Holocene is usually recognized in the Gulf of Mexico by the disappearance, Last Appearance Datum (LAD), of the cool-water species *Globorotalia inflata* below and the coincident reappearance of warm-water *Globorotalia menardii* group above the boundary. However, at the Mississippi Fan sites, the high sedimentation rate during the late Wisconsin (associated with density flows) diluted the planktonic fauna, so that occurrences of *Gl. inflata,* usually common to abundant in Wisconsin glacial sediments elsewhere in the Gulf of Mexico, are rare. The selection of the Z/Y boundary is further complicated by a long transition zone that contains rare occurrences of *Gl. menardii* at some

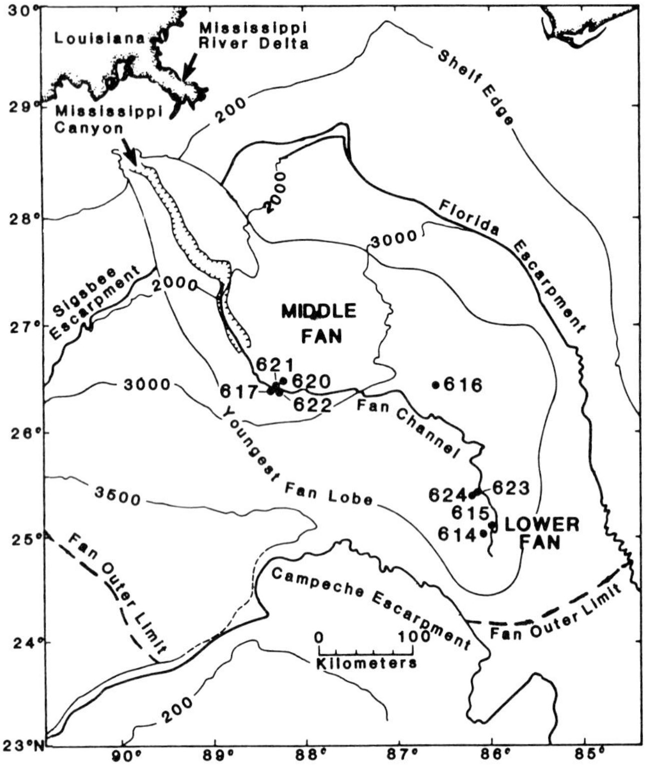

Figure 1. Generalized map showing the location of the Mississippi Fan sites drilled on DSDP Leg 96. Site 616 is not discussed in this paper.

sites. The base of Zone Z at the Mississippi Fan sites is, therefore, placed near the common reoccurrence of the warm-water species *Gl. menardii* and *Gl. tumida.*

Zone Y: The late Wisconsin glacial (Ericson Zone Y) in the Gulf of Mexico is typically dominated by low-diversity benthic foraminiferal faunas and the common to abundant cool-water species *Gl. inflata*. However, the Mississippi Fan sites lacked this assemblage, and *Cibicides wuellerstorfi* and *Melonis pompilioides*, usually associated with bathyal water depths, were not present (possibly because of higher turbidity conditions on the sea floor). The base of Zone Y is marked by the disappearance (LAD) of *Globorotalia flexuosa* and the coincident dominance of *Emiliania huxleyi* over *Gephyrocapsa spp.* above the boundary [3]. Williams [4] correlates the X/Y boundary to isotope stage 5b (Fig. 2).

Zone X: The *Globorotalia flexuosa* Zone (Wisconsin interstadial) in the Gulf of Mexico is recognized by the common occurrence of *Gl. flexuosa* and an abundance of the warm-water *Gl. menardii* complex. There is usually an increase in the diversity of the entire foraminiferal fauna. Generally, *Gl. inflata* is absent in Zone X, but is common to abundant in Zone W (early Wisconsin glacial). A change from dominant warm-water planktonics above to dominant cool-water planktonics and the absence of *Gl. flexuosa* below marks the X/W boundary.

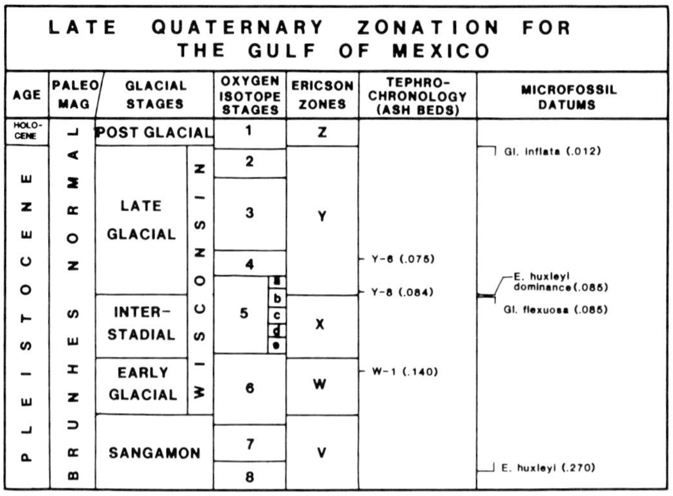

LATE QUATERNARY ZONATION FOR THE GULF OF MEXICO
AGE
PALEO MAG
GLACIAL STAGES
OXYGEN ISOTOPE STAGES
ERICSON ZONES
TEPHRO-CHRONOLOGY (ASH BEDS)
MICROFOSSIL DATUMS
HOLOCENE
PLEISTOCENE
BRUNHES NORMAL
POST GLACIAL
LATE GLACIAL
INTER-STADIAL
EARLY GLACIAL
WISCONSIN
SANGAMON
1
2
3
4
5
a
b
c
d
e
6
7
8
Z
Y
X
W
V
Y-6 (.075)
Y-8 (.084)
W-1 (.140)
Gl. inflata (.012)
E. huxleyi dominance (.085)
Gl. flexuosa (.085)
E. huxleyi (.270)

Figure 2. Late Quaternary zonation for the Gulf of Mexico. Modified after Rabek and others [10] and Williams [4]. The dates used in column 6 and 7 are in Ma.

Zone W: The uppermost portion of the early Wisconsin glacial was encountered only at Site 615. The W/V boundary was not penetrated at any of the Mississippi Fan sites.

Depositional Processes

Once the general age zonation was established for the fan sites, the main biostratigraphic interpretation dealt with paleoenvironmental analysis, using reworked and indigenous foraminifers to assist in determining sedimentation patterns and local correlations between sites. Occurrences of Quaternary radiolarians and tasmanitids (cysts of pelagic algae) also assisted in local correlations.

Sandy units are generally devoid of bathyal benthic foraminifers, but abraded shallow-water benthic foraminifers are dispersed throughout the coarser clastic sections. Genera such as *Ammonia, Elphidium, Hanzawaia,* and *Amphistegina* were probably derived from the Louisiana Continental Shelf and transported as clasts. Some clay units, from channel Sites 621 and 622, contain well-preserved shallow-water specimens which were probably transported as part of a density flow.

Reworked late Cretaceous planktonic foraminifers and radiolarians are also associated with the coarser clastic silty and sandy units. A study of late Wisconsin terrace deposits in Missouri by Thompson [5] found reworked Cretaceous species which are similar to those in the Leg 96 fan sites. This supports an upper Mississippi River Valley source for the reworked late Cretaceous planktonic foraminifers. Chalk clasts with well-preserved Cretaceous foraminifers were found at several fan sites. These occurrences may explain the excellent state of preservation of some late Cretaceous planktonics in the same samples as highly abraded specimens.

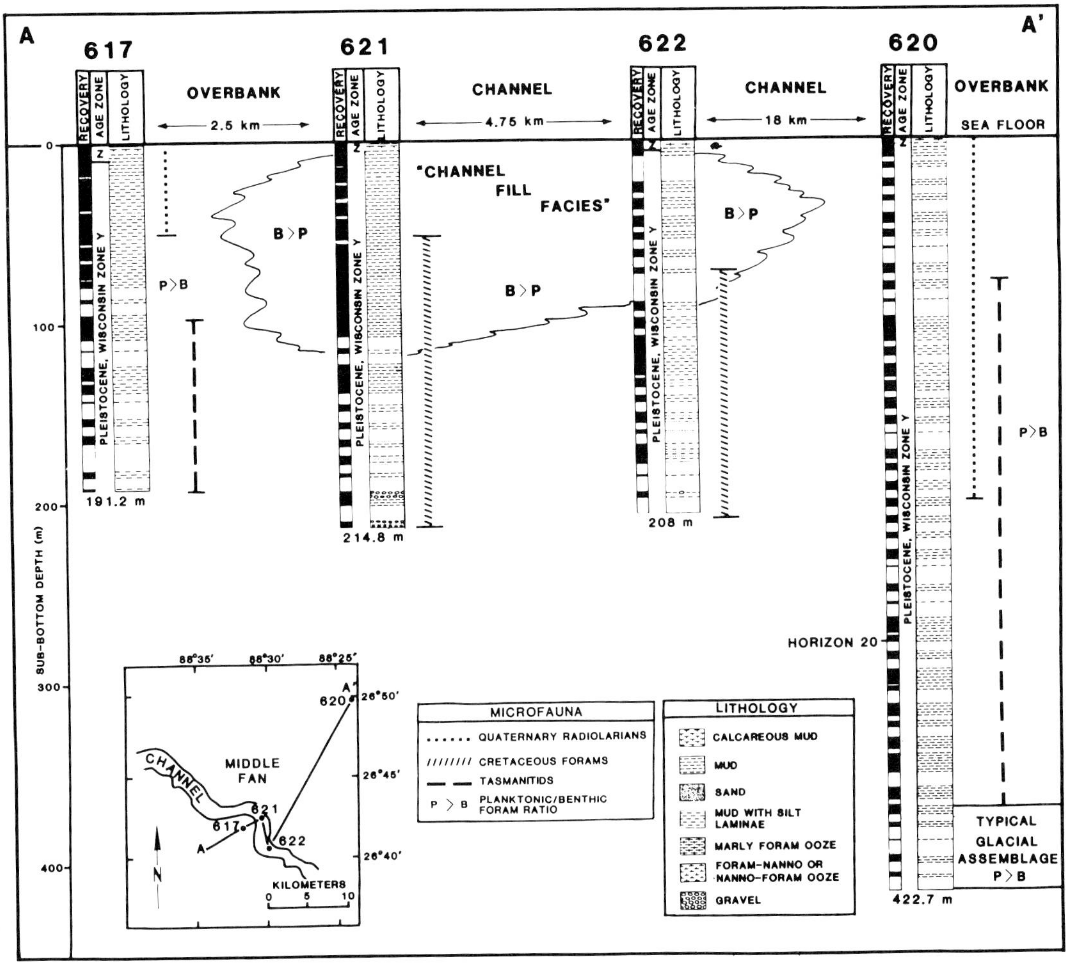

Figure 3. Cross-section across the middle-fan sites. The "channel fill facies" as well as the reworked Cretaceous foraminifers and radiolarians are confined to Sites 621 and 622. Common occurrences of tasmanitids and Quaternary radiolarians are found only in the overbank sites.

Middle-Fan Sites

Five sites were cored on the middle fan. Site 616 was selected to core the Walker-Massingill slump [6] and will not be discussed in this paper. The four other sites which cored the sinuous channel region are shown in the cross-section (Fig. 3). Sites 617 and 620 are overbank sites while Sites 621 and 622 are in the channel (Stelting and others, Chapter 41). All four sites penetrated sediments no older than the late Wisconsin glacial (Ericson Zone Y).

Channel Sites 621 and 622 have a thick (1.5 m) Holocene interval underlain by a clay-rich "channel fill facies." Channel Site 621 is dominated (from 5 to 110-m subbottom) by an *Elphidium-Ammonia-Hanzawaia* assemblage with abundant organic debris, rare planktonic foraminifers, and the absence of bathyal benthics. The sandy section below 110 m shows a decrease in foraminifers and an increase in reworked Cretaceous planktonics such as *Heterohelix*, *Hedbergella*, and *Globigerinelloides*. Site 622 has a similar "channel fill facies" from 30 to 90-m subbottom with *Ammonia* and *Elphidium* as the dominant shallow-water genera (Fig. 3).

In contrast, overbank Sites 617 and 620 have a more abundant planktonic foraminiferal fauna; shallow-water benthics and reworked Cretaceous planktonics are absent. Planktonic faunas are most abundant at Site 620, which is 18 km from the center of the channel; bathyal benthics such as *Cibicides wuellerstorfi* and *Melonis pompilioides* occur below 370-m subbottom with common planktonics and the cool-water species *Globorotalia inflata*.

Tasmanitids, cysts of pelagic chlorophyllous algae from a low-salinity littoral environment [7], are present only in the overbank sites, indicating that the cysts may be deposited from the water column rather than from a density flow.

The absence of shallow-neritic benthic foraminifers at the two overbank sites implies that the benthics found in Sites 621 and 622 were part of density flows of which the coarse-grained fraction was restricted to the channel. The sediments at the overbank sites appear to have been deposited by very fine material overflowing the channel and by a significant amount of fine sediment settling out from the water column.

Quaternary radiolarians are rare but persistent in the upper part of Zones Y and Z at the overbank sites, but absent at the channel sites. This relationship also supports the idea of rapid bottom sedimentation at the channel sites with a very slight contribution to sediment accumulation from the water column. The increase in abundance of tasmanitids, restricted to Zone Y, may coincide with high meltwater runoff during the late Wisconsin glacial.

The graveliferous deposit at the bottom of Site 621 is dominated by chert clasts, some up to 4.0-cm in length. Impressions of crinoids and brachiopods of Carboniferous age in the chert imply that these clasts are probably derived from the upper Mississippi River Valley.

Lower-Fan Sites

Four sites were cored on the lower fan. Sites 623 and 624 are located in a buried channel fill/overbank complex, while Sites 614 and 615 are located near the distal ends of the channel terminations (Fig. 1) (O'Connell and others, Chapter 43). All sites (except site 615) encountered sediments no older than Ericson Zone Y (late Wisconsin glacial). However, the deepest site (Site 615) penetrated the Wisconsin interstadial (Ericson Zone X) and bottomed in Ericson Zone W (early Wisconsin glacial) (Fig. 4).

Sites 623 and 624 have a thin Holocene (Zone Z) planktonic ooze, 0.25-m thick, underlain by a section dated as Ericson Zone Y. These late Wisconsin glacial sediments have a very sparse Pleistocene foraminiferal fauna with rare shallow water (neritic) benthic species. Reworked Cretaceous calcareous nannoplankton, foraminifers, and radiolarians occur throughout the late Wisconsin glacial and are associated with high deposition rates. Rare Quaternary radiolarians occur in the Holocene and uppermost portion of Zone Y. These occurrences are interpreted as indicating a reduction of the sedimentation rate for these intervals. This relationship is similar to that of overbank Sites 617 and 620 on the middle fan.

Sites 614 and 615 are located approximately 40-km south of Sites 623 and 624. The Holocene ooze is 0.5-m thick at Sites 614 and 615. Ericson Zone Y (late Wisconsin glacial) is characterized by a very sparse planktonic foraminiferal assemblage and rare, commonly abraded, shallow-water (neritic) benthic foraminifers associated with the coarser clastic intervals. Abraded *Amphistegina*, bryozoa, and pelecypod fragments are dispersed throughout this section.

The Wisconsin interstadial (Ericson Zone X) occurs at Site 615 at 485-m subbottom. This contact (Core 48, core catcher) is marked by an abrupt lithologic change with a brown terrigenous clay above and gray, carbonate-rich nannofossil ooze below. Since the remainder of the cores in the carbonate recovered almost 100% of the ooze, it is assumed that the sample from the core catcher of Core 48 was taken at the *bottom* of the cored interval rather than at the top as is the normal convention.

The carbonate is a calcareous nannofossil-foraminiferal ooze, 30-m thick. The upper 29 m can be divided into two intervals based on size-sorting of the clasts. The upper 10 m contains a nannofossil ooze mixed with juvenile planktonic foraminifers grading downward to common adult-sized planktonic foraminifers greater than 149 μm. The lower 19 m grades from a normal-size planktonic foraminiferal ooze downward to a carbonate gravel at the base which contains some rare chert pebbles.

Common-to-abundant *Gl. menardii* and rare-to-common *Gl. flexuosa* in the carbonate indicate warm-water conditions. Dispersed throughout is a mixture of late Cretaceous (*Globotruncana spp.*), Miocene and Pliocene planktonic foraminifers and Pliocene, Miocene, and Eocene calcareous nannofossil species. Occurrences of calcareous algae (*Lithothamnion*), bryozoa, coral fragments, barnacle plates, and abraded *Amphistegina gibbosa* suggest transport as a debris flow from a neritic environment, possibly the Florida escarpment to the east.

The lowermost 1 m of the carbonate ooze contains a warm-water Pleistocene foraminiferal fauna dominated by *Gl. menardii* and occurrences of *Gl. flexuosa* with no reworked Pliocene or older foraminifers. It appears that this interval may be in place and represent normal hemipelagic deposition.

Below 515-m subbottom, the fauna changes to a planktonic ooze dominated by *Gl. inflata* with only rare occurrences of *Gl. menardii*. This cool-water fauna is interpreted as part of the lower Wisconsin glacial interval (Ericson Zone W).

Sedimentation Rates

Holocene: A summary of the nondecompacted sedimentation rates is shown in Figure 5. The sedimentation rate for the Holocene (Ericson Zone Z) ranges from 2.1 to 12.5 cm/1000 yr and is based on a 0.012 Ma date for the base of the Holocene (Y/Z boundary). The thickest accumulation of Holocene sediment is found in channel Sites 621 and 622 on the middle fan; the thinnest accumulation is at Sites 617, 623, and 624.

Since the sinuous central channel is a depression (25 to 45-m deep), the Holocene planktonic ooze could accumulate to greater depths than at the overbank sites. "Sand waves" were seen on EDO deep-towed side-scan sonar [8] and were originally thought to be formed in terrigenous sand by bottom currents. However, bottom cores at channel Sites 621 and 622 show that the sand is bioclastic, composed predominantly of planktonic and bathyal benthic foraminiferal tests.

Samples of the upper few centimeters at each site were collected and stained with Rose-Bengal dye to determine whether they contained living benthic foraminifers. None were found. The absence of stained (living) foraminifers may result from a wave generated by the Hydraulic Piston

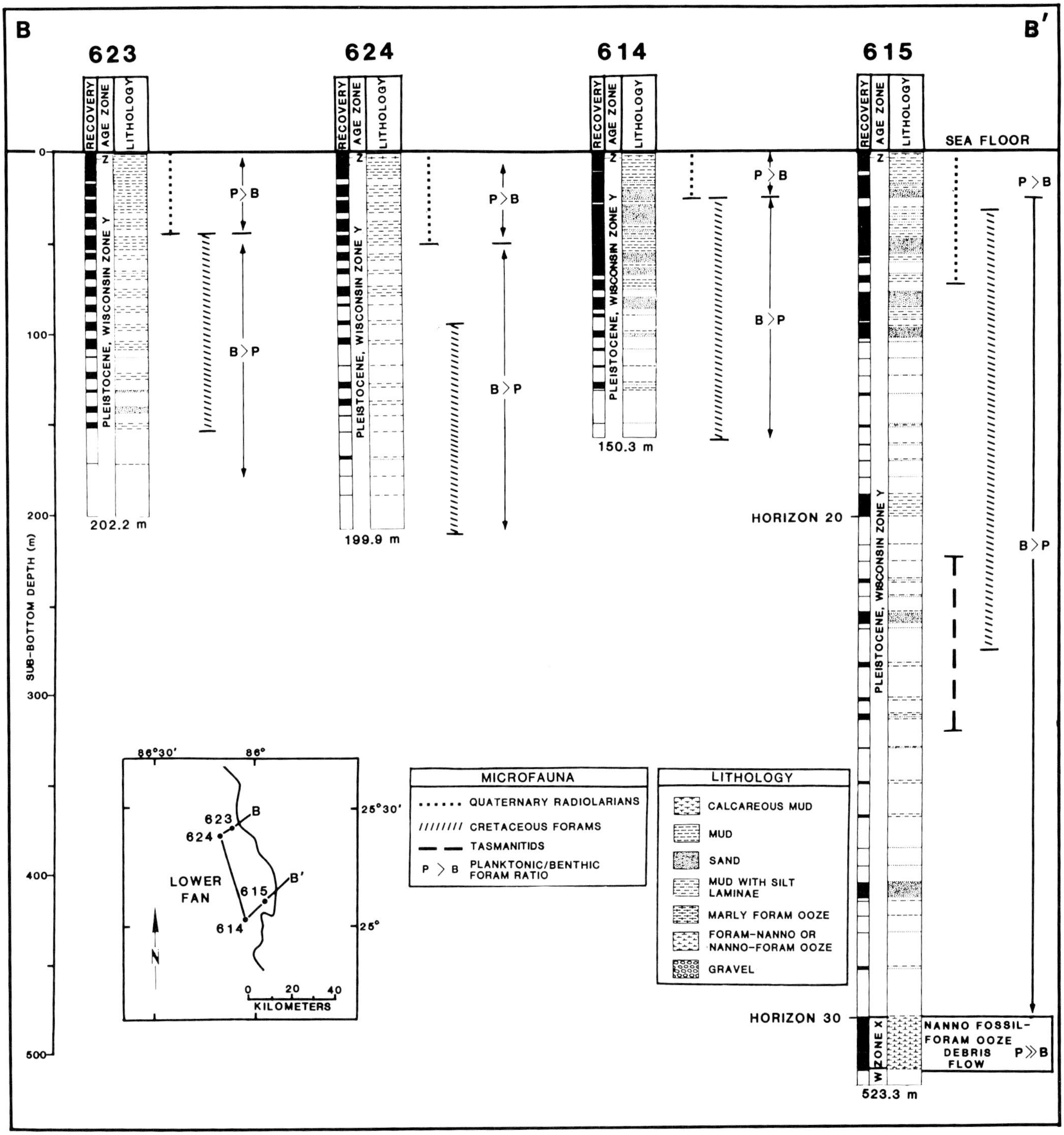

Figure 4. Cross-section across the lower-fan sites. Quaternary radiolarians occur near the top of each corehole, with reworked Cretaceous foraminifers and radiolarians confined to the sandier sections below. The common tasmanitids occurrence in Site 615 projects below the total depth of the other sites. The Wisconsin interstadial (Zone X) is penetrated at 485-m subbottom.

Corer (HPC) which, as it approaches the sea floor, "blows away" the upper few centimeters of sediment, or by bottom currents that scour the sea floor and remove the sediment containing the living foraminifers. Sedimentation rates presented for the Holocene, therefore, are minimum values.

Late Wisconsin Glacial: The sedimentation rate for the late Wisconsin glacial (Ericson Zone Y) is calculated using a date of 0.085 Ma for the Y/X boundary (coincident with the LAD of *Gl. flexuosa*). This date was chosen because of Y-8 ash (dated at 0.084 Ma [9], and occurring just above

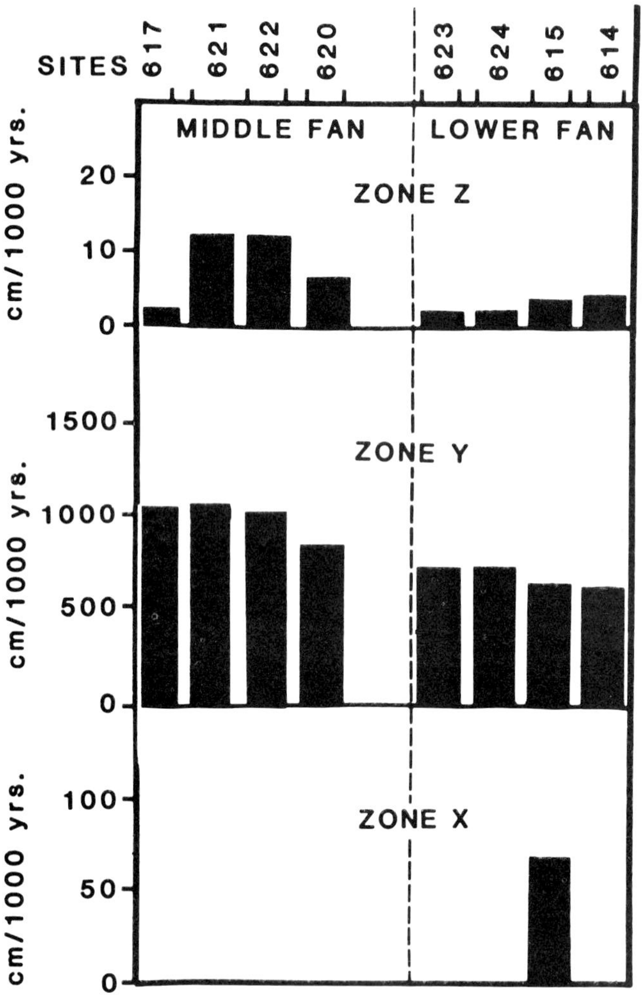

Figure 5. Undecompacted sedimentation rates of Mississippi Fan sediments, Leg 96. The sedimentation rates for Ericson Zone Y are based on seismic projections to Horizon 30 (Y/X boundary). Site 615 shows the actual rate based on the penetrated thickness of Zone Y.

the Y/X boundary and in Isotopic Stage 5b [4] in the Gulf of Mexico) (Fig. 2). The duration of Zone Y is 0.073 Ma. The Y/X boundary, as found in Site 615, coincides with seismic "Horizon 30" which is mapped throughout the Mississippi Fan (Bouma and others, Chapter 21). A seismic projection down to Horizon 30 was made at each site on the Mississippi Fan and a sedimentation rate was calculated on the basis of the projected thickness of Zone Y. These seismic projections are used in Figure 5. Sedimentation rates range from 6.4 to 10.7 m/1000 yr. Highest values occur at middle-fan sites and lowest values occur at lower-fan sites.

Wisconsin Interstadial: The sedimentation rate for the Wisconsin interstadial (Ericson Zone X) is calculated using a date of 0.127 Ma for the X/W boundary (close to the isotopic stage boundary 5/6, [4]. Zone X was encountered only at Site 615 in the lower fan and shows a sedimentation rate of 71 cm/1000 yr. The actual rate may be much lower if the upper 29 m of the Wisconsin interstadial are turbidite or debris flow deposits.

Ericson Zone V was not penetrated at any of the Mississippi Fan sites.

Summary

1) Relative sedimentation rates can be inferred from abundances of planktonic foraminifers and occurrences of indigenous bathyal benthic faunas. The presence of tasmanitids may be related to increased meltwater intervals and the Quaternary radiolarians conversely related to normal marine conditions.

2) Sandy units are generally devoid of bathyal benthic foraminifers. They contain common, reworked, late Cretaceous planktonic foraminifers and radiolarians derived from the upper Mississippi River Valley.

3) Shallow (neritic) reworked benthic foraminifers present in the channel sites and lower fan were derived from the Louisiana Continental Shelf. These occurrences support the conduit model for transport of coarser clastics over 560 km to the lower fan.

4) The overbank sites are characterized by lower sedimentation rates, occurrences of bathyal benthic faunas, and the absence of reworked Cretaceous foraminifers and radiolarians.

5) Calcareous bank assemblages in Site 615 were probably derived from the Florida Escarpment and Continental Shelf. This turbidite-debris flow may have occurred as several pulses of sedimentation. This zone is equivalent to seismic Horizon 30.

6) The nondecompacted sedimentation rates for the fan sites range from 2 to 13 cm/1000 yr for Zone Z, 6 to 11 cm/1000 yr for Zone Y, and 71 cm/1000 yr for Zone X.

Acknowledgments

The author thanks Chevron USA and W.P.S. Ventress for supporting my participation on DSDP Leg 96 and for providing the technical assistance needed to prepare this manuscript. I am obliged to the following people: R. L. Fleisher and Paula J. Quinterno for critically reviewing this paper and providing comments that improved the text, Gisela Galjour for typing the manuscript, Debbie Koffskey for preparing the illustrations.

References

[1] Ericson, D. B., and Wollin, G., 1968. Pleistocene climates and chronology in deep-sea sediments. Science, v. 162, pp. 1227–1234.

[2] Kennett, J. P., and Huddleston, P., 1972. Late Pleistocene paleoclimatology, foraminiferal biostratigraphy and tephrochronology, western Gulf of Mexico. Quaternary Research, v. 2, pp. 38–39.

[3] Thierstein, H. R., and others, 1977. Global synchroneity of Late Quaternary coccolith datums: validation by oxygen isotopes. Geology, v. 5, pp. 400–404.

[4] Williams, D., 1984. Correlation of Pleistocene marine sediments of the Gulf of Mexico and other basins using oxygen isotope stratigraphy. In: N. Healy-Williams (ed.), Recent Advances in Pleistocene Stratigraphy Applied to the Gulf of Mexico. International Human Resources Development Corporation Press, Boston, MA, pp. 65–118.

[5] Thompson, T. L., 1983. Late Cretaceous marine foraminifers from Pleistocene fluviolacustrine deposits in eastern Missouri. Journal of Paleontology, v. 57, pp. 1304–1310.

[6] Walker, J. R., and Massingill, J. V., 1970. Slump features on the Mississippi Fan, northeastern Gulf of Mexico. Geological Society of America Bulletin, v. 81, pp. 3101–3108.

[7] Williams, G. L., 1978. Dinoflagellates—acritarchs and tasmanitids. In: B. W. Haq and A. Boersma (eds.), Introduction to Marine Micropaleontology. Elsevier, Amsterdam, pp. 293–326.

[8] Bouma, A. H., Stelting, C. E., and Coleman, J. M., 1983–1984. Mississippi Fan: internal structure and depositional processes. Geo-Marine Letters, v. 3, pp. 147–153.

[9] Ledbetter, M., 1984. Late Pleistocene tephrochronology in the Gulf of Mexico region. In: N. Healy-Williams (ed.), Recent Advances in Pleistocene Stratigraphy Applied to the Gulf of Mexico. International Human Resources Development Corporation Press, Boston, MA, pp. 119–148.

[10] Rabek, K., Ledbetter, M. T., and Williams, D. F., 1985. Tephrochronology of the western Gulf of Mexico for the last 185,000 years. Quaternary Research (in press).

CHAPTER 40

Drilling Results on the Middle Mississippi Fan

Charles E. Stelting, Kevin T. Pickering, Arnold H. Bouma, James M. Coleman, Michel Cremer, Laurence Droz, Audrey A. Meyer-Wright, William R. Normark, Suzanne O'Connell, Dorrik A. V. Stow, and DSDP Leg 96 Shipboard Scientists

Abstract

The middle-fan area of the youngest Mississippi fanlobe is a convex-shaped aggradational body with a sinuous, migratory channel located along its apex. The middle-fan/lower-fan boundary corresponds to a change from the sinuous channel pattern to one of lateral switching and abandonment. Sites drilled in the mid-fan channel indicate an upward-fining fill, commencing with gravel and ending with clay. The overbank deposits are primarily muddy with some silty and very-fine sandy turbidites. Sedimentation rates during the Late Wisconsin age for both the channel fill and overbank areas are about 12 m/1000 yr.

Introduction

The middle-fan area of the youngest Mississippi fanlobe commences at the base of slope at a water depth of about 2000 to 2400 m and merges with the lower fan at about 3000 to 3200 m (Bouma and others, Chapter 21). While the upper fan area is basically a conduit with some overall aggradational characteristics, the middle fan is an active aggradational system (Bouma and others, Chapters 21, 36). The fanlobe in the middle-fan area is lenticular in cross-section, with a maximum width of about 200 km and a maximum thickness of about 400 m. Morphologically, the mid-fan consists of a channel-levee complex centered along the apex of the fanlobe and is flanked by extensive overbank deposits.

The mid-fan channel is highly sinuous in nature, asymmetric in cross-section, and bounded morphologically by levees. Near the mid-fan drill sites, the channel width is 3 to 4 km and its bathymetric relief ranges from 25 to 45 m. Downfan, the dimensions of the channel and its sinuosity decrease.

Analyses of side-scan sonar images (GLORIA, Sea MARC I, and EDO) show a wide variety of morphologic features (Fig. 1) [1–3]. Lineations and bedforms suggestive of sand-sized deposits are present on the surficial channel floor [3]. Drilling indicated that this sand-sized material consists of foraminiferal ooze. Features that are morphologically similar to fluvial ridge and swale structures are observed on both side-scan sonar records and high resolution seismic reflection profiles (Figs. 1 and 2). Several other side-scan sonar images adjacent to the channel also encourage a comparison with fluvial morphology (Fig. 1). Away from the channel, side-scan sonar images show only low-relief irregularities [1,3].

Reconstructions of channel development, using watergun and airgun seismic reflection profiles, suggest several depositional episodes in both the channel and overbank environments (Stelting and others, Chapter 41). The lowermost part of the present channel fill corresponds to the top of a zone of acoustically high amplitude reflections on seismic profiles. This zone is much wider than the present channel floor, suggesting lateral migration during the most active periods of fanlobe construction (Fig. 3).

Four sites were drilled in the upper region of the middle fan; two in the present channel and two outside the channel (Fig. 1). Site 621 is located in the axis (thalweg) of the channel on the outer side of a meander; Site 622 is on the inner side ("pointbar") of the next meander loop downchannel. Site 617 is in one of the swales on the inside of the meander loop about 4.8-km southwest of Site 621. Site 620 is located outside the zone with ridge and swale features, in overbank deposits about 18-km northeast of the channel.

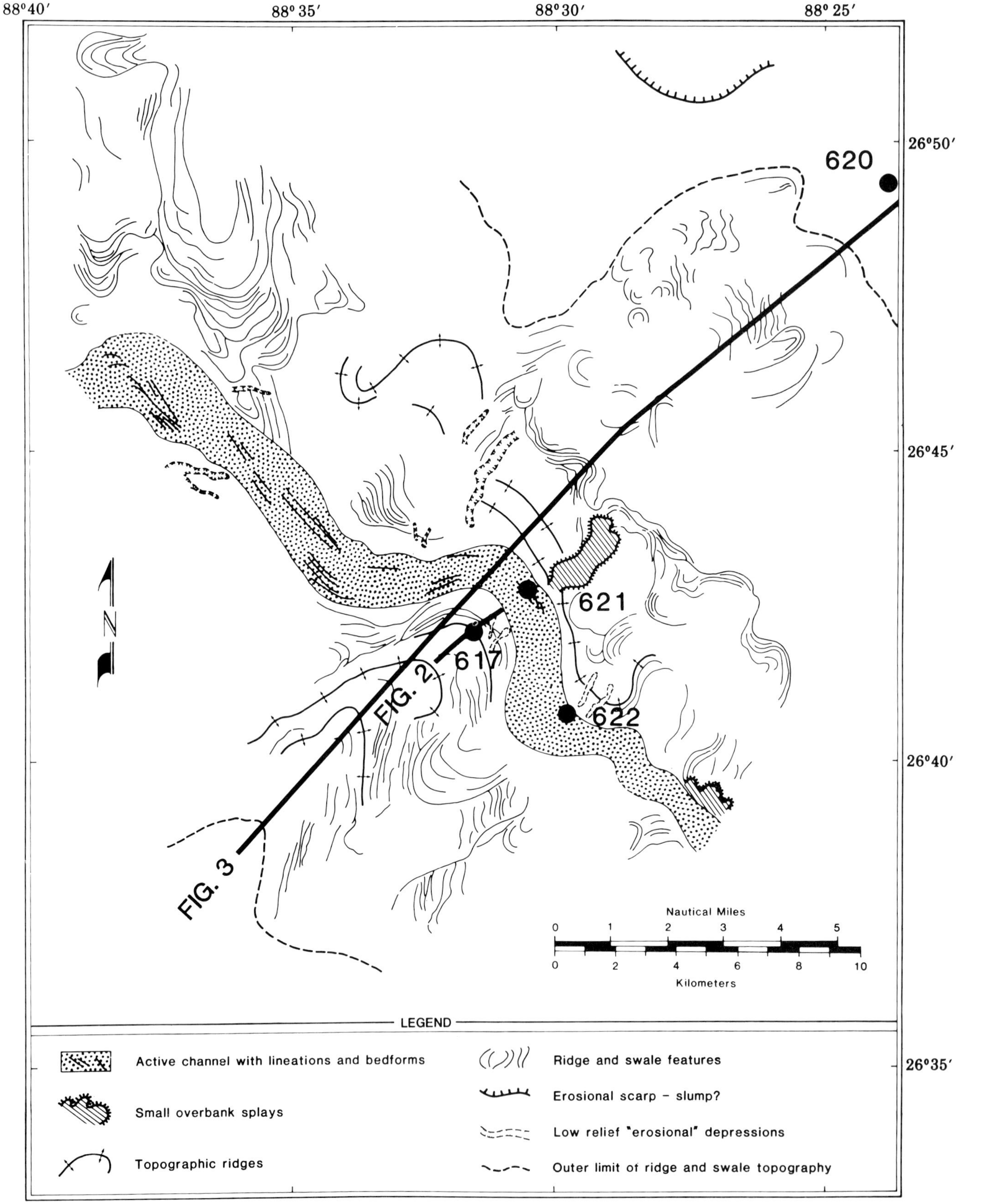

Figure 1. Morphologic features of the middle-fan region as mapped from side-scan sonar. Locations of the DSDP drill sites and seismic reflection profiles (shown in Figures 2 and 3) are indicated.

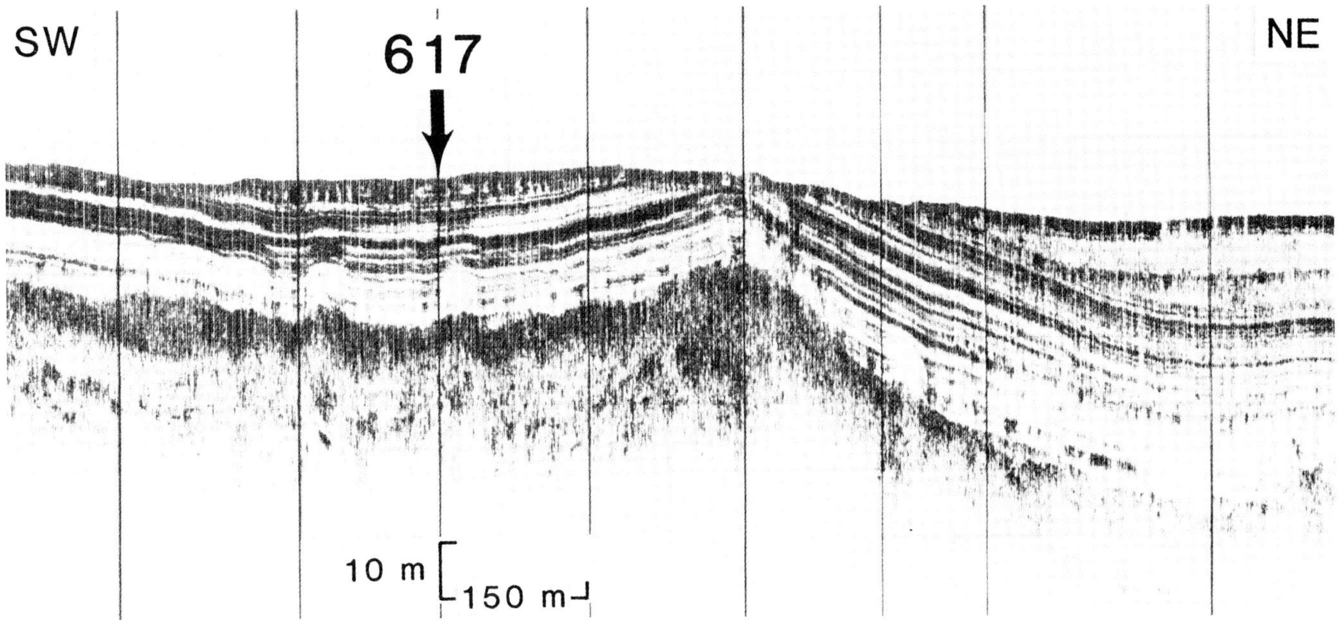

Figure 2. High resolution (4.5 kHz) EDO deep-towed seismic reflection profile over ridge and swale topography near Site 617. Conformable fill, truncation of reflectors over the top of the ridges, and a deeper seismically opaque zone are shown. See Figure 1 for location. (Printed by permission of D. B. Prior; Louisiana State University, Coastal Studies Institute).

Seismic Stratigraphy

Single- and multichannel seismic reflection profiles over the middle fan show a chaotic alternation of semitransparent to strongly reflective, irregular, more or less continuous to discontinuous, parallel reflectors [4,5]; high amplitude reflectors beneath the modern channel correspond to the lower channel fill (Fig. 3) (Stelting and others, Chapter 41). Generally, the higher the resolution of the seismic system, the more distinct these different seismic facies become (Bouma and others, Chapter 36).

Over the ridge and swale area, near Site 617, a very high resolution 4.5-kHz record from the EDO deeptow system shows erosion over the tops of the ridges. Acoustically,

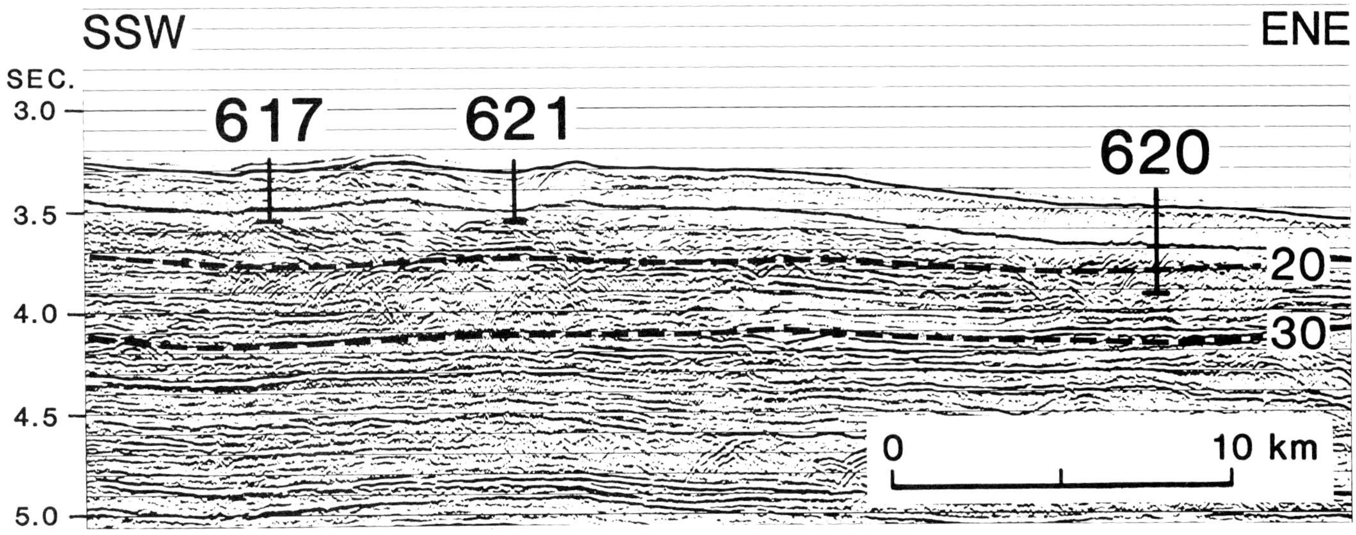

Figure 3. Multichannel seismic reflection profile (Line MC 11-A) over the middle-fan region. Major seismic reflectors "20" and "30", seismic facies (including high amplitude reflectors beneath modern channel), and the location of Sites 617, 621, and 620 are shown. See Figure 1 for location. (Printed by permission of R. T. Buffler; University of Texas, Institute for Geophysics).

sediments filling the swales generally show a conformable pattern. Both the ridges and the swales are underlain by a seismically opaque zone on these 4.5-kHz records (Fig. 2).

Sedimentary Characteristics

Sediments recovered from the four mid-fan sites can be divided into eight sedimentary facies. The characteristics, mineralogy, and texture of each of these sedimentary facies are described in more detail by Stow and others (Chapter 38) and Roberts and Thayer (Chapter 47). Several of these facies are minimally represented in the mid-fan cores (i.e., the calcareous facies comprise < 0.5% of total core recovered; Table 1 in Stow and others). To simplify the discussion of lithologic characteristics, we consider only the following five groups of sedimentary facies: 1) gravels and pebbly muds, 2) sands and silts, 3) silty muds, 4) silt-laminated muds, and 5) muds and clays. These basic sedimentary facies are identical to Stow's descriptions in Chapter 38 except for the combination of gravels with muddy gravels and pebbly muds and the elimination of oozes, muddy oozes, and calcareous muds.

Because the Mississippi Fan is commonly characterized as a dominantly muddy system, the most surprising facies is the one with gravels and pebbly muds at the base of the channel sites (Sites 621 and 622). The clean gravels (Fig. 4A), recovered at the bottom of Site 621, contain clasts ranging from about 2-mm to 2.5-cm in length and appear to be both clast-supported and normally graded. It is possible, however, that the gravels were washed clean of interpebble sands during the raising of the core barrel, thus creating a minor artificial sorting of the gravels. The pebbly muds (Fig. 4B) are poorly sorted, with the maximum clast sizes up to 3.4 cm (long axis) at Site 621 and 1.5 cm at Site 622. These sediments form structureless beds in which pebbles and sand granules are matrix-supported.

Sands and silts (Fig. 4C) comprise an important sedimentary facies in the lower part of the two channel sites, but they are nearly absent from the overbank sites. These predominantly sandy silts occur in beds from 0.5-cm to 3.0-m thick and are typically interbedded with muds or silty muds.

The characteristics of the sands and silts differ between the two channel sites. At Site 621 (thalweg site), they commonly occur as thin, inclined, and contorted layers (0.5 to 10-cm thick) interbedded with mud. At Site 622 (pointbar site), they comprise beds (0.8 to 3.0-m thick) of very finely laminated or structureless, relatively clean, sandy silt.

Silty muds occur as distinct interbeds in all four mid-fan sites and comprise about 6% of the recovered sediments. They are distinguished from the mud and clay facies by an increased abundance of sand and silt, but are not distinguishable from silt-laminated muds on well logs. The silty muds occur in beds ranging from a few centimeters to a meter in thickness and are typically normal graded.

Silt-laminated muds contain common to very abundant silt layers ranging from very thinly laminated (< 1 mm) to very thinly bedded (1 to 3 cm) (Figs. 4D,E). The silt-laminated muds are one of the major facies at the mid-fan sites, constituting about 38% of the recovered section. It is the dominant facies at Site 617 (82%), and comprises about 24% of the total recovered section in the channel sites (generally occurring only in the lower part).

Muds and clays are the dominant facies on the mid-fan, comprising about 52% of the total recovered section at the four sites. They are the dominant lithology at the overbank sites, are common in the upper parts of the channel fill, and occur as thin interbeds in the lower part of Site 621.

Vertical Lithological Characteristics

Core recovery at the four middle-fan drill sites was generally good in the upper 80 to 90 m, but decreased drastically below that depth (Chapter 36). A suite of well logs run at all sites (except Site 617) allowed us to fill in the gaps produced by the incomplete core recovery. Wire line logs can provide information on lithologies, but not on sedimentary facies. Therefore, the following discussion of lithological characteristics as well as the sedimentary columns in Figures 5 and 6 present interpretative lithological information resulting from correlative studies between actual core lithologies and well-log data.

A thin cover of Holocene ooze and calcareous muds (10 to 25-cm thick) occurs at the top of all four drill sites. These calcareous sediments overlie detrital muds, silts, and coarser sediment of Late Pleistocene age. All sites drilled in the middle fan penetrated only into the Upper Wisconsin glacial stage (Ericson Zone Y [6]; Kohl and others, Chapter 39).

Channel Sites

The two channel sites (Sites 621 and 622) are similar in overall lithologic properties, each showing a basically fining-upward sequence (Fig. 5). Both holes bottomed in coarse-grained sediment (about 13 m of gravel and sand at Site 621 and about 15 m of interbedded pebbly mud and mud at Site 622). Above the basal gravelly sections, there is a thick section (about 100 m) of interbedded muds and sands that generally fine upward. The lower part of this interbedded section consists of interbedded silty muds and sands (to about 141-m subbottom at Site 621 and to about 134-m subbottom at Site 622). Clayey mud is the dominant lithology in the upper part of the interbedded section at Site 621 (about 141 to 86-m subbottom). The mud is slightly siltier in the same interval at Site 622 (about 134 to 92-m subbottom) and contains few sand layers. Above the interbedded interval, muds and clays are dominant and are rarely interrupted by thin bedded silts and sands. The basic

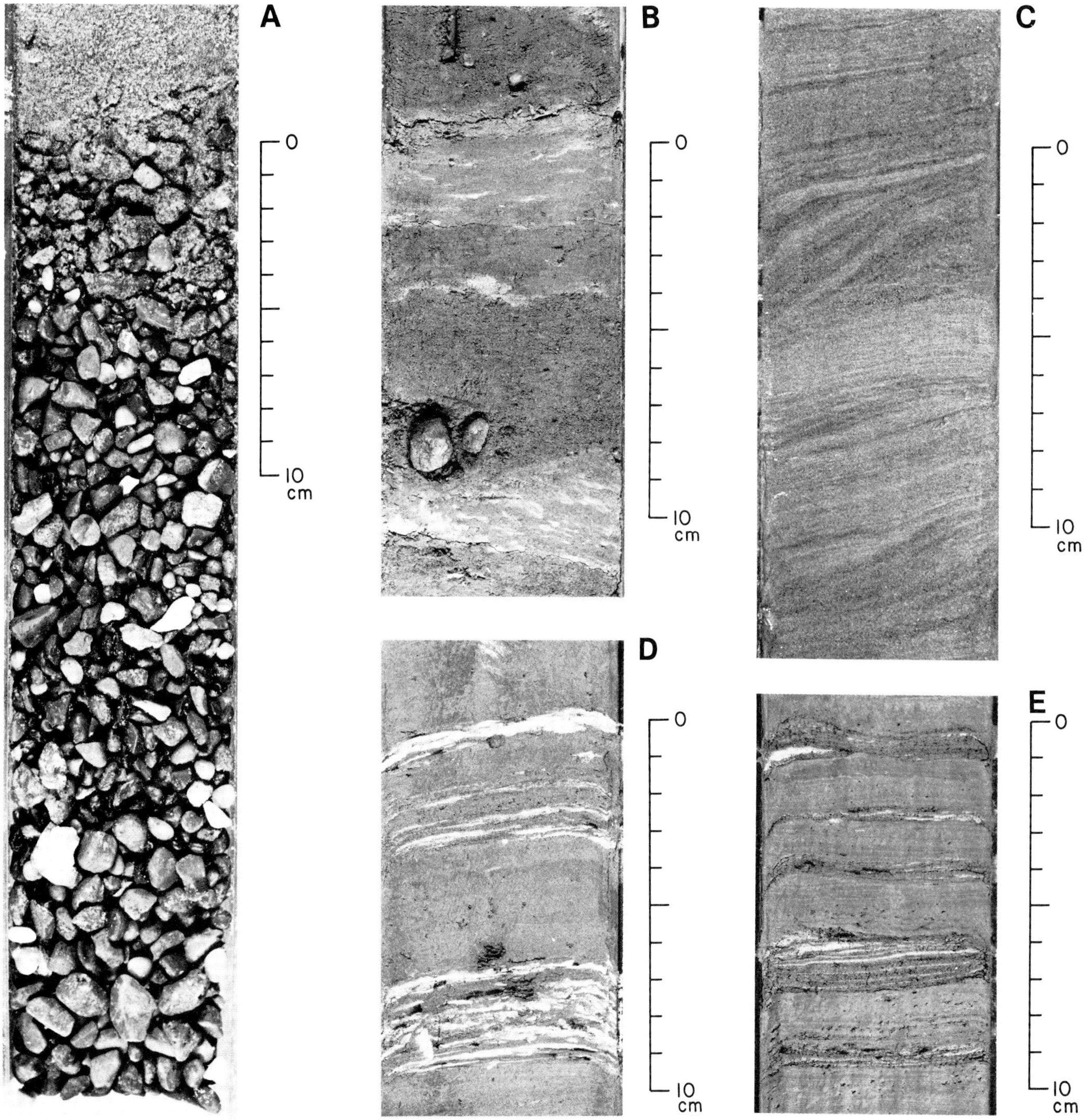

Figure 4. Photographs showing characteristics of selected sedimentary facies (subbottom depths correspond to top of each photograph). (A) Gravel overlain by sand (Site 621; 213.9 m). (B) Pebbly mud (Site 622; 197.7 m). (C) Sandy silt (Site 622; 156.2 m). (D) Silt-laminated mud (Site 621; 158.7 m). (E) Silt-laminated mud (Site 622; 93.4 m).

difference between these two sites is that the sandy section at Site 622 is about 40 to 50-m thicker than at Site 621 (Fig. 5). This difference is common in fluvial systems in which the pointbar deposits have a thicker sand section than the thalweg fill.

The approximately 210-m section cored at each of the two channel sites probably represents deposition during the later phase of aggradation of the youngest fanlobe. The gravel, pebbly mud, and sand recovered at the bottom of each hole correspond to the top of the seismically, high amplitude, reflection zone (Fig. 3), suggesting that this seismic zone characterizes coarse-grained sediment and that such material could be interpreted as channel-lag deposits (Chapter 41). The fining-upward trend [7] probably

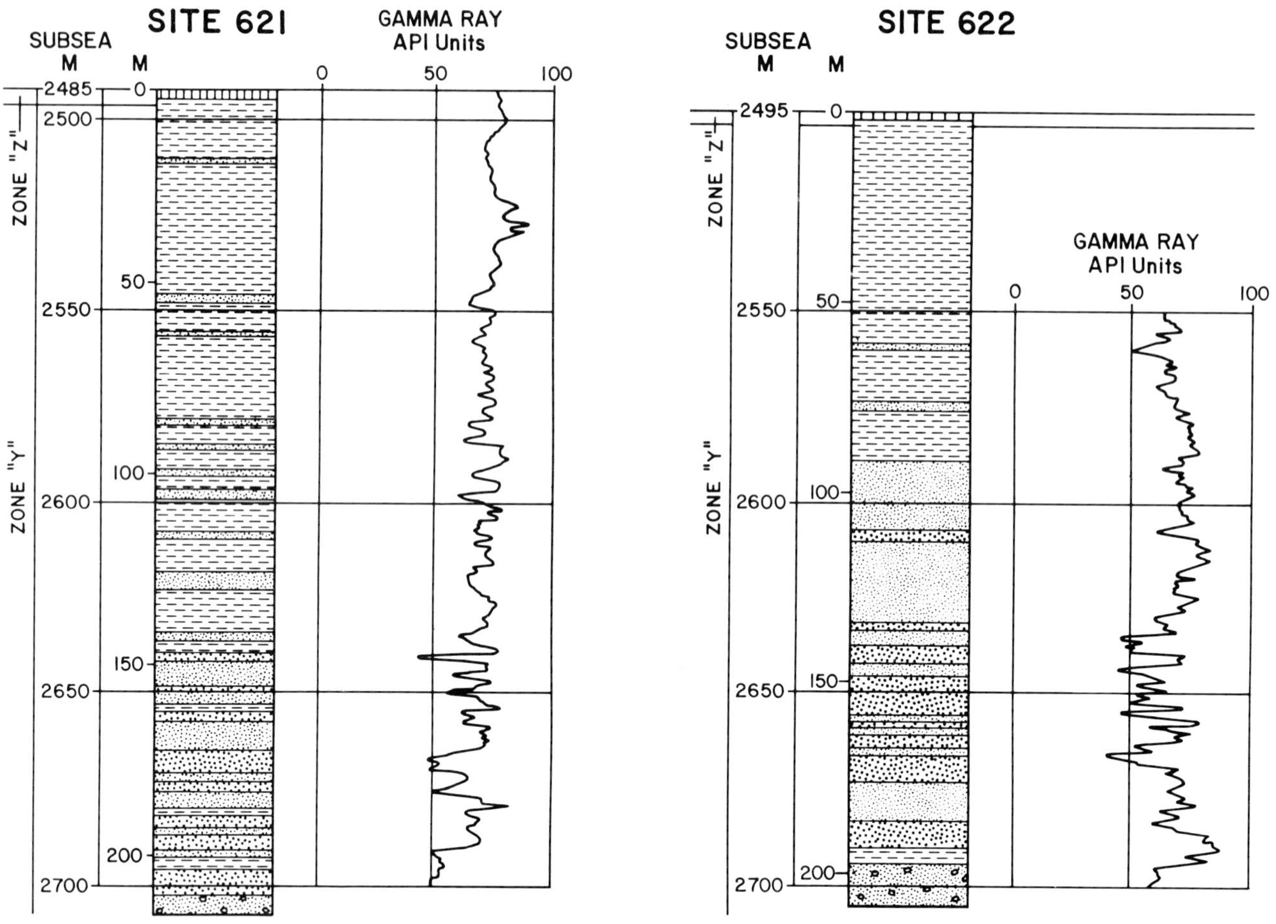

Figure 5. Interpretative lithostratigraphic summary of channel sites (Sites 621 and 622) showing age, lithology, and gamma-ray log response. See Figure 6 for legend. Summaries based on recovered sediment and well log data; actual core recovery and observed lithologies given in reference 9.

indicates a change in the depositional mode (waning phase) and, finally, abandonment or inactivity of the system as the sea level rose to its present high stand.

Overbank Sites

The overbank sites (Sites 617 and 620) are not directly correlatable because of their differing distances from the channel. Site 617 (4-km southwest of the channel) was drilled to a depth of 191 m and contains somewhat coarser sediment than was recovered at Site 620 (18-km northeast of the channel). Site 617 was proposed to examine the type of sediments that constitute the ridge and swale structures and to discover how much coarse material (turbidites) was deposited directly adjacent to the channel. Site 620 was expected to consist of numerous thin-bedded turbidites, thought to characterize overbank or interchannel deposits in studies on ancient turbidite systems [8]. The drill sites, however, did not support these working hypotheses (Chapter 36).

The lowermost 108 m (191 to 83-m subbottom) of Site 617 (Fig. 6) represents a minor coarsening-upward sequence of dominantly silt-laminated muds, with the frequency of silt laminae increasing from a minimum of 6% near 162-m subbottom to 18% of the total volume at 85-m subbottom. This lower part is overlain by a 38-m interval of silt-laminated mud (83 to 45-m subbottom); abundant silt laminae averages 35% of the total volume. The uppermost 45 m comprises a weakly defined fining-upward sequence, beginning with silt-laminated muds (45 to 18-m subbottom), in which the silt laminae decrease in frequency upward and are topped by structureless muds and clays. The silt laminae are typically very thin with sharp bases. Deformation in the form of inclined and contorted layers is most prevalent at this site.

It was intended to drill through at least the upper two fanlobes at Site 620 to combine time-stratigraphy with lithological characteristics of overbank deposits. At 422-m subbottom depth, however, friction on the drill string forced abandonment of deeper objectives. Because the originally intended penetration at this site was 750 m, a rotary drilling/

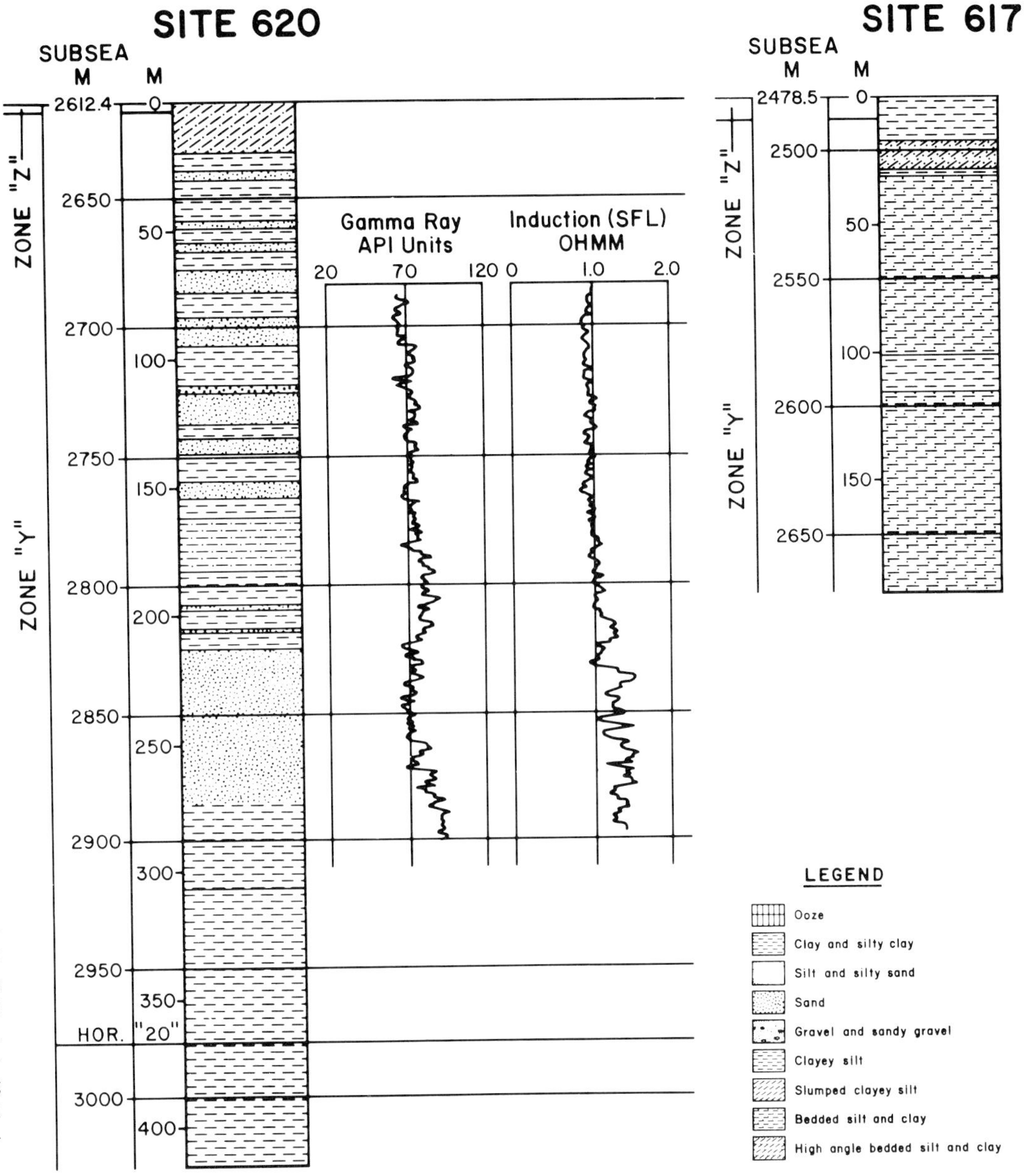

Figure 6. Interpretative lithostratigraphic summary of overbank sites (Sites 617 and 620) showing age, lithology, and gamma-ray and induction log responses (Site 620). Site 620 summary based on recovered sediments and well log data; see reference 9 for actual core recovery and observed lithologies.

coring system was used. Accordingly, the cores recovered were severely disturbed and sedimentary structures were not preserved (Chapter 36). Compositionally, the sediments are mostly muds and clays with rare silt laminae or very thin sandy turbidites (Fig. 6).

Two major slightly coarsening-upward lithological intervals, based primarily on well logs, can be suggested for Site 620. The lowermost interval extends from at least 289 m (the base of the logged interval) to 217-m subbottom; it might extend to the base of the cored interval at 422-m subbottom since no drastic lithological changes were observed below the well-logged interval. This basal interval starts with a clayey mud which changes rather abruptly into a silty mud at about 258-m subbottom. The well-log response suggests variability in the silt and clay contents of the mud (Fig. 6). Above 237-m subbottom, the silt component becomes dominant.

The second interval (from 217 to about 65-m subbottom) shows a similar, minor, coarsening-upward trend, but is lithologically more variable than the lower interval. The lower part of this upper section is composed mainly of clayey muds. Above about 155-m subbottom, the interval consists of interbedded silty muds and clayey muds with some coarse silt to fine sand intercalations.

The uppermost 65 m of the hole were extensively disturbed by the coring, and no adequate observations can be made on the relationship of this section to the underlying intervals.

Conclusion

Analyses of seismic reflection profiles, side-scan sonar images, and cores from the four drill sites show that the middle fan is the thickest part of the youngest fanlobe and that it is aggradational in nature. It is lenticular in cross-section, with a channel-levee complex along its apex. The channel has a sinuous course and is migratory in nature. The deepest core recovered from the channel fill contained a gravel-pebbly mud-sand lag deposit that is visible on seismic reflection profiles as a zone of acoustically high amplitude reflections. This high amplitude zone is several times wider than the present channel visible on the seafloor, suggesting a decrease in the size of the channel with time. The channel fill is primarily a fining-upward sequence.

The levee and overbank deposits are typically fine-grained, with only a minor amount of thin silty and sandy turbidites which rapidly decrease in number away from the channel.

Total sedimentation rates of the youngest fanlobe, both in the channel and overbank deposits, average nearly 12 m/1000 years (Kohl and others, Chapter 39). Although the present channel floor is only 30 to 45-m below its levees, seismic reconstructions suggest that the channel was two to three times deeper during active transport. Because of the high sediment accumulation rates and the reconstructed size of the channel, it is assumed that very large, together with smaller, density flows (debris flows and turbidity currents) moved through the channel. It is likely that the size of the flows decreased with time. Sand-sized and coarser material was basically transported below the levee crests and, therefore, was deposited within the channel margins, whereas enormous amounts of finer-grained sediments moved over the levees and were deposited in the overbank areas.

References

[1] Garrison, L. E., Kenyon, N. H., and Bouma, A. H., 1982. Channel systems and lobe construction in the Mississippi Fan. Geo-Marine Letters, v. 2, pp. 31–39.

[2] Kastens, K. A., and Shor, A. N., 1985. Depositional processes of a meandering channel on the Mississippi Fan. American Association of Petroleum Geologists Bulletin, v. 69, pp. 190–202.

[3] Prior, D. B., Adams, C. E., and Coleman, J. M., 1983. Characteristics of a deep-sea channel on the middle Mississippi Fan as revealed by a high-resolution survey. Gulf Coast Association of Geological Societies Transactions, v. 33, pp. 389–394.

[4] Moore, G. T., and others, 1978. Mississippi Fan, Gulf of Mexico—physiography, stratigraphy, and sedimentation patterns. In: A. H. Bouma, G. T. Moore, and J. M. Coleman (eds.), Framework, Facies and Oil-Trapping Characteristics of the Upper Continental Margin. American Association of Petroleum Geologists Studies in Geology No. 7, pp. 155–191.

[5] Stuart, C. J., and Caughey, C. A., 1976. Form and composition of the Mississippi Fan. Gulf Coast Association of Geological Societies Transactions, v. 26, pp. 333–343.

[6] Ericson, D. B., and Wollin, G., 1968. Pleistocene climates and chronology in deep-sea sediments. Science, v. 162, pp. 1227–1234.

[7] Mutti, E., and Ricci Lucchi, F., 1972. Le torbiditi dell'Appennino settentrionale: introduzione all' analisi di facies. Memorie della Societa Geologica Italiana, pp. 161–199.

[8] Nelson, C. H., and others, 1978. Thin-bedded turbidites in modern submarine canyons and fans. In: D. J. Stanley and G. Kelling (eds.), Sedimentation in Submarine Canyons, Fans, and Trenches. Dowden, Hutchinson & Ross, Stroudsburg, PA, pp. 177–189.

[9] Leg 96 Scientific Party, 1984. Challenger drills Mississippi Fan. Geotimes, v. 29, no. 7, pp. 15–18.

CHAPTER 41

Migratory Characteristics of a Mid-Fan Meander Belt, Mississippi Fan

Charles E. Stelting and DSDP Leg 96 Shipboard Scientists

Abstract

Seismic reflection profiles across the most recent Mississippi Fan mid-fan channel reveals high amplitude reflectors in the lower part of the channel fill. Subsequent drilling results show that these high amplitude reflectors correspond to coarse-grained channel-lag deposits. These lag deposits can be divided into at least three distinct channel-fill units which show channel migration with time. Isopach maps indicate that the dimensions of the channel-lag accumulations are up to 6.5-km wide, slightly more than 200-m (250-msec) thick, and that the northernmost meander belt has migrated about 2.0-km laterally, 1.2-km downfan, and has climbed 175-m (220-msec) stratigraphically. Evolution of the meander belt shows characteristics similar to meandering fluvial systems.

Introduction

The concept of sinuous channel systems on submarine fans is not new. The presence of a meandering channel and associated fluviatile-type morphologic features were reported from the La Jolla Fan in the early 1950's [1]. These features and similar ones on other West Coast fans were described in greater detail nearly two decades later [2]. Stratigraphic traps in sinuous submarine channels have also been recognized by explorationists in subsurface production areas [3,4].

Renewed interest in sinuous submarine channels during the past few years has resulted from bathymetric surveys using SEABEAM echo sounders and swath mapping with long-range and medium-range side-scan sonar systems (GLORIA, Sea MARC). These surveys have revealed sinuous (or meandering) channels on the Amazon, Cap Ferret, Mississippi, Nile, Orinoco, and Rhone Fans [5–10] (Chapters 15, 17, 21, 22).

In addition to the single, sinuous, leveed channel revealed by GLORIA on the Mississippi Fan [7], high resolution seismic reflection profiles collected simultaneously show zones of high amplitude reflectors directly beneath and adjacent to the modern channel margins. A re-examination of previously collected seismic data showed that these reflectors occur near the base of the channel fill across most of the youngest fanlobe.

Numerous questions were raised about the meaning of the sinuous channel and the high amplitude reflectors. Is the sinuosity characteristic of total fanlobe growth, or is it a phenomena which evolved late in the construction of an individual fanlobe? Do the fluvial-type morphologic features merit comparison with subaerial fluvial systems? Are the high amplitude reflectors representative of geologic features or are they seismic artifacts (e.g., focusing effect of the acoustic signal)? How are they related to channel development and what information can they provide about the processes that were active in this part of the fan?

With these questions in mind, additional high resolution seismic reflection profiles, subbottom profiles, and side-scan sonar data were collected for selection of DSDP Leg 96 drill sites [11]. Four sites were selected within and adjacent to the northernmost mid-fan meander belt to evaluate the lithologies and the depositional processes responsible for the formation of the observed channel/overbank morphologic features. Two of the sites (Sites 621 and 622) were located within the margins of the present channel (Fig. 1). Cores recovered from the upper part of the high amplitude reflector zone indicate that these reflectors correspond to coarse-grained sediments (Stelting and others, Chapter 40). These sediments are interpreted to be channel-lag deposits overlain by lateral accretion deposits laid down in a migratory submarine-fan channel.

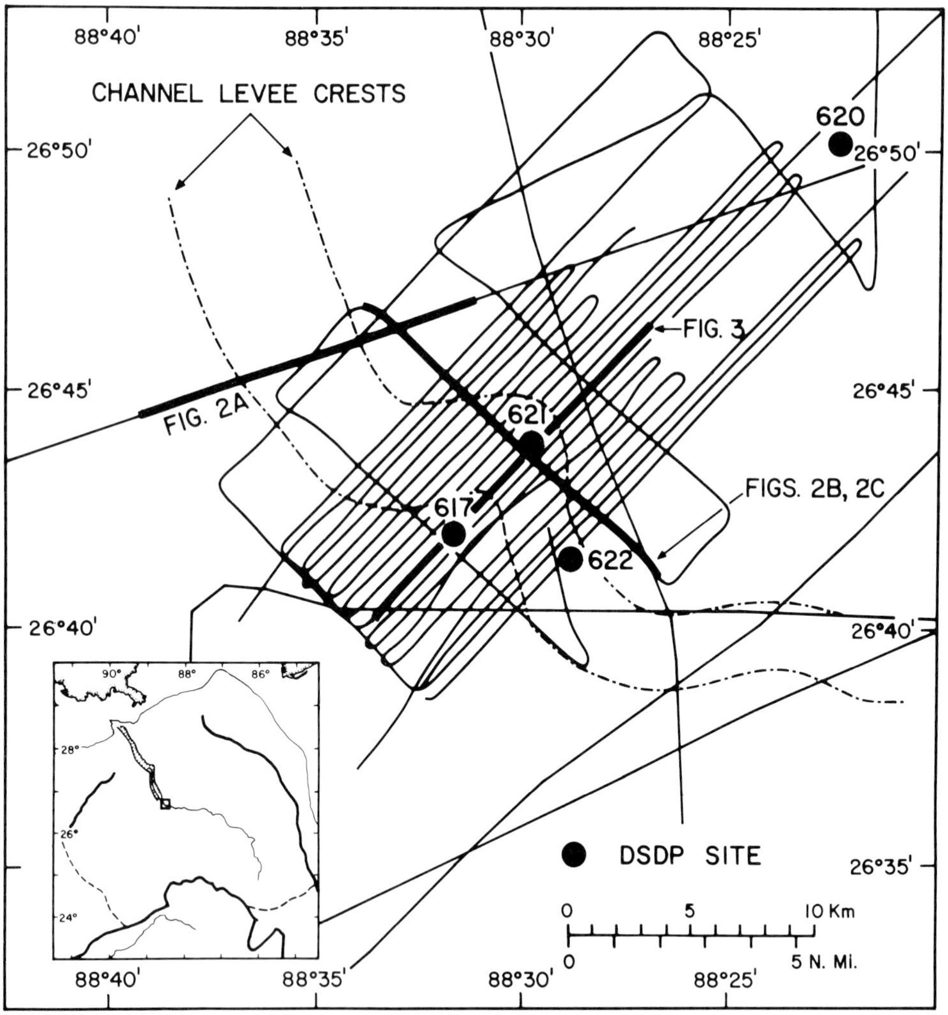

Figure 1. Location map showing seismic tracklines over the northernmost mid-fan meander belt and DSDP drill sites. Location of seismic reflection profiles (Figs. 2 and 3) shown by bold lines. Inset map shows location relative to overall channel trend (modified from Bouma and others, Chapter 36).

In this chapter, we evaluate the high amplitude reflectors and relate their distribution to the growth patterns of a mid-fan meander belt.

Seismic Stratigraphy

Prior to 1983, available seismic reflection data consisted primarily of deep penetration regional survey lines. The GLORIA survey (80 cu. in. airgun) and the DSDP Leg 96 site survey (80 cu. in. watergun) were the first two surveys designed to collect data over specific depositional environments on the Mississippi Fan [7,11]. These high resolution seismic reflection surveys, together with the DSDP mid-fan cores, comprise the primary data.

Mid-Fan Meander Belt

The Mississippi Fan is made up of at least seven distinct fanlobes ([12]; Bouma and others, Chapter 21). The youngest fanlobe is the depositional unit overlying seismic Horizon "20." This fanlobe is about 200-km wide and has a maximum thickness of 400 m at the mid-fan channel/overbank complex (Fig. 3A of Chapter 21).

The modern channel narrows slightly as it passes from the upper-fan channel onto the aggradational middle-fan region (2000 to 2400-m water depth) where it assumes a highly sinuous pattern. The channel in the vicinity of the mid-fan drill sites is about 3-km wide with a bathymetric relief (channel floor to levee crest) of 20 to 45 m. Below the initial meander loop, near Site 622, the channel bends to the east and maintains an easterly direction to the lower fan (Fig. 1, inset).

Seismic stratigraphic and facies studies near the mid-fan drill sites utilized several different seismic systems including high resolution systems (Figs. 2 and 3) and multichannel systems (Fig. 3 of Chapter 40). As illustrated in these figures, each acquisition system produces a different resolution and signature for the seismic facies. Therefore, site survey data are used to describe the seismic facies because of its proximity to the drill sites and the dense coverage over the mid-fan meander belt (Fig. 1).

Essentially, four seismic facies are observed in the vicinity of the mid-fan channel-overbank complex (Figs. 2 and 3). The seismic facies are characterized by the following reflector types: Type 1 = high amplitude, low continuity reflectors; Type 2 = semitransparent reflection zones with very short, hummocky, medium amplitude reflectors; Type 3 = semitransparent to transparent reflection zones with relatively more continuous, curvilinear, low amplitude reflectors; and Type 4 = hummocky, discontinuous, medium amplitude, subparallel reflectors. Reflector Types 1 to 3 are associated with the channel system, whereas Type 4 occurs in the overbank areas.

All four reflector types were penetrated at the four DSDP mid-fan sites. An interactive evaluation of the core and seismic data provided for a lithological correlation to each of the seismic facies (Fig. 3). The high amplitude reflector facies (Type 1) corresponds to gravels and sands deposited at the base of the channel. The semitransparent reflectors (Type 2) generally indicate sand and silt deposited within the channel margins. The semitransparent to transparent reflectors (Type 3), making up the upper channel fill, correspond to muds and clays. The discontinuous, subparallel, medium amplitude reflectors (Type 4) in the overbank area are typified by clays and silts.

The composite interpretation shown in Figure 3B is based on the evaluation of numerous seismic reflection profiles. The reflector horizons (A through F) represent the various channel shapes during aggradation of the youngest fanlobe. The three lower units thin away from the channel and downlap onto older surfaces (unpublished data). The upper units also thin away from the channel.

Channel-Lag Deposits

The hummocky (concave side down) to relatively flat, high amplitude reflectors constitute the most prominent seismic facies in the mid-fan region. These reflectors, interpreted to correspond to coarse-grained channel-lag deposits, occur at the base of the most recent fanlobe channel system. They also occur within the channel complex of the underlying fanlobe (Fig. 2A) and have been described as the primary criteria for identification of channel systems in older Mississippi fanlobes (Feeley and others, Chapter 37). Furthermore, high amplitude reflectors have been associated with channel deposits on other submarine fans [13–15].

Detailed examination of the seismic reflection profiles indicates that: 1) the base of the high amplitude reflections is offset to the side of the channel and climbs upsection to directly beneath the modern channel; and 2) the zone of high amplitude reflectors can be divided into at least three distinct units. These seismic units correspond to the channel-lag units labelled A, B, and C in Figures 2C and 3B.

Seismically, the channel-lag units are typified by short, discontinuous, high amplitude reflectors, with extremely short, lower amplitude reflectors becoming more abundant toward the margins. This change in seismic character may suggest a decrease in grain size toward the channel margins. Erosion during deposition of each unit is suggested by the truncation of individual high amplitude reflectors against each other. This implies that each unit is made up of many depositional events.

Channel-Lag Depositional Units

Isopach maps for each of the three channel-lag depositional units show the geometries and lateral distribution of each unit and document the development of the sinuosity of the modern channel system (Fig. 4). The following discussion regarding the characteristics of the individual units and the extent and direction of migration refers only to the northernmost meander belt (Fig. 1).

Channel-Lag Unit A

The lowermost unit unconformably overlies seismic Horizon "20." Truncation of the underlying short, discontinuous reflectors suggests that Unit A cut down into older deposits (Figs. 2C and 3B), by possibly as much as 30 msec (about 23 m). Unit A is typically thicker than 60 msec and is 4.0 to 4.5-km wide (Fig. 4A). The isopach map shows that this relatively linear unit thickens downfan. The thickening might be an artifact from using different seismic systems or could represent a rapid accumulation of sediment where the channel trend swings to the east. The thalweg (contours > 70 msec) curvature is generally opposite to the modern channel thalweg.

The isopach map reveals that the initial channel trended slightly more to the east-southeast than does the modern one. Near the center of the meander belt, the initial channel deflected sharply to the south and completed three meander loops in the next 6 km. The sinuosity of these meanders decreases downfan from 1.4 to 1.2. Generally, the contour gradients are broader on the convex side of the bends, suggesting a thicker accumulation of sediments in the direction of migration.

Channel-Lag Unit B

The isopach map shows that Unit B is characterized by a low sinuosity and conforms more to the shape of the modern channel than does Unit A (Fig. 4B). Unit B is 3.0 to

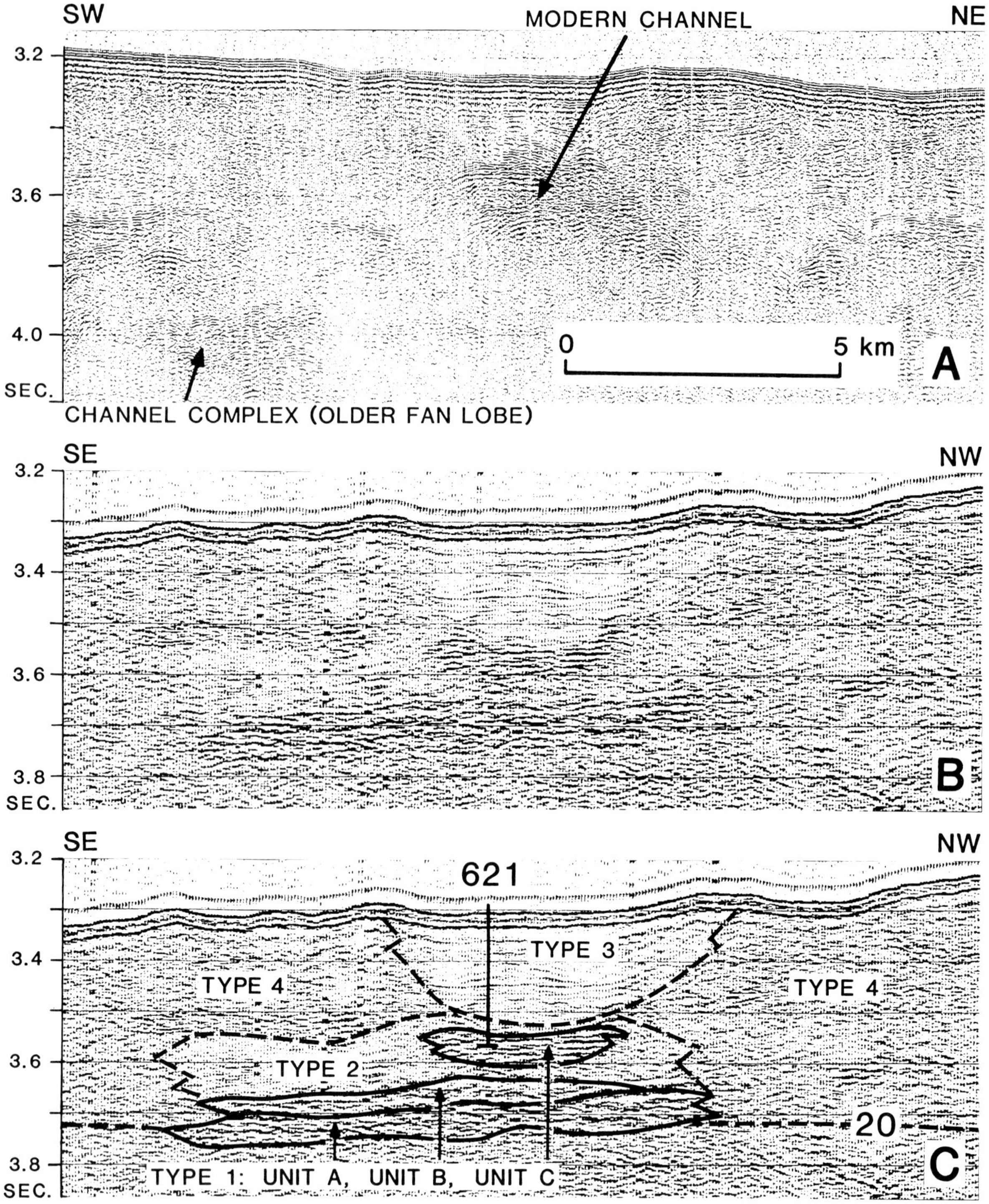

Figure 2. High resolution seismic reflection profiles showing seismic facies in the mid-fan meander belt area. (A) Airgun record (USGS Line 21) showing high amplitude reflectors beneath modern channel and underlying fanlobe channel system. (B) Dip-oriented watergun record (R/V Conrad Line 2003). (C) Annotated watergun record showing seismic facies, channel-lag units, and position of Site 621. Vertical scale in seconds (two-way travel time). See Figure 1 for locations.

3.3-km wide in the northwest and widens to 4.0 km in the southeast. The line drawing (Fig. 3B) suggests that the width of this unit is probably controlled by the channel relief established by the previous unit.

Unit B unconformably overlies Unit A, with the boundary between the two units commonly being marked by an undulatory (erosional) surface. This unit is thinner than the other channel-lag units in the northwestern portion of the study area, but thickens downfan. Active lateral migration dominates the channel growth pattern between Units A and

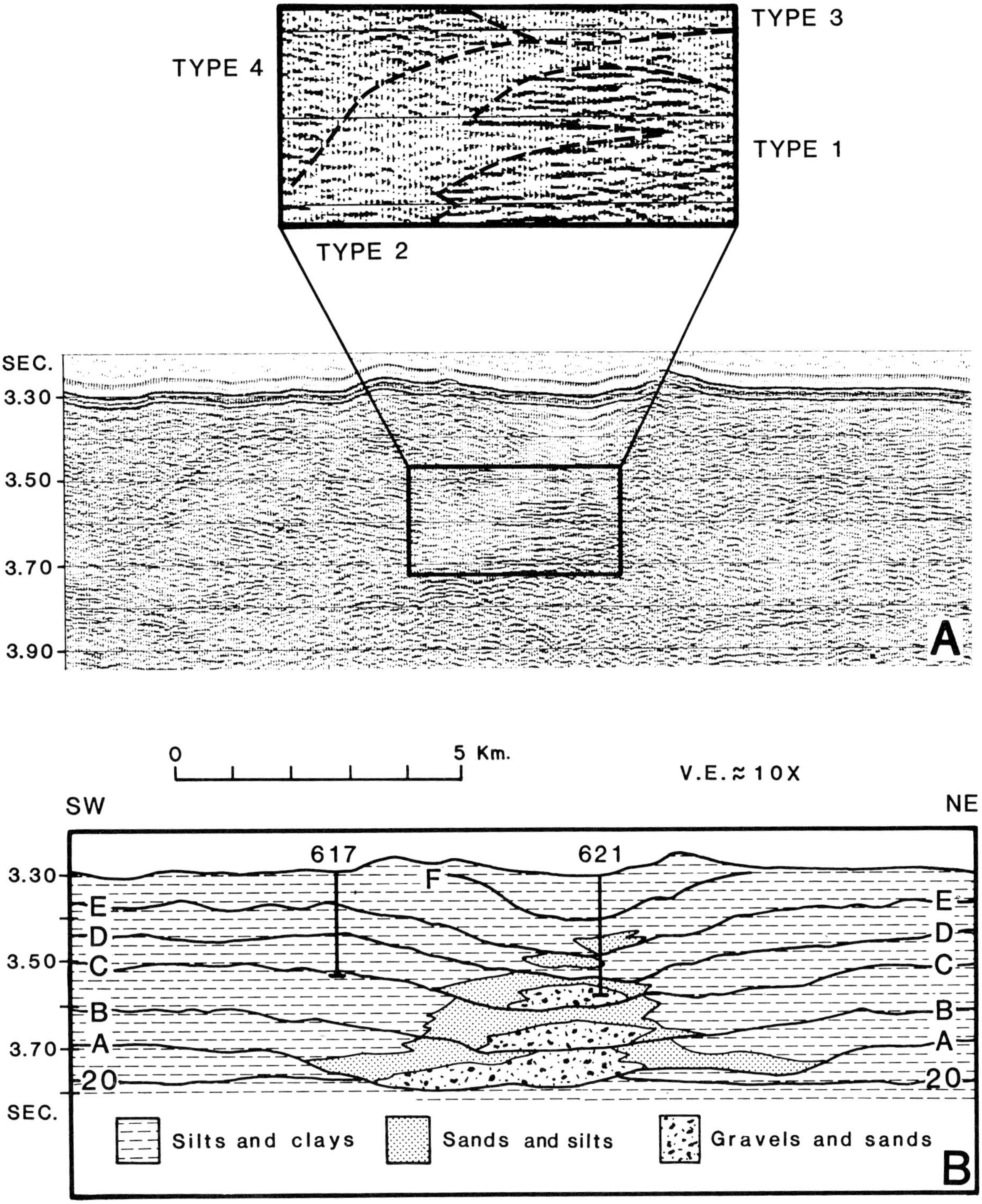

Figure 3. (A) Watergun seismic reflection profile (R/V Conrad Line 1017). Top figure is an enlargement of blocked part of A showing detailed character of seismic reflector types. Vertical scale in seconds (two-way travel time). See Figure 1 for location. (B) Composite interpretation showing seismic unit boundaries, probable channel margins, inferred lithologies of mid-fan channel-overbank complex, and location of Sites 617 and 621.

B, with the thalweg (contours > 40 msec) having migrated about 1.2-km laterally and 1.1-km downfan. While Unit A displayed three meander loops, this unit is characterized by one major bend (sinuosity 1.2) and a minor meander (sinuosity 1.1) positioned over the third bend of Unit A. This change in channel shape has reduced the effective length of the channel in this area by about 20%.

Unit B is characterized by an overall decrease in seismic reflector amplitude, especially in the northwestern half of the meander belt. Very short, discontinuous reflectors are

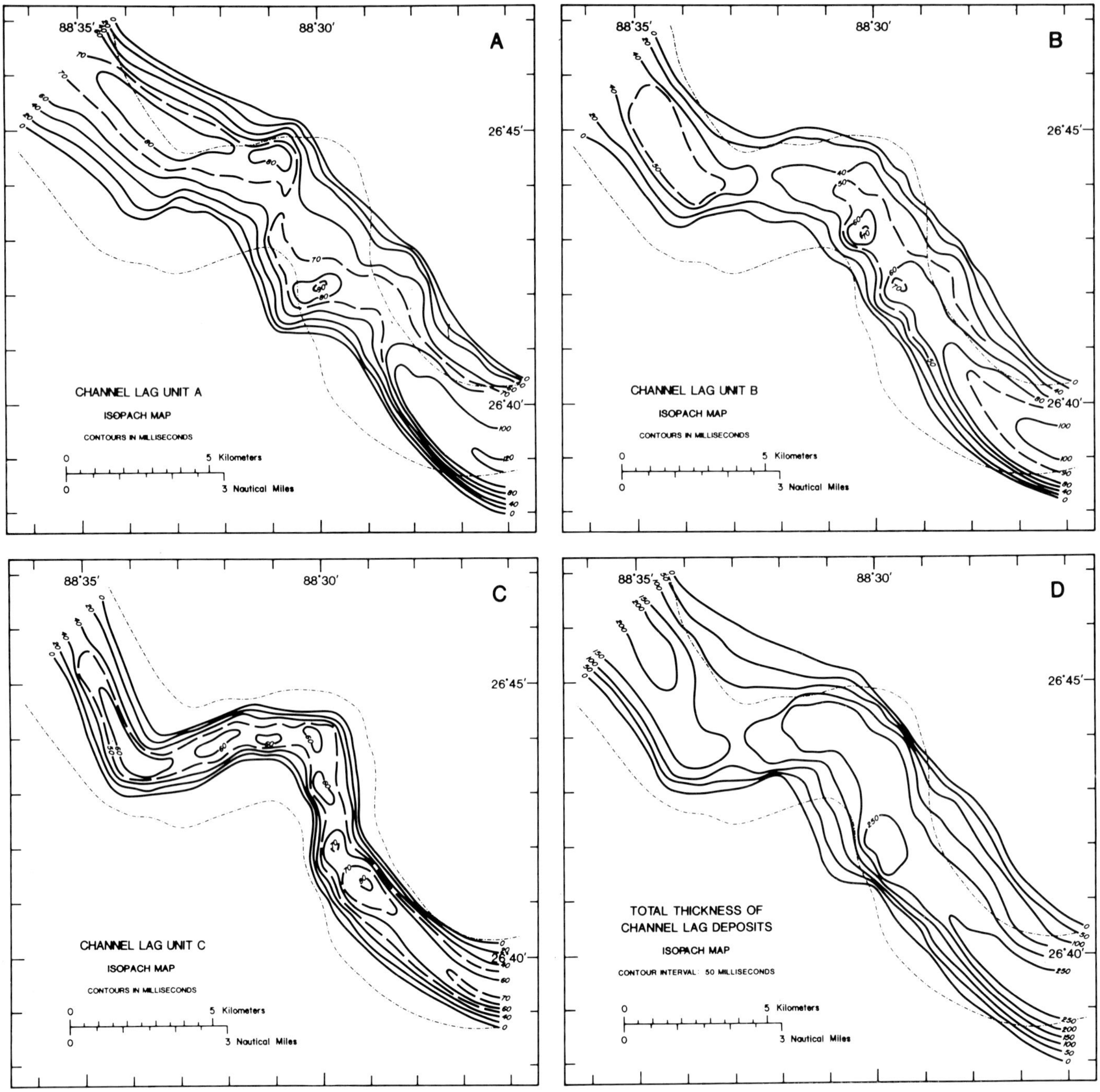

Figure 4. Isopach maps of the channel-lag deposits (high amplitude reflectors). (A) Channel-lag Unit A (oldest unit). (B) Channel-lag Unit B. (C) Channel-lag Unit C (youngest unit). (D) Total thickness of the three channel-lag units. Levee crests of the modern channel (dash-dot line) are shown for reference. For an approximate depth conversion; 20 msec equals about 15 m (50 ft); 50 msec equals about 38 m (126 ft).

commonly interspersed in this unit, which may imply improved grain size homogenity or reduced bed continuity relative to the lower unit.

Channel-Lag Unit C

The youngest of the three major channel-lag depositional units, Unit C, is more narrow than the present channel margins and essentially lies directly underneath the present thalweg (Fig. 4C). The unit is 1.8 to 2.5-km wide and is relatively uniform in thickness. Unlike the other units, Unit C does not lie directly on the next older unit, but is separated from the top of Unit B by a thin zone of semitransparent to medium amplitude, discontinuous reflectors about 30 to 40 ms thick (Figs. 2C and 3B). This gap could possibly be the result of a brief waning stage occurring after deposition of Unit B, during which time the sediments deposited within the channel were more uniform in grain size.

As in Unit B, erosion during migration of the thalweg (contours > 50 msec) continued to smooth out the geometries established during the initial phase of channel development. Lateral migration occurred only in the northwest and central areas and ranges from 0.3 to 0.8 km. Sinuosity along the thalweg is about 1.4.

The high amplitude reflectors of this unit tend to be more continuous than the previous units. This is perhaps an artifact of the tracklines which run more obliquely to the channel trend than in the older units.

Channel-Lag Deposits—Cumulative Thickness

The total thickness and lateral distribution of the three channel-lag units are shown in Figure 4D. The aforementioned interval between Units B and C is interpreted to consist predominantly of sands and silts and is included in the total thickness calculations. The lateral extent of the channel-lag deposits ranges from a maximum of 6.5 km (in the northwest) to a minimum of 4.3-km downfan of the first meander loop (Unit A). The total thickness of the lag deposits is relatively uniform, but is slightly thicker to the northwest and southeast.

The lateral distribution of each unit, as indicated by the individual isopach maps, shows that the channel migrated to the southwest in the northern area, to the northeast in the central portion, and again to the southwest in the southern region. Migration of the thalweg to its present position is denoted by the isopach thickening under the modern channel. In total, the channel thalweg migrated laterally a distance of 1.5 to 2.0 km, downfan a distance of 1.2 km, and climbed about 220 msec (175 m) stratigraphically during the most active period of fanlobe aggradation.

Comparison of the geometries depicted by the isopach maps and seismic reflection profiles shows that the sinuosity of the channel, as observed at the seafloor, appears to have been established during the earlier, more active periods of fanlobe aggradation. The growth pattern of the channel-overbank complex of the youngest fanlobe (in the northernmost mid-fan meander belt) can be characterized by active lateral migration between Units A and B and reduced lateral migration between Units B and C.

Discussion

Several sediment transport mechanisms have been proposed recently for the sinuous channel pattern and fluvial-type morphologic features of the most recent Mississippi Fan mid-fan channel-overbank complex. These include: debris flows; a combination of debris flows, turbidity currents, and locally generated currents; and fluvial-type processes [11,12,16]. The model discussed herein follows the lead of Bouma and others ([12], Chapter 21), who suggest that sea-level fluctuations caused sediment failure on the upper slope and at the shelf edge, producing slump masses that transformed into debris flows and, in turn, into turbidity currents as they proceeded downslope.

The external controls that may have had a direct effect on the flow characteristics of the gravity flows as they passed from the upper-fan channel to the middle fan are unknown. However, the reduction in gradient at the base of slope (upper/middle fan boundary) probably was an important factor in producing a more rapid deposition in the middle-fan area. In addition, pre-existing fan topography, such as the bathymetric depression between older fanlobes (L. E. Garrison, personal communication, 1983), or a central channel in debris flow deposits, as suggested by Prior and others [16], could also have been factors, although they probably acted more as steering controls (i.e., controlled position of channel trend).

Erosional events within the entire channel-lag complex imply that each unit is comprised of numerous depositional events. It is highly possible that both deposition and erosion within the channel complex were directly influenced by retrogressive slumping upslope (see Chapter 21). The size of these slumps would likely diminish as either the slump scarps approached a more stable gradient or as sea level rose. Either case would likely cause a volumetric and grain size reduction in subsequent gravity flows and would affect depositional processes on the middle fan. Accordingly, we believe that the presence of several distinct channel-lag depositional units together with episodes of erosion between and within each unit suggest that the growth pattern of, at least, the youngest fanlobe was influenced by reinitiation of major slope failure periods and/or several cycles of sea-level change.

Constriction of the channel margins during aggradation of the youngest fanlobe is suggested by the seismic profiles. The channel margins at the end of Unit A deposition were about 10-km wide. By the end of Unit C deposition, channel width had decreased to less than 6 km and has since been further reduced to about 3 km for the present channel (Fig. 3B). The amount of channel constriction between succeeding seismic units appears to have increased after deposition of the channel-lag units. This may reflect a change in the sediment transport/depositional mode from dominantly lateral migration (or accretion) during more active periods of fanlobe aggradation to mainly infilling of the channel complex during less active or waning periods of aggradation. Such a change in the depositional mode may be the result of a decrease in the grain size and/or amount of sediment supplied as sea level rose to its present high stand.

Conclusion

The growth pattern of the northernmost mid-fan meander belt on the Mississippi Fan is likely the result of channelized debris flows and turbidity currents that were active during periods of lowered sea level. The sinuosity of earlier channel patterns is still reflected in the channel presently visible at the seafloor. These patterns are the product of lateral and downfan migration of the meander belt during the earlier phase of fanlobe construction. The resulting geometries and lateral facies changes appear to be analogous to a meandering, aggradational, subaerial fluvial system.

Migration of the channel could have been accomplished by random scouring of unconsolidated channel margins. However, the isopach maps and lateral facies changes, inferred from seismic reflection profiles, suggest that channel migration may have resulted from erosion of the concave margin and deposition on the convex margin. At the minimum, the products of channelized flow in the deep-sea environment seem to be similar to the products of channelized flow in a subaerial fluvial system.

References

[1] Menard, H. W., and Ludwick, J. C., 1951. Applications of hydraulics to the study of marine turbidity currents. In: J. L. Hough (ed.), Turbidity Currents and the Transportation of Coarse Sediment to Deep Water. Society of Economic Paleontologists and Mineralogists Special Publication No. 2, pp. 2–13.

[2] Normark, W. R., 1970. Growth patterns of deep-sea fans. American Association of Petroleum Geologists Bulletin, v. 54, pp. 2170–2195.

[3] Payne, M. W., 1976. Basinal sandstone facies, Delaware Basin, west Texas and southeast New Mexico. American Association of Petroleum Geologists Bulletin, v. 60, pp. 517–527.

[4] Lovick, G., 1983. Exploration methods for submarine fan deposits: a west central Texas model. Oil & Gas Journal, Feb. 7, 1983, pp. 89–93.

[5] Damuth, J. E., and others, 1983. Distributary channel meandering and bifurcation patterns on the Amazon deep-sea fan as revealed by long-range side-scan sonar (GLORIA). Geology, v. 11, pp. 94–98.

[6] Coumes, F., and others, 1979. Etude des éventails detritiques profonds du golfe de Gascogne. Analyse geomorphologique de la carte bathymetrique du canyon Cap Ferret et de ses abords. Bulletin Société Géologique France, (7) t. 21, pp. 563–568.

[7] Garrison, L. E., Kenyon, N. H., and Bouma, A. H., 1982. Channel systems and lobe construction in the Mississippi Fan. Geo-Marine Letters, v. 2, pp. 31–39.

[8] Kenyon, N. H., Stride, A. H., and Belderson, R. H., 1975. Plan views of active faults and other features on the lower Nile cone. Geological Society of America Bulletin, v. 86, pp. 1733–1739.

[9] Belderson, R. H., and others, 1984. A 'braided' distributary system on the Orinoco deep-sea fan. Marine Geology, v. 56, pp. 195–206.

[10] Bellaiche, G., and others, 1984. Detailed morphology, structure and main growth pattern of the Rhone deep-sea fan. Marine Geology, v. 55, pp. 181–193.

[11] Kastens, K. A., and Shor, A. N., 1985. Depositional processes of a meandering channel on the Mississippi Fan. American Association of Petroleum Geologists Bulletin, v. 69, pp. 190–202.

[12] Bouma, A. H., Stelting, C. E., and Coleman, J. M., 1983/84. Mississippi Fan: internal structure and depositional processes. Geo-Marine Letters, v. 3, pp. 147–153.

[13] Normark, W. R., 1970. Channel piracy on Monterey deep-sea fan. Deep-Sea Research, v. 17, pp. 837–846.

[14] Damuth, J. E., and others, 1983. Age relationships of distributary channels on the Amazon deep-sea fan: Implications for fan growth patterns. Geology, v. 11, pp. 470–473.

[15] Cremer, M., 1983. Approches sédimentologique et geophysique des accumulations turbiditiques; L'évential profound du Cap-Ferret (Golfe de Gascogne), la serie des grès d'Annot (Alpes de Haute Provence). Ph.D. Dissertation, L'Universite de Bordeaux I, France, 300 p.

[16] Prior, D. B., Adams, C. E., and Coleman, J. M., 1983. Characteristics of a deep-sea channel on the middle Mississippi Fan as revealed by a high-resolution survey. Gulf Coast Association of Geological Societies Transactions, v. 33, pp. 389–394.

CHAPTER 42

Drilling Results on the Lower Mississippi Fan

Suzanne O'Connell, Charles E. Stelting, Arnold H. Bouma, James M. Coleman, Michel Cremer, Laurence Droz, Audrey A. Meyer-Wright, William R. Normark, Kevin T. Pickering, Dorrik A. V. Stow, and DSDP Leg 96 Shipboard Scientists

Abstract

The sinuous migratory channel of the middle Mississippi Fan changes to a pattern of frequent lateral shifting and abandonment on the lower fan, with one channel being active at any given time. Borings in the upfan part of the lower fan recovered vertical successions of thin channel fill and overbank deposits. Near the distal end of the fan, the channels often bifurcate before merging with a low relief area of assumed laterally extensive "sheet sand" deposition. A significant amount of all the sand-sized sediment that came from the source area onto the Mississippi Fan during the Late Pleistocene was apparently transported to the lower fan.

Introduction

The transition from the middle- to the lower-fan region of the youngest Mississippi fanlobe is found at a water depth of approximately 3100 m; the lower fan merges to the southeast with the Florida Abyssal Plain at water depths around 3400 m ([1]; Bouma and others, Chapter 21). Geometrically, this aggradational lower-fan system is broad and slightly convex-shaped. It is about 350 to 400-km wide with maximum thicknesses (<250 m) under the central portion of the lower fanlobe.

The middle fan/lower fan boundary coincides with a change in character of the channel-overbank complex. Morphologically, the middle fan is dominated by a single, highly sinuous, migratory channel. In contrast, the lower-fan morphology is characterized by numerous small channels that may result from frequent lateral shifting, filling, and abandonment. Lateral to the channel complex, the upfan part of the lower fan consists of broad, low relief areas presumed to be extensive overbank deposits.

Side-scan sonar images (Sea MARC I, Site Survey Cruise, Lamont-Doherty Geological Observatory) of the northern part of the lower fan show one major, slightly sinuous channel with oblong features and a geometry that resembles fluvial mid-channel islands. A zone with several, indistinct sub-parallel echoes on the Sea MARC images trends semiparallel to the modern channel and is interpreted to represent relict morphology from several older, abandoned channels (Fig. 1). Apparently only one channel was active at any given time. The number of relict channels suggest that they were relatively short-lived. Bifurcation is commonly observed near the downfan terminus of the channels (Fig. 1C). Farther downfan, channel systems cannot be recognized on side-scan sonar images. This downfan area is presumed to result from unconfined spreading of density flows resulting in the deposition of "sheet sands" (i.e., sand lobes [2]).

The character of the seismic reflection profiles on the lower fan differs from those of the middle fan in that the seismically high amplitude reflectors (which represent the coarse-grained channel-lag deposits) and major erosional surfaces with onlapping reflectors are absent (Stelting and others, Chapter 41). The seismic facies in the lower fan are primarily characterized by sets of parallel to sub-parallel, discontinuous reflectors which become more continuous downfan.

Four sites were drilled on the lower fan, two (Sites 623 and 624) in the area of channel abandonment and two (Sites 614 and 615) near the channel terminations. These two areas are separated by a distance of about 70 km (Fig. 1). Drilling details of these sites are presented in Table 1 of Chapter 36.

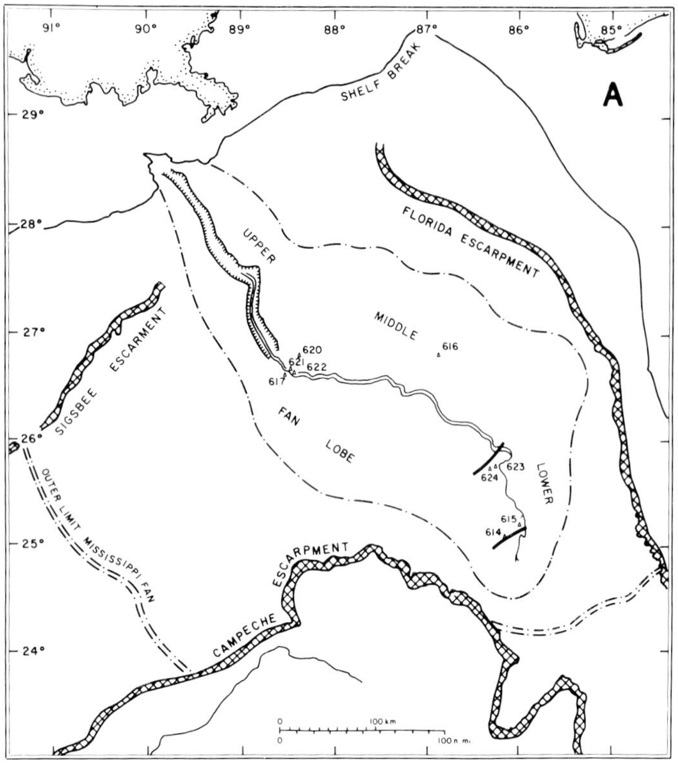

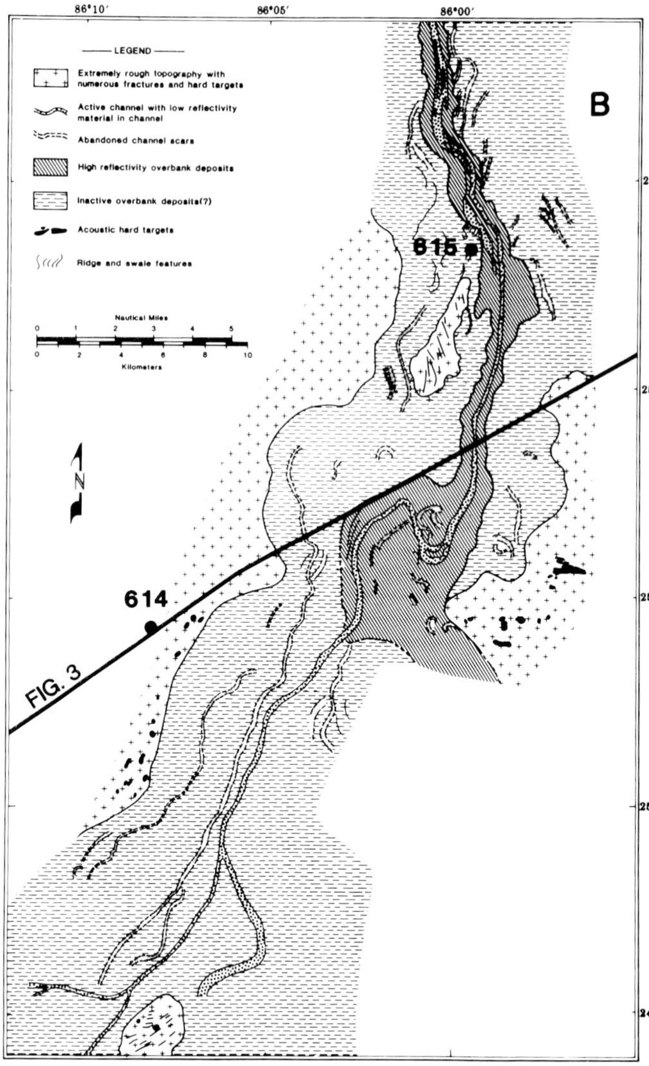

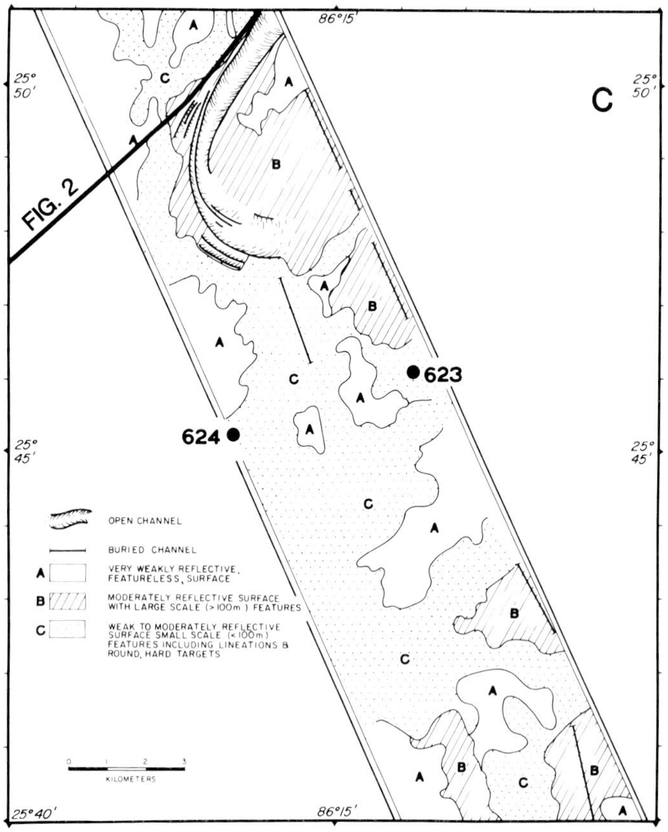

Figure 1. (A) Map showing youngest Mississippi fanlobe and the geographic positions of the lower-fan drill sites. Location of seismic profiles (Figs. 2 and 3) indicated by heavy solid lines. (Modified from Bouma and Coleman, Chapter 36). (B) Morphologic features in the distal lower-fan area as mapped from side-scan sonar. Locations of DSDP sites and the position of seismic profile in Figure 3 indicated. (C) Side-scan sonar map showing location of Sites 623 and 624 and location of seismic profile in Figure 2.

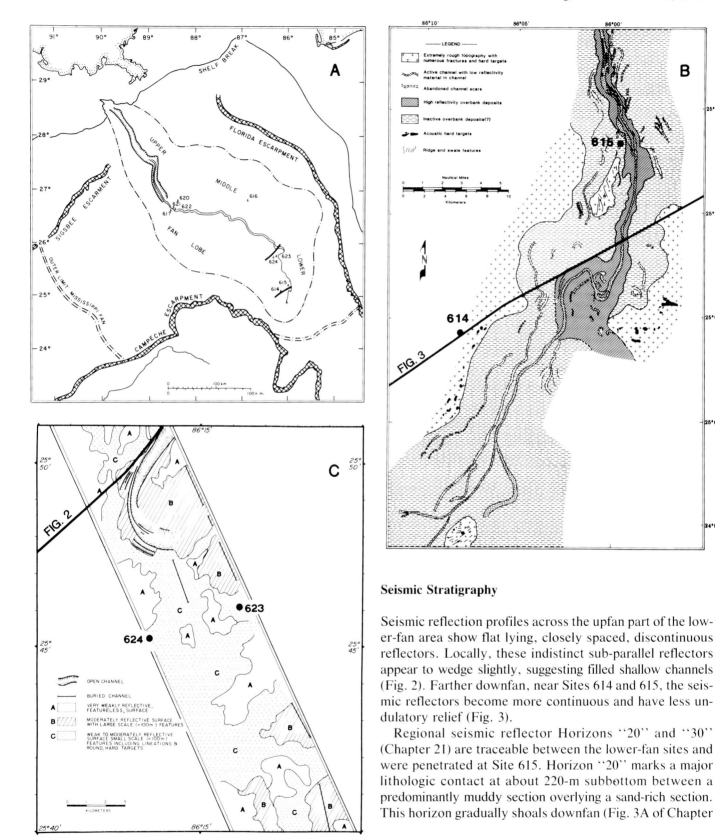

Seismic Stratigraphy

Seismic reflection profiles across the upfan part of the lower-fan area show flat lying, closely spaced, discontinuous reflectors. Locally, these indistinct sub-parallel reflectors appear to wedge slightly, suggesting filled shallow channels (Fig. 2). Farther downfan, near Sites 614 and 615, the seismic reflectors become more continuous and have less undulatory relief (Fig. 3).

Regional seismic reflector Horizons "20" and "30" (Chapter 21) are traceable between the lower-fan sites and were penetrated at Site 615. Horizon "20" marks a major lithologic contact at about 220-m subbottom between a predominantly muddy section overlying a sand-rich section. This horizon gradually shoals downfan (Fig. 3A of Chapter

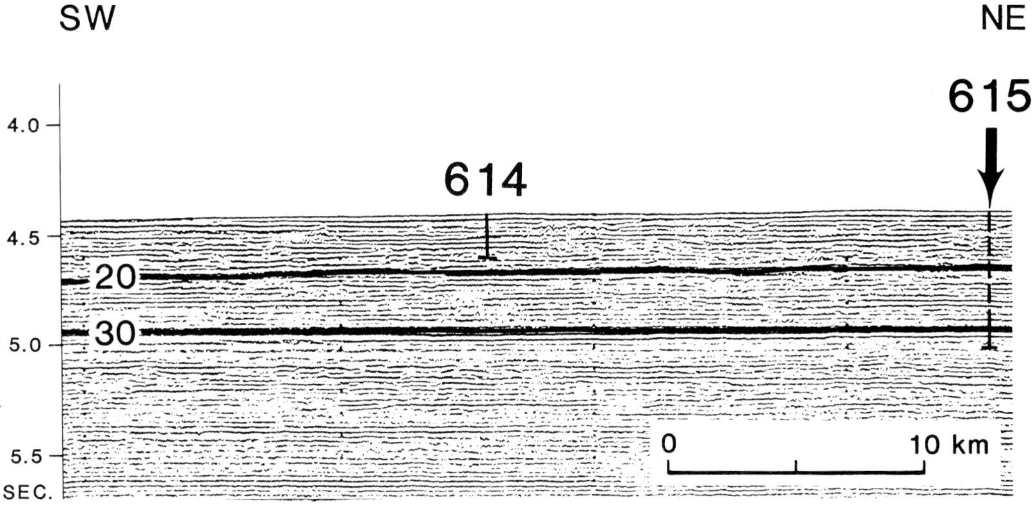

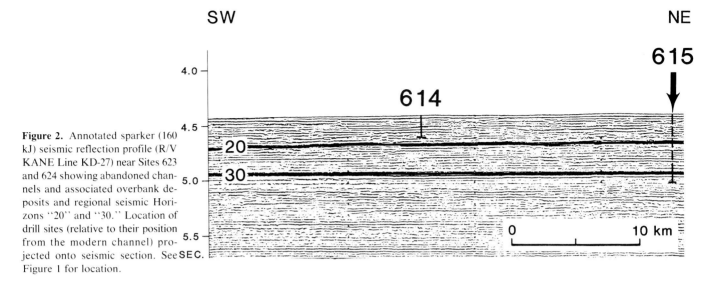

Figure 2. Annotated sparker (160 kJ) seismic reflection profile (R/V KANE Line KD-27) near Sites 623 and 624 showing abandoned channels and associated overbank deposits and regional seismic Horizons "20" and "30." Location of drill sites (relative to their position from the modern channel) projected onto seismic section. See Figure 1 for location.

21). Horizon "30" corresponds to an abrupt change from interbedded terrigenous sands, silts, and muds to a carbonate ooze (Chapters 38, 39).

Sedimentary Characteristics

Sediments recovered from the lower-fan drill sites can be grouped into five major sedimentary facies: 1) sands and silts, 2) silty muds, 3) silt-laminated muds, 4) clays and muds, and 5) oozes and calcareous muds. These facies follow Stow and others (see Chapter 38 for more detailed descriptions for each of the sedimentary facies).

Sands and coarse silts (Figs. 4A and B) typically occur in beds from 10 cm to at least 12 m in thickness. Sedimentary structures are not observed in thicker beds. The lack of sedimentary structures could be the result of several factors: flow-in during the coring operations, destruction of fabric during drilling, or flowage of the watery sediment while raising the core barrel. The sediments of these layers are moderately to poorly sorted. Contacts with fine-grained layers are generally sharp and occassionally show some scoured or loaded bases (Figs. 4D and E). Texturally, this coarse-grained facies is highly variable; sand/silt ratios range from 90:10 for the sandy sediments (Fig. 4A) to 1:99 for the silty sediments (Fig. 4E).

The silty muds form beds that range in thickness from about 5 cm to a few meters. They are moderately to poorly sorted and occur both as structureless layers and as beds that exhibit indistinct grading. These silty muds make up less than 10% of the recovered sediments. At Sites 623 and 624, silty muds mainly occur below 125-m subbottom and

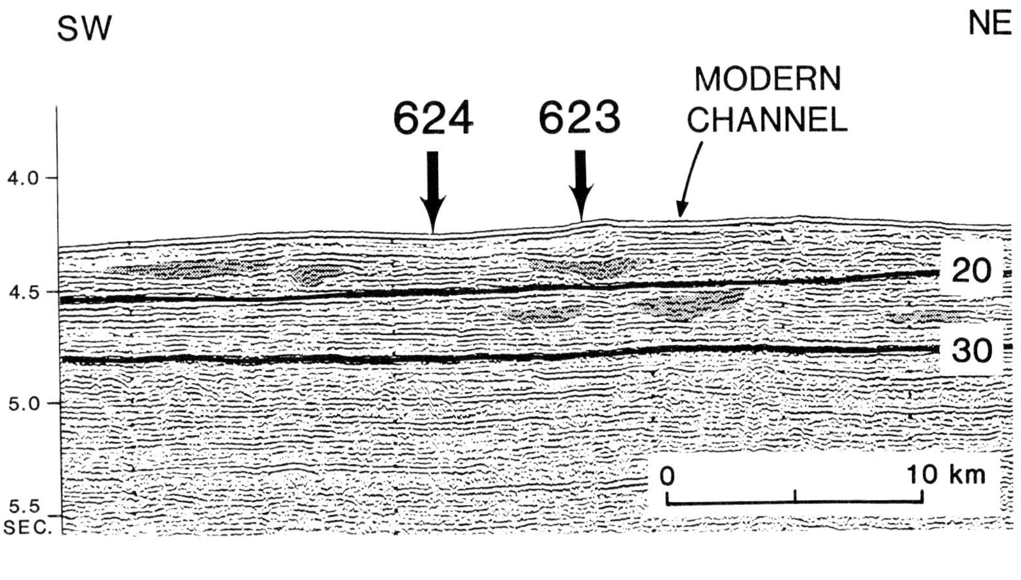

Figure 3. Annotated sparker (160 kJ) seismic reflection profile (R/V KANE Line KD-10) near Sites 614 and 615 showing seismic character typical of the distal portion of the lower fan, regional Horizons "20" and "30," position of Site 614, and relative position of Site 165. See Figure 1 for location.

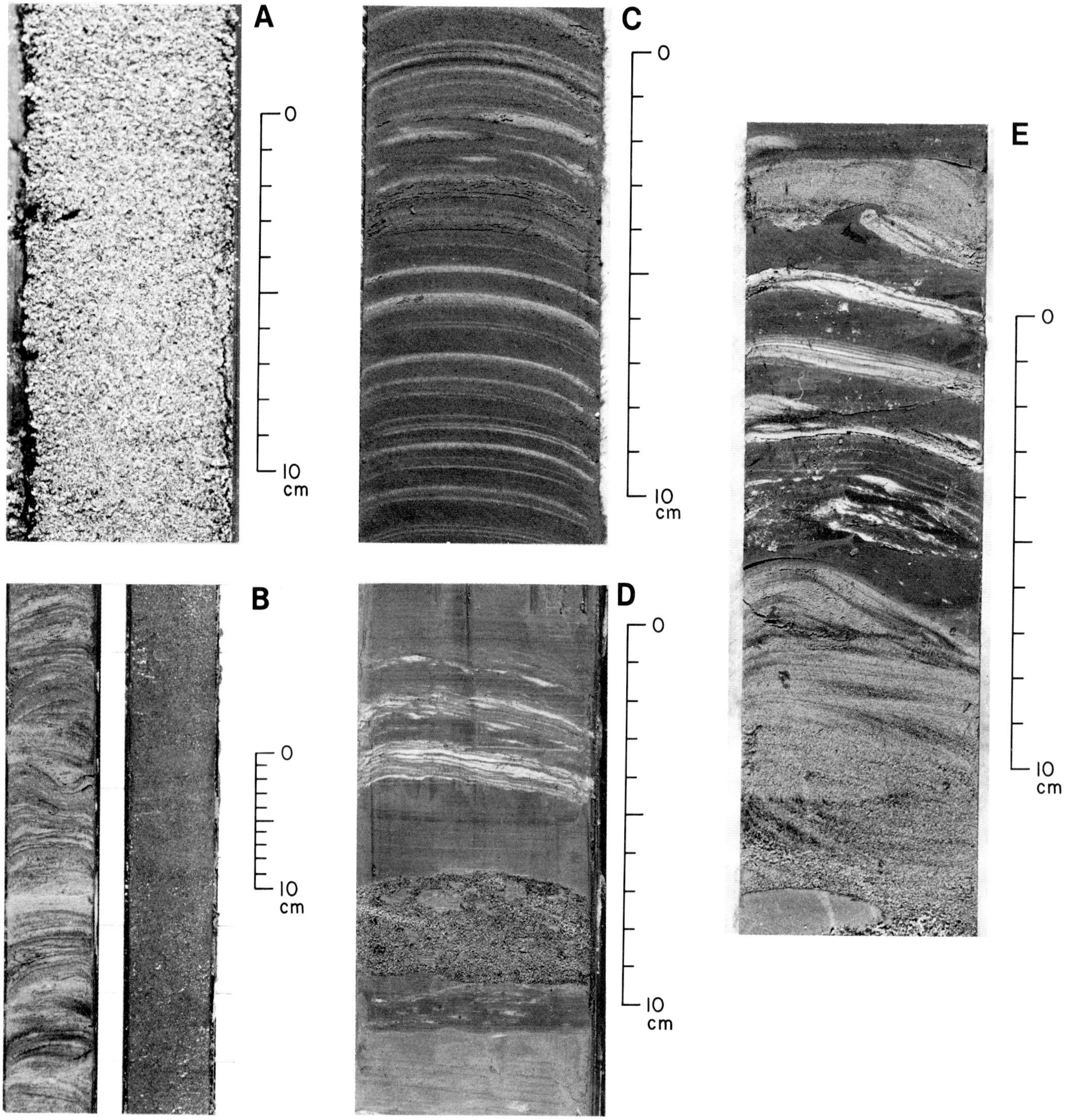

Figure 4. Photographs showing characteristics of selected sedimentary facies (subbottom depths correspond to the top of each photograph). (A) Medium-grained sand (Site 615; 135.4 m). (B) Finely laminated silt (Site 615; 98.3 m) and massive fine-grained sand (Site 615; 99.8 m). (C) Silt-laminated muds (Site 615; 90.2 m). (D) Interbedded mud, sand, and silt-laminated mud (Site 623; 50.5 m). (E) Interbedded mud, silt-laminated mud, and finely laminated silt (Site 624; 169.2 m).

are commonly in association with coarser-grained layers (Fig. 5). At Sites 614 and 615, these silty muds are interspersed with the other sedimentary facies (Fig. 6).

Silt-laminated muds (Figs. 4C through E) are present in intervals ranging from a few cm to over 10 m in thickness. The silts occur as beds or lenses (1-mm to about 5-cm thick) within visually structureless, slightly banded or laminated muds. The silt laminae are normally graded and show compositional sorting. The bottom contact is usually sharp, and the upper contact may be either sharp or gradational. Both commonly display sedimentary structures typical of fine-grained turbidites [3] (Fig. 4C). Convolute bedding and dis-

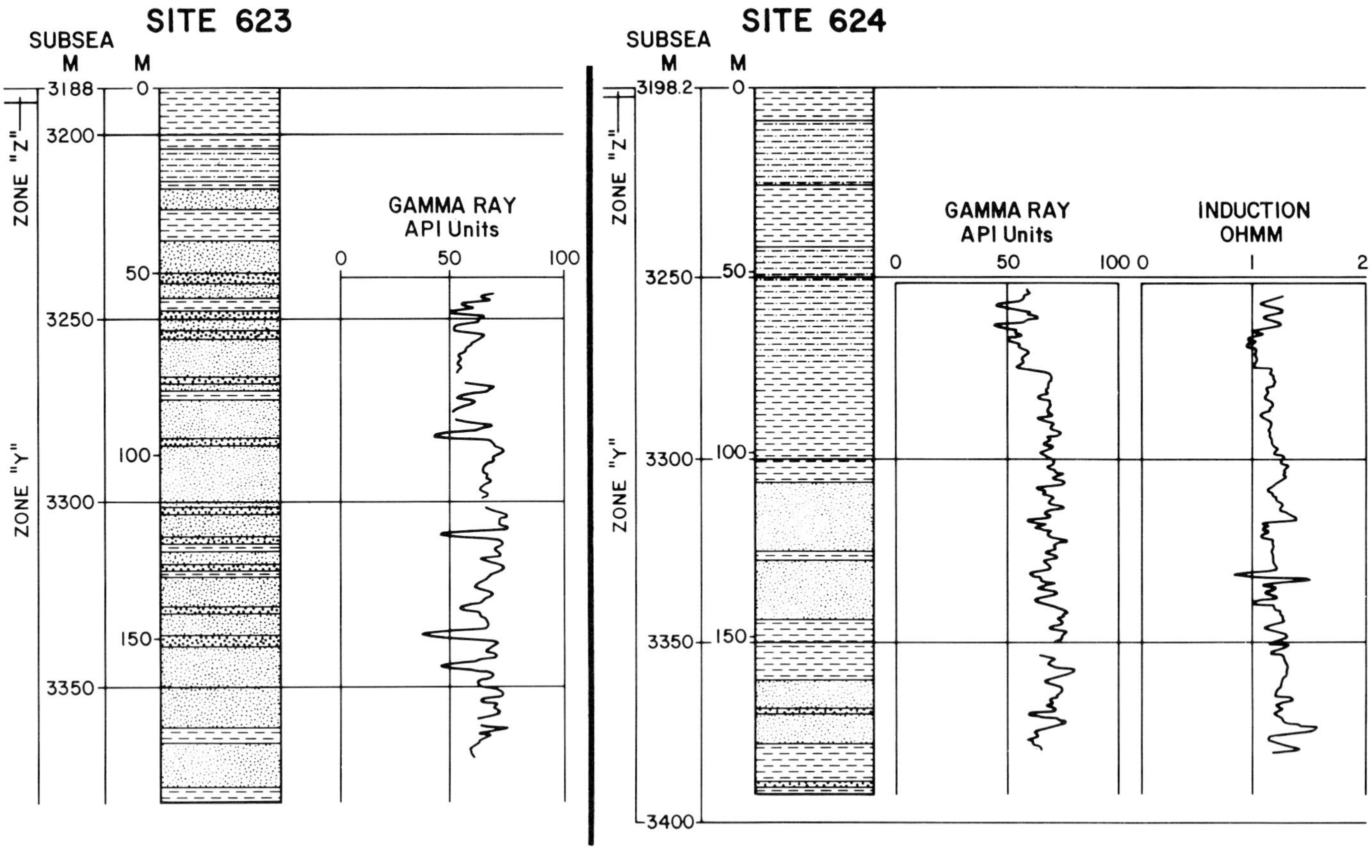

Figure 5. Interpretative lithostratigraphic summary of Sites 623 and 624 (based on recovered sediments and well-log data) showing age, lithology, and gamma-ray and induction-log responses in these lower-fan channel, levee, and overbank deposits. See Figure 6 for legend. Actual core recovery and observed lithologies given in reference 5.

continuous laminations are commonly observed. At Sites 623 and 624, silt-laminated muds form the dominant facies, comprising about 60 to 65% of the recovered material (Fig. 5). At Sites 614 and 615, this facies comprises about 10% of the recovered sediment and occurs as beds ranging from a few to tens of centimeters in thickness (Fig. 6).

The clay and mud sedimentary facies is comprised of sediments that form visually homogeneous layers; they are commonly color-banded or mottled. The color bands range from a few millimeters to a few centimeters in thickness. This facies is an important component at Sites 623 and 624 (30 to 40% of the recovered sediment), but is a relatively minor component at Sites 614 and 615 (10 to 15% of the recovered sediment).

Oozes and calcareous muds occur as a thin cover (25 to 50-cm thick) at the top of all four drill sites. This facies typically consists of foraminiferal ooze at the top and grades downward to a calcareous mud. The basal 29 m of Site 615 consists of a nannofossil-foraminiferal ooze with some chalk gravel-sized fragments at the base (Fig. 6) (see also Chapters 38 and 39).

Additional details on the textural and mineralogical aspects of these sedimentary facies are discussed by Stow and others (Chapter 38) and by Roberts and Thayer (Chapter 47). Sedimentary structures are described from X-ray radiographs by Coleman and others (Chapter 44), and from thin section by Cremer and others (Chapter 45).

Vertical Lithological Characteristics

Actual core recovery at the four lower-fan drill sites was generally good in the upper 100 m but was poor below that depth (Chapter 36). The following discussion of lithological characteristics and their intrepretative summaries in Figures 5 and 6 are the result of *interactive* studies between recovered core lithologies and well-log data (run at all sites except Site 614). Because wire line logs can be used to obtain lithological data but not the discussed sedimentary facies, lithological terms are used in Figures 5 and 6.

The sedimentary column at each of the four drill sites consists of a thin cover of Holocene foraminiferal ooze underlain by a thick series of detrital sands, silts, and muds deposited during Late Pleistocene (Ericson Zone Y [4]). The lithological characteristics of these terrigeneous deposits differ between the upfan part of the lower fan (channel abandonment area; Sites 623 and 624) and the more

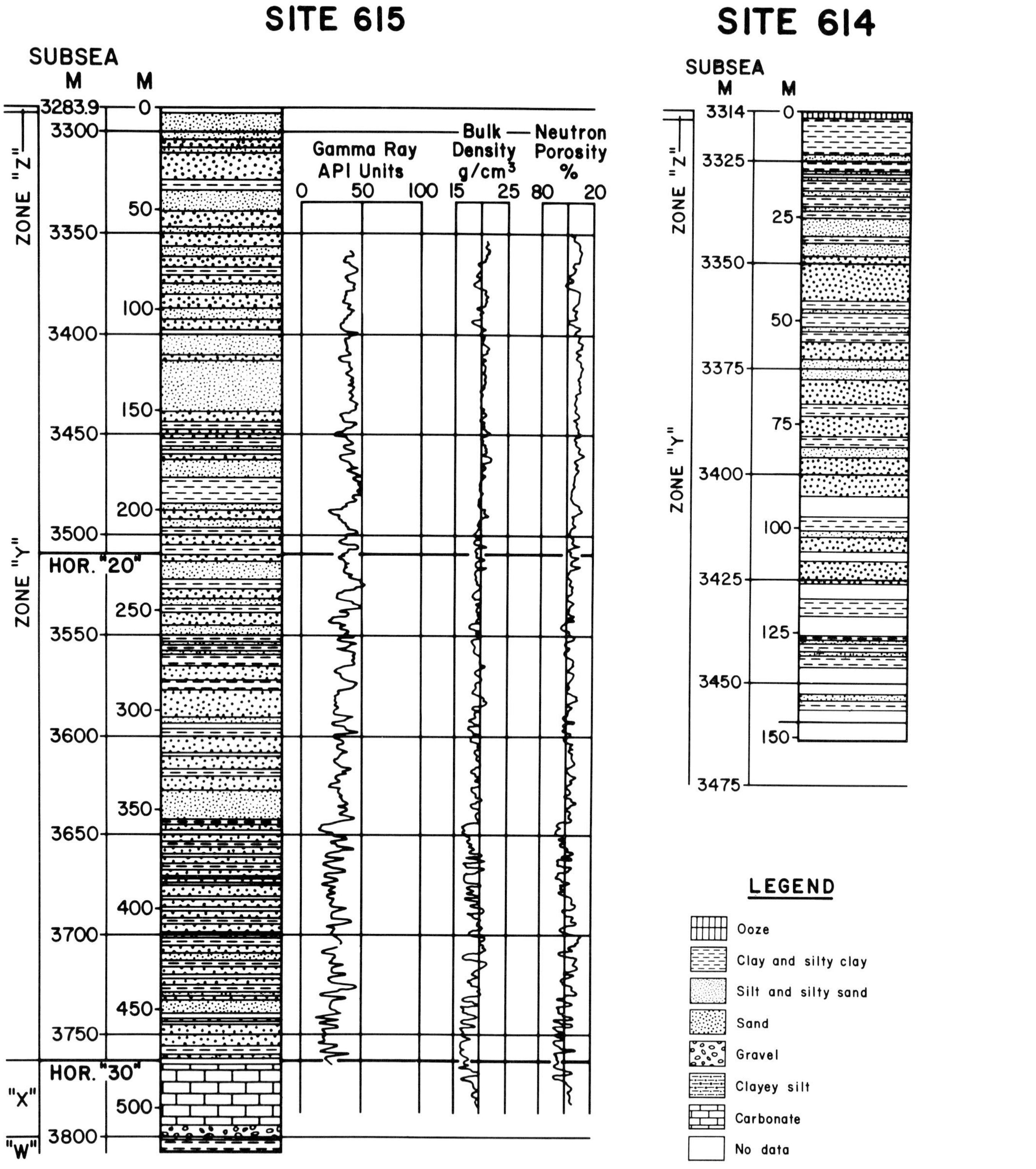

Figure 6. Interpretative lithostratigraphic summary of Sites 614 and 615 showing age, lithology, and well-log responses in these "sheet sand" deposits. Site 615 summary based on recovered sediments and well-log data; see reference 5 for actual core recovery and observed lithologies.

distal lower-fan area (sheet flow depositional area; Sites 614 and 615).

Channel Abandonment Area

Site 623 (202-m penetration) is located just west of the most recent (or most prominent) channel surveyed on the lower fan. Volumetrically, silt-laminated muds are the dominant sediments observed at this site. They extend from the bottom of the hole (202-m subbottom) to about 45-m subbottom and are commonly interbedded with sands and silts and silty-muds (Fig. 5). Above 45-m subbottom, the silt-laminated muds are interbedded with clays and muds, which in turn become the dominant facies above 25-m subbottom.

The vertical succession can be interpreted in several ways: as one major fining-upward sequence, as two fining-upward sequences with the boundary located at about 92-m subbottom, or as multiple 10 to 30-m thick fining-upward sequences (Fig. 5). The latter interpretation follows the jagged well-log response best. Multiple fining-upward sequences could be produced by a vertical succession of channel fill, levee, and overbank deposits.

Site 624 (200-m penetration) was drilled about 4.8-km west of Site 623. The sediments recovered are similar to those at Site 623, although sediments at Site 624 are generally finer-grained (predominantly silt-laminated muds interbedded with clays and muds). Fewer and thinner channel-fill sands are observed at this site (Fig. 5).

Sheet Flow Depositional Area

Site 614, approximately 600-km down-channel from the shelf break, is the most distal location drilled on the fan during leg 96 (Fig. 1). It was cored to a depth of 150-m subbottom and penetrated only the youngest fanlobe (Figs. 3 and 6). The lowermost 37 m (150 to 113-m subbottom) consists of clays and muds; this may be a function of poor recovery, particularly in sandier intervals (Chapter 36). Sand is the dominant lithology from 113 to 25-m subbottom. These relatively massive sands are commonly interbedded with silts and muds. The upper 25 m of the section is comprised primarily of clays and muds and silt-laminated muds.

Site 615, located 21 km to the northeast, was drilled to a depth of 523-m subbottom and penetrated the upper two fanlobes as well as an underlying carbonate ooze (Fig. 6). The average percentage of sand at this site is very high (about 50%) and unconsolidated, thus resulting in poor core recovery.

The youngest fanlobe lies above seismic Horizon "20" and extends from the seafloor to about 220-m subbottom (Figs. 4 and 6). Sediments in this zone are comprised of interbedded sands and silts, silty muds (or silt-laminated muds), and clays and muds. Sandy sediments are dominant (see below). The second fanlobe forms the section from 220 to 475-m subbottom (between seismic Horizons "20" and "30"). The same sediment types are interbedded in the underlying fanlobe, but sand is even more abundant than in the youngest fanlobe.

The two Late Wisconsin glacial fanlobes (Ericson Zone Y [4]; see Kohl and others, Chapter 39) are underlain by a 29-m thick carbonate that consists of one or more fining-upward sequences of nannofossil-foraminiferal ooze with some small chalk gravel-sized fragments at the base. This carbonate interval is interpreted to be the product of a debris flow, probably originating from the Florida Platform (see Chapter 39). The carbonate unit is bounded on top and bottom by detrital muds. Stratigraphically, it represents the Wisconsin interstadial (Ericson Zone X).

Well logs obtained at Site 615 complemented the lithologic information available from the limited core material. Primarily using the gamma-ray log, a total net sand percentage of 41% was calculated for the youngest fanlobe, whereas for the underlying fanlobe, net sand percentage was 64%. The gamma-ray logs frequently show a blocky pattern for the sands with maximum bed thicknesses of as much as 10 m (Fig. 6).

Conclusions

Downfan changes in the Mississippi Fan channel system are observed in the sediments, side-scan sonar images and 3.5-kHz and seismic reflection profiles. In the upfan part of the lower fan, the channel changes from a single, sinuous, migratory channel to a shallow channel system in which the channels become filled and abandoned. Recently filled channels are identifiable on side-scan sonar images and many filled channels can be observed in the seismic records. At the two drill sites in this area (Sites 623 and 624), the sediments are predominately silt-laminated muds that are commonly interbedded with sands and silts and silty-muds. These sediments are considered to be characteristic of overbank deposition and numerous thin channel fills. The more distal lower fan shows very low channel relief in an apparently bifurcating system. Sediments recovered in this area (Sites 614 and 615) are composed dominantly of sand layers (ranging from a few centimeters to several meters in thickness) that might be continuous over large sections of the lower fan (so-called "sheet-flow deposits"). The drilling results demonstrate that the Mississippi Fan is a sand-efficient system in which much of the sand is transported to the lower-fan region.

References

[1] Bouma, A. H., Stelting, C. E., and Coleman, J. M., 1983/84. Mississippi Fan: internal structure and depositional processes. Geo-Marine Letters, v. 3, pp. 147–153.

[2] Mutti, E., and Ricci-Lucchi, F., 1972. Le torbiditi dell'Appennino settentrionale: introduzione all' analisi di facies. Memorie della Societa Geologica Italiana, v. 11, pp. 161–199.

[3] Stow, D. A. V., and Piper, D. J. W., 1984. Deep-water finegrained sediments: facies models. In: D. A. V. Stow and D. J. W. Piper (eds.), Fine-Grained Sediments: Deep-Water Processes and Facies. Geological Society of London Special Publication 14. Blackwell Scientific Publications, Oxford, pp. 611–645.

[4] Ericson, D. B., and Wollin, G., 1968. Pleistocene climates and chronology in deep-sea sediments. Science, v. 162, pp. 1227–1234.

[5] Leg 96 Scientific Party, 1984. Challenger drills Mississippi Fan. Geotimes, v. 29, no. 7, p. 15–18.

CHAPTER 43

Consolidated Characteristics and Excess Pore Water Pressures of Mississippi Fan Sediments

William R. Bryant and DSDP Leg 96 Shipboard Scientists

Abstract

Examination of the geotechnical properties of the Quaternary sediments from the middle and lower portions of the Mississippi Fan shows that the sediments are normally consolidated to overconsolidated to a depth of 40 to 50 below the seafloor. Beyond that depth, the sediments become highly underconsolidated and exhibit high excess pore water pressures. The state of underconsolidation is attributed to the rapid rates of accumulation and the low permeability of smectite clays.

Introduction

An extensive geotechnical sampling program was carried out during Deep Sea Drilling Project Leg 96 on the Mississippi Fan. Of the nine sites drilled, eight produced excellent quality cores using the Advance Piston Corer (APC) and the Extended Core Barrel at Sites 614, 615, 616, 617, 621, 622, 623, and 624. The excellent quality of the cores was also attributed to the absence of methane and other gases. The cores taken at Site 620 were recovered by the use of rotary drilling and were highly disturbed. Duplicate APC cores collected at Sites 615, 616, 617, and 623 were used exclusively for geotechnical investigations. Analyses of these dedicated geotechnical cores are underway at this time. The preliminary results of the onshore geotechnical analyses and the data collected on board are reported in this chapter.

The majority of sediments drilled on the Mississippi Fan were from the late Wisconsin (Ericson's Y zone, [1]). Because of the high sedimentation rates in this zone, over 11 m/1000 yr for the middle fan and over 5 m/1000 yr for the lower fan [Kohl and others, Chapter 39], all sediments were expected to be underconsolidated.

Underconsolidation is the condition in which the sediments and pore waters are in disequilibrium with the overburden stress, resulting in pore water pressures in excess of hydrostatic pressure. A state of total underconsolidation is where the pore water pressures would be equal to the total vertical effective stress plus the hydrostatic pressure (geostatic pressure). A state of normal consolidation and overconsolidation is where the pore water pressures are equal to the hydrostatic pressure. Overconsolidation is the state in which the fabric of the sediments is capable of supporting a load in excess of the existing overburden. Overconsolidation results from unusual bonding or the removal of pre-existing overburden by erosional processes and slumping.

High rates of sediment accumulation and low permeabilities are the usual causes of underconsolidation. The permeabilities of the clays on the Mississippi Fan are extremely low. These clays are composed of 70 to 80% smectites, and have coefficients of permeabilities in the range of 10^{-6} cm/sec (10^{-3} darcys) for the near-surface sediments and 10^{-11} cm/sec (10^{-8} darcys) for sediments at 550-m subbottom depth.

The primary aim of this chapter is to define the degree of consolidation and pore water pressure conditions of Mississippi Fan sediments and determine what such conditions may mean to the sediment column. Excess pore water pressures can have a profound influence on the in situ shear strength and other geotechnical properties. The effective stress ($\bar{\sigma}$) principal states that the effective stress is equal to the total stress (σ) minus the pore pressure (u):

$$\bar{\sigma} = \sigma - u. \quad (1)$$

The effective stress state more closely correlates with sediment behavior than either total stress or pore water pressure. An increase in the effective stress should cause the sediment particles to shift into a denser packing, while equal

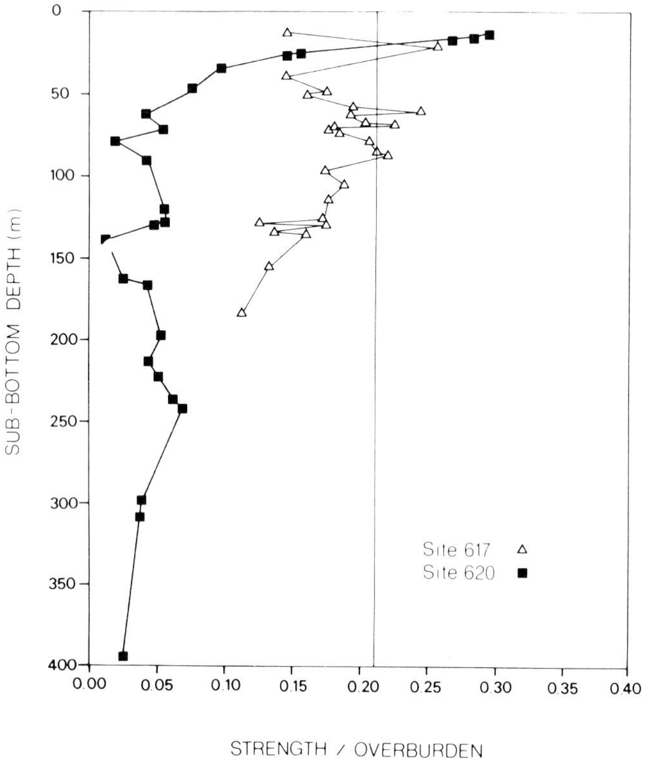

Figure 1. Undrained shear strength/effective overburden stress ratios for Sites 617 and 620. Sediments with a ratio of less than 0.22 are underconsolidated.

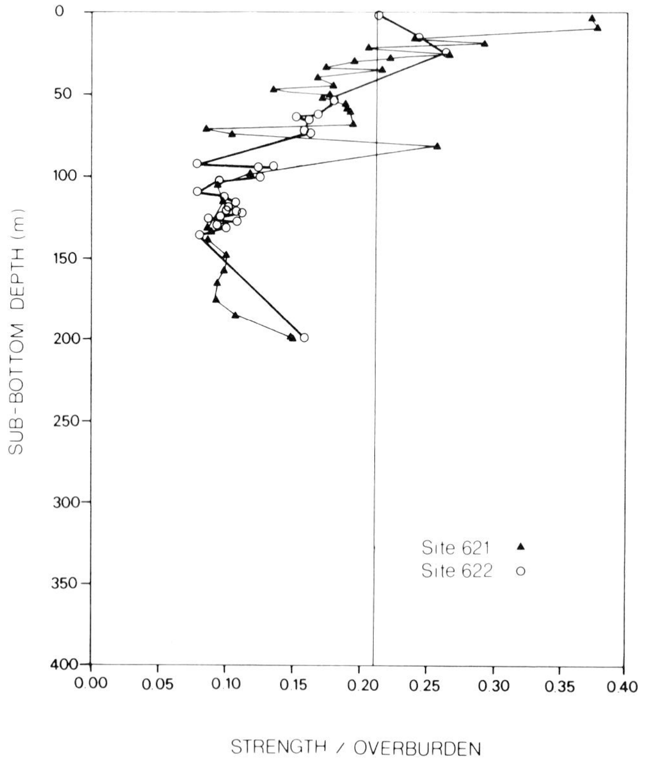

Figure 2. Undrained shear strength/effective overburden stress ratios for Sites 621 and 622. Sediments with a ratio of less than 0.22 are underconsolidated.

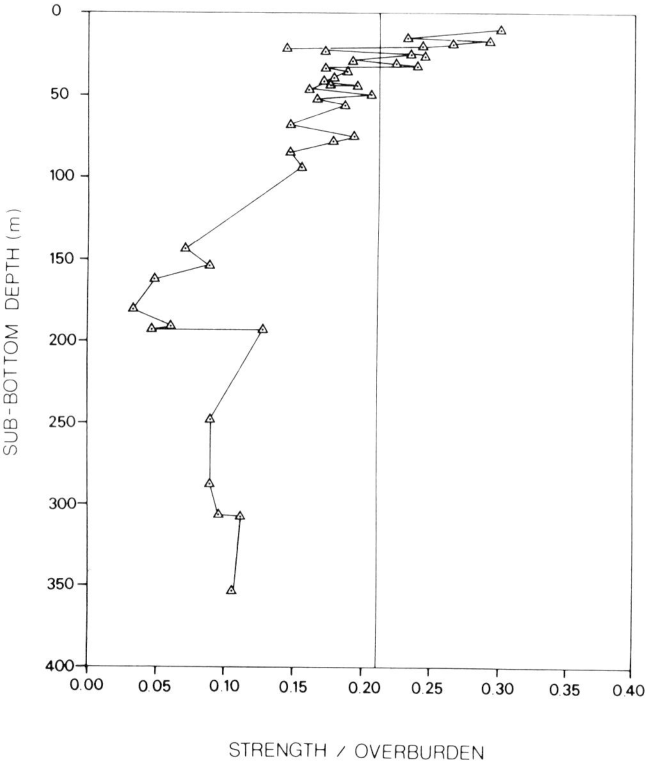

Figure 3. Undrained shear strength/effective overburden stress ratios for Site 616. Sediments with a ratio of less than 0.22 are underconsolidated.

increases in the total stress and the pore water pressure would keep the effective stress constant and would have little or no effect on the particle packing [2].

The dramatic fashion in which the shearing resistance of a sediment can be effected by excess pore water pressure is shown by examination of the following equation:

$$s_u = c + \bar{\sigma} \tan \phi \quad (2)$$

where s = undrained shear strength, c = cohesion, $\bar{\sigma}$ = vertical effective stress [see (1)], and ϕ = angle of internal friction.

As the pore water pressure increases, the value of the vertical effective stress decreases. For the condition where a state of total underconsolidation exists, the value of the effective stress ($\bar{\sigma}$) is zero, in which case the shear strength (s_u) is then equal only to the cohesion (c) of the sediment. Under these conditions, the shearing resistance of the sediments remains constant with depth, and the deposits are in an unstable condition on almost any slope.

Overpressured formations, which are very common on the Gulf Coast, are the result of high excess pore water conditions resulting from compaction (consolidation), disequalibrium, and other conditions. In highly underconsolidated sediments, such as certain Mississippi Fan sedi-

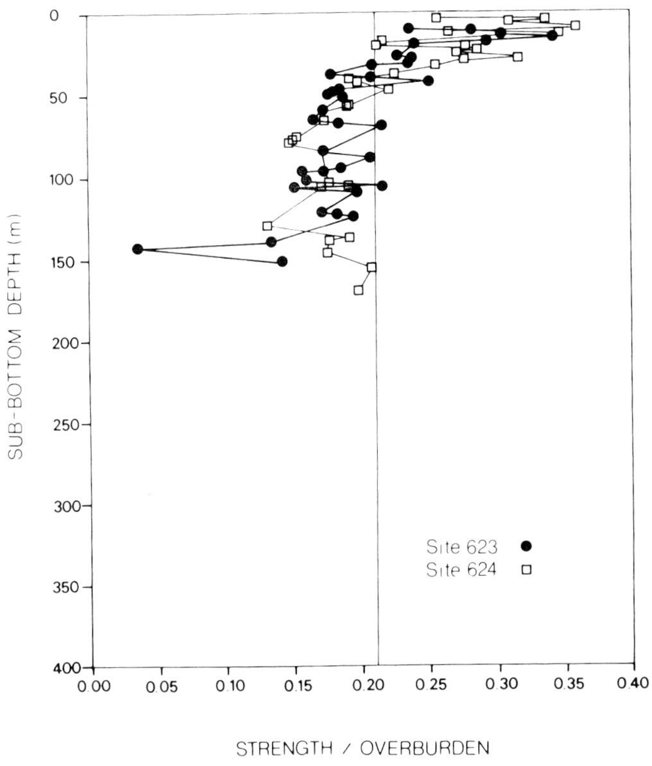

Figure 4. Undrained shear strength/effective overburden stress ratios for Sites 623 and 624. Sediments with a ratio of less than 0.22 are underconsolidated.

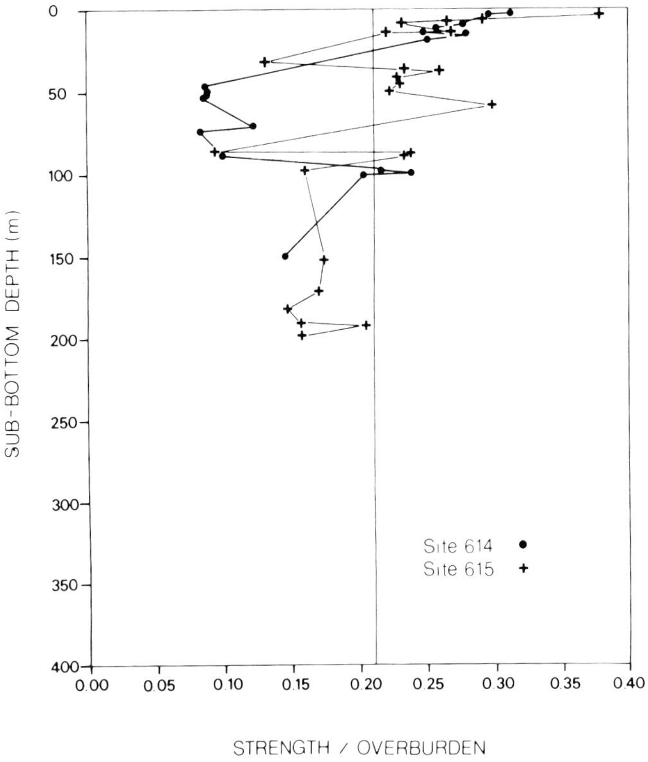

Figure 5. Undrained shear strength/effective overburden stress ratios for Sites 614 and 615. Sediments with a ratio of less than 0.22 are underconsolidated.

ments, overpressured conditions can be encountered at fairly shallow depths below the seafloor.

Consolidation Characteristics Inferred from Sediment Shear Strength

The most direct and positive way of determining the degree of consolidation for normally consolidated and underconsolidated sediment conditions is to measure the in situ pore water pressures. Only in rare cases was that possible at the majority of DSDP sites drilled throughout the world. Another method to approximate the conditions of consolidation is to examine the relationship between shear strength and overburden stress.

Skempton [3] devised a method based on empirical relationships for predicting the shear strength/effective overburden stress ratio for normally consolidated marine sediments. This relationship is given in a simple equation:

$$S_u / \bar{\sigma}_o = 0.11 + 0.0037\ I_p, \qquad (3)$$

where S_u = shear strength, $\bar{\sigma}_o$ = effective overburden stress, and I_p = plasticity index.

Figures 1, 2, and 3 show the ratio of shear strength to effective overburden stress plotted against subbottom depth for sediments of the middle fan at Sites 616, 617, 621, 622, and 620. Figures 4 and 5 depict this ratio for the lower-fan sediments at Sites 614, 615, 623, and 624.

The plots include a vertical line with a value of 0.22. This line represents the limits derived from Skempton's relationship, as expressed by equation 3, for normally consolidated sediments with a plasticity index of 0.25. All values of the shear strength/overburden stress ratio less than 0.22 are thus assumed to be underconsolidated and those with values greater than 0.22 are considered to be normally consolidated to overconsolidated.

Characteristics of Middle-Fan Sediments

The sediments at Site 617 are overbank deposits consisting of thin, fine-grained turbidite sequences. Figure 1 shows that the upper 60 m of the deposits at this site are most likely underconsolidated, the 60 to 90 m interval normally consolidated, and the sediments below the 90 m level underconsolidated with the degree of underconsolidation increasing with depth in the core.

Site 620 was located in the middle-fan area, approximately 18-km northeast of the channel in overbank sedi-

Table 1. Depth of Normally Consolidated Sediments, Mississippi Fan

	Site	Sediment Normally Consolidated to a Depth of and Underconsolidated Beyond
Middle Fan	617	90 m
	620	25 m
	621	40–50 m
	622	40–50 m
	616	40 m
Lower Fan	614	30–40 m
	615	65 m
	623	40 m
	624	40 m

ment. Figure 1 indicates that the clay and silty clays recovered from this site are normally consolidated in the upper 25 m of the section and highly underconsolidated at all depths below the 25-m level. Most sediments from Site 620 were highly disturbed as a result of rotary drilling to recover cores.

Sites 621 and 622 were drilled in the channel 18-km southwest of Site 620. Both sites contained clay and mud to a depth of 135 m below the seafloor. Below the 135-m level, the sediments become sandy with the amount of sand increasing with depth and finally turning to loose sand and gravel at approximately 200-m subbottom depth. Figure 2 shows that at both channel sites, the sediments became underconsolidated at a depth of 40 to 50 m, with the degree of underconsolidation increasing with depth.

Site 616 was drilled in the eastern margin of the youngest fanlobe. The upper 90-m section consisted of fine-grained muds and silty clays with extremely steep dipping beds (ranging up to 65°), suggesting disturbance by mass movement. Below this disturbed zone the sediments are dominantly silt-laminated muds and minor sands.

Figure 3 shows that the sediments at Site 616 are normally consolidated to a depth of 40 m, slightly underconsolidated to a depth of 95 m, and highly underconsolidated below a depth of 95 m. It is unusual that a large Holocene or late Pleistocene slump would contain normally consolidated sediments. It is also unusual that a massive slump should contain well-preserved primary sediment structures and that the only indication of mass movement is the appearance of inclined strata. The sediment directly below the slump presents all the evidence of being rapidly loaded, which supports the idea that a slump mass had been implaced at Site 616.

Characteristics of Lower-Fan Sediments

Sites 623 and 624 were drilled adjacent to the main channel in the lower fan. The strength/overburden stress ratios of the sediments (terrigenous clays and muds; silty muds and silts; and silty sands and sands) at both sites (Fig. 4) indicate that the sediments are normally consolidated in the upper 40-m level and slightly underconsolidated below the 40-m level.

Sites 614 and 615 were drilled in the lower fan near the ends of the channel and its associated depositional lobe. Large amounts of sand were recovered from both sites. Sand made up an estimated 47% of the upper 150 m of the sediment column of Site 614 and 47% at Site 615. The remaining sediments consisted of clays and silty clays, underlain by 29 m of nannofossil ooze at the bottom of the hole of Site 615. Figure 5 shows that the sediments at Site 614 are normally consolidated to a depth of 30 to 40 m, underconsolidated between 40 and 90 m, and normally consolidated at a depth of 100 m. The sediment becomes underconsolidated again at a depth below 110 m. At Site 615, the sediments are normally consolidated to a depth of 65 m and slightly underconsolidated below a depth of 64 m.

Examination of the relationship between the shear strength and overburden stress indicates that the following conditions of the degree of consolidation on the modern lobe of Mississippi Fan exist (Table 1). The shear strength/overburden stress relationships only give indications of the degree of consolidation. They are, however, extremely useful because of the large number of points that can be tested. Trends of consolidation characteristics can be generated for a given sediment section. The trends can then be used to determine the sampling and testing scheme for more definitive investigations.

Consolidation Characteristics Inferred from Consolidation Tests

A standard method used to determine the conditions of consolidation and the amount of excess pore water pressures that exists in a sediment section is to determine the preconsolidation pressure by the analysis of the void ratio to vertical effective stress relationships. This is determined by a consolidation test using an oedometer.

Consolidation Tests

One-dimensional consolidation tests were performed on Leg 96 sediments using Anteus backpressure consolidometers. The standard consolidation test involves the incremental loading of a relatively thin, laterally confined sediment sample. Axial strain-time relationships are obtained by measuring the change in sample height during the test. A complete description of this technique was given by Lambe and others [4]. A variation of this method involves the application of sufficient backpressure to redissolve gas bubbles and completely saturate the consolidation sample. Lowe [5] discussed the benefits of this method. A serious drawback to both of these methods is that more than 15 days are required to complete a test.

Skempton [3] defined consolidation as the result of all processes that cause the progressive transformation of an argillaceous sediment from a soft clay to a shale. These processes include: 1) interparticle bonding, 2) dessication, 3) cementation, and 4) the squeezing out of pore water under the increasing weight of overburden. The effect of each process on consolidation will vary through time. Because of the relatively young age of the Mississippi Fan sediments, their consolidation state is basically a function of pore water loss as a result of the weight of the overburden.

Consolidation tests are used to evaluate the sediment response to an applied load. These results provide insight into the relative degree of consolidation that the sediment has experienced in situ under the imposed load of the sedimentary column. The state of consolidation is determined using the ratio of preconsolidation stress ($\bar{\sigma}_c$) to the present effective overburden stress ($\bar{\sigma}_o$). Preconsolidation stress, defined as the maximum effective stress that the sediment has experienced is usually calculated using the graphical reconstruction technique of Casagrande [6]. Various investigators, including Cooling and Skempton [7], Bishop and others [8], Schmertmann [9], and the Geotechnical Consortium [10] have concluded that the Casagrande method is inadequate under certain circumstances to define the preconsolidation stress when applied to curves from oedometer (consolidation) tests. Other procedures for the determination of preconsolidation pressures have been proposed by Burmister [11] and Schmertmann [9].

Recent results of a study conducted at Texas A&M University to determine preconsolidation pressures of marine sediments indicate that the Casagrande method underestimates preconsolidation pressures by 35% or more. Other methods to determine preconsolidation pressure were devised on the basis of the rebound characteristics of the sediments and modification of the Casagrande method. A detailed discussion of these alternate methods are given by the Geotechnical Consortium [10].

The ratio of the preconsolidation stress to the effective overburden stress yields the overconsolidation ratio (OCR $= \bar{\sigma}_c/\bar{\sigma}_o$). A sediment is considered normally consolidated if the present effective overburden stress is the greatest ever imposed. Thus, for a normally consolidated sediment, the preconsolidation stress equals or closely approximates the effective overburden stress ($\bar{\sigma}_c = \bar{\sigma}_o$). A sediment is overconsolidated if it has been consolidated under a stress that exceeds the present effective overburden stress ($\bar{\sigma}_c < \bar{\sigma}_o$). Sediment that has not fully consolidated under the present overburden stress is underconsolidated. For underconsolidated sediments, the pore water pressure exceeds the hydrostatic pressure, and the preconsolidation stress will be less than the effective overburden stress ($\bar{\sigma}_c < \bar{\sigma}_o$). Thus, sediments with an overconsolidated ratio of approximately 1.0 are considered normally consolidated, greater than 1.0 are overconsolidated, and less than 1.0 are underconsolidated.

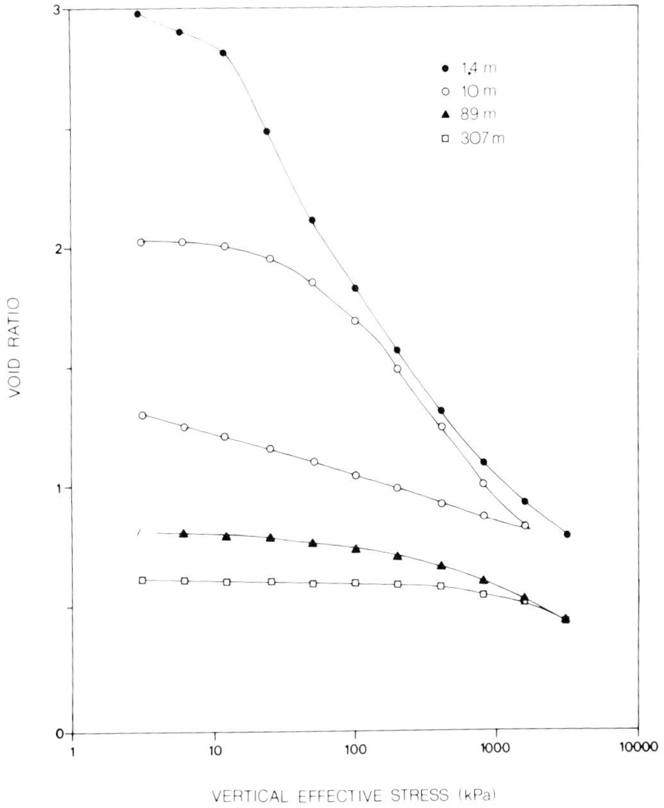

Figure 6. Void ratio log of vertical effective stress plots of sediments from Site 615 at 1.4, 10, 89, and 307-m below the seafloor.

Results of Consolidation Tests

Forty-one consolidation tests were performed on samples from the nine Mississippi Fan sites. Twenty-three of the tests were performed in the Goetechnical Laboratory at Texas A&M University. The remaining tests were run by Dr. A. Wetzel of the Universität Tubingen, West Germany. An example of the results of the consolidation tests for Site 615 is shown in Figure 6. The four curves shown in Figure 6 were generated from samples taken at 1.4, 10, 89, and 307 m in the core from Site 615. The sample taken at the 1.4-m level had an initial void ratio of approximately 3.00 and a final void ratio of 0.80 at a vertical effective stress (load) of 3200 kPa. This sample is a good example of the large amount of volume reduction (decrease in void ratio) that high water-content sediments undergo with relatively small increases in vertical effective stress (load). These sediments undergo as much reduction in void ratio under a load of 100 kPa as they do under a subsequent load of 3200. This may in part explain the conditions expressed by the shear

Table 2. Consolidation Characteristics and Pore Water Pressures of Mississippi Fan Sediments

	Site	Depth (m)	Effective Overburden Stress $\bar{\sigma}_o$ kPa	Minimum and Maximum Preconsolidation Stress $\bar{\sigma}_c$ kPa	OCR for High Values of $\bar{\sigma}_c$	Consolidation Characteristics	Minimum Pore Water Pressure kPa	Maximum Pore Water Pressure kPa
Lower Fan	614	7	35	40–46	1.31	OC	0	0
		100	805	800	.99	NC	0	0
	615	1.4	15	16	1.06	NC	0	0
		10	45	30–60	1.30	OC	0	15
		89	711	100–410	.57	UC	301	511
		307	2664	200	.07	HUC	2464	2464
	623	5.9	30	33	1.13	NC	0	0
		95.9	671	200–530	.78	UC	141	471
	624	12.8	70	20–43	.61	UC	27	50
		56.3	366	130–230	.62	UC	136	236
		137.8	970	200–500	.51	HUC	470	770
Middle Fan	616	4.4	20	20	1.00	NC	0	0
		75.9	532	100–400	.75	UC	132	432
		105	783	500–700	.63	UC	83	653
	617	5.9	30	30	1.00	NC	0	0
		61.7	445	200–410	.92	NC	35	245
		124	945	240–560	.59	HUC	705	385
	620	119.9	751	250–450	.59	HUC	301	501
	621	35.2	444	130–460	1.06	NC	0	314
		138.8	1100	400–600	.54	HUC	500	700
	622	4.9	26	22–40	1.53	OC	0	4
		73	53	240–600	1.12	NC	0	291

N = normally consolidated; UC = underconsolidated; HUC = highly underconsolidated; OC = overconsolidated

strength/overburden relationship where the sediments become underconsolidated at a depth of approximately 40-m subbottom or under a vertical effective stress of 200 kPa.

Figure 6 also shows that if all the sediment samples were subjected to sufficient stress, their final void ratio would be similar. The rebound portion of the curve for the sample taken at 10 meters indicates that these sediments are fairly elastic and are subject to an increase in void ratio of up to 40% with the release of the vertical effective stress.

The results of consolidation tests are by custom displayed on the void ratio log of vertical effective stress diagrams. The results of the consolidation tests conducted at Texas A&M University are listed in Table 2. The values listed for the preconsolidation stress ($\bar{\sigma}_c$) cover a range from the absolute minimum to the absolute maximum that can be obtained by using four different methods for determining the preconsolidation stress.

Figures 7 through 12 are plots (line connected with dots) of the calculated in situ vertical effective stress (overburden pressure) under hydrostatic conditions (UHC) as a function of depth at the nine drill sites. Also plotted on the figures is the range of the absolute minimum and maximum values of the preconsolidation stress ($\bar{\sigma}_c$) as determined from the consolidation tests. The difference between the preconsolidation stress and the vertical effective stress is the approximate amount of pressure in excess of the hydrostatic pressure that one would expect in the underconsolidated sediments. For example, at Sites 617 and 620 (Fig. 9), the excess pore water pressures at the 120 to 124-m depth level would range from a maximum of 705 kPa to a minimum of 385 kPa in excess of hydrostatic or from 40 to 67% of the vertical effective stress (overburden pressure) under hydrostatic conditions (UHC). The shaded areas of each figure represent the range of the vertical effective stress in the presence of excess pore water pressures.

During the drilling operations at Site 620, a pressure of approximately 2460 kPa (350 PSI) was required to maintain circulation at the 395-m depth level. Even at this pump pressure, the drill string became stuck. Downhole pressures of 2460 kPa at 395-m below the seafloor translate to an excess pore water pressure (abnormal pressure) equal to 78% of vertical effective stress (UHC). The one consolidation sample tested on Site 620 sediments (Table 2) shows that at a depth of 119 m, the sediments are in a state of high underconsolidation. The value of the overconsolidation ratio (OCR) ranged from 0.33 to 0.59. An OCR value of 0.33 means that the downhole excess pore water pressure would be 67% of the total vertical effective stress (UHC). Projecting the underconsolidation trend with depth infers that one could easily account for excess pore water pressures equal to 78% of the vertical effective stress at the 395-m level.

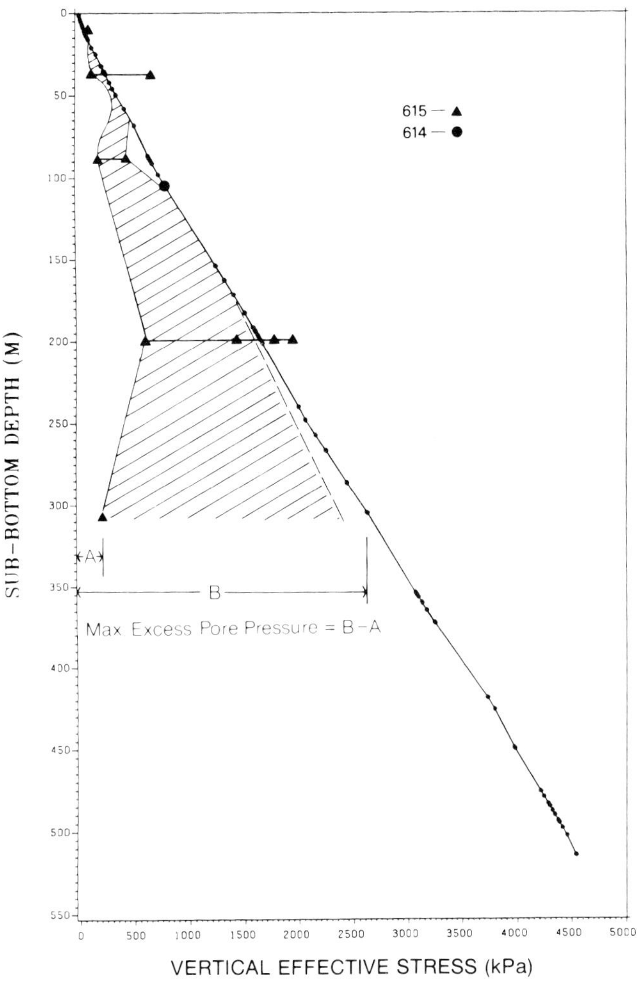

Figure 7. A plot (the dotted line) of the vertical effective stress (overburden pressure) assuming hydrostatic conditions as a function of depth at Sites 615 and 614. The shaded area is the range of the minimum and maximum excess pore water pressures. The left-hand boundary of the shaded area is the maximum excess pore water pressure found at any particular level. The triangles and circles represent the minimum and maximum values of the preconsolidation stress as determined from the consolidation tests. The vertical effective stress in the presence of excess pore water pressures would fall somewhere in the shaded area.

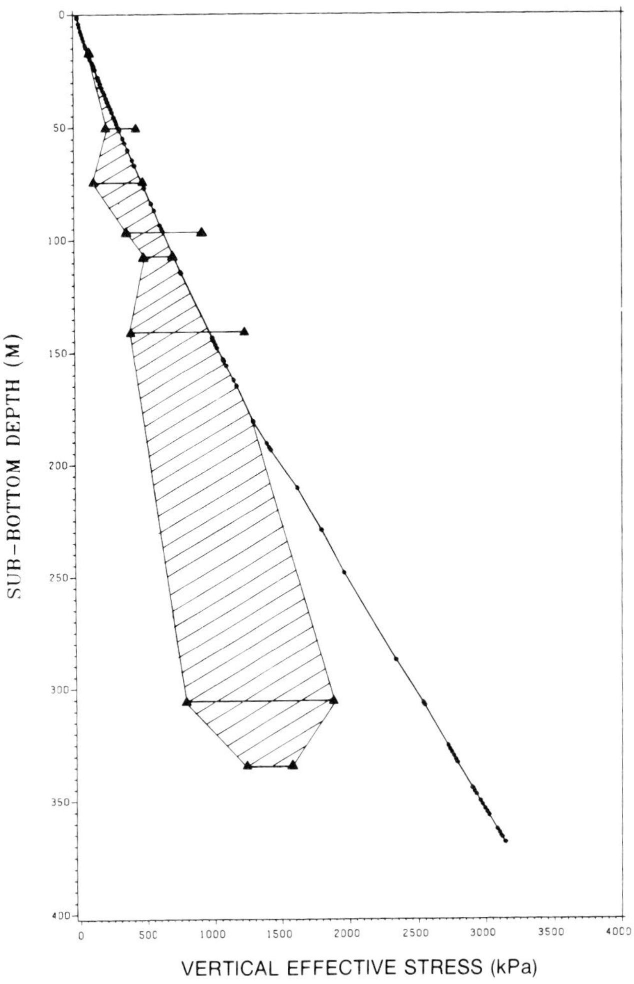

Figure 8. A plot (the dotted line) of the vertical effective stress (overburden pressure) assuming hydrostatic conditions as a function of depth at Site 616. The shaded area is the range of the minimum and maximum excess pore water pressures. The left-hand boundary of the shaded area is the maximum excess pore water pressure found at any particular level. The triangles represent the minimum and maximum values of the preconsolidation stress as determined from the consolidation tests. The vertical effective stress in the presence of excess pore water pressures would fall somewhere in the shaded area.

A minimum OCR value of 0.07 was measured for sediments from 307-m subbottom at Site 615 (Fig. 7). This degree of underconsolidation suggests that downhole conditions at Site 615 are such that the excess pore water pressures almost equal the vertical effective stress. This conclusion, however, is based on only one sample, which may have been highly disturbed. In general, the middle-fan sediments show a higher degree of underconsolidation than do lower-fan sediments. This trend probably results from lower rates of accumulation for the lower-fan sediments as well as the more rapid draining of the pore water in the sediments through higher permeability sands which are more common to the lower fan.

Table 2 lists the minimum and maximum excess pore water pressures determined for the Mississippi Fan sites. A summary of other related geotechnical properties, presenting the seafloor values and vertical gradients, are also listed in Table 3.

Conclusions

The results of the oedometer tests show similar conditions of consolidation for all sites as expressed by the shear strength/overburden relationship. In general, the upper sediments (40 to 65 m) are normally consolidated and, in

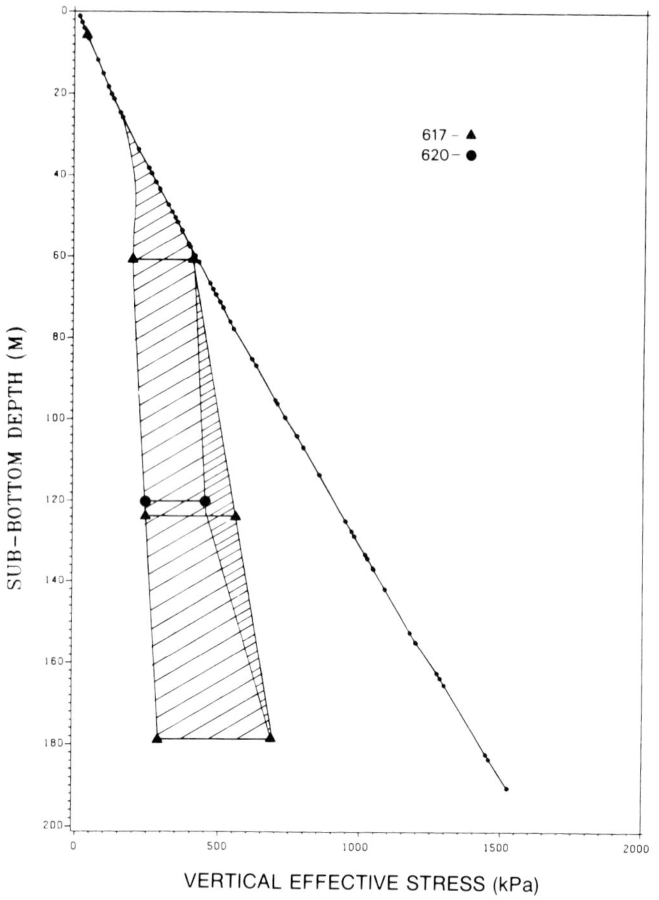

Figure 9. A plot (the dotted line) of the vertical effective stress (overburden pressure) assuming hydrostatic conditions as a function of depth at Sites 617 and 620. The shaded area is the range of the minimum and maximum excess pore water pressures. The left-hand boundary of the shaded area is the maximum excess pore water pressure found at any particular level. The triangles and circles represent the minimum and maximum values of the preconsolidation stress as determined from the consolidation tests. The vertical effective stress in the presence of excess pore water pressures would fall somewhere in the shaded area.

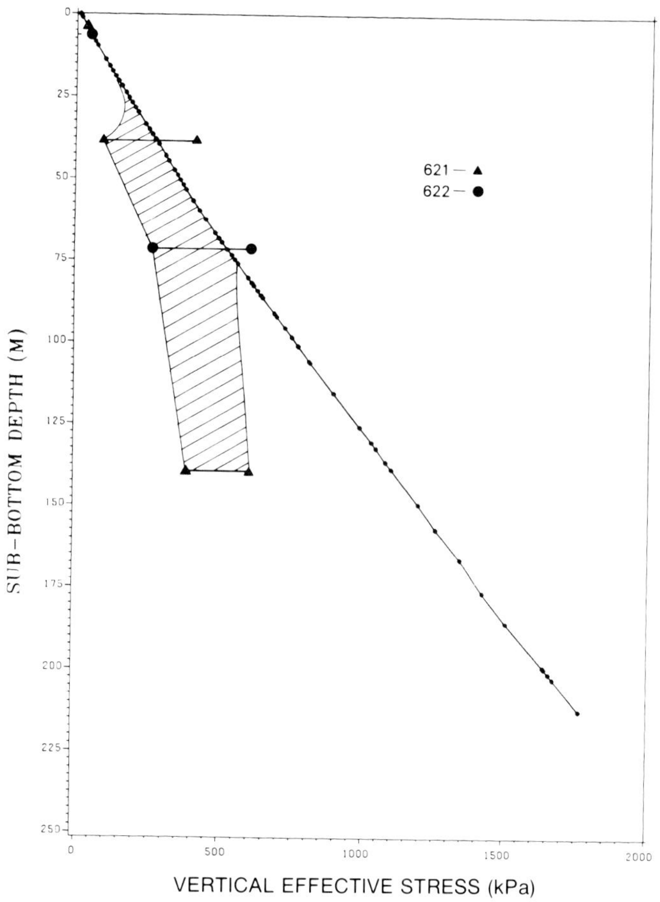

Figure 10. A plot (the dotted line) of the vertical effective stress (overburden pressure) assuming hydrostatic conditions as a function of depth at Sites 621 and 622. The shaded area is the range of the minimum and maximum excess pore water pressures. The left-hand boundary of the shaded area is the maximum excess pore water pressure found at any particular level. The triangles and circles represent the minimum and maximum values of the preconsolidation stress as determined from the consolidation tests. The vertical effective stress in the presence of excess pore water pressures would fall somewhere in the shaded area.

some cases, overconsolidated. The sediments below that level are underconsolidated, and the degree of underconsolidation increases with depth in the section. The significance of these findings are that at depths below approximately 40 to 65 m, the fan sediments exhibit excess pore water pressures which sustain the state of underconsolidation. If such conditions continue at depth, and there is no reason to believe otherwise, highly abnormal pressure conditions are expected to exist at almost all depths within the fan, a condition similar to those found on the shelf and upper slopes of the Gulf of Mexico.

Table 3 shows that almost all the gradients of the geotechnical properties change by an order of magnitude at approximately the 40- to 60-m depth in the cores. This change in gradients takes place at the same depth as the transition from normally consolidated to underconsolidated conditions at the various sites. The explanation for this is not known at the time of publication of this article.

It appears that the major diagenic process at work on the Mississippi Fan is fine-grained sediment consolidation (compaction). Consolidation results in the rapid expulsion of pore water and a reduction of porosity (void ratio). The reduction of porosity results in a decrease in the volume of the sediment mass and, thus, a change in the geometry of that mass. Thus, the total geometry of the fan and the geometry of individual sediment sequences are constantly changing with time. If sedimentation was to cease at the present time, the fan would experience approximately a 15

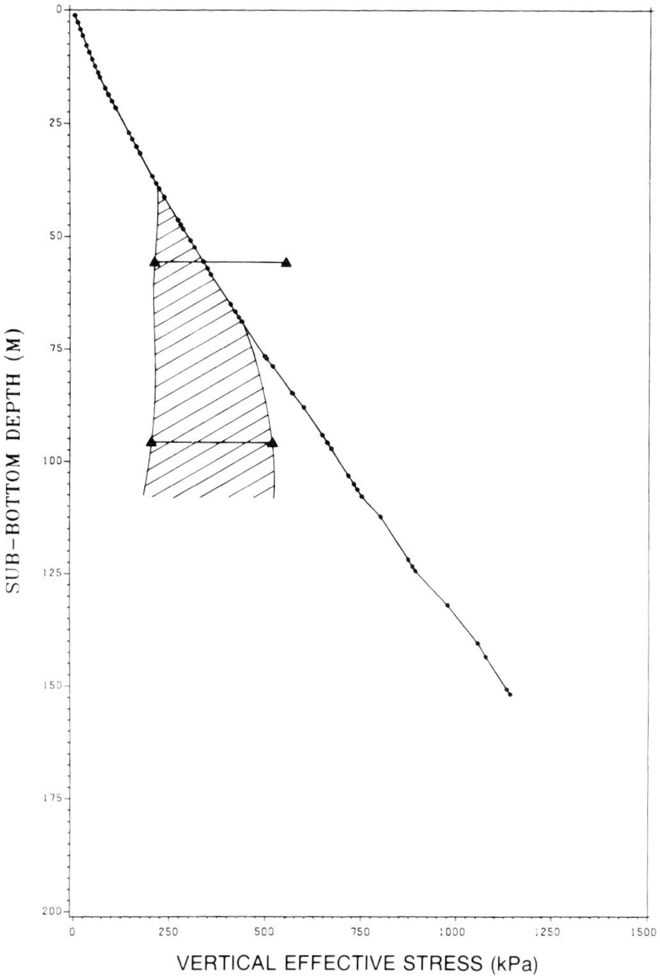

Figure 11. A plot (the dotted line) of the vertical effective stress (overburden pressure) assuming hydrostatic conditions as a function of depth at Site 623. The shaded area is the range of the minimum and maximum excess pore water pressures. The left-hand boundary of the shaded area is the maximum excess pore water pressure found at any particular level. The triangles represent the minimum and maximum values of the preconsolidation stress as determined from the consolidation tests. The vertical effective stress in the presence of excess pore water pressures would fall somewhere in the shaded area.

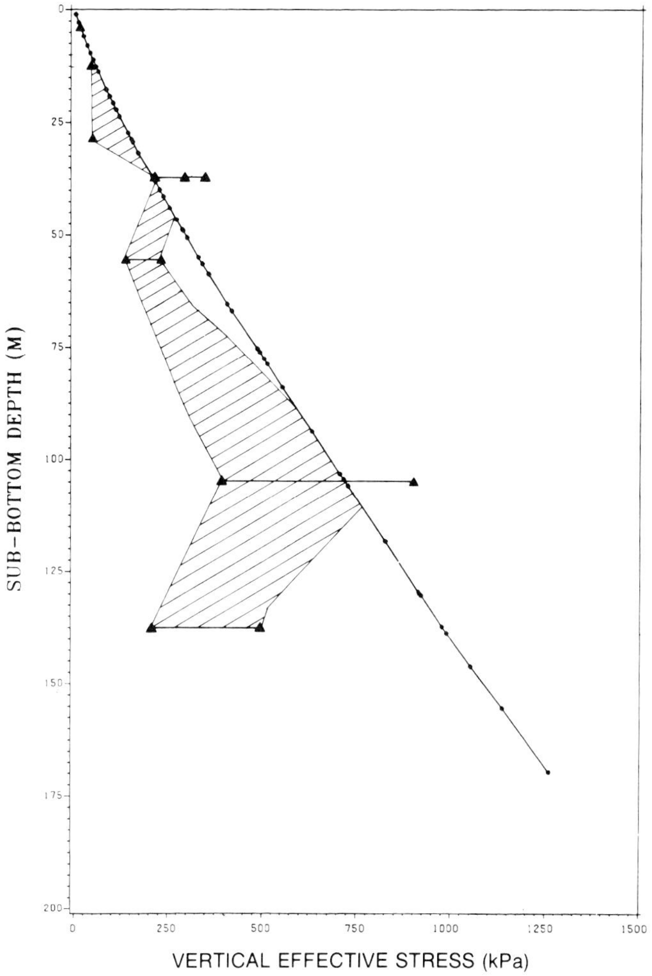

Figure 12. A plot (the dotted line) of the vertical effective stress (overburden pressure) assuming hydrostatic conditions as a function of depth at Sites 624. The shaded area is the range of the minimum and maximum excess pore water pressures. The left-hand boundary of the shaded area is the maximum excess pore water pressure found at any particular level. The triangles represent the minimum and maximum values of the preconsolidation stress as determined from the consolidation tests. The vertical effective stress in the presence of excess pore water pressures would fall somewhere in the shaded area.

to 20% decrease in volume through the subsequent consolidation process.

The time required for the fan to achieve a totally normally consolidated state would range from millions of years to perhaps hundreds of millions of years depending on the permeability, the coefficient of consolidation, and the thickness of the sediments. Because the coefficient of consolidation (the rate at which it consolidates) of the sediments on the fan probably vary over wide ranges, the process of consolidation and the time required to achieve an equilibrium state would vary greatly. Thus, not only would the amount of reduction in volume of the fan vary with time, but the amount of volume reduction would not be equal throughout the various parts of the fan. Unequal volume adjustments would result in a fan with a variable geometry through time.

Geologic reconstructions and fan models should take in consideration the effects of the consolidation process which are reflected in the porosity of the sediments. It is interesting to note that the subsidence resulting from sediment loading of the crust probably proceeds at a faster rate than does the subsidence resulting from the consolidation process. Both events may prove to be equally important in the interpretation of marine fans.

Table 3. Seafloor Geotechnical Properties and Gradients of Mississippi Fan Sediments

	Site	*Bulk Density (seafloor) g/cm^3	Bulk Density Gradient g/cm^3 per m	Dry Water Content (seafloor) %	Dry Water Content (gradient) % per m	Shear Strength (seafloor) kPa	Shear Strength (gradient) kPa per m	Pososity %	Porosity Gradient % per m
Lower Fan	614	1.55	.009	90	−1.75 (0–40 m) +.076 (40–130 m)	6	.96	72	−.75 (0–40 m) +.076 (40–130m)
	615	1.47	.011 (0–40 m) .0002 (40–450 m)	105	−1.3 (0–50 m) −.02 (40–450 m)	5	1.37	75	−.05
	623	1.45	.0036	106	−2.04 (0–25 m) −0.24 (25–150 m)	7.5	1.47	75	−.24
	624	1.42	.008 (0–50 m) .0005 (50–150 m)	116	−1.64 (0–25 m) .275 (25–170 m)	3	—	74	−.40 (0–50 m) .07 (50–150 m)
Middle Fan	616	1.45	.006 (0–60 m) .001 (60–368 m)	100	−.90 (0–60 m) −.051 (60–368 m)	6	1.07 (0–60 m) 1.1 (60–368 m)	72	−0.18 (0–60 m) .0617 (60–368 m)
	617	1.57	.006 (0–35 m) .0001 (35–190 m)	84	−1.0 (0–35 m) −.096 (35–190 m)	7	1.34 (0–35 m) .84 (35–190 m)	70	−0.42 (0–35 m) −.058 (35–190 m)
	620	ALL SAMPLES DISTURBED							
	621	1.54	.002 (0–70 m) .0007 (70–212 m)	60	−1.39	2	1.29 (0–70 m) 1.15 (70–212 m)	89	−.0667
	622	1.62	.0019	72	−.231	12	1.25	67	−.130

High rates of accumulation of low-permeability sediments such as seen on the Mississippi Fan result in a sediment mass with a low effective stress. Such a mass is susceptible to slope failure and mass movements. A slumping and a detachment or decollement similar to that found on the Mexican Ridges may well be the future for the Mississippi Fan.

References

[1] Ericson, D. B., and Wollin, G., 1968. Pleistocene climates and chronology in deep-sea sediments. Science, v. 162, pp. 1227–1234.

[2] Lambe, T. W., 1951. Soil Testing for Engineers. John Wiley & Sons, Inc., New York, 165 pp.

[3] Skempton, A. W., 1970. The consolidation of clays by gravitational compaction. Quarterly Journal of Geology Society of London, v. 125, pp. 373–411.

[4] Lambe, T. W., and Whitman, R. V., 1969. Soil Mechanics. John Wiley & Sons, Inc., New York, 553 pp.

[5] Lowe, J., III, 1974. New concepts in consolidation and settlement analysis. Journal of the Geotechnical Engineering Division, American Society of Civil Engineers, v. 100, no. GT6, Proceedings, Paper 10623, pp. 574–612.

[6] Casagrande, A., 1936. The determination of the preconsolidation load and its practical significance. Proceedings, First International Conference on Soil Mechanics and Foundation Engineering, v. 3, pp. 60–64.

[7] Cooling, L. F., and Skempton, A. W., 1942. A laboratory study of London clay. Journal of the Institution of Civil Engineers, London, v. 17, pp. 251–256.

[8] Bishop, A. W., Webb, D. L., and Lewis, P. I., 1965. Undisturbed samples of London Clay from the Ashford Common Shaft: strength-effective stress relationship. Geotechnique, v. 15, pp. 1–31.

[9] Schmertmann, J. H., 1955. The undisturbed consolidation behavior of clay. Transactions American Society of Civil Engineers, v. 120, pp. 1201–1227.

[10] Geotechnical Consortium, in press. Geotechnical properties of Northwest Pacific pelagic clays, Deep Sea Drilling Project Leg 86, Hole 576A. *In:* Burckle, L., Heath, G., and others (eds.), Initial Reports of the Deep Sea Drilling Project, v. 86. U.S. Government Printing Office, Washington, D. C.

[11] Burmister, D. M., 1951. The application of controlled test methods in consolidation testing. Consolidation Testing of Soils, American Society for Testing and Materials, STP 126, pp. 83–91.

CHAPTER 44

X-ray Radiography of Mississippi Fan Cores

J. M. Coleman, A. H. Bouma, H. H. Roberts, P. A. Thayer, and DSDP Leg 96 Scientific Party

Abstract

Thin sediment slabs were collected from Mississippi Fan cores for analysis by X-ray radiography. Minor sedimentary structures and large-scale features, when considered with lithologic data, help characterize the various depositional environments of the fan. Parallel bedding, ranging in thickness from less than 1 mm to several centimeters, is the most common small-scale structure. Most of these thin beds are normally graded, ranging from either sand to silt or from silt to very fine clay; concentrations of forams at the top of the fine clays are common. Micro-cross stratification, distorted bedding, and micro-fracturing are rather common; reworked plant material, mica, and volcanic shards are abundant. Diagenetic inclusions, while present, are not especially common. Bioturbation is extremely rare.

Introduction

Nine sites were drilled in the Mississippi Fan during the DSDP Leg 96 cruise (Fig. 1). Sediment slabs (8-mm thick, 7.5-cm wide, and 30-cm long) were sampled onboard, sealed in Plexiglas frames, and transported to the laboratory for analysis by X-ray radiography. The minor sedimentary structures and inclusions as well as the large-scale structures, when considered with detailed lithologic descriptions, were found to be indicative of the various depositional environments in the Mississippi Fan. The DSDP drill sites are divided into six groups: middle-fan channel (Sites 621 and 622), middle-fan overbank (Sites 617 and 620), slump-marginal fan slope (Site 616), lower-fan channel (Site 623), lower-fan overbank (Site 624), and channel-mouth depositional lobes (Sites 615 and 614).

Middle Fan

The middle-fan channel borings (Sites 621 and 622) show a fining-upward channel-fill sequence commencing with a basal lag gravel composed of quartzitic sandstones, faceted chert, and polycrystalline quartz up to 3 cm in diameter. The basal lag grades upward into quartz-rich sands containing chert granules, mica, and woody organic fragments. The entire unit is capped by fine-grained silts and clays containing minor amounts of methane gas. Fauna is generally rare, but reworked shallow water benthic fauna was found in the channel fill. Most of the thicker sands contain little evidence of any sedimentary structures, possibly as a result of liquefaction during coring operations. Planar thin and thick laminae and small-scale cross-laminations are the major structures observed in the radiographs of undisturbed sections. It is remarkable that scour at the base of the thin sands is not apparent. Some evidence of small-scale distorted layers (a, Fig. 2A) was seen in a number of samples and is assumed to be real rather than the result of coring disturbance. Multidirectional tilted beds, possibly indicating larger-scale cross-bedding, were sometimes found. The most common structures in the finer-grained units between the thicker sands were alternations of thinly laminated sands (a, Fig. 2B) grading upward into clays (b, Fig. 2B). Graded units of variable dimensions are the most frequently observed structure in the channel fill; the graded unit generally displays a sharp lower base (a, Fig. 2C). The base of the sand is relatively sharp, often displaying a scour relationship with the underlying units. The sand displays small-scale micro-cross lamination (b, Fig. 2C) containing abundant scattered wood fragments often referred to as "coffee-grounds." In the upper finer-grained "passive" channel fill, thin-graded laminations (a, Fig. 2D) containing iron and carbonate-cemented inclusions (b, Fig. 2C) are the most abundant structures. These diagenetically formed inclusions occur in repetitive thin zones within the finer-grained clays. Microfractures and small faults (c, Fig. 2D) are common.

The overbank sites (Sites 617 and 620) contain primarily fine-grained sediments with only minor amounts of sand.

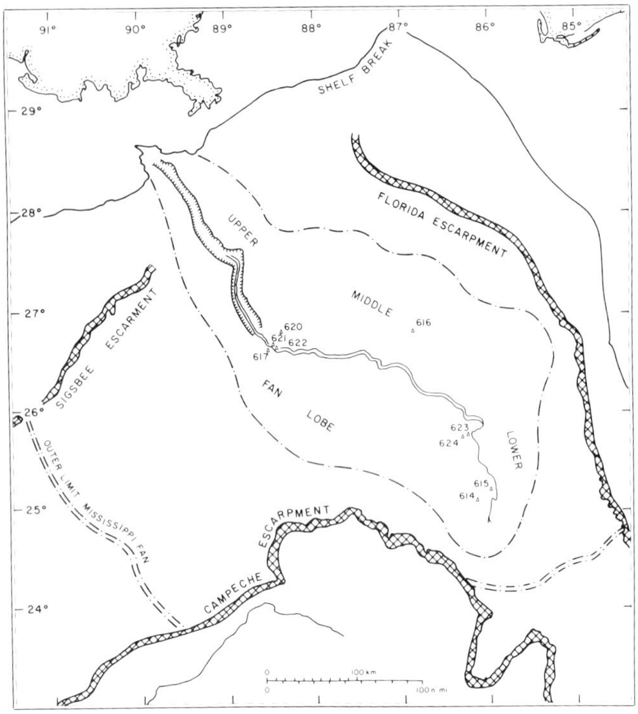

Figure 1. Location map showing the DSDP Leg 96 drill sites.

Bedding is generally thin, microfauna is rare but more abundant than in the channel, and the overall sequences can be described as thin-bedded turbidites. One of the most common sedimentary structures is variable-thickness graded bedding. The X-ray radiographs illustrated in Figures 2E and F are typical of the finer-grained overbank deposits. These parallel laminations (a, Figs. 2E and F) range in thickness from a few millimeters to more than 10 cm. They are generally characterized by a sharp basal contact of fine sand or silt, grading to a thinly laminated clay, and capped by a homogeneous clay that often contains scattered microfauna and microburrows. In some instances, the basal sand displays extremely thin, parallel laminations that in themselves are graded. A few small diagenetically formed inclusions (b, Fig. 2E), primarily iron and carbonate-cemented silts, are present in some of the near-surface cores. The sample slabs containing only fine silts and clays (Fig. 2F) display extremely thin, graded laminations, generally grading from fine silts or coarse clays to extremely fine clays.

One of the characteristic features in all of the overbank cores is the presence of distorted structures ranging in scale from several meters (tilted and microfractured bedding) to small-scale convoluted laminations (Fig. 3A). These types of bedding disruptions are undoubtedly local and are probably associated with the extremely high sedimentation rates of the overbank sediments. In the coarser-grained units (very fine sand and coarse silt), stranded ripples, small load features, and occasional clay clasts are present. Although bioturbation is evident, it is not common and is generally scattered throughout the cores.

Slump-Marginal Fan Area

Site 616 was drilled on the margins of the youngest fanlobe in a region described as a massive "slump" [1]. The upper 96 m of the boring is composed primarily of clay and thin silt stringers that display sections with high dip angles up to 65° (Fig. 3B). Although the bedding planes have variable steep dip angles, microfractures are generally not abundant. Within the "slump mass," dipping units are often separated by parallel bedded sequences, possibly implying that the material failed as a series of smaller slumps rather than one massive feature. The radiograph in Figure 3C illustrates the base of one of the dipping units and shows that high distortion and fracturing (a, Fig. 3C) are present. Although the origin of the slump has not been ascertained, Coleman and others [2] described a massive shelf-edge failure off the Mississippi Delta from the late Pleistocene that can be traced down the slope to within a few tens of kilometers of the drill site.

Lower Fan

In the lower fan, the main channel narrows considerably and displays much less sinuosity. The levees bordering the channel are subdued, and it is apparent that flow was not as confined in this region as in the middle fan. Site 623 was drilled on the banks of the active channel in a region where the seismic reflection and side-scan sonar data indicated the presence of many abandoned channels, tending to record a shifting sequence of channels through relatively short periods of time. The core shows alternating units of channel fill (fining upward sand and silt units) and overbank sediments (alternating fine sands, silts, and clays). The sands are always sharp based, commonly show minor scouring into underlying clays, have small-scale cross-laminations (a, Fig. 3D), and contain both small clay chips and larger clay clasts. The larger clay clasts (a, Fig. 3E) are generally angular and show little or no distortion of laminations. The cross-laminated sands (Fig. 3D) contain abundant mica (high absorption) and transported organics (low absorption) along the bedding planes. The silts tend to be graded and arranged in numerous thin cycles (a, Fig. 4A). The coarser-graded units contain sand at the base, grading upward into fine-grained clays (b, Fig. 4A), and commonly contain abundant scattered wood fragments (c, Fig. 4A; low absorption) and mica. These graded units vary in thickness from a few centimeters up to 20 cm. Micro-cross laminations and thin parallel laminations are common in the silts.

At Site 624, drilled several kilometers away from the

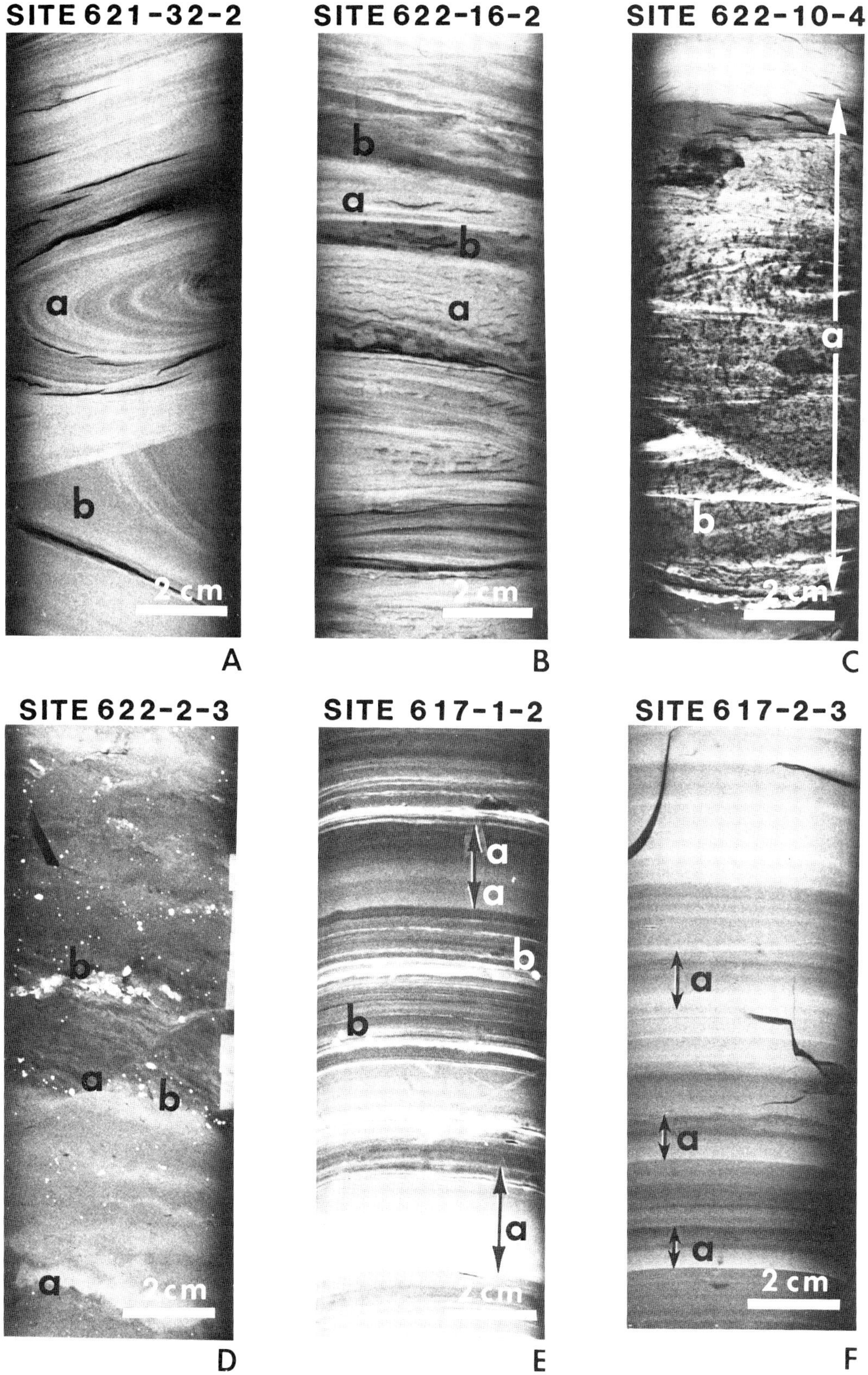

Figure 2. Radiographs of Mississippi Fan cores. (A) Site 621-32-2 (201 m). (B) Site 622-16-2 (136 m). (C) Site 622-10-4 (95 m). (D) Site 622-2-3 (8 m). (E) Site 617-1-2 (2 m). (F) Site 617-2-3 (3.5 m). Numbers between parentheses indicate approximate subbottom depth in meters.

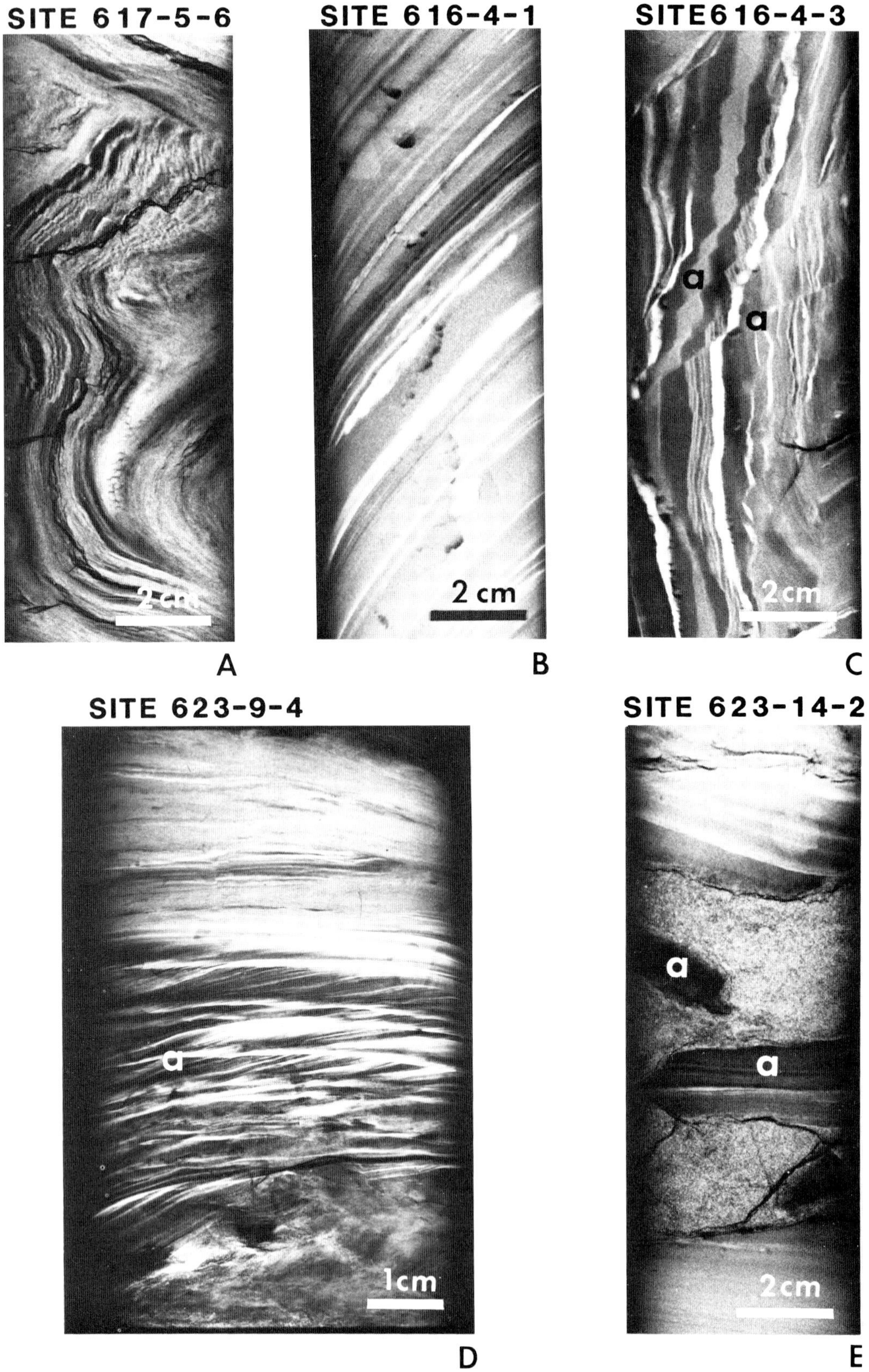

Figure 3. Radiographs of Mississippi Fan cores. (A) Site 617-5-6 (39 m). (B) Site 616-4-1 (26 m). (C) Site 616-4-3 (29 m). (D) Site 623-9-4 (78 m). (E) Site 623-14-2 (123 m).

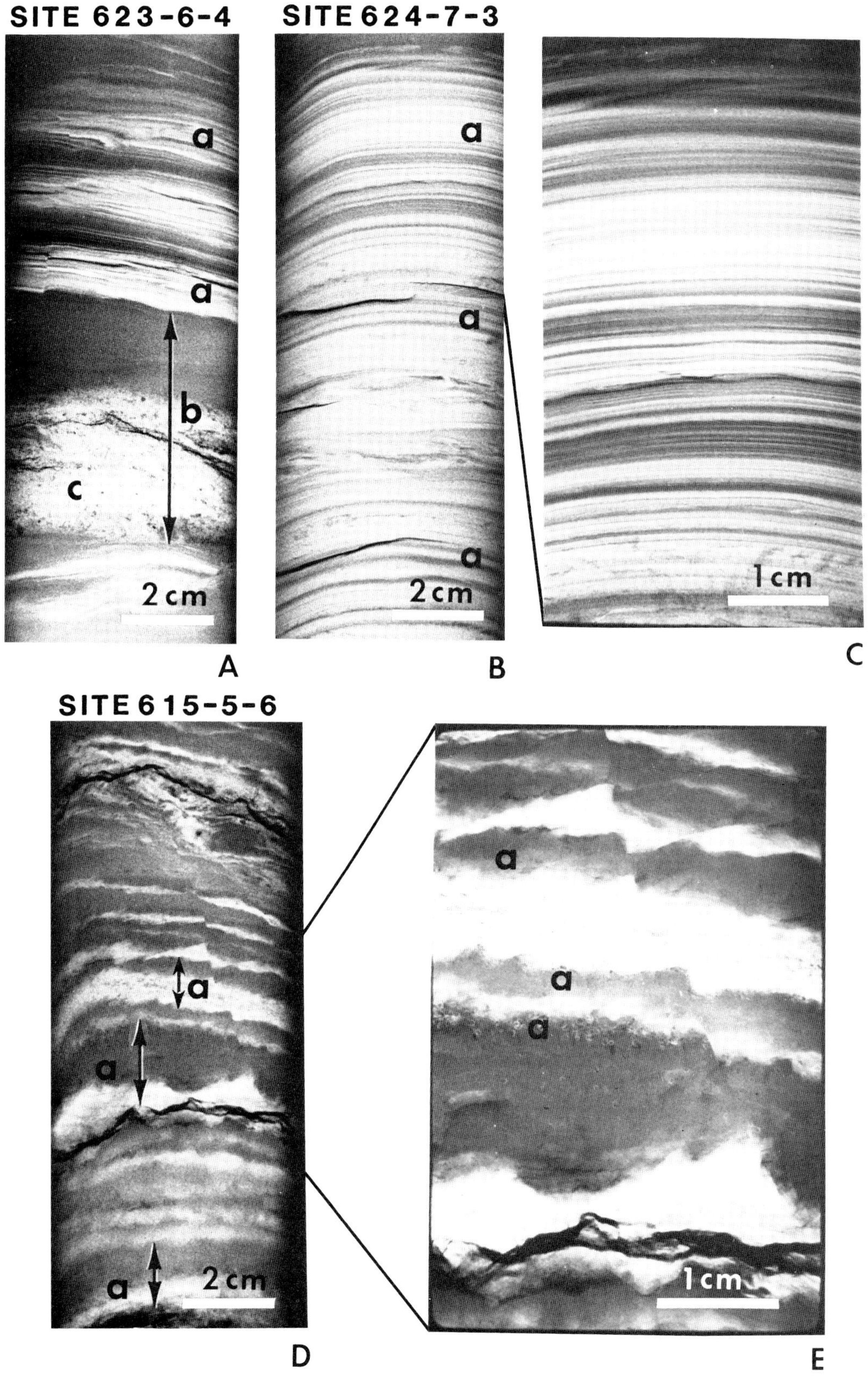

Figure 4. Radiographs of Mississippi Fan cores. (A) Site 623-6-4 (52 m). (B) Site 624-7-3 (58 m). (C) Detail of B. (D) Site 615-5-6 (38 m). (E) Detail of D.

channel complex, the overall sediment size is considerably finer, laminations are much thinner, and no major sand units were encountered. The fine silts and clays are very thinly laminated (a, Fig. 4B), with thicknesses ranging from only a fraction of a millimeter to a few centimeters. The thin, normally graded parallel laminations are characterized by a lack of any basal scouring. Figure 4C is an enlargement of the radiograph shown in Figure 4B, illustrating the delicate nature of the laminations. A high percentage of the core displays this type of bedding. Each unit generally consists of a thinly parallel laminated or micro-cross laminated, sharp-based silt grading upward into low-absorption clays that often shows microburrowing at the top. Small-scale contorted bedding, only a few centimeters in thickness, is also common. Transported organic debris is abundant in the silts and fine sands.

Two sites (Sites 615 and 614) were drilled near the terminus of the channel in the lower fan. The channel is barely discernible on the side-scan sonar images, and this area represents the point at which the sediments are no longer confined to the channel but are free to spread laterally. The sediments in the core contain a considerable amount of sand; of the two fanlobes of Upper Wisconsin age cored at Site 615, the older lobe contains 65% net sand, while the younger lobe contains 41% net sand. The coarser units are dominantly composed of medium- to fine-grained massive sands containing thin silt and clay beds. The sands are generally sublitharenites composed of quartz, with subordinate feldspars, rock fragments, micas, and heavy minerals. There are minor amounts of reworked foram tests, glauconite, shell debris, and transported organic debris. Contacts are sharp and often display microscouring at the base. Except for clay clasts (1 to 5 cm in size) and well-defined horizons of transported organic debris, these coarse units have little internal structure. The interbedded fine sand, silt, and clay units are generally graded and occur in repetitive cycles a few millimeters to tens of centimeters in thickness.

One of the more common attributes of the graded units in these cores, as well as of many of those in the overbank settings, is the presence of a thin concentration of microfauna, mainly forams, at the top of the fine-grained graded units. Figure 4D illustrates a core from Site 615, showing the repetitive graded units (a, Fig. 4D); some microfracturing was present in the core. In Figure 4E (an enlargement of a part of Fig. 4D), the concentration of forams (a, Fig. 4E) can be seen at the top of the low-absorption clay. The presence of this foram concentration, containing primarily planktonic species, indicates a break in the rapid deposition of the thin graded units that are barren of microfauna. The cores containing only fine-grained silts and clays are very thinly bedded, as illustrated in Figure 5A. Figures 5B and C show the details of this thin lamination, the graded units in radiograph 5B being composed of finer-grained sediments than in 5C. In some instances, grading on a scale of 1 to 2 mm could be detected on the radiographs, and this grading was confirmed by scanning electron microscopic examination. Although most of the thin graded laminations are parallel (a, Fig. 5B), there is some evidence of minor scour associated with the deposition (a, Fig. 5C). In cores containing alternations of sands, silts, and clays, the graded sands are very characteristic. Figures 5D (a) and E (a) illustrate one such sand unit; the unit commences with micro-cross laminated sands (b, Fig. 5E) grading upward into a thin massive or parallel bedded sand (c, Fig. 5E) containing scattered organic fragments. Overlying the sand unit is a concentration of organic debris (d, Fig. 5E) containing sand and mica particles. Often this feature grades upward into a micro-cross laminated silt, which then grades into a homogeneous clay, and is normally capped by a thin (generally less than 1 or 2-mm thick) foram zone. Figure 5F illustrates a similar sequence, but the basal micro-cross lamination is not present. The high-absorption inclusions at the top of the graded unit (a, Fig. 5F) are concentrations of mica flakes.

Conclusions

The study of microstructures in the cores from the Mississippi Fan provides considerable information on some of the processes that were active during deposition of the fan lobes. The most characteristic feature is the abundance of graded units of variable thickness and varying grain size. Some of these units consist of sand grading to silt, while others consist of silt grading to very fine-grained clay. Based on average sedimentation rates computed from faunal boundaries (Kohl and others, Chapter 39, this volume), it is highly probable that more than one such fine-grained graded unit, and possibly as many as five or six, were deposited each year. Deposition of the fine clays at least suggests a period of hours or days for accumulation before introduction of the sediment forming the overlying graded unit.

The presence of a foram concentration capping only selected units (ranging from every third to every tenth) tends to support the concept that the processes responsible for delivering fine-grained shelf-derived sediment to the fan operates on an average of several times per year, but geologically continuously over relatively long periods of time (the late Wisconsin glacial stages). The coarser-grained graded beds, which lack burrows and forams, might be more the result of pulsations in a turbidity current. A set of such beds could be deposited in a very short time. Deposition rates were extremely high, as little evidence of burrowing organisms was observed in the cores. The lack of extensive scouring at the base of the sands supports the

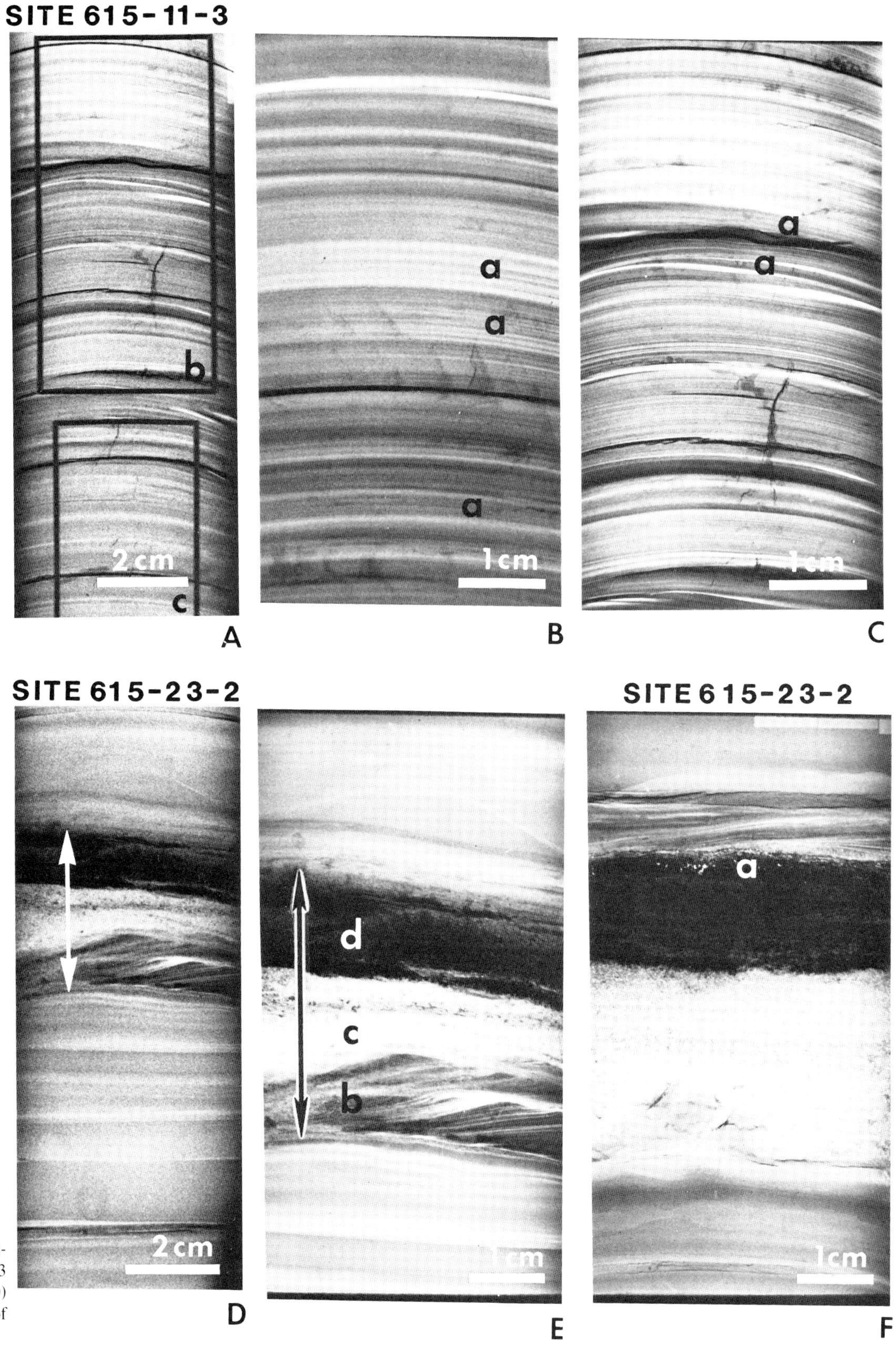

Figure 5. Radiographs of Mississippi Fan cores. (A) Site 615-11-3 (91 m). (B and C) Details of A. (D) Site 615-23-2 (202 m). (E) Detail of D. (F) Site 615-23-2 (202 m).

concept of rapid suspension settling rather than highly erosive density flows. Traction currents, however, were operative, as many of the thin basal sand and silt units display micro-cross stratification.

In the middle fan, the coarser debris carried by the density flows was confined to the channel, and only the finer-grained sediments escaped the channel to be deposited as overbank sediments. A wide range of grain sizes, including high volumes of reworked organic debris and mica, were transported long distances from their source and incorporated into the coarser sediments. Many of the minor structure associations cannot be explained at the present time, and considerably more work on the cores is required to determine the details of the depositional processes.

References

[1] Walker, J. R., and Massingill, J. V., 1970. Slump features on the Mississippi fan, northeastern Gulf of Mexico. Geological Society of American Bulletin, v. 81, pp. 3101–3108.

[2] Coleman, J. M., Prior, D. B., and Lindsay, J. F., 1983. Deltaic influences on shelf edge instability processes. In: D. J. Stanley and G. T. Moore (eds), The Shelf Break, Critical Interface on Continental Margins. Society of Economic Paleontologists and Mineralogists Special Publication 33, pp. 161–199.

CHAPTER 45

Thin-Section Studies, Mississippi Fan

Michel Cremer, Laurence Droz, William R. Normark, Suzanne O'Connell, Kevin T. Pickering, Charles E. Stelting, Dorrik A. V. Stow, and DSDP Leg 96 Shipboard Scientists

Abstract

Thin sections from fine-grained sediments collected during DSDP Leg 96 on the Mississippi Fan were analyzed to better understand the characteristics of sedimentary structures observed during the shipboard core descriptions and to look for finer-scale structures not visually resolvable. The silt, laminated silt and mud, and mud facies show sedimentary structures (oblique stratification, parallel lamination, lenses, and normal and inverse grading) that are suggestive of single fine-grained turbidites, multiple thin turbidites (cyclic fluctuations), and deposits from nepheloid layers.

Introduction

The Mississippi Fan, cored during Deep Sea Drilling Project Leg 96, consists mainly of terrigenous sediments derived from the Mississippi River system that were transported to the basin by density currents during the late Wisconsin low sea-level stage. In addition to samples selected for grain size and detailed mineralogical studies, a few vertical samples were analyzed in thin sections to determine the microsedimentary structures and grain fabric and to infer the depositional processes from these observations.

Sample Preparation

The samples collected consisted of quarter-rounds of cores, 3-cm wide and 6 to 20-cm long. They were taken mainly from unconsolidated muds, muddy silts, and silts that were sufficiently firm to be removed from the core liner while preserving their sedimentary structures. The samples were cut to obtain a slice about 1-cm thick, the remaining material being further sampled for grain size analyses. The slices were freeze-dried and then impregnated with an epoxy resin. The impregnated slices were thin-sectioned as carried out for hard rocks.

Sedimentary Facies

A selected number of thin sections will be discussed from three of the common facies: coarse silts, laminated silts and muds, and muds (Stow and others, Chapter 38). These sections are generally representative of all examples studied from these facies.

Coarse Silt Facies

Sample 622-21-1, 62 to 74 cm (subbottom depth 172.62 to 172.74m) was obtained from a sandy-silty core and was selected from the middle of a sequence described as fining-upward from sand to silt. This 60-cm thick sequence comprises numerous black laminae throughout and irregular dark brown-colored mud beds in its upper part. The bottom of the thin section (Fig. 1a) shows coarse well-sorted silt. Elongated grains emphasize a crude plane stratification. Grain size and mineralogic variations permit the definition of several laminae rich in lignite particles. No general vertical trend could be observed. The middle of the section (Fig. 1b) consists of a mud with fine silt laminae. The boundaries of this muddy zone with the underlying and overlying silt are generally sharp, and this structural characteristic can be interpreted as representing the destruction of an initial single bed. Toward the top of the thin section (Fig. 1c), there is an increase in the number of mud intervals that occur as isolated bleds, discontinuous layers, or lenses interbedded with silt. The silt in the upper part of the section is very similar to that observed in the lower part, except for a slight decrease of grain size (Md: 40 → 35 μm).

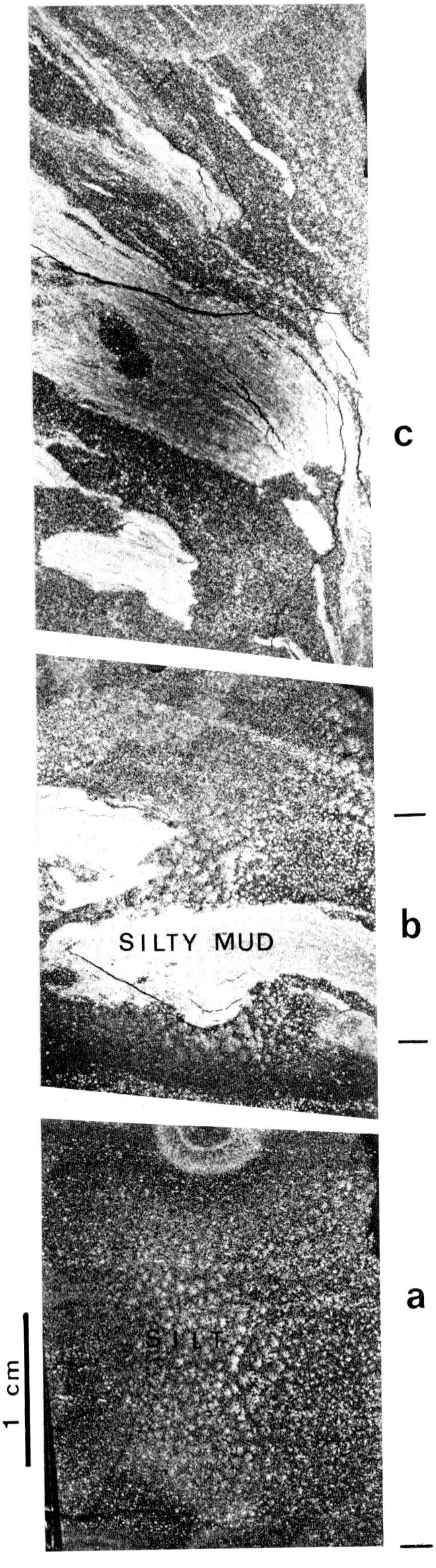

Figure 1. Coarse silts and laminated mud layers. Sample 622-21-1, 62 to 74 cm. The illustrations are negatives of thin sections; thus transparent grains (quartz) are dark areas, opaque grains (lignite) are white, and silty layers are darker than muddy layers.

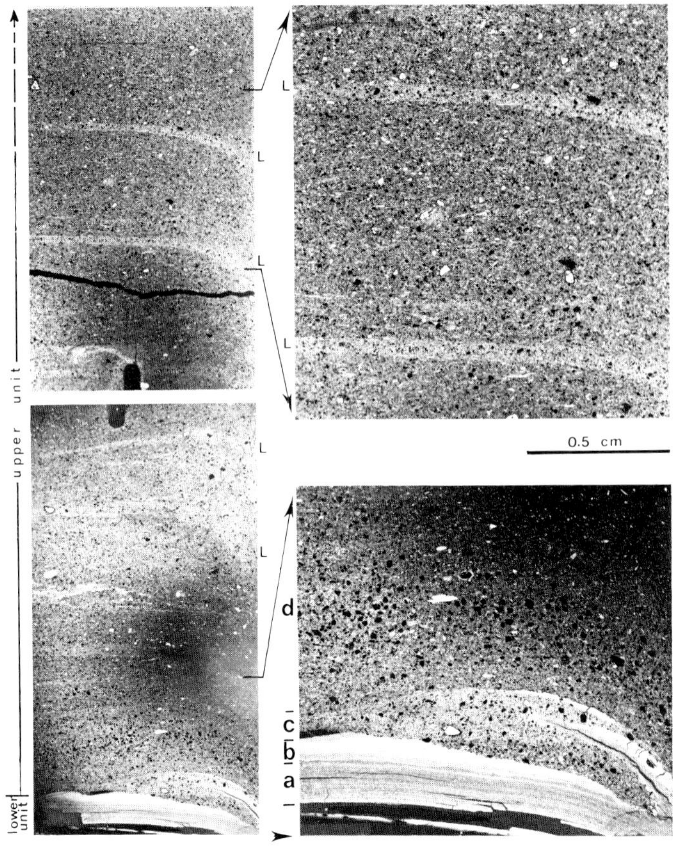

Figure 2. Uppermost and lower parts of two coarse silt turbidites. Sample 615-33-2, 29 to 42 cm. From base to top: grading, bed forms, quartz concentration, and spaced laminae (L). The illustrations are negatives of thin sections.

Comments: The observed sedimentary structures may suggest partial sediment flow during coring or sampling, or the succession of several turbidites. Nevertheless, the observed fining-upward sequence and the minor variation between silt layers imply a single turbidite rather than a succession of turbidites. Because the silty muds show a bedding more or less parallel to the silt beds, it is difficult to argue that they are rip-up clasts that were deposited simultaneously with the silt.

Sample 615-33-2, 29 to 42 cm (subbottom depth 306.49 to 306.62 m) was taken from a 80-cm long core section that consists of two units of coarse silt grading upward into silty mud. Each unit contains several beds separated by darker laminae; the thickness of beds (average 1.5 cm) decreases upward. The thin section comprises the upper part of the lower unit and the bottom part of the upper unit (Fig. 2).

The top of the lower unit contains a graded sequence from a fine silt with mud laminae (Fig. 2a) to a graded silty mud (Fig. 2b). Although the base of the upper unit is sharp, it molds rather than erodes the wavey top of the lower unit. The upper unit consists of a matrix of muddy silt ($\simeq$30% grains < 2 μm) with coarser (up to 400 μm) grains of quartz and lignite. Near its base, a discontinuous mud lamina con-

taining coarse quartz (Fig. 2c) shows oblique stratification. It is overlain by a 1-cm thick bed (Fig. 2d) that has a concentration of coarse grains in its center. Darker-colored laminae appear and become more distinct and regularly spaced in the upper part of the thin section. These laminae of equal thickness separate beds of identical character. Moreover, studies of grain size do not show a significant difference between beds and laminae. The characteristic that highlights the thin laminae seems to be related to grain fabric; a denser grain arrangement gives rise to a darker laminae (lighter in the illustration). One can also observe a larger proportion of coarser grains in the center of these laminae, similar to that observed near the base of the upper unit.

Comments: The general homogeneity of the sediment in the upper unit supports the interpretation of a single turbidite (the bottom part of the thin section being the top of another turbidite). The presence of coarse grains in a finer matrix suggests rapid deposition. The darker laminae may be related to a change in bed form similar to the ripple shape in the lower part of this unit. It is more likely, however, that the structure results from changes in flow behavior with phases of rapid deposition and phases where the partially discharged turbidity current forms a traction carpet [1].

Laminated Silt and Mud Facies

Sample 614-3-1, 120 to 126 cm (subbottom depth 19.26 to 19.60 m) came from a section of silty mud containing thin silt beds and numerous colored laminae. It contains a 2-cm thick silt layer.

The base of the thin section is a fine-grained mud (Fig. 3a) that shows planar lamination in polarized light. Overlaying this is a mud bed (Fig. 3b) with discontinuous silt layers at its base and scattered coarse grains. It grades upward, through interbedded silts and muds (Fig. 3c) and into laminated silts (Fig. 3d). The interbedded silts and muds exhibit an oblique, wavey stratification; the mud laminae that contain lignite grains pass laterally into silt laminae. The overlaying silts are finely laminated and well-sorted. This silt section (Fig. 3d) contains two layers (layers 1 and 2) in which the proportion of mud and lignite grains increases; the lower layer (layer 1) exhibits oblique stratification. The general fining-upward continues in the upper part of the section; silty mud with silt laminae (Fig. 3e) changes into a graded mud (Fig. 3f) and then into a foraminifera-bearing mud (Fig. 3g). Although this mud is very fine-grained (90% clay), it is still laminated as can be observed in cross-polarized light. A microfault was observed in the silt bed that does not reach the upper muddy beds.

Comments: While structures described as fine-grained turbidites are present in this section [2,3], it remains difficult to define its limits. The muddy interval underneath the silt bed could be interpreted as resulting from a different event.

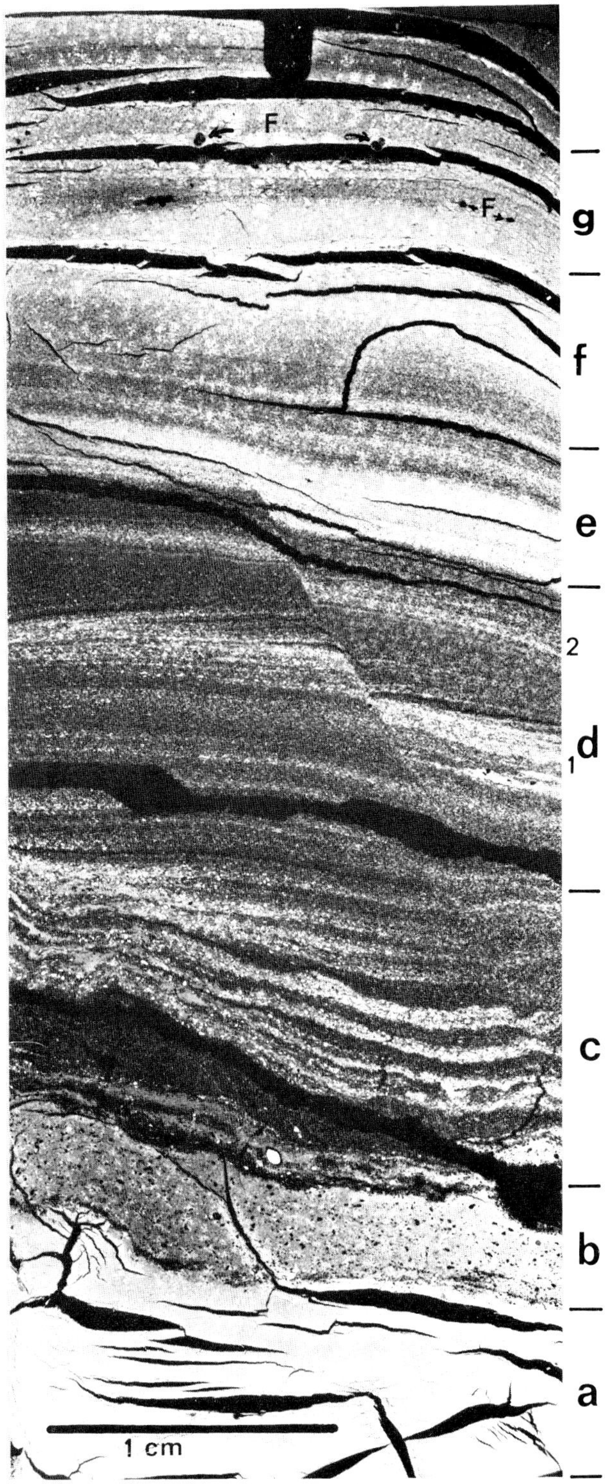

Figure 3. Fine-grained turbidite. Sample 614-3-1, 120 to 126 cm. Note the muddy bed with scattered coarse quartz grains at the bottom and foraminifers (F) at the top. The illustrations are negatives of thin sections.

This mud either represents a rapid deposition of suspended sediments (crude lamination, poor sorting) prior to a traction phase (well-defined laminae, good sorting) or deposition from the head and then the body of a turbidity cur-

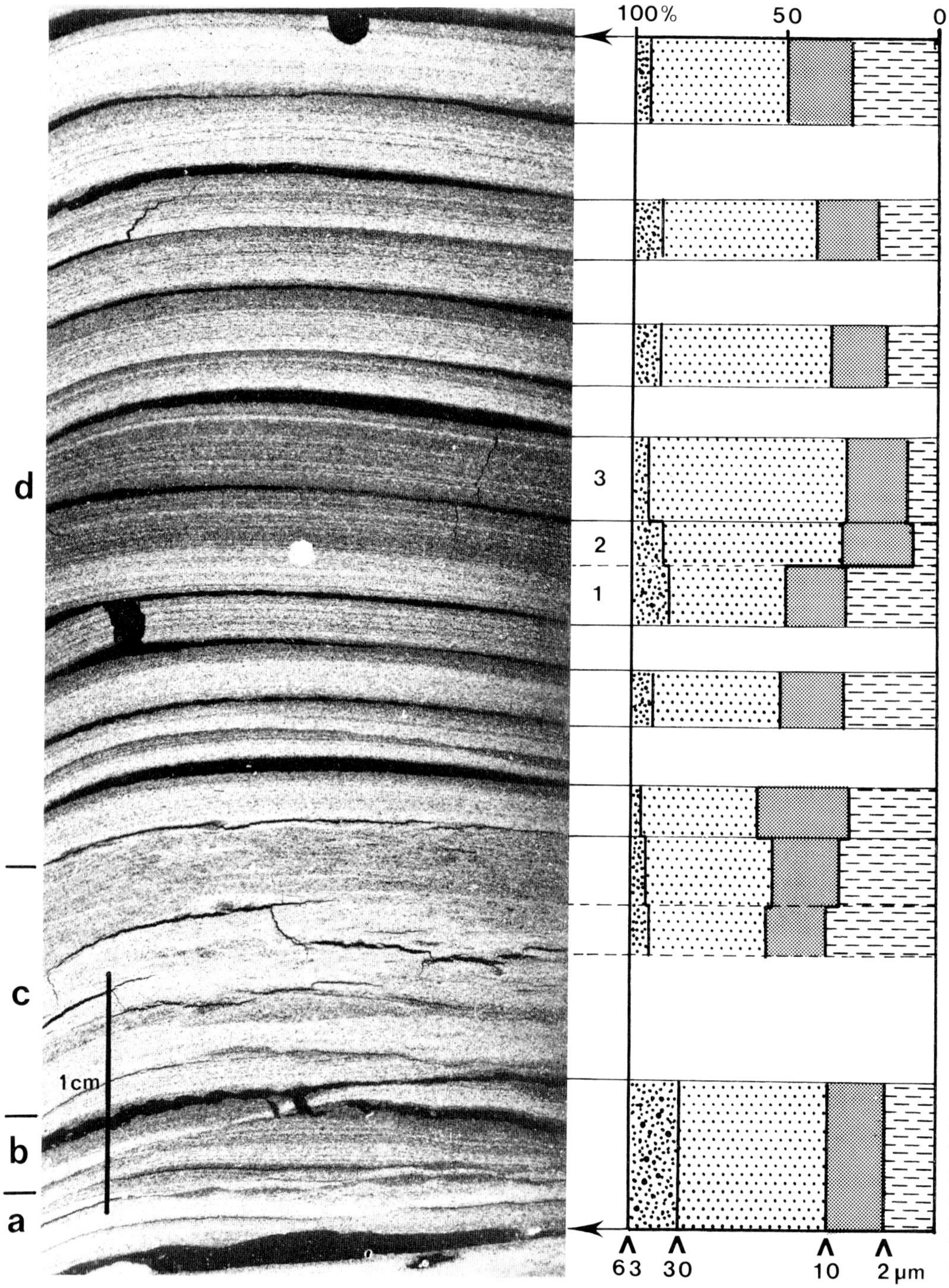

Figure 4. T_c and T_d divisions of a turbidite with grain size analyses of beds. Sample 614-11-4, 98 to 104 cm. Note the inverse grading of individual beds in the upper part. The illustrations are negatives of thin sections.

rent. At the top of the thin section, the presence of pelagic foraminifers in a very fine mud may characterize a hemipelagic interval that overlies the turbidite. However, this mud still contains plane lamination and no bioturbation. Thus, either the deceleration of the turbidity current was sufficiently long to allow some pelagic forms to settle out or this mud represents deposits from some type of nepheloid layer emplaced rapidly enough to limit bioturbation.

Sample 615-11-4, 98 to 104 cm (subbottom depth 91.68 to 91.74 m) was taken from a section of interbedded silts and silty muds that consists of a unit of about 20 individual beds that can be interpreted as a single depositional event.

The overall characteristics of the sediment do not vary from the base to the top of the thin section (Fig. 4). However, the sedimentary structures are emphasized by an alternating grain size between clean silt laminae (Md: 15 μm, % < 2 μm : 10) and muddier laminae (Md: 10 μm, % < 2 μm : 30). In the lower part of the thin section, we observed wavy lamination in place beds (Fig. 4a) overlain by cross-lamination defining ripples (Fig. 4b) and a bed with less developed parallel lamination (Fig. 3c). The upper half of the thin section (Fig. 4d) shows only fine and planar laminae displaying clearly defined beds. The majority of these beds are inversely graded, starting with a muddy silt with scattered coarse grains at its base and topped by a clean silt. The grain size distribution within a bed thick enough to be subsampled (Fig. 4, 1-2-3) confirms this observation; the proportion of finer and coarser grains decreases upward and the sorting increases.

Comments: Based on the general homogeneity of the sediment and the observed sedimentary structures, we interpret this section as a $T_{c\text{-}d}$ turbidite sequence despite the fact that inversely graded beds are not a recognized characteristic of the T_d division. Their formation appears to be linked to depositional sorting in the boundary layer at the base of the turbidity current [4,5].

Mud Facies

Sample 616-8-2, 117 to 123 cm (subbottom depth 66.07 to 66.13 m) was collected from a fine mud that shows only thin laminae and dispersed silt pockets. In addition, the stratification dips as a result of mass movement in the upper 100 m. The sediment is a fine mud; 70% of the grains are clay-sized and less than 5% are coarser than 10 μm. In the thin section (Fig. 5), one can observe several thin layers of mud, either homogenous, graded, or bearing discontinuous silt laminae. No distinct vertical trend is visible. One fine silty bed (Md: 10 μm, 2-mm thick) very well-sorted and having a sharp base and an undulated upper boundary with the overlying graded mud, was observed. In spite of their fineness, the muds are laminated and reorientated grains along oblique or vertical planes are also noticed.

Comments: The characteristics of this sample are more related to several very fine-grained turbidites than to the upper part of a single turbidite. The absence of biogenic components perhaps indicates a high frequency of flows. The orientation of grains along oblique and vertical planes appears to have been induced by fracturing related to mass movement.

Sample 616-32-6, 124 to 130 cm (subbottom depth 351.84 to 351.90 m) was taken from a visually homogenous muddy core that did not reveal any sedimentary structure except for some thin dark-colored laminae which are rather regularly spaced. Coring disturbance was an initial explanation for this structure (shear planes). The first attempt to impregnate these muds failed, probably because of their low porosity. After humidification, impregnation was successful and a thin section was made. The sediment is almost homogenous from base to top. It is a fine terrigenous mud with scattered coarser grains (up to 50 μm); stratification is roughly parallel. Darker laminae noted in visual descriptions of the sample cannot be clearly observed in the thin section, although slight variations in grain size and fabric that define indistinct bedding are emphasized in cross-polarized light.

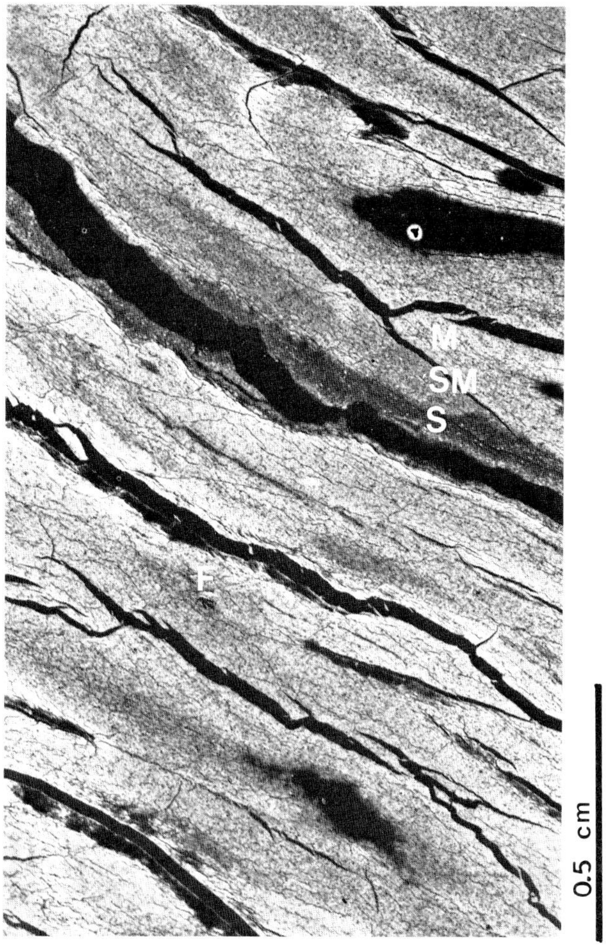

Figure 5. Detail of thin silty mud turbidites. Sample 618-8-2, 117 to 123 cm. F = reorientated grains along fractures; M = mud; S = thin silt bed; SM = silty mud.

Comments: One cannot define the origin of these "homogenous muds" on the basis of only one sample. Some remarks may, nevertheless, be made. The sediment is not completely homogenous. The bedding, with an average grain size larger than that in the upper part of the earlier described fine-grained turbidite (sample 614-3-1, 120 to 126 cm), and the absence of a biogenic component probably characterize a deposit originating from a density current. Moreover, no shear planes have been observed. The cracks made during the preparation of the thin section have various directions. One can establish a similarity between the structures observed in these muds and those described from sample 615-33-2, 29 to 42 cm (i.e., a fine matrix with scattered coarse grains and spaced fine laminae).

Consequently, the structures observed may represent dynamic processes that take place during deposition of a thick muddy turbidite, rather than a coring disturbance or a succession of thin muddy turbidites.

Conclusions

These observations of a limited number of fine sediment thin-sections already reveal a great variety of small-scale sedimentary structures and grain fabric. Some features are characteristic of well-defined sequences, such as those published by Piper, Stow, and Bouma [2,3,6]. Others are not yet understood. Additional samples should be studied to answer three major questions: 1) Where are the upper and lower limits of a single turbidite event? 2) Can we differentiate very fine-grained turbidites from the deposits of nepheloid layers? 3) Is it possible to recognize sedimentary structures which characterize a distinct depositional setting in a submarine fan?

References

[1] Lowe, D. R., 1982. Sediment gravity flows: II. Depositional models with special reference to the deposits of high-density turbidity currents. Journal of Sedimentary Petrology, v. 52, pp. 280–297.
[2] Piper, D. J. W., 1978. Turbidite muds and silts on deep-sea fans and abyssal plains. In: D.J. Stanley and G. Kelling (eds.), Sedimentation in Submarine Canyons, Fans and Trenches. Dowden, Hutchinson and Ross, Stroudsburg, PA, pp. 163–176.
[3] Stow D. A. V., and Shanmugan, G., 1980. Sequence of structure in fine-grained turbidites: comparison of recent deep sea and ancient flysh sediment. Sedimentary Geology, v. 25, pp. 23–42.
[4] Hesse, R., and Chough, S. K., 1980. The northwest Atlantic Mid-Ocean Channel of the Labrador Sea: II. Deposition of parallel laminated levee-muds from the viscous sublayer of low density turbidity currents. Sedimentology, v. 27, pp. 697–711.
[5] Stow, D. A. V., and Bowen, A. Y., 1978. Origin of lamination in deep-sea, fine-grained sediments. Nature, v. 274, pp. 324–328.
[6] Bouma, A. H., 1962. Sedimentology of Some Flysh Deposits: A Graphic Approach to Facies Interpretation. Elsevier, Amsterdam, 168 pp.

CHAPTER 46

Geochemistry of Mississippi Fan Sediments

Mahlon C. Kennicutt II, Ronald C. Pflaum, Debra DeFreitas, James M. Brooks, and DSDP Leg 96 Shipboard Scientists

Abstract

Organic matter in late Pleistocene Mississippi Fan fine-grained sediments averages 0.9% and is predominantly terrestrial in origin as defined by carbon isotopic composition, the presence of plant biowaxes, nitrogen content, and lithologic associations. Organic matter from all fan sites is similar in isotopic and chemical composition as well as in concentration. The sediment is generally nongaseous but when gas is present, it is predominantly microbial methane. Gas chromatographic and fluorescence analyses suggest the presence of petroleum at all sites; possibly migrated. The depth of the onset of sulfate reduction varies, being shallowest in the middle-fan channel sites.

Introduction

The distribution of organic matter derived from terrestrial and marine sources in deltaic sequences can be used to infer paleo-oceanographic conditions. Organic matter composition also has initial control over whether oil or gas is generated from a given sedimentary unit [1]. Because of the different economic considerations for oil versus gas production, it is important to understand the factors that control the distribution of organic matter in marine sediments. Attempts to differentiate marine and terrestrial organic matter are primarily based on chemical (n-alkanes, lignin, elemental composition) and isotopic ($\delta^{13}C$) differences [2–8].

Differentiation of organic matter sources on the basis of carbon isotopic composition is complicated by: 1) the overlap of the isotopic composition of marine and terrestrial plants; 2) the effects of water temperature on the fractionation of carbon in plankton; 3) the alteration in the isotopic composition of organic matter during degradation and diagenesis; 4) migration of petrogenic hydrocarbons into shallower sediments; 5) anomalous isotopic compositions produced in localized (closed) environments; and 6) the incorporation of recycled organic matter into recent sediments. Despite these complications, carbon isotopic compositions are useful in understanding the distribution of organic matter in recent sediments.

The presence of long-chain alkanes ($>$ n-C_{23}) with a strong odd-over-even carbon preference is a useful indicator of a terrestrial source [9–11]. Surface cuticle waxes, which prevent evaporation in higher plants, are esters of long-chain acids and alcohols and produce long-chain normal alkanes when degraded. These higher plants appear to be the only source of normal alkanes with chain lengths greater than n-C_{25} in recent sediments and crude oils. Oils containing high percentages of long-chain paraffins are waxy in nature, have high pour points, and are generally associated with nearshore sedimentation, especially in deltaic settings [12]. Carbon-to-nitrogen ratios are also useful in differentiating sources because marine organisms are enriched in nitrogen relative to terrestrial plants [13–15].

This report summarizes geochemical data from the Mississippi Fan (DSDP Leg 96). Mainly sediments of late Pleistocene age were recovered. Organic and inorganic geochemical parameters that were measured include % organic carbon, $\delta^{13}C$ of the organic matter, total nitrogen content, extractable organic matter, total carbon dioxide, sulfate concentration, gaseous hydrocarbon content, pH, total alkalinity, concentration of the unresolved complex mixture, pristane/phytane ratio, and carbon preference index. These data are used to document early diagenetic transformations and determine the relative distribution of marine and terrigenous organic matter in late Pleistocene sediments of the Mississippi Fan.

Table 1. The Ranges for Geochemical Parameters Measured in Late Pleistocene Mississippi Fan Sediments

Parameter[4]	Lower Fan[1]	Middle Fan[2]	Fan Margin[3]
$\delta^{13}C_{O.M.}$	−25.5 to −27.7	−25.0 to −27.9	−26.3 to −27.1
C/N	11.1 to 170.0	8.5 to 90.0	14.3 to 90.0
% Org. C	0.1 to 1.7	0.5 to 2.1	1.1 to 1.2
EOM (ppm)	3.8 to 222.2	31.5 to 214.1	24.7 to 90.6
ΣCO_2 (mg C/L)	28.1 to 99.3	29.9 to 255.9	38.8 to 155.7
$SO_4^=$ (mM)	0.0 to 32.6	0.0 to 38.1	25.3 to 38.9
$\delta^{13}C$-CH_4 (‰)	No Significant Gas Accumulation	−72.0 to −82.4[5]	No Significant Gas Accumulation
CH_4 (%)	—	1.3 to 78.5	—
C_1/C_2	—	25,300 to 39,200	—
CO_2 (%)	—	0.00 to 0.24	—
UCM (ppm)	3.5 to 53.9	3.1 to 69.3	2.3 to 52.2
CPI	2.4 to 5.8	2.8 to 5.7	2.3 to 5.3
Pristane/Phytane	0.36 to 1.72	0.54 to 3.30	0.88 to 2.16

[1]Sites 614, 615, and 623; [2]Sites, 617, 620, 621, and 622; [3]Site 616; [4]$\delta^{13}C_{O.M.}$ = $\delta^{13}C$ (‰ relative to PDB) of the sedimentary organic matter; C/N = organic carbon to total nitrogen ratio on a % basis; % Org. C. = precent organic carbon; EOM = hexane extractable organic matter; ΣCO_2 = total interstitial CO_2; $SO_4^=$ = interstitial sulfate; $\delta^{13}C$-CH_4 = $\delta^{13}C$ (vs. PDB) of the CH_4; C_1/C_2 = ratio of methane to ethane on a % basis; UCM = unresolved complex mixture; CPI = carbon preference index of normal alkanes from C_{23} to C_{32} (odd/even); [5]Sites 620 and 621 only.

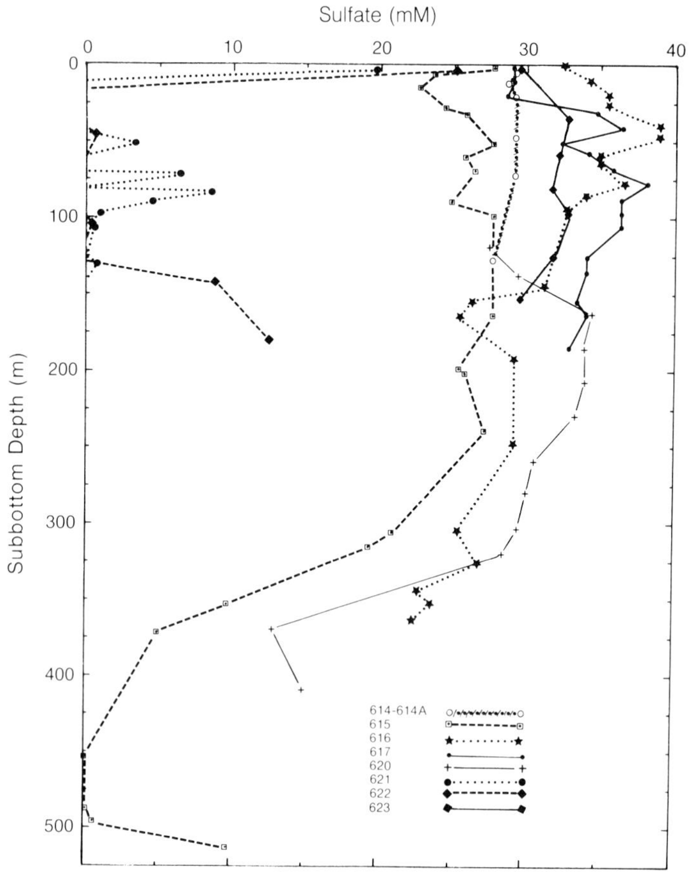

Figure 1. Interstitial water sulfate profiles for Mississippi Fan sites.

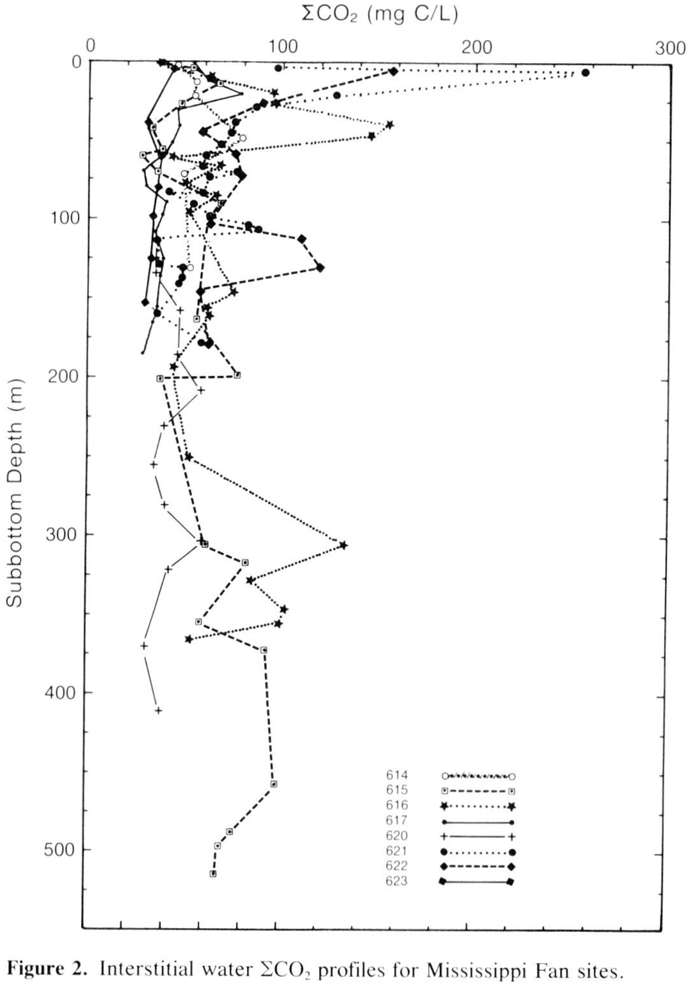

Figure 2. Interstitial water ΣCO_2 profiles for Mississippi Fan sites.

The Chemistry of Late Pleistocene Mississippi Fan Sediments

The ranges of the parameters measured in individual samples at each site are summarized in Table 1. Individual site data and methods are reported in detail elsewhere [16].

Bulk Parameters

Organic carbon content (% Org. C) ranged from 0.1 to 2.1%, averaged 0.9%, and was monotonously distributed vertically and areally. These values from the Mississippi Fan are typical of shallow water deltaic sequences [6,7], but differ substantially from typical deep water marine sediments. The carbon isotopic composition of the organic matter ranged from -25.0 to -27.9‰. This composition is indicative of a predominantly terrestrial source for the organic matter at all locations [2–8]. The low nitrogen contents, as illustrated by high C/N ratios (i.e., C/N ranging from 8.5 to 170), and lithologic data [Chapters 38, 40, 42, 47 this volume] support the carbon isotopic interpretation of a predominantly terrestrial origin for the organic matter [13–15]. In addition, significant amounts of wood fragments and lignitic material were frequently observed.

Pore Water Chemistry

Sulfate ($SO_4^{=}$) and total CO_2 (ΣCO_2) concentrations range from 0.0 to 38.9 (mM) and 28 to 256 mgC/liter, respectively (Figs. 1 and 2). Sulfate reduction, as evidenced by sulfate concentrations lower than seawater (average 29.6 mM), was observed at all locations except for one lower fan site (Site 614). Total alkalinity often showed a three- to six-fold increase over that of seawater. Total alkalinities ranged from 3.8 to 22.4 meq/liter, and pH values ranged from 6.1 to 7.6. Total alkalinity and pH were frequently lower in sands than in clays. Pore water salinities were relatively constant and near that of seawater (range 32.0 to 36.5 ‰). A slight decrease in salinity with depth in the sediment column was commonly observed. Pore water salinity was generally higher in sands than in clays, possibly because of clay dewatering. If this is correct, then compaction of the sediments began very early in its burial history.

Gaseous Hydrocarbons

Gas pockets in cores from the middle fan (Sites 620 and 621) contained methane, trace amounts of ethane, and low concentrations of carbon dioxide. The C_1/C_2 ratios averaged 31,600 (range 25,300 to 39,200) and $\delta^{13}C$-CH_4 ranged from -72.0 to -82.4‰ (vs. PDB). These values are consistent with a biological production process [17]. Lower-fan and fan-margin sediments were generally nongaseous, with only local, minor gas pockets observed. When the gas content was sufficient for analysis, its molecular composition was suggestive of a biogenic origin. The absence or near absence of gaseous hydrocarbons at most locations on the Mississippi Fan may result from several factors including: 1) high sedimentation rates that dilute biodegradeable organic matter (derived from either overlying water or riverine inputs) with clays or refractory organic matter (derived from slumped or continental-derived organic matter); 2) coarse-grained sediments that permit rapid loss of methane to the overlying water column; 3) temperatures and pressures at the drill sites that suppress microbial metabolic activity; 4) early thermal diagenesis of organic matter would be low level as a result of the low temperatures and geothermal gradient; and 5) microbial sulfate reduction can inhibit microbiological methane production.

One, or a combination, of these parameters could result in the low level of gaseous hydrocarbons observed. Slow deposition rates that favor aerobic microbial activity were not observed at any fan locations. At many locations, interstitial-water sulfate levels remained near seawater values so that a sulfate-free zone where methanogenic bacterial activity might have occurred was never observed. High C_1/C_2 ratios, low interstitial-water sulfate levels, and the relatively high alkalinities are all consistent with a microbiological source for the methane. The nonhydrocarbon gases observed at several sites were either nitrogen (possibly from nitrate-reducing micro-organisms), carbon dioxide, or air.

Extractable Organic Matter

Hexane extractable organic matter (EOM) ranged from 3.8 to 222 ppm (dry wt.). No regular trend with depth was observed. The distribution and quantity of the EOM was typical of shallow water deltaic sequences. Significant amounts of plant biowaxes were present in all samples (Fig. 3). The carbon preference index of normal alkanes from C_{23} to C_{32} (range 2.3 to 5.8) is indicative of a terrestrial source of the organic matter. Nonindigenous, high molecular weight, thermogenic hydrocarbons in the sediments (Fig. 4) were indicated by a complete suite of normal alkanes from n-C_{15} to n-C_{32}, the presence of significant amounts of pristane and phytane, and the unresolved complex mixture (UCM). The subbottom depth of the samples, the young age of the sediments, and the vertical trends observed suggest that these hydrocarbons most likely originated from upward migration of petroleum from deeper sources and not from pollution in the overlying water column.

Total scanning fluorescence analyses of the sediment extracts suggested the presence of significant quantities of two-, three-, and four-ring aromatic compounds (Fig. 5).

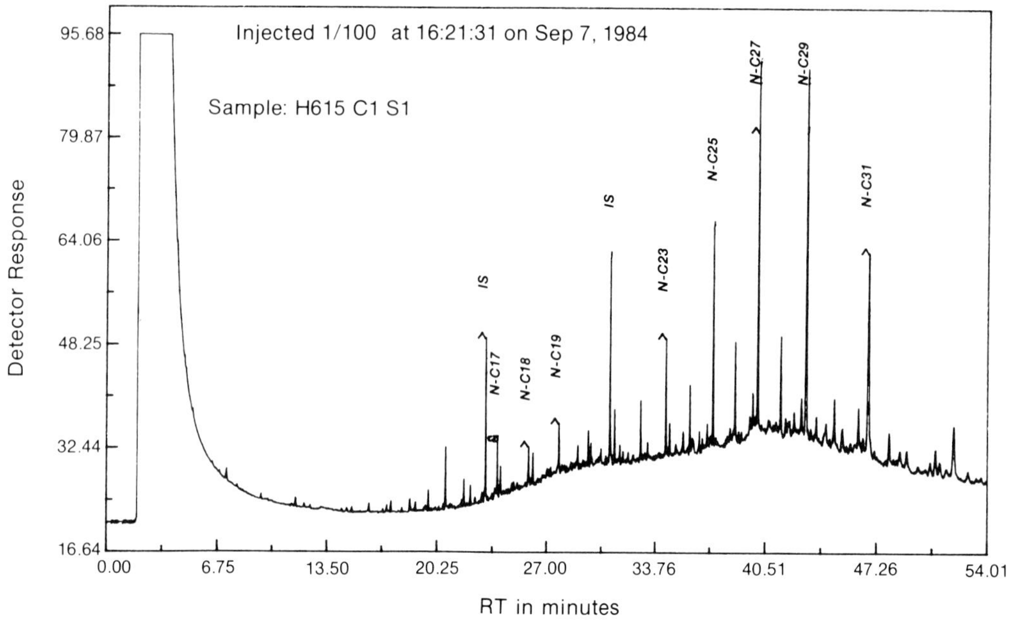

Figure 3. Capillary gas chromatogram from DSDP Site 615 (1.37-m subbottom) showing significant amounts of the n-C_{27} to n-C_{31} plant waxes.

This is consistent with the gas chromatographic analyses that suggest the presence of petrogenic or thermogenic hydrocarbons. Perylene was detected at all sites.

Conclusions

Geochemical parameters measured at the DSDP sites on the Mississippi Fan are similar to those reported from deltaic sequences. The organic matter present in these sediments is chemically and isotopically similar to terrigenous material. The qualitative and quantitative distribution of organic matter is similar at all fan locations. The sediments were generally nongaseous, but when gas was detected, it was primarily of a biogenic origin. Gas chromatographic and fluorescence analyses detected significant amounts of petroleum hydrocarbons, presumably resulting from the upward migration of petroleum from deeper sources. Sediments generally showed little if any sulfate reduction. Channel sediments had zero or near-zero sulfate levels, suggesting that methanogenesis was occurring.

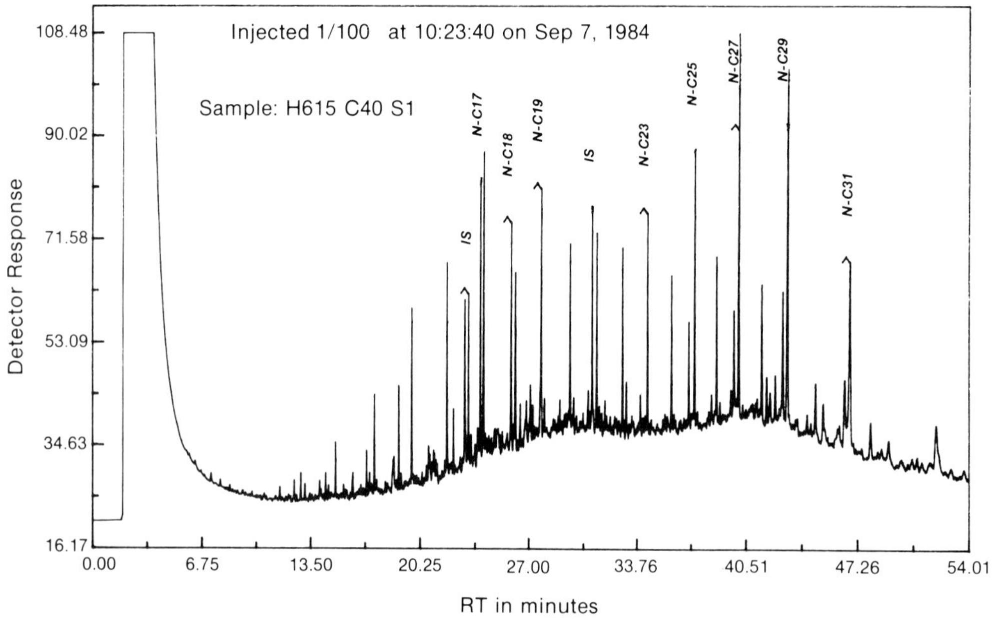

Figure 4. Capillary gas chromatogram from DSDP Site 615 (372-m subbottom) showing thermogenic hydrocarbons as evidenced by the UCM and normal alkane distributions.

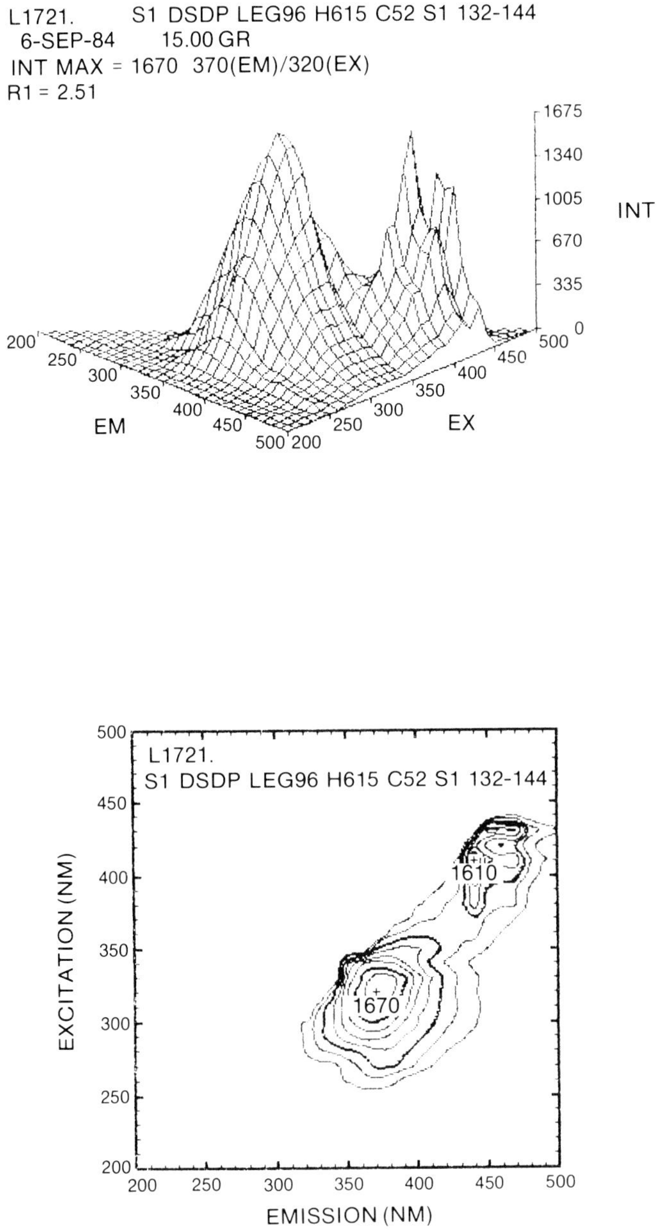

Figure 5. Total scanning fluorescence spectra from DSDP Site 615 (515-m subbottom) showing a thermogenic hydrocarbon signature consisting of significant amounts of three- and four-ring aromatics.

Acknowledgments

This work was supported by the Marine Chemistry Program of NSF through Grant No. OCE-8301538. Instrumentation support was provided by the Center for Energy and Mineral Resources at Texas A&M University.

References

[1] Barker, C., 1979. Organic geochemistry in petroleum exploration. American Association of Petroleum Geologists Continuing Education Course Note Series 10, 159 pp.

[2] Hedges, J. I., and Parker, P. L., 1976. Land derived organic matter in surface sediments from the Gulf of Mexico. Geochimica Cosmochimica Acta, v. 40, pp. 1019–1029.

[3] Hedges, J. I., 1975. Lignin compounds as indicators of terrestrial organic matter in marine sediments. Ph.D. Thesis, University of Texas at Austin.

[4] Gardner, W. S., and Menzel, D. W., 1974. Phenolic aldehydes as indicators of terrestrially derived organic matter in the sea. Geochimica Cosmochimica Acta, v. 38, pp. 813–822.

[5] Sackett, W. M., and Thompson, R. R., 1963. Isotopic organic carbon composition of recent continental derived clastic sediments of eastern Gulf coast, Gulf of Mexico. American Association of Petroleum Geologists Bulletin, v. 47, pp. 525–531.

[6] Sackett, W. M., 1964. The depositional history and isotopic organic carbon composition of marine sediments. Marine Geology, v. 2, pp. 173–185.

[7] Gearing, P., Plucker, F. E., and Parker, P. L., 1977. Organic carbon stable isotope ratios of continental margin sediments. Marine Chemistry, v. 5, pp. 251–266.

[8] Hunt, J. M., 1970. The significance of carbon isotope variations in marine sediments. In: G. D. Hobson and G. C. Spears (eds.), Advances in Organic Geochemistry. Pergammon, Oxford, pp. 27–35.

[9] Farrington, J. W., and Tripp, B. W., 1977. Hydrocarbons in western North Atlantic sediments. Geochimica Cosmochimica Acta, v. 41, pp. 1627–1641.

[10] Farrington, J. W., and Meyers, P. A., 1975. Hydrocarbons in the Marine Environment. In: G. Eglington (ed.), Environmental Chemistry, vol. 1, ch. 5. Specialist Periodical Report, The Chemical Society, London, pp. 109–136.

[11] Gearing, P., and others, 1976. Hydrocarbons in 60 northeast Gulf of Mexico sediments. A preliminary survey. Geochimica Cosmochimica Acta, v. 40, pp. 1005–1017.

[12] Hedberg, H. D., 1968. Significance of high wax oils with respect to the genesis of petroleum. American Association of Petroleum Geologists Bulletin, v. 52, pp. 736–750.

[13] Prahl, F. G., Bennett, J. T., and Carpenter, R., 1980. The early diagenesis of aliphatic hydrocarbons and organic matter in sedimentary particulates from Dabob Bay, Washington. Geochimica Cosmochimica Acta, v. 44, pp. 1967–1976.

[14] Müller, P. J., 1977. C/N ratios in Pacific deep-sea sediments: Effect of inorganic ammonium and organic nitrogen compounds sorbed by clays. Geochimica Cosmochimica Acta, v. 41, pp. 765–776.

[15] Meyers, P. A., and others, 1984. Organic geochemistry of suspended and settling particulate matter in Lake Michigan. Geochimica Cosmochimica Acta, v. 48, pp. 443–452.

[16] Kennicutt, M. C. II, and others, 1985. Non-volatile organic matter at DSDP sites 614-623, DSDP Leg 96. Initial Reports of the Deep Sea Drilling Project (in press).

[17] Bernard, B. B., 1978. Light hydrocarbons in marine sediments. Ph.D. Thesis, Texas A&M University, 144 pp.

CHAPTER 47

Petrology of Mississippi Fan Depositional Environments

Harry H. Roberts and Paul A. Thayer

Abstract

Radiography, scanning electron microscope, thin-section, and X-ray diffraction analyses provided initial data on petrographical characteristics of Mississippi Fan deposits. Channel gravels are composed of polycrystalline quartz and faceted chert. Sands from channels and channel terminations (sheet sands) consist of feldspathic litharenites, sublitharenites, and subarkoses composed of quartz with various subordinate amounts of feldspar, mica, and heavy minerals plus reworked foraminifer tests, shell debris, glauconite, and abundant wood fragments. Finely laminated clays, silts, and fine sands from the upper channel fill and overbank contain authigenic clays and other diagenetic products, as well as authigenic gypsum, pyrite, calcite, and dolomite.

Introduction

Nine Mississippi Fan borings have provided samples from which the sedimentologic/petrographic characteristics of the fan's major depositional environments can be described. These samples came from channel fill sequences (Sites 621 and 622), overbank deposits (Sites 617 and 620), and slumped marginal-fan sediments (Site 616) of the middle fan. Both channel fill (Site 623) and overbank deposits (Site 624), as well as sandy channel-mouth sheet sands (Sites 614 and 615), were sampled from the lower fan. A variety of analytical techniques were used in this study. X-ray radiographs of thin sediment slabs (6-mm thick) proved to be exceptionally valuable to describe macroscopic sedimentary structures and inclusions, as well as to determine locations for detailed sampling and analysis. One hundred and eight sediment slabs were radiographed from the DSDP borings (Chapter 44). From these slabs and other samples, representative subsamples were selected for thin-section, scanning electron microscope (SEM), X-ray diffraction (XRD) (mineralogy), and total organic carbon analyses. This paper represents a summary of the initial results. Other details of the sedimentary facies are described by Stow and others in Chapter 38.

Sands

With the exception of a few very thin sands in the overbank deposits, sands and gravels are confined to the channel fill and sheet sands at the channel terminations. Macroscopic investigations of the cores and their X-ray radiographs indicate that several styles of deposition are represented in the coarse fraction deposits (Figs. 1A through D). At the base of both borings in the middle-fan channel fill (Sites 621 and 622), gravels grade upward into sands with thin clay interbeds. This active channel fill sequence is overlain by a passive fill composed of thinly stratified clays, silty clays, and silts (Figs. 1E and F). Thicker sands of the active fill contain very few sedimentary structures, a condition that may be a product of liquefaction associated with the coring process. Thinner sands (<0.5 m) of the mid-fan channel fill are commonly graded, with abundant organic remains sometimes organized into parallel units of nearly 100% woody fragments ("coffee grounds"). The graded units display sharp basal contacts, but do not generally indicate significant scour into underlying sediments. It is not uncommon for the sands and silts in these graded packets to have small-scale cross-laminations that grade upward into parallel bedded silts and clays (Figs. 1A and B). Other sands cut across finer-grained and parallel laminated interbeds of silts, silty clays, and clays. These coarse units are generally structureless (Figs. 1D and E).

The lower-fan channel deposits nearly always appear to be sharp-based, with evidence of scour into underlying units. As in the middle fan, these channel sands sometimes cut across bedding in finer-grained underlying sediments. Angular clay-rich rip-up clasts are typical components of

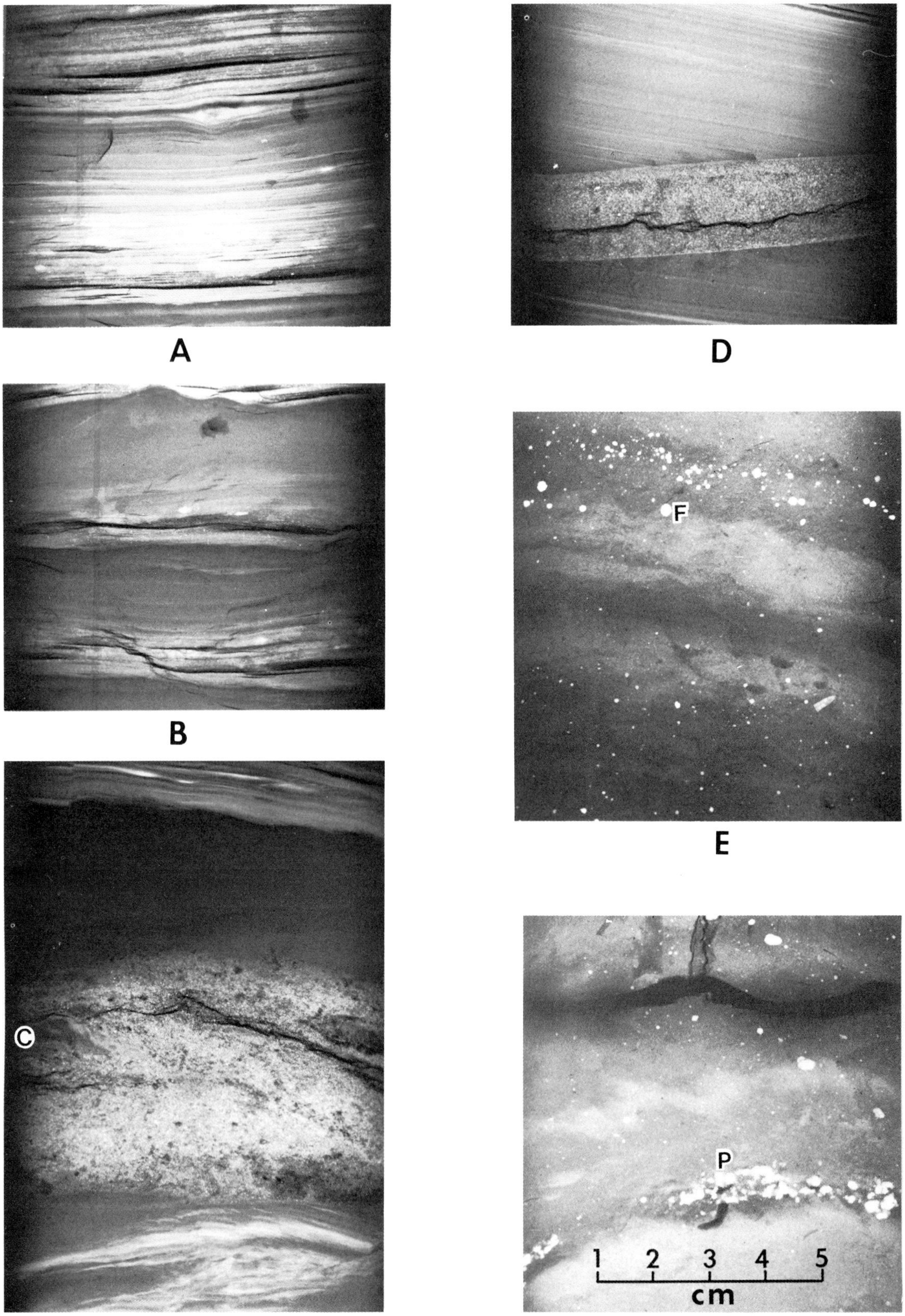
A
B
C
C
D
E
F
F
P
1 2 3 4 5
cm

these channel sands, which generally have little or no internal structure. It is not uncommon for the sands to contain abundant scattered woody organic particles as well as mica. The thinner sand and silt units (~2 to 15 cm) may display several types of stratification including: 1) very small-scale cross-laminations, 2) cyclic graded packets, and 3) massive to faintly parallel bedded units. Additional details of sedimentary structures are described by Coleman and others in Chapter 44.

In terms of total sand and coarse silt, the lower-fan sheet sands (Sites 614 and 615) contained the highest percentage. Specifically, Site 615 contained 65% net sand in the lower 274 m and 41% net sand in the upper 202 m [1]. The thicker sands tend to be massive and sharp-based. They are commonly associated with horizontally bedded thin silt and clay-graded interbeds, as well as units of woody organic remains that may be several centimeters thick. The graded interbeds usually occur in numerous cycles a few millimeters to a few centimeters thick. Small-scale cross-laminations may occur in the thin sand and silt units, which usually have sharp but nonscouring bases with only a small thickness of coarse sediment that grades into clay. A few planktonic foraminifera tests are sometimes found at the tops of these units [2].

Sand composition has thus far been evaluated by studying approximately 100 thin-sections of samples selected from specific depositional environments and their various coarse sedimentation units as just described. Composition of the grain framework and relative pore space was determined by counting 300 points/thin section. Grain size was estimated by measuring long diameters of all grains encountered in the point count procedure. Figure 2 shows how sands encountered in the DSDP sites can be classified using a ternary diagram with end members being quartz, feldspar, and rock fragments [3]. Most sands are feldspathic litharenites, sublitharenites, or subarkoses. A few sands qualify as lithic arkoses, and very few are true arkoses. Generally, the coarser sands contain more rock fragments and, hence, fall within the feldspathic litharenite field.

Within the grain framework, quartz accounts for 40 to 60% of the total. The dominant variety is strained monocrystalline quartz, with subordinate amounts of polycrystalline types (Figs. 3A through C). Unstrained monocrystalline quartz accounts for approximately 1% of the framework. These grains commonly display overgrowths that appear worn and abraided, which suggests that they are reworked. Polycrystalline quartz increases in abundance with increasing grain size. It is most abundant in the medium- to coarse-grained sands, but this variety never becomes the dominant quartz type.

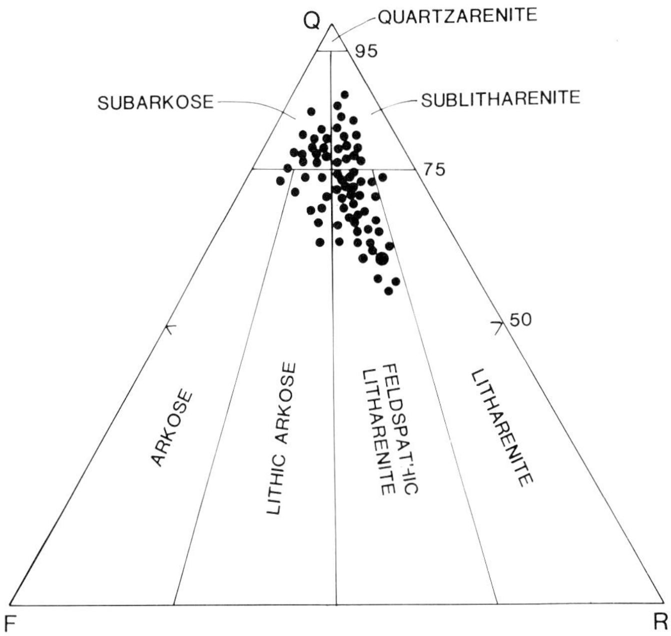

Figure 2. Framework composition of Mississippi Fan sands [3]. The larger dot is the average composition of modern Mississippi River sand [4].

Feldspars comprise approximately 7 to 15% of the sand-sized constituents and average about 10%. Of the feldspars, plagioclase is the most abundant variety in the fine and very fine sands, with microcline and orthoclase being subordinate. However, in the medium and coarse sands, microcline is usually dominant. Some of the plagioclase displays oscillatory zoning, which is indicative of a volcanic provenance. Quite often plagioclase grains show evidence of dissolution in the grain core and along the twin lamellae. These effects of dissolution are interpreted as being inherited from the source area and probably do not represent the products of recent diagenesis.

Rock fragments account for 10 to 25% of the framework grains. Sedimentary rock fragments form more than 50%

Figure 1. Selected intervals from X-ray radiographs of DSDP Leg 96 cores. The scale (shown in F) is the same for each radiograph. (A) Finely graded sequence from fine sand to clay. The high density, light areas in coarser units represent calcite-cemented grain clusters (Site 622, 94.2-m subbottom). (B) Graded bedding typical of middle to upper channel fill as well as overbank sediments. Pelagic foram tests sometimes found at top of mud unit (Site 622, 94.3-m subbottom). (C) Rather structureless sand unit with sharp bottom and top contacts. Note the clay clasts (C) and dark spots that represent disseminated woody organic fragments (Site 623, 50.5-m subbottom). (D) Sand unit with sharp base and top that cuts across silty clay and clay laminations (Site 623, 123.1-m subbottom). (E) Passive channel fill composed of poorly defined graded silty clays and clays. Note the numerous high density features (white spots). Some of these white areas represent foram tests (F) filled with calcite (Site 622, 10.3-m subbottom). (F) Passive channel fill showing a variety of high density inclusions (white areas), some of which are pyrite (P) in the form of framboid clusters (Site 621, 10.1-m subbottom).

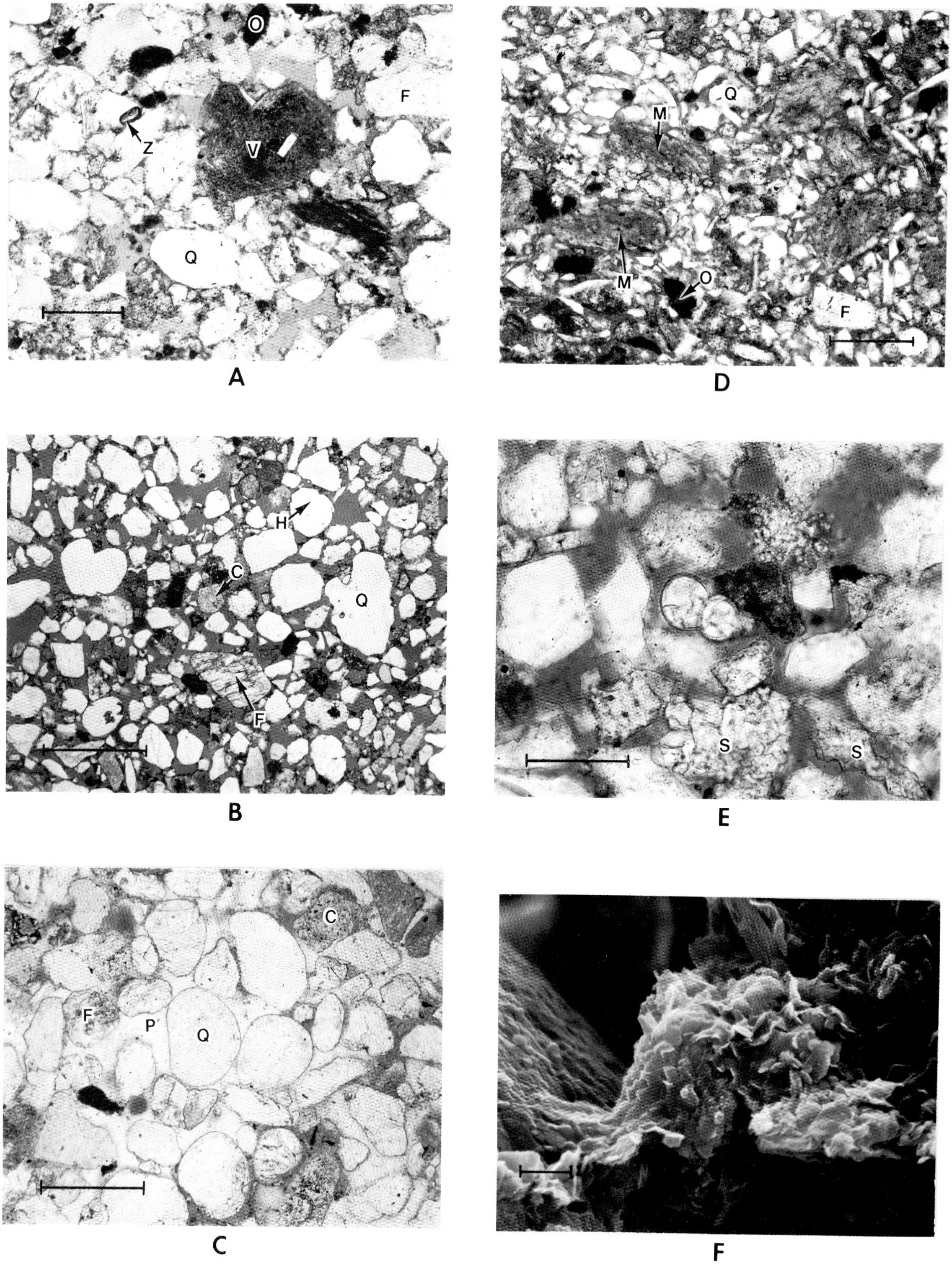
A
D
B
E
C
F

of the total rock fragment population. Of the sedimentary rock fragments, chert is the most abundant (Fig. 3B). Cherts sometimes contain euhedral dolomite rhombs, suggesting a provenance characterized by carbonate terrain. In many of the cherts, however, the carbonates have been dissolved, leaving pore spaces. This dissolution process was probably the result of subaerial weathering.

Next in abundance to chert are sandstone rock fragments (Fig. 3D). The sand-sized clasts tend to be rounded to well-rounded, while larger mud clasts commonly display angular morphologies (Fig. 1C). Most of the mudstone rock fragments show evidence of being relatively soft at the time of deposition and, therefore, are interpreted as having an intrabasinal origin. A few of the mudstone rock fragments are much more indurated and may be extrabasinal.

Other sedimentary rock fragment types include carbonates and indurated terrigenous sandstone. The carbonates all tend to be well-lithified (micrites to sparites), rounded, and occasionally fossiliferous. All are interpreted as having an extrabasinal origin. Quartz-rich sandstone rock fragments cemented with calcite are relatively rare. Microscopic inspection of washed coarse fraction samples, as well as the occurrence of high density clots on X-ray radiographs in sand units, especially thin-graded sands (Figs. 1A and B), suggest that these grain types can have an intrabasinal origin.

Remaining rock fragment types include metamorphics, volcanics, and minor plutonics. Combined, these varieties account for no more than 3% of the total rock fragment population. The metamorphics are primarily schist and phyllite, while the volcanics consist chiefly of euhedral plagioclase laths in an aphanitic groundmass.

Accessory constituents of the sands include mica, glauconite, dolomite, heavy minerals, detrital opaque minerals, and woody fragments. Mica is found most commonly in the fine sands and coarse silts. Muscovite is most abundant, with minor occurrences of biotite. Glauconite grains are found only in trace amounts, while dolomite rhombs are quite common, especially in the very fine sands and silts (Fig. 4). Some dolomite rhombs display rounding, which suggests transport. Most dolomite, however, is euhedral or occurs in grain clusters with no evidence of transport or reworking. Heavy minerals consist of green hornblende, epidote, garnet, pyroxene, as well as zircon and tourmaline. Apatite, kyanite, staurolite, monazite, and sphene are present but rare. Detrital opaques include magnetite, ilmenite, leucoxcene, and pyrite.

Woody organic fragments are found throughout the sands (Fig. 2), but they are most abundant in the fine and very fine sands and silts. These organic particles display a size range from medium silt to very coarse sand. The woody organic grains are generally black in thin section (inertite) and, in rare cases, show a golden-brown color. Rock evaluation pyrolysis data from 13 sand samples from various depositional settings in the fan show total organic carbon (TOC) values that range from 0.7 to 7.9% (average 2.6%). The hydrogen index of these samples was low ($\bar{X} = 56$, range 41 to 79), and the oxygen index was high ($\bar{X} = 212$, range 131 to 608). These data indicate that the detrital organics are in an oxidized state and have limited potential as a hydrocarbon source. However, shrinkage and advanced stage oxidation of organics may serve to enhance pore space. Perhaps this process accounts for the oversized pores commonly observed in ancient sands.

Other than woody grains, biogenic components rarely form more than a small part of the total grain framework. Broken and rounded skeletal debris, including echinoderm, mollusk, and foram tests, are the most common. Mollusk fragments are typically bored and have micrite rims, while forams tend to be filled with calcite (Figs. 1D and F). Nasselarian radiolarians are present and usually show evidence of reworking. Rare occurrences of broken silico flagellates were observed.

Authigenic constituents of the sands consist primarily of dolomite, clays, pyrite, and traces of gypsum (Fig. 4). As previously discussed, the dolomite occurs as euhedral rhombs or clusters of rhombs. Authigenic clays, identified by Energy Dispersive X-ray analyses (EDX) and clay morphology, were found growing on a substrate of detrital clay (Fig. 2F) as well as on other grain surfaces. Both illite and smectite were identified. Pyrite occurs most commonly as framboids (Fig. 4E) or clusters of framboids. It is also fre-

Figure 3. Thin-section and SEM photomicrographs of Mississippi Fan sands. (A) Fine-grained, moderately sorted, feldspathic litharenite composed of quartz (Q), felsdspar (F), and volcanic (V) and metamorphic (M) rock fragments. Grain with high relief in upper left is zircon (Z). Middle-fan channel (Site 621, 186.6-m subbottom). Porosity: 18%. Plane-polarized light. Scale bar = 200 μ. (B) Medium-grained, well-sorted sublitharenite composed of quartz (Q), feldspar (F), and chert (C), with accessory hornblende (H). Lower-fan channel/levee complex (Site 623, 107.6-m subbottom). Intergranular porosity (dark gray areas): 34%. Plane-polarized light. Scale bar = 1000 μ. (C) Coarse-grained, well-sorted, feldspathic litharenite composed of quartz (Q), feldspar (F), and chert (C). Lower-fan channel/levee complex (Site 623, 140.2-m subbottom). Intergranular porosity (P): 32%. Plane-polarized light. Scale bar = 1000 μ. (D) Fine-grained, moderately sorted, feldspathic litharenite composed of subangular quartz (Q) and feldspar (F), along with larger, rounded mudstone fragments (M) and organic debris. Lower-fan channel/levee complex (Site 623, 140.6-m subbottom). Intergranular porosity (dark gray areas): 21%. Plane-polarized light. Scale bar = 200 μ. (E) Recrystallized planktonic foraminifer in coarse silt. Dark gray areas are pores. Grains with high relief are broken skeletal fragments (S). Lower-fan channel-mouth lobe (Site 614A, 41-m subbottom). Plane-polarized light. Scale bar = 100 μ. (F) SEM photomicrograph of authigenic smectite growing on pore-bridging detrital clay in a fine sand (Site 615, 17.4-m subbottom). Scale bar = 3 μ.

A
D
B
E
C
F

quently found filling foram tests. Gypsum occurs as large poikilotopic crystals (Fig. 4F) that engulf terrigenous grains.

Figure 5 illustrates the relationship between size, sorting, and porosity of the Mississippi Fan sands. Size and sorting were measured from thin sections. Two methods were used to estimate porosity. When samples were interpreted as undisturbed by the coring and sample preparation processes (retained internal structure, pore-bridging clay, etc.), porosity was simply estimated by point counting. Of the 50 undisturbed sand samples, porosity values ranged from 5 to 35% (most in the range 20 to 30%). If samples were disturbed, porosity was estimated from Figure 5 on the basis of size and sorting characteristics. Figure 5 illustrates that most porosities fall within the 30 to 40% range, which is probably a more reasonable estimate of the actual porosity values. Porosities derived from thin-section analyses are typically lower than those measured by other standard methods (R. M. Sneider, 1984, personal communication).

In general, the sands range in size from fine to coarse, but most are fine to very fine and are moderately well-sorted (Fig. 5). Because of the small grain size of most sands, the individual grains tend to be subangular (Fig. 3). As grain size increases, rounding also increases. Accordingly, sands of the Mississippi Fan can be classified as immature to submature [3], which is characteristic of modern Mississippi River sands [4]. However, the average Mississippi River sands appear to have more rock fragments (Fig. 2), a disparity that may be related to selective sorting during transport into the deep Gulf of Mexico.

Silts and Clays

Fine-grained sediments of the Mississippi Fan are concentrated in the middle-fan passive channel fill and in the overbank deposits of both the middle and the lower fan. These sediments are composed of clays, silty clays, and silts organized into a variety of sedimentary structures, the most common of which are graded beds of various dimensions.

Aside from the sediment deformational features that are common to the overbank deposits of the middle fan, the most common structures of the overbank facies in both the upper and the lower fan are graded parallel laminations. Commonly, the silts and silty clays that comprise these laminations are rich in woody organic particles. In thin-section, the organic remains are bedded (Fig. 4A). One of the characteristic features of these typically fine-grained deposits is the thin nature of most graded units. Lower-fan overbank deposits as well as fine-grained interbeds of the channel-mouth sheet sands may have up to 15 to 20 thick micrograded units per centimeter [Coleman and others, Chapter 44]. In addition to these features, small clay clasts, stranded ripples, a few microcross-laminated silts, and small load features have been observed. Bioturbation is rare, but was noted in some of the X-ray radiographs of overbank deposits. Burrowing usually occurs at the clay-rich top of a graded unit.

The clay-rich passive channel fills of the middle fan contain repetitive occurrences of poorly defined, graded units composed of silty clay to clay and silt to clay sediments. Loading in these fine-grained sequences has caused slight deformation of the stratigraphy, primarily in the form of fractures and small offsets of laminae.

One hundred and ten X-ray diffraction analyses of clays from all the major depositional environments of the fan as well as the two intraslope basins (Sites 618 and 619) suggest considerable variability in relative percentages of smectite, illite, and kaolinite. The dominant clay mineral is smectite, with lesser amounts of illite and kaolinite. Relative percentages between these clay mineral species as determined by peak areas are similar to relative percentages found on the modern Mississippi River delta front and upper continental slope [6]. At this early stage of investigation into clay mineral trends in the Mississippi Fan, it is premature to characterize significant variations within and between depositional environments. However, we feel confident that a significant difference exists between the average clay mineral signature of the intraslope basins (Sites 618 and 619) as compared with that of the Mississippi Fan (Fig. 6). The intraslope basin sediments display increases in illite and kaolinite at the expense of smectite compared with the averaged clay mineral composition of the fan. As suggested for similar variations between shelf and undisturbed slope sediments opposite the modern Mississippi Delta [6], this trend toward increased kaolinite and illite probably indicates the significant influence of a pelagic clay component. Fan clays are closer to those currently being deposited by the Mississippi River.

Within the silts and sands, SEM work has revealed the frequent occurrence of authigenic clays, both illite and smectite. Figure 3D shows authigenic smectite, as interpreted by EDX as well as clay morphology, growing on a detrital clay substrate that has formed a bridge between two fine sand grains. Authigenic clays have been found more frequently in the silts, although most clay in the silts and sands is allogenic. The allogenic clays occur as coats

Figure 4. Thin-section and SEM photomicrographs of Mississippi Fan fine-grained sediments. (A) Layered organic debris (black grains) in coarse silt. Middle-fan channel (Site 622, 103-m subbottom). Plane-polarized light. Scale bar = 200 μ. (B) SEM photomicrograph of dolomite rhomb with authigenic illite growing on the grain (Site 614, 126.2-m subbottom). Scale bar = 16 μ. (C) SEM photomicrograph of authigenic gypsum in a clay matrix (Site 615, 197.3-m subbottom). Scale bar = 2.5 μ. (D) SEM photomicrograph of calcite growing on detrital grain (Site 615, 135.1-m subbottom). Scale bar = 7 μ. (E) SEM photomicrograph of a pyrite framboid typical of those found throughout the silt-clay facies (Site 615, 51.6-m subbottom). Scale bar = 5 μ. (F) Euhedral crystal of authigenic gypsum in claystone from lower-fan channel/levee complex (Site 623, 78.7 m subbottom). Plane polarized light. Scale bar = 200 μ.

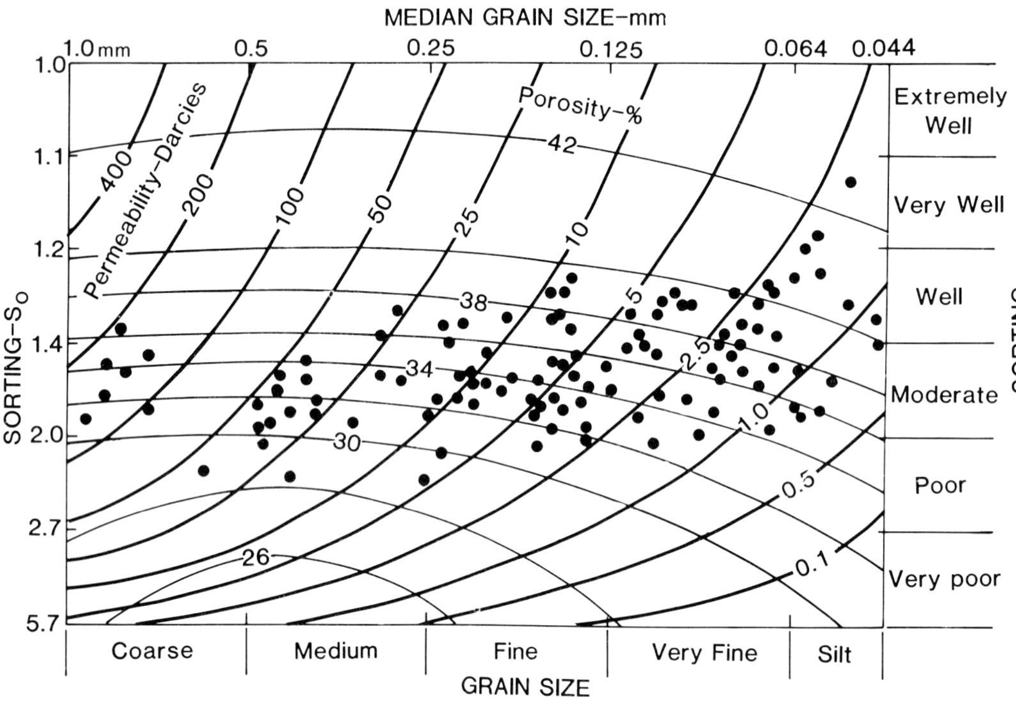

Figure 5. The relationship between size, sorting, porosity, and permeability of unconsolidated Mississippi Fan coarse silts and sands as determined from thin section [5, as modified by R. M. Sneider, 1984, personal communication].

on framework grains, interparticle pore fills, and detrital bridges between grains. Authigenic clays are of particular interest because of the implications concerning deterioration of reservoir properties.

Although small amounts of gypsum have been found as an authigenic constituent of the sands, much more of this mineral is present in the silt and clay fractions (Figs. 4C and F). Both thin-section and SEM examination of the silts and clays indicate that gypsum occurs as single poikilotopic crystals and clusters of crystals that penetrate the surrounding fine-grained clay-rich sediment. In relatively clay-free silts, the gypsum crystals overgrow adjacent silt grains.

Other common authigenic minerals in the silt and clay fraction include calcite and pyrite. The calcite occurs in several different formats, including cemented silt grain clusters (Figs. 1A and B), cavity fills in foram tests (Figs. 1C, 1D, and 3E), and free precipitates on grain surfaces (Fig. 4D). Pyrite also occurs as a cavity fill in foram tests, but framboidal pyrite (as small framboids or clustered framboids) is the most commonly encountered form (Fig. 4F).

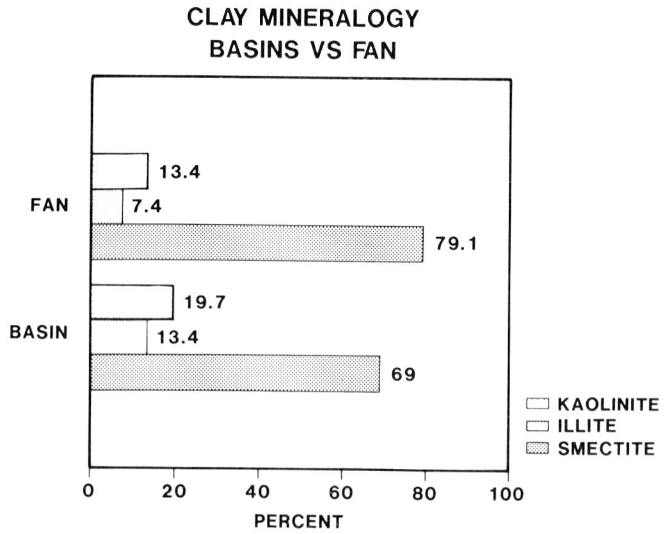

Figure 6. Bar graph of average clay mineral content of the Mississippi Fan sediments as contrasted with the intraslope basins. Data are based on peak areas derived from 110 XRD analyses.

Conclusions

Initial sedimentologic/petrographic studies of the DSDP Leg 96 sediments have led to the following conclusions:

1. Sands are concentrated in the middle- and lower-fan channel fills as well as in the lower-fan sheet sands. Silts and clays occur at overbank deposits, passive channel fills, and interbeds associated with the coarser facies. Graded bedding of various thicknesses constitutes the most common sedimentary structure in all environments.

2. Sands are dominantly feldspathic litharenites, sublitharenites, and subarkoses, with measured porosities in thin-section that average 20 to 30% and woody organic contents of up to 7.9% TOC.

3. Authigenic minerals occur in both the sand and the silt-clay fractions, but are most common in the fine-grained

sediments. Smectite and illite, dolomite, pyrite, and gypsum are the most frequently encountered authigenic minerals. The latter three are most typical of the silt-clay fraction.

4. At this stage in their depositional history, the sands are clean, have high porosities, show only minor pore-reducing diagenetic effects, and therefore appear to have good reservoir potential.

5. Petrology of the Mississippi Fan sediments suggests that there are few indicators within small samples (e.g., cuttings or sidewall samples) that would indicate a deep-water setting for these deposits.

Acknowledgments

Many of the samples discussed in this chapter came from the slabs prepared for X-radiography by A. H. Bouma and J. M. Coleman onboard the D/V *Glomar Challenger*. We appreciate the access to that material. Dr. Robert K. Olson of AMOCO provided rock eval pyrolysis data on selected sand samples. AMOCO Production Company supplied thin sections, SEM analyses, and some drafting as well as photography support. Dr. Syed Ali of Gulf Oil Corporation arranged for SEM work to be performed on a few of our samples. Celia Harrod helped with drafting, and Kerry Lyle did the photographic work.

References

[1] Bouma, A. H., and Coleman, J. M., 1984. Drilling program on Mississippi Fan - DSDP Leg 96 (abstract). In: Programs and Abstracts, 5th Annual Research Conference, Gulf Coast Section Society of Economic Paleontologists and Mineralogists, December 2-5, 1984, Austin, Texas, pp. 8–22.

[2] Roberts, H. H., and Thayer, P. A., 1984. Sedimentology of depositional environments in DSDP Leg 96 borings (abstract). In: Program and Abstracts, 5th Annual Research Conference, Gulf Coast Section Society of Economic Paleontologists and Mineralogists, December 2-5, 1984, Austin, Texas, pp. 86–87.

[3] Folk, R. L., 1968. Petrology of Sedimentary Rocks. Hemphill's, Austin, Texas, 170 p.

[4] Potter, P. E., 1978. Petrology and composition of modern big river sands. Journal of Geology, v. 86, pp. 423–449.

[5] Beard, D. C., and Weyl, P. K., 1973. Influence of texture on porosity and permeability of unconsolidated sand. American Association of Petroleum Geologists Bulletin, v. 5, pp. 349–369.

[6] Roberts, H. H., 1985. Clay mineralogy of contrasting mudflow and distal shelf deposits on the Mississippi River delta front. Geo-Marine Letters (in press).

VII

Submarine Fans and Related Turbidite Systems

Conclusions

CHAPTER 48

Comments and New Directions for Deep-Sea Fan Research

William R. Normark, Neal E. Barnes, and Arnold H. Bouma

Abstract

Comparison of fan descriptions presented in this volume demonstrates the major problems in developing general models that incorporate modern fans and ancient turbidite sequences. Therefore, attempts to develop a unifying fan model are premature at the present time. The most pressing need is a refinement of definitions of the primary common characteristics of and the terminology used to describe submarine turbidite systems. The incorporation of results obtained by the Deep Sea Drilling Project Leg 96 on the Mississippi Fan provides the first comprehensive collection of stratigraphic information from a modern submarine fan. The drilling results also help to bridge the gap in the comparison between modern and ancient turbidite systems because for the first time we obtained detailed vertical sections through a modern submarine fan.

Introduction

After viewing the descriptions of the submarine fans and related turbidite systems presented in this volume, it should be clear to the reader that the editors did not attempt to coerce the individual authors to standardize terminology or interpretations. Our effort to provide a more or less uniform set of bathymetric and morphometric maps for several modern submarine fans, rather than presenting an attempt to define commonality of parameters, emphasizes many of the descriptive problems noted in Chapter 2. In addition to differences reflecting the wide variety of submarine fan features, most of the problems stem from a lack of agreement among the authors regarding the definition of the diagnostic characters of submarine fans. This inability to define even the basic features and divisions of submarine fans became obvious during our attempts to assemble the data table for the wall chart [1; Chapter 3], which also accompanies this publication. Some creative editing was required just to provide entries for some of the pigeon holes in the table. Thus, as a famous cartoonist once observed, "We have met the enemy and he is us" [2, p. 5].

The comments that follow are just a few that we noted while editing the chapters and compiling the morphometric maps and the table of characteristics for the wall chart [1]. In general, these items are in addition to the problems identified in the opening chapter of the special issue of *Geo-Marine Letters* [3] and Chapter 2 of this volume.

Modern Fans

The lack of an accepted set of definitions and terms to describe modern submarine fans is clearly evident in the morphometric maps and in the various descriptions of those fans. The number of fan divisions varies widely among the examples presented—none, two, three, or four divisions. The use of four divisions is applied to those submarine fans for which the canyon is considered as a separate division or for which the abyssal plain is treated as a separate entity from the fan. For some fans, the authors included ponded basin plains or abyssal plain equivalents as part of the lower fan division (e.g., Monterey, San Lucas). For others, the abyssal plain is not considered to be related to the fan (e.g., Amazon, Astoria, Cap Ferret). For the purposes of compiling the table for the wall chart [1], we included the basin plain or abyssal plain as part of the fan if the fan was the only or major source of clastic sediment for the plain.

The morphometric maps also show that the characters ascribed to lower- and middle-fan divisions vary widely. For some fans, the absence of channels is a major factor in identifying lower-fan divisions (e.g., Navy, San Lucas). For other fans, channels occur in all divisions (e.g., Amazon, Indus, Magdalena, Mississippi). The divisions, likewise, are not consistent with changes in gradient of the fan surface. Furthermore, there is a tendency for authors to confine their attention to those fan areas for which the best or sufficient data are available; this tends to produce skewed fan divisions that can cause interpretational problems especially when using short review articles such as those reported here.

Ancient Fans

Few of the authors describing ancient fans or turbidite systems have sufficient outcrop control to produce a schematic map that is somewhat comparable with the morphometric maps for modern fans. Extreme differences in extent and type of exposures, together with the lack of detailed stratigraphy for correlation purposes, make it difficult to standardize descriptions of ancient fans or provide comparable maps. At the present, one could argue that few of the ancient fans actually display many of the diagnostic characters that are used to define modern fans. Indeed, one is tempted to say that the "classic" ancient fans that provided so much data for determining fan facies may not be "true" fans.

Comparisons

The compilation of modern fans and ancient turbidite systems emphasizes the differences in shape, size, and margin setting between these two groups and within each group. Of the modern fans described, about half developed on active margins and the other half on passive margins. In contrast, all but two of the ancient examples formed on active margins. This difference is not surprising because turbidites deposited in active marginal basins are more likely to be exposed in continental areas than is the case with oceanic crust with its overlying sediment as would be generally required for passive margin systems.

The difference in margin setting is clearly reflected in the shapes of modern fans. In passive margin settings, the basins normally are much larger and have lower seafloor gradients than is the case in active margin settings. Consequently, the submarine fans that develop on a passive margin are less influenced by basin shape. Such fans can build out more equidimensionally, the depositional processes are more important than basin size in controlling the final shape, and the downfan direction is commonly in seaward direction. The sediment deposited on passive margin fans typically has a low sand/clay ratio, has probably been transported over long distances to the coast, and then was deposited in a deltaic shelf-upper slope setting from which the sediment is released by such mechanisms as slope failure.

Active margin basins typically are smaller, elongate, and subparallel to the strike of the margin. The source of the sediment may be a nearby mountain range, resulting in a much higher sand/clay ratio. The smaller basins consequently have a strong effect on the shape of the developing submarine fan, and the commonly steeper gradients influence the movement of the sediment. Downfan directions can be more parallel to the margin if the basin axis dips in one direction, even if more than one source is involved (i.e., line source). A number of small "radial" fans can develop, each having a seaward downfan direction, and these can coalesce giving the impression of a longitudinal basin fill.

A more comprehensive examination of modern fans commonly shows that each fan is a composite of several individual bodies (fanlobes). As discussed for the Mississippi Fan (Chapters 21, 36, 37), a fanlobe consists mainly of a channel-overbank system that is of fan-scale and not just growth within a radial segment of a fan; however, the morphologic and lithological characteristics of fanlobes from other fans may differ from the Mississippi Fan, depending on basin setting and sand/clay ratio of the input materials.

Another important method for comparing modern submarine fans and ancient turbidite systems included in this volume concerns the basic controls on fan dimensions. The use of seismic reflection profiles provides a direct method of calculating area, thickness, and volume for modern fans. Several workers have carefully defined that part of the sediment accumulation forming the true "modern" fan as we know it (Chapter 3, this volume). Thus, only the upper part of the basin fill might be included in the calculations. Some differences in dimensions between fans may actually be the result of using different seismic systems with differing degrees of resolution of reflectors. For ancient turbidite systems, limited exposures might be expected to lead to underestimation of fan volume. The table on the wall chart [1], however, shows that using paleogeographic reconstructions together with the measured-section estimates for fan thickness give surprisingly large volumes for many ancient fans, often much larger than expected from comparison with modern fans in similar settings. We suggest that many of the comparisons of volumes for modern and ancient fans presented in this volume can be quite misleading.

Conclusions

The main conclusions we can draw from editing the contributions to this volume are:

1) An attempt to compare modern and ancient fans provides a good basis for understanding the major problem areas in the development of general models for turbidite and deep-sea fan sedimentation.

2) It seems advisable, at this point, to discourage attempts to develop a unifying fan model by cross-multiplying characters from different modern and ancient fans. There may be more similarities than these contributions indicate. The

insufficient knowledge of those characters make cross-multiplication a misleading game.

3) It is time to concentrate on defining the primary common characters of deep-sea fans that can be used in mapping both modern and ancient fans. Simplification of the number of terms presently in use should be a first step. Further, more stringent requirements should be used for determining subdivisions of fans, and no fan should be used as the basis for a model unless the entire deposit is well surveyed.

4) One submarine fan has now been drilled, and although more drill sites are needed to answer many of the pertinent questions, a considerable increase in existing knowledge has resulted from that program. A similar drilling program on a fan from an active margin setting should be the next step before initiating a large-scale drilling program. Another step toward needed information is the use of more compatible data collections from many fans.

References

[1] Barnes, N. E., and Normark, W. R., 1983/84. Diagnostic parameters for comparing modern and ancient submarine fans. Geo-Marine Letters, v. 3, enclosed map.

[2] Kelly, W., 1974. Pogo Revisited. Simon & Schuster, New York, p. 5.

[3] Normark, W. R., Mutti, E., and Bouma, A. H., 1983/84. Problems in turbidite research: a need for COMFAN. Geo-Marine Letters, v. 3, pp. 53–56.

Index

D

E

P

R

S

T

U

V

W

X

Y

Z